FROM GENE TO PROTEIN: INFORMATION TRANSFER IN NORMAL AND ABNORMAL CELLS

MIAMI WINTER SYMPOSIA—VOLUME 16

1. *W. J. Whelan and J. Schultz, editors:* HOMOLOGIES IN ENZYMES AND METABOLIC PATHWAYS *and* METABOLIC ALTERATIONS IN CANCER, 1970
2. *D. W. Ribbons, J. F. Woessner, Jr., and J. Schulz, editors:* NUCLEIC ACID– PROTEIN INTERACTIONS *and* NUCLEIC ACID SYNTHESIS IN VIRAL INFECTION, 1971
3. *J. F. Woessner, Jr., and F. Huijing, editors:* THE MOLECULAR BASIS OF BIOLOGICAL TRANSPORT, 1972
4. *J. Schultz and B. F. Cameron, editors:* THE MOLECULAR BASIS OF ELECTRON TRANSPORT, 1972
5. *F. Huijing and E. Y. C. Lee, editors:* PROTEIN PHOSPHORYLATION IN CONTROL MECHANISMS, 1973
6. *J. Schultz and H. G. Gratzner, editors:* THE ROLE OF CYCLIC NUCLEOTIDES IN CARCINOGENESIS, 1973
7. *E. Y. C. Lee and E. E. Smith, editors:* BIOLOGY AND CHEMISTRY OF EUCARYOTIC CELL SURFACES, 1974
8. *J. Schultz and R. Block, editors:* MEMBRANE TRANSFORMATION IN NEOPLASIA, 1974
9. *E. E. Smith and D. W. Ribbons, editors:* MOLECULAR APPROACHES TO IMMUNOLOGY, 1975
10. *J. Schultz and R. C. Leif, editors:* CRITICAL FACTORS IN CANCER IMMUNOLOGY, 1975
11. *D. W. Ribbons and K. Brew, editors:* PROTEOLYSIS AND PHYSIOLOGICAL REGULATION, 1976
12. *J. Schultz and F. Ahmad, editors:* CANCER ENZYMOLOGY, 1976
13. *W. A. Scott and R. Werner, editors:* MOLECULAR CLONING OF RECOMBINANT DNA, 1977
14. *J. Schultz and Z. Brada, editors:* GENETIC MANIPULATION AS IT AFFECTS THE CANCER PROBLEM, 1977
15. *F. Ahmad, T. R. Russell, J. Schultz, and R. Werner, editors:* DIFFERENTIATION AND DEVELOPMENT, 1978
16. *T. R. Russell, K. Brew, H. Faber, and J. Schultz, editors:* FROM GENE TO PROTEIN: INFORMATION TRANSFER IN NORMAL AND ABNORMAL CELLS, 1979

MIAMI WINTER SYMPOSIA–VOLUME 16

FROM GENE TO PROTEIN: INFORMATION TRANSFER IN NORMAL AND ABNORMAL CELLS

edited by

Thomas R. Russell   Keith Brew
University of Miami School of Medicine
Harvey Faber   Julius Schultz
The Papanicolaou Cancer Research Institute

Proceedings of the Miami Winter Symposium, January 1979
Sponsored by The Department of Biochemistry
University of Miami School of Medicine, Miami, Florida
Symposium Director: W. J. Whelan

and by

The Papanicolaou Cancer Research Institute, Miami, Florida
Symposium Director: J. Schultz

1979

ACADEMIC PRESS   New York London Toronto Sydney San Francisco
A Subsidiary of Harcourt Brace Jovanovich, Publishers

ACADEMIC PRESS RAPID MANUSCRIPT REPRODUCTION

574.1924 45
F92
116599
Jan.1981

ACADEMIC PRESS, INC.
111 Fifth Avenue, New York, New York 10003

United Kingdom Edition published by
ACADEMIC PRESS, INC. (LONDON) LTD.
24/28 Oval Road, London NW1 7DX

Library of Congress Cataloging in Publication Data
Main entry under title:

From gene to protein.

(Miami winter symposia ; v. 16)
"Proceedings of the Miami winter symposium,
January 1979, sponsored by the Department of
Biochemistry, University of Miami School of
Medicine . . . and by the Papanicolaou Cancer
Research Institute, Miami . . . "
    1. Protein biosynthesis—Congresses.
2. Cellular control mechanisms—Congresses.
3. Genetic regulation—Congresses.   I. Russell,
Thomas Robert, 1945–   II. Miami, University
of, Coral Gables, Fla. Dept. of Biochemistry.
III. Papanicolaou Cancer Research Institute.
IV. Series. [DNLM: 1. Genetics, Biochemical—
Congresses. 2. Proteins—Metabolism—Congresses.
W3 MI202 v. 16 1979 / QU4.3 F931 1979]
QP551.F93   574.1'9245   79-19255
ISBN 0-12-604450-3

PRINTED IN THE UNITED STATES OF AMERICA

79 80 81 82   9 8 7 6 5 4 3 2 1

CONTENTS

SPEAKERS, CHAIRMEN, AND DISCUSSANTS   xix
PREFACE   xxiii

How to Live with a Golden Helix   1
   Francis Crick

The Ovomucoid Gene: Organization, Structure, and Regulation 15
   Bert W. O'Malley, Joseph P. Stein, Savio L.C. Woo,
   James F. Catterall, Ming-Jer Tsai, and Anthony R. Means
     DISCUSSION: F.H.C. Crick

Structural Organization and Expression of Ovalbumin and
Related Chicken Genes 55
   P. Chambon, C. Benoist, R. Breathnach, M. Cochet,
   F. Gannon, P. Gerlinger, A. Krust, M. LeMeur,
   J. P. LePennec, J. L. Mandel, K. O'Hare, and F. Perrin
     DISCUSSION: R. Hamilton, F. H. C. Crick,
     A. J. Shatkin, B. W. O'Malley, A. Peterson, D. Apirion,
     O. Smithies, and J. Bag

An Endonuclease-Sensitive Region in Simian Virus 40
Chromatin 83
   Walter A. Scott, Dianne J. Wigmore, and
   Richard W. Eaton
     DISCUSSION: J. Sygusch, P. Chambon, and J. Alwine

Expression of Cloned Viral and Chromosomal Plasmid-
Linked DNA in Cognate Host Cells 99
    Charles Weissmann, Ned Mantei, Werner Boll, Robert
    F. Weaver, Neil Wilkie, Barklie Clements, Tadatsugu
    Taniguchi, Albert Van Ooyen, Johan Van Den Berg, Mike
    Fried, and Kenneth Murray
        DISCUSSION: W. Marzluff, P. Reddy, J. Lebowitz,
        P. Chambon, B. W. O'Malley, and F. H. C. Crick

The Arrangement, Rearrangement, and Origin of
Immunoglobulin Genes 133
    Philip Leder, Jonathan G. Seidman, Edward Max, Yutaka
    Nishioka, Aya Leder, Barbara Norman, and Marion Nau
        DISCUSSION: C. Weissmann, P. Reddy, and
        F. W. Putnam

The Structure and Expression of Normal and Abnormal
Globin Genes 149
    R. A. Flavell, G. C. Grosveld, F. G. Grosveld,
    R. Bernards, J. M. Kooter, P. F. R. Little , and E. De Boer
        DISCUSSION: P. Leder

Genomic Clones from Unfractionated DNA 167
    O. Smithies, J. L. Slightom, P. W. Tucker, and
    F. R. Blattner
        DISCUSSION: P. Leder, J. Slightom, M. Philipp, and
        P. Chambon

The Processing of Messenger RNA and the Determination of
Its Relative Abundance 187
    R. P. Perry, U. Schibler, and O. Meyuhas
        DISCUSSION: A. Travers, P. Chambon, P. Leder,
        K. Gupta, C. M. McGrath, and C. Weissmann

Steps in Processing of mRNA: Implications for Gene
Regulation 207
    J. E. Darnell, Jr.
        DISCUSSION: C. Weissmann, R. P. Perry, T. Hunt, and
        S. Zain

5'-Capping and Eukaryotic mRNA Function 229
    Aaron J. Shatkin, Nahum Sonenberg, and Yasuhiro
    Furuichi
        DISCUSSION: R. P. Perry, K. Gupta, A. Shatkin,
        T. Hunt, and P. Chambon

Splice Patterns of Adenovirus and Adenovirus-SV40 Mosaic
RNA's 245
    Heiner Westphal, Geoffrey R. Kitchingman, Sing-Ping
    Lai, Charles Lawrence, Tony Hunter, and Gernot Walter

Processing of Bacteriophage T7 RNAs by RNase III 261
    F. William Studier, John J. Dunn, and Elizabeth
    Buzash-Pollert
        DISCUSSION: M. Philipp and C. Weissmann

Regulation of Promoter Selection by RNA Polymerases 271
    A.A. Travers, R. Buckland, and P.G. Debenham
        DISCUSSION: T. Hunt, M. F. Miles, K. Nath, and
        J. Lebowitz

Synthesis and Maturation of Transmembrane Viral
Glycoproteins 297
    Harvey F. Lodish, Dyann Wirth, and Mary Porter
        DISCUSSION: A. Schultz, A. Shatkin, T. Hunt, and
        K. S. McCarty

The Control of Protein Synthesis in Rabbit Reticulocyte
Lysates 321
    Tim Hunt
        DISCUSSION: J. E. Darnell, E. Kaminskas, and
        V. Ernst

Determinants in Protein Topology 347
    Gunter Blobel
        DISCUSSION: R. P. Perry, B. R. Rabin, and
        H. F. Lodish

Precursors in the Biosynthesis of Insulin and Other Peptide
Hormones 361
    Shu J. Chan, Christoph Patzelt, John R. Duguid, Paul
    Quinn, Allen Labrecque, Barbara Noyes, Pamela Keim,
    Robert L. Heinrikson, and Donald F. Steiner
        DISCUSSION: G. Blobel and T. Hunt

Biosynthesis and Processing of Glycoproteins 379
    E. C. Heath, S. A. Brinkley, R. C. Das, and T. H. Haugen
    DISCUSSION: H. F. Lodish, A. Hasilik, and
    B. S. Hartley

The Enzymatic Conversion of Membrane and Secretory
to Glycoproteins 407
    William J. Lennarz
    DISCUSSION: E. G. Krebs, S. Kornfeld, R. K. Keller,
    and E. C. Heath

Oligosaccharide Processing during Glycoprotein Biosynthesis 425
    Ira Tabas, Ellen Li, Mark Michael, and Stuart Kornfeld
    DISCUSSION: B. S. Hartley, A. Hasilik, and
    J. Lebowitz

Structural Basis of the Regulation of Galactosyltransferase 433
    Keith Brew, Richard H. Richardson, and Sudhir K. Sinha
    DISCUSSION: B. S. Hartley, S. Kornfeld, G. Blobel,
    R. K. Keller, and E. G. Krebs

Specificity Considerations Relevant to Protein Kinase
Activation and Function 449
    Tung-Shiuh Huang, James R. Feramisco, David B. Glass,
    and Edwin G. Krebs
    DISCUSSION: R. Scott and D. F. Steiner

The Role of Protein Phosphorylation in the Coordinated
Control of Intermediary Metabolism 463
    P. Cohen, N. Embi, G. Foulkes, G. Hardie, G. Nimmo,
    D. Hylatt, and S. Shenolikar
    DISCUSSION: H. L. Segal, E. G. Krebs, M. Singh, and
    F. Ahmad

Protein Phosphatase C: Properties, Specificity, and Structural
Relationship to a Larger Holoenzyme 483
    Ernest Y.C. Lee, James H. Aylward, Ronald L. Mellgren,
    and S. Derek Killilea
    DISCUSSION: H. L. Segal, T. Hunt, and P. Cohen

Regulation of Pyruvate Dehydrogenase by Phosphorylation/
Dephosphorylation 501
    Philip J. Randle, Nancy J. Hutson, Alan L. Kerbey, and
    Peter H. Sugden
    DISCUSSION: E. G. Krebs and P. Cohen

Allosteric Control of *E. coli* Glutamine Synthetase is Mediated
by a Bicyclic Nucleotidylation Cascade System 521
    E. R. Stadtman, P. B. Chock, and S. G. Rhee
    DISCUSSION: P. Cohen and A. R. Peterson

Poly(ADP-Ribose) and ADP-Ribosylation of Proteins 545
    O. Hayaishi, K. Ueda, M. Kawaichi, N. Ogata, J. Oka,
    K. Ikai, S. Ito, Y. Shizuta, H. Kim, and H. Okayama
    DISCUSSION: R. Leif, E. Kun, E. R. Stadtman,
    W. J. Whelan, B. Lesser, and P. Cohen

FREE COMMUNICATIONS

Identification of Two Collagen RNAs Larger than α1 and α2
Collagen mRNAs which Are Decreased in Transformed Cells 567
    Sherrill L. Adams, James. C. Alwine, V. Enrico
    Avvedimento, Mark E. Sobel, Tadashi Yamamoto, Benoit
    de Crombrugghe, and Ira Pastan

Processing of RNA in Bacteria 568
    David Apirion, Basanta K. Ghora, Greg Plautz, Tapan
    K. Misra, and Peter Gegenheimer

The Protein Moieties of the Nonpolysomal Cytoplasmic (Free)
mRNP Complexes of Embryonic Chicken Muscle 569
    J. Bag and B.H. Sells

Complex Oligosaccharides Assembled on Lipid Intermediates:
Studies with UDP-N-Acetylglucosamine and GDP-Mannose 570
    David S. Bailey, Mathias Dürr, and Gordon Maclachlan

Is There a Different Mode of DNA Synthesis in Normal and
Transformed Human Diploid Fibroblasts? 571
    David E. Berry, James W. Rawles, and James M. Collins

A Study of the DNA Region Coding for Initiation of Precursor
rRNA Transcription in *Lytechinus variegatus* 572
    D. Bieber, N. Blin, and D.W. Stafford

A Lectin Resistant Variant of Mouse Fibroblasts with Altered
Cell Surface Galactose 573
    D.A. Blake, W.S. Stanley, B.P. Peters, E.H.Y. Chu, and
    I.J. Goldstein

Alterations in the Rate of Protein Synthesis of Polysomes
Extracted from Fibroblasts of Patients with Duchenne
Muscular Dystrophy and Carriers of the Disease 574
    Léa Brakier-Gingras, Marie Boulé, and Michel Vanasse

Biosynthesis of the Small Nuclear RNAs in a Cell-Free System
from Mouse Myeloma Cells 575
    Anne Brown, Chi-Jiunn Pan, and William F. Marzluff

A Shared Control Mechanism for Regulation of Adenovirus and
SV40 Gene Expression 576
    Timothy Carter, Rebecca Blanton, and David Cosvin

In Vitro Synthesis of Virus-Specific RNA Using Transcriptional
Complex Isolated from Adenovirus-Infected Cells 577
    N.K. Chatterjee and M.H. Sarma

Gene Expression of Mammalian Lens Tissue: Coexistence and
Properties of Poly(A)-Containing and Poly(A)-Lacking 10S and
14S mRNA Species 578
    John H. Chen

Gene Expression of Mammalian Lens Tissue: Is Lens mRNA
Bicistronic? 579
    John H. Chen

The Expression of α-Fetoprotein Gene in Normal and Neoplastic
Rat Liver 580
    J.F. Chiu, L. Belanger, P. Commer, and C. Schwartz

Nuclear Phosphoprotein Phosphatase from Calf Liver 581
    Ming W. Chou, Eng L. Tan, and Chung S. Yang

Chemically Induced Gene Expression: The Role of Histone
Hyperacetylation in Transformation by SV40 Minichromatin 582
    Bennett N. Cohen and Thomas E. Wagner

Transcription of the Histone mRNA Precursor during the Cell
Cycle 583
    Dave Cooper, Chi Jiunn Pan, and William F. Marzluff

Characterization of a Cytoplasmic Poly(A)·Protein Complex:
Endproduct or Byproduct of Messenger RNA Metabolism? 584
    Peter L. Davies

Factors which Influence the N^{α}-Acetylation of ACTH and
Fragments of ACTH 585
    Jack E. Dixon and Terry A. Woodford

Instability of Higher Plant Chromosomal DNA Molecularly
Cloned in E. coli via Recombinant Plasmids 586
    Virginia K. Eckenrode, Harald Friedrich, Ron T. Nagao,
    Sidney R. Kushner, and Richard B. Meagher

Translation of Endogenous Non-IgG and IgG Messenger
RNA in Human Leukemic Plasma Cells after Treatment with
Several Alkylating Agents 587
    B. Emmerich, R. Maurer, G. Schmidt, and J. Rastetter

The in situ Phosphorylation of the α-Subunit of eIF-2 in
Reticulocyte Lysates Inhibited by Heme Deficiency, dsRNA,
GSSG, or the Heme-Regulated Inhibitor 588
    Vivian Ernst, Daniel Levin, and Irving M. London

Purification of Arginyl- and Lysyl-tRNA Synthetases from
Rat Liver: Evidence of Glycoproteins 589
    Raymant L. Glinski, Jr., Philip C. Gainey, Thomas
    P. Mawhinney, and Richard H. Hilderman

Mammalian Cell Nuclear RNA Ligase Activity 590
    Norma Ornstein Goldstein

Multiple Forms of α2u Globulin in the Rat Studied *in Vivo*
and in Primary Cultures of Adult Hepatocytes 591
    Laura Haars and Henry Pitot

Biosynthesis of Lysosomal Enzymes in Diploid Human
Fibroblasts 592
    Andrej Hasilik and Elizabeth F. Neufeld

Isolation and Partial Characterization of β-like Globin Genes
from Total Genomic Goat DNA 593
    J.R. Haynes, P. Rosteck, Jr., J. Robbins, G. Freyer,
    D.J. Burks, M. Cleary, and J.B. Lingrel

Regulation of Metallothionein Synthesis in Cultured Mammalian
Cells: Induction, Deinduction, and Superinduction 594
    C.E. Hildebrand and M.D. Enger

A Class of Highly Repeated DNA Sequences Ubiquitous
to the Human Genome 595
    Catherine M. Houck, Frank P. Rinehart, and Carl
    W. Schmid

Regulation of HMG-CoA Reductase and Reductase Kinase by
Reversible Phosphorylation 596
    T.S. Ingebritsen, R.A. Parker, and D.M. Gibson

Structure and Expression of the Alpha Fetoprotein Gene 597
    M.A. Innis and D.L. Miller

Control of eIF-2 Phosphorylation in Rabbit Reticulocyte
Lysate 598
    Rosemary Jagus and Brian Safer

Organization of Immunoglobulin Kappa Light Chain Genes in
Germ Line and Somatic Cells 599
    Rolf Joho, Irving Weissman, Philip Early, and Leroy Hood

Cloning of Fragments of the Sea Urchin Histone Gene Cluster
with SV40 DNA as a Vector 600
    John S. Kaptein and George C. Fareed

Regulation by Metabolites of Inactivating and Additional
Phosphorylation Reactions in the Pig Heart Pyruvate
Dehydrogenase Complex 601
    Alan L. Kerbey, Peter H. Sugden, and Philip J. Randle

Cyclic AMP Stimulation of Ribosomal Phosphoprotein
Phosphatase Activity 602
    Robert Kisilevsky and Margaret Treloar

Unusual Genome Organization of *Neurospora crassa* 603
    Robb Krumlauf and George Marzluf

Cloning and Characterization of Alleles of the Natural
Ovalbumin Gene Created by Mutations in the Intervening
Sequences 604
    Eugene C. Lai, Savio L. C. Woo, Achilles Dugaiczyk,
    Donald A. Colbert, and Bert W. O'Malley

Differential Influence of mRNA Cap Metabolism on Protein
Synthesis in Differentiating Embryonic Chick Lens Cells 605
    Gene C. Lavers, Gregory Gang, and E. M. Ciccone

Molecular Studies of Retrodifferentiation Phenomena in
Proliferation-Competent Adult Hepatocyte Cultures 606
    H. Leffert, J. Bonner, J. Brown, T. Moran, J. Sala-Trepat,
    S. Sell, H. Skelly, and K. Thomas

Role of Protein Synthesis in the Replication of the "Killer
Virus" of Yeast 607
    Michael J. Leibowitz

Effects of the Catalytic Subunit of Protein Kinase II from
Reticulocytes and Bovine Heart Muscle on Protein
Phosphorylation and Protein Synthesis in Reticulocyte
Lysates 608
    Daniel Levin, Vivian Ernst, and Irving M. London

Functional Heterogeneity in the RNAs of TYMV Virions 609
  D. Lightfoot and R. Clark

Subcellular Distribution of SV40 T Antigen Species in Various
Cell Lines 610
  Samuel W. Luborsky and K. Chandrasekaran

Hormonal Regulation of Protein Biosynthesis in Rat
Epididymal Fat Cells: Stimulation by Insulin, Inhibition by
Fasting, and Permissive Effects of Adrenal Glucocorticoids 611
  Robert T. Lyons and Donald T. Young

Reversal of Posttranslational Tyrosylation of Tubulin 612
  Todd M. Martensen and Martin Flavin

Salt-Dependent Transcriptional Specificity of T7 RNA
Polymerase 613
  William T. McAllister and Anthony D. Carter

Differential Sensitivity of Inhibition of Cyclic AMP
Phosphodiesterase in the Thymus during Murine
Leukemogenesis 614
  Lawrence A. Menahan

In Vivo Subnuclear Distribution of Larger than Tetrameric
Polyadenosine Diphosphoribose in Rat Liver 615
  Takeyoshi Minaga and Ernest Kun

Processing and Transport of RNA following Mitogenic
Stimulation of Human Lymphocytes 616
  M. F. Mitchell, E. Bard, and J. G. Kaplan

Identification of Drosophila 2S rRNA as the 3'-Part of 5.8S
rRNA 617
  George Pavlakis, Regina Wurst, and John Vournakis

Facilitation by Pyrimidine Nucleosides and Hypoxanthine
of MNNG Mutagenesis in Chinese Hamster Cells 618
  A. R. Peterson, Hazel Peterson, J. R. Landolph, and
  Charles Heidelberger

..

Noncoding Inserts in Hemoglobin and Simian Virus 40 Genes
Are Bounded by Self-Complementary Regions 619
  Manfred Philipp, Robert Bernholc, Avi Rubinsztejn, and
  Darrell Ballinger

A Variant of Polyoma Virus Altered in Its Processing of Late
Viral RNA 620
  B. J. Pomerantz, C. E. Blackburn, E. K. Thomas,
  V. A. Merchant, and J. D. Hare

Protease-Mediated Morphological Changes Induced by
Treatment of Transformed Chick Fibroblasts with a Tumor
Promoter: Evidence for Direct Catalytic Involvement of
Plasminogen Activator 621
  James P. Quigley

Recovery of a Transforming Virus from AKR Mouse Tumor
Tissue by DNA Transfection 622
  John M. Rice, Anna D. Barker, and Anthony J. Dennis

Synthesis and Posttranslational Modification of Retrovirus
Precursor Polyproteins Encoded by *gag* and *env* Genes 623
  A. M. Schultz and S. Oroszlan

Avian Myeloblastosis Virus (AMV) Linear Duplex DNA:
In Vitro Enzymatic Synthesis and Structural Organization
Analysis 624
  Robert A. Schulz, Jack G. Chirikjian, and Takis S. Papas

Studies on the DNA-Directed *in Vitro* Synthesis of Bacterial
Elongation Factor TU 625
  Tanya Schulz, Carlos Spears, and Herbert Weissbach

In Vitro Transcription of Ad2 DNA by Eukaryotic RNA
Polymerases II. I. Kinetics of Formation of Stable Binary
Complexes at Specific Sites 626
  S. Seidman, F. Witney, and S. Surzycki

Characterization of Human Fetal γ Globin DNA 627
  Jerry L. Slightom and Philip W. Tucker

Phosphorylation and Inactivation of Glycogen Synthase
by a Ca^{2+} CDR-Dependent Kinase 628
    Thomas R. Soderling and Ashok Srivastava

A Polypeptide in Initiation Factor Preparations That
Recognizes the 5'-Terminal Cap in Eukaryotic mRNA 629
    Nahum Sonenberg and Aaron J. Shatkin

Location and Sequence of a Putative Promoter for the B Gene
of Bacteriophage S13 630
    J. H. Spencer, E. Rassart, and K. Harbers

Regulation of Protein Synthesis: Role of NAD^+ and Sugar
Phosphates in Lysed Rabbit Reticulocytes 631
    R. J. Suhadolnik, J. M. Wu, C. P. Cheung, A. Bruzel, and
    J. Mosca

The Effect of Cortisol on Glucocorticoid "Resistant"
Thymocytes 632
    E. Aubrey Thompson, Jr.

The Effects of *Streptococcus pneumoniae* Infection on
Hepatic Messenger RNA Production 633
    William L. Thompson and Robert W. Wannemacher, Jr.

Identification of Putative Precursors of Ovalbumin and
Ovomucoid Messenger RNAs 634
    Ming-Jer Tsai, J. L. Nordstrom, D. R. Roop, and
    B. W. O'Malley

Purification of mRNA Guanylyl Transferase from HeLa
Cells 635
    S. Venkatesan and B. Moss

Emergence of Resistance to Glucocorticoid Hormones in
Cancer Cells: Detection of Differences in Proteins Synthesized
by Glucocorticoid-Sensitive and -Resistant Mouse P1798
Lymphosarcoma Cells 636
    Bruce P. Voris, Mary L. Nicholson, and Donald A. Young

Complexity and Diversity of Polysomal and Informosomal
mRNA from Chinese Hamster Cells 637
    Ronald A. Walters, Paul M. Yandell, and M. Duane Enger

Characterization of a Virion-Associated RNA Polymerase
from Killer Yeast 638
    J. Douglas Welsh and Michael J. Leibowitz

In Vitro Transcription of Ad2 DNA by Eukaryotic RNA
Polymerases II. II. Transcription from Stable Binary
Complexes 639
    F. Witney, S. Seidman, and S. Surzycki

Mouse Fetal Hemoglobin Synthesis in Murine
Erythroleukemic Cells 640
    N. C. Wu and R. M. Zucker

Spectroscopic Probes for Study of Gene Transcription 641
    L. R. Yarbrough, Michael Baughman, and
    Joseph G. Schlageck

Positive Transcriptional Control of Inducible Galactose
Pathway Enzymes in Yeast 642
    James G. Yarger and James E. Hopper

Localization and Characterization of the 5' End of
Adenovirus-2 Fiber mRNA 643
    B. S. Zain, J. Sambrook, A. Dunn, and R. J. Roberts

SPEAKERS, CHAIRMEN, AND DISCUSSANTS

F. Ahmad, Papanicolaou Cancer Research Institute, Miami, Florida

J. Alwine, National Cancer Institute, Bethesda, Maryland

D. Apirion, Washington University Medical School, St. Louis, Missouri

J. Bag, Memorial University, St. John's Newfoundland, St. John's, Canada

G. Blobel, The Rockefeller University, New York, New York

R. E. Block, Papanicolaou Cancer Research Institute, Miami, Florida

Z. Brada, Papanicolaou Cancer Research Institute, Miami, Florida

K. Brew, University of Miami School of Medicine, Miami, Florida

P. Chambon, Strasbourg Institute of Molecular Biology, Strasbourg, France

P. Cohen, University of Dundee, Dundee, Scotland

F. H. C. Crick, The Salk Institute, La Jolla, California

J. E. Darnell, The Rockefeller University, New York, New York

V. Ernst, Massachusetts Institute of Technology, Cambridge, Massachusetts

H. Faber, Papanicolaou Cancer Research Institute, Miami, Florida

R. A. Flavell, The University of Amsterdam, Amsterdam, The Netherlands

K. Gupta, University of Alabama Medical Center, Birmingham, Alabama

R. Hamilton, Pennsylvania State University, Hershey, Pennsylvania

B. S. Hartley, Imperial College, London, England

A. Hasilik, National Institutes of Health, Bethesda, Maryland

O. Hayaishi, Kyoto University Faculty of Medicine, Kyoto, Japan

E. C. Heath, University of Iowa, Iowa City, Iowa

Names in bold indicate speakers at the conference.

T. Hunt, University of Cambridge, Cambridge, England

E. Kaminskas, Mount Sinai Medical Center, Milwaukee, Wisconsin

G. J. Kaplan, University of Ottawa, Ottawa, Ontario, Canada

R. K. Keller, University of South Florida, Tampa, Florida

S. Kornfeld, Washington University School of Medicine, St. Louis, Missouri

E. G. Krebs, University of Washington, Seattle, Washington

E. Kun, University of California, San Francisco, California

J. Lebowitz, University of Alabama, Birmingham, Alabama

P. Leder, National Institutes of Health, Bethesda, Maryland

E. Y. C. Lee, University of Miami School of Medicine, Miami, Florida

M. Leibowitz, Rutgers Medical School, Piscataway, New Jersey

R. C. Leif, Papanicolaou Cancer Research Institute, Miami, Florida

W. J. Lennarz, The Johns Hopkins University School of Medicine, Baltimore, Maryland

B. Lesser, University of Calgary, Calgary, Alberta, Canada

H. E. Lodish, Massachusetts Institute of Technology, Cambridge, Massachusetts

W. Marzluff, Florida State University, Tallahassee, Florida

K. S. McCarty, Duke University, Durham, North Carolina

C. M. McGrath, Michigan Cancer Foundation, Detroit, Michigan

M. F. Miles, Northwestern University, Chicago, Illinois

K. Nath, Merck Sharp & Dohme Research Laboratories, Rahway, New Jersey

B. W. O'Malley, Baylor College of Medicine, Houston, Texas

R. P. Perry, Fox Chase Institute for Cancer Research, Philadelphia, Pennsylvania

A. R. Peterson, University of Southern California Cancer Center, Los Angeles, California

M. Philipp, Lehman College of the City University of New York, Bronx, New York

F. W. Putnam, Indiana University, Bloomington, Indiana

B. R. Rabin, University College, London, England

P. J. Randle, University of Oxford, Oxford, England

P. Reddy, National Cancer Institute, Bethesda, Maryland

T. R. Russell, University of Miami School of Medicine, Miami, Florida

A. Schultz, Frederick Cancer Research Center, Frederick, Maryland

J. Schultz, Papanicolaou Cancer Research Institute, Miami, Florida

R. Scott, Mayo Clinic/Foundation, Rochester, Minnesota

W. A. Scott, University of Miami School of Medicine, Miami, Florida

H. L. Segal, State University of New York, Amherst, New York

A. J. Shatkin, The Roche Institute of Molecular Biology, Nutley, New Jersey

M. Singh, Veterans Administration Medical Center, Dallas, Texas

J. Slightom, University of Wisconsin, Madison, Wisconsin

O. Smithies, University of Wisconsin, Madison, Wisconsin

E. R. Stadtman, National Institutes of Health, Bethesda, Maryland

H. K. Stanford, University of Miami, Coral Gables, Florida

D. F. Steiner, University of Chicago, Chicago, Illinois

F. W. Studier, Brookhaven National Laboratory, Upton, New York

J. Sygusch, University of Sherbrooke, Canada

A. Travers, MRC Laboratory of Molecular Biology, Cambridge, England

C. Weissmann, University of Zurich, Zurich, Switzerland

H. Westphal, National Institutes of Health, Bethesda, Maryland

W. J. Whelan, University of Miami School of Medicine, Miami, Florida

S. Zain, Cold Spring Harbor Laboratory, Cold Spring Harbor, New York

PREFACE

This volume is the sixteenth in the continuing series published under the title "Miami Winter Symposia." In January 1969, the Department of Biochemistry of the University of Miami and the University-affiliated Papanicolaou Cancer Research Institute organized the first of these symposia. This is the eleventh year in which the symposia have been held.

As topics for the Miami Winter Symposia, we select areas of biochemistry in which recent progress offers new insights into the molecular basis of biological phenomena. Until 1977, we organized two symposia. The first, sponsored by the Department of Biochemistry, emphasized the basic science aspects of the chosen topic, while the second, sponsored by the Papanicolaou Cancer Research Institute, dealt with the application of this research to the cancer problem. The proceedings of each symposia were published in separate volumes. With cancer research becoming increasingly concerned with basic cellular mechanisms, the division of the symposia into basic research and cancer-related research became rather academic. For this reason the 1978 meeting was organized as a single symposium and the proceedings were published in a single volume, a practice that we have continued this year. This year's symposium is entitled "From Gene to Protein: Information Transfer in Normal and Abnormal Cells." The broad nature of the subject enabled us to bring together, as speakers, scientists from technically diverse fields who have a common interest in the expression and processing of genetic information at the levels of both nucleic acids and proteins. To bring forward as much of the recent work as possible, short communications are also presented in this volume.

Associated with the symposia is the Feodor Lynen Lecture, named in honor of the Department of Biochemistry's distinguished visiting professor. Past speakers have been George Wald, Arthur Kornberg, Harland G. Wood, Earl W. Sutherland, Jr., Luis F. Leloir, Gerald M. Edelman, A.H.T.

Theorell, Paul Berg, and James D. Watson. This year the Lynen lecture was given by Francis H.C. Crick. These lectures have provided insights into the history of discovery, and have included the personal and scientific philosophies of our distinguished speakers. The Lynen lecturer for 1980 will be Fred Sanger, and the symposium will deal with the mobilization and reassembly of genetic information.

Our arrangement with the publishers is to achieve rapid publication of the symposium proceedings, and we thank the speakers for their prompt submission of manuscripts. Our thanks also go to the participants whose interest and discussions provided the interactions that bring a symposium to life and to the many local helpers, faculty, and administrative staff who have contributed to the success of the present symposium. Special gratitude should be accorded to the organizers and coordinators of the program: W. J. Whelan (joint director with J. Schultz), Sandra Black, and Olga Lopez, and to Virginia Salisbury who did the major job of assembling the typescripts for the Papanicolaou Cancer Research Institute.

The financial assistance of several Departments of the University of Miami School of Medicine, namely, Oncology, Ophthalmology, Pathology, and Pediatrics, as well as of Dr. Emanuel M. Papper, Vice President for Medical Affairs and Dean of the University of Miami School of Medicine, the Dermatology Foundation of Miami, the Graduate School of the University of Miami, the Howard Hughes Medical Institute, Abbott Laboratories, Eli Lilly and Company, Hoffmann-La Roche, Inc., North American Biologicals, and Smith Kline & French Laboratories is gratefully acknowledged.

Keith Brew
Harvey Faber
Thomas R. Russell
Julius Schultz

HOW TO LIVE WITH A GOLDEN HELIX

Francis Crick[1]

The Salk Institute
La Jolla, California

The Lynen Lecture is traditionally an occasion for a broad overview of one's life and scientific interests, or at least some part of them; an occasion for reminiscences of the past together with a few speculations about the future. I shall try to conform to this agreeable format as far as I can but I have a difficulty. I can hardly avoid saying something about the discovery of the structure of DNA, if only because audiences invariably show such a keen interest in it, although I notice that they are usually less concerned with the scientific aspects than the psychological ones. "What did it feel like?" they ask. An embarrassing question, since after almost 25 years I have some difficulty in recalling in detail the curious combination of excitement, euphoria, and skepticism which we experienced at the time.

My main problem is that the discovery has been described in print several times already, not only by Jim Watson in that rather breathless fragment of his autobiography he called "The Double Helix" (perhaps "Lucky Jim" would have been a better title), but also, in a more sober, detailed and scholarly way by Bob Olby ("The Path to the Double Helix"). At least one TV documentary has been made about it. Leaving more ephemeral effusions, whether girlish or soured, on one side we shall soon have an excellent account by Horace Judson. His book, "The Eighth Day of Creation," to be published in the spring of 1979, covers not only the discovery of the double helix but also the search for the genetic code and the three-dimensional structure of proteins. Very well researched, scientifically accurate and written in

[1]Supported by Ahmanson Foundation, The Eugene and Estelle Ferkhauf Foundation, The J. W. Kieckhefer Foundation, and The Samuel Roberts Noble Foundation, Inc.

a lively and readable style, it reveals more about the way molecular biology was done and about the people who did it than any other account I know.

Some of you may have already seen three extracts from it which appeared last year in the New Yorker. If you enjoyed these, I think you will like the more extended account in the book itself.

What more can I add? Before the whole thing gets out of hand and becomes an academic cottage industry I think a dose of cold water would do no harm. No doubt it is fascinating to read just how a scientific discovery is made; the misleading experimental data, the false starts, the long hours spent chewing the cud, the darkest hour before the dawn, and then the moment of illumination, followed by the final run down the home straight to the winning post.

And what a cast of characters! The Brash Young Man from the Middle West, the Englishman who talks too much (and therefore must be a genius since geniuses either talk all the time or say nothing at all), the older generation, replete with Nobel Prizes, and best of all, a Liberated Woman who appears to be unfairly treated. And in addition, what bliss, some of the characters actually quarrel, in fact almost come to blows. The reader is delighted to learn that after all, in spite of science being so impossibly difficult to understand, SCIENTISTS ARE HUMAN, even though the word "human" more accurately describes the behavior of mammals rather than anything peculiar to our own species, such as mathematics. Surely the script must have been written, not in heaven, but in Hollywood.

Unfortunately a closer study shows that real life is not always exactly like a soap opera. Not everybody was competing madly, with one eye on Stockholm. In actual fact there was a considerable amount of cooperation mixed in with the inevitable competition. The major opposition Rosalind Franklin had to cope with was not from her scientific colleagues, nor even from King's College, London (an Anglican foundation, it should be noted, and therefore inherently biased against women) but from her affluent, educated and sympathetic family who felt that scientific research was not the proper thing for a normal girl. Rosalind's difficulties and her failures were mainly of her own making. Underneath her brisk manner she was oversensitive and, ironically, too determined to be scientifically sound and to avoid short cuts. She was rather too set on succeeding all by herself and rather too stubborn to accept advice easily from others when it ran counter to her own ideas. She was proffered help but she would not take it. The soap opera has many other distortions and simplifications. I need not elaborate further. The plain fact that science is largely an intellectual pursuit, that it involves an enormous amount of hard, often grinding, work (both theoretical and experimental), that it is based upon an immense body of closely interlocking facts and theories, much of which must be thoroughly mastered before any progress at all can be made — all this tends to be submerged in the popular mind beneath those personal aspects which ordinary people relate to more easily. It is certainly an excellent idea to kill the stereotype of the cold, impersonal scientist in the white coat — such people do exist but they are as dull in science as they are in life — but we must not let the public think that because they understand some of our motives they thereby understand what science is about. The most surprising characteristic of modern western society is that in spite of being largely based on science and technology, the average citizen under-stands so little about the scientific enterprise. It is not only that elementary scientific facts are not known (the shape of H_2O, for example) but there is an almost complete lack of any scientific overview, a lack of any description, even in outline, of what is well established, what we still have to discover, and how we hope to go about discovering it.

Let me return for a moment to the discovery of the double helix. I think what needs to be emphasized is that the path to the discovery was, scientifically speaking, fairly common-place. What was important was not the way it was discovered but the object discovered — the structure of DNA itself. One can see this by comparing it with almost any other scientific discovery. Misleading data, false ideas, problems of personal interrelationships occur in much if not all scientific work. Consider, for example, the discovery of the basic structure of collagen. It will be found to have all these elements. The characters are just as colorful and diverse. The facts were just as confused and the false solutions just as misleading. Competition and friendliness also played a part in the story. Yet nobody has written even one book about "The Race for the Triple Helix." This is surely because, in a very real sense, collagen is not as important a molecule as DNA.

Of course this probably depends upon what you consider important. Before Alex Rich and I worked (quite by accident, incidentally) on collagen we tended to be rather patronizing about it. "After all," we said, "there's no collagen in plants." After we got interested in the molecule we found ourselves saying, "Do you realize that one-third of all the protein in your body is collagen?" But however you look at it, DNA <u>is</u> more important than collagen, more central to biology, and more significant for further research. So, as

I have said before: it is the molecule which has the glamor, not the scientists.

There was in the early fifties a small, somewhat exclusive biophysics club at Cambridge, called the Hardy Club, named after a Cambridge zoologist of a previous generation who had turned physical chemist. The list of those early members now has an illustrious ring, replete with Nobel Laureates and Fellows of the Royal Society, but in those days we were all fairly young and most of us not particularly well-known. We boasted only one F.R.S. — Alan Hodgkin — and one member of the House of Lords — Victor Rothschild. Jim was asked to give an evening talk to this select gathering. The speaker was customarily given dinner first at Peterhouse. The food there was always good but the speaker was also plied with sherry before dinner, wine with it, and, if he was so rash as to accept it, drinks after dinner as well. I have seen more than one speaker struggling to find his way into his topic through a haze of alcohol. Jim was no exception. In spite of it all he managed to give a fairly adequate description of the main points of the structure and the evidence supporting it but when he came to sum up he was quite over-come and at a loss for words. He gazed at the model, slightly bleary-eyed. All he could manage to say was, "It's so beautiful, you see, so beautiful!" But then, of course, it was.

What then, do Jim Watson and I deserve credit for, if anything? There are certain technical points which are sometimes overlooked. It took courage (or rashness, according to your point of view) and a degree of technical expertise to put firmly on one side the difficult problem of unwinding the double helix and to reject a side-by-side structure. Such a model was suggested by Gamow, not long after ours was published, and it has been suggested again more recently by two other groups of authors. It is less well-known that in 1953 we very briefly considered a four-stranded model — the structure eventually published by McGavin — and had the good sense to reject that also. But these are small points. If we deserve any credit at all it is for persistance and the willingness to discard ideas when they became untenable. One reviewer thought that we can't have been very clever because we went on so many false trails, but that is the way discoveries are usually made. Most attempts fail not because of lack of brains but because the investigator gets stuck in a cul-de-sac or gives up too soon. We have also been criticized because we had not perfectly mastered all the very diverse fields of knowledge needed to guess the double helix but at least we were <u>trying</u> to master them all, which is more than can be said for some of our critics.

However I don't believe all this amounts to much. The
major credit I think Jim and I deserve, considering how early
we were in our research careers, is for selecting the right
problem and sticking at it. It's true that by blundering
about we stumbled on gold but the fact remains that we were
looking for gold. Both of us had decided, quite independently
of each other, that the central problem in molecular biology
was the chemical structure of the gene. Muller had pointed
this out as long ago as the early twenties and many others
had done so since then. What both Jim and I sensed was that
there might be a short cut to the answer, that things might
not be quite as complicated as they seemed. Curiously
enough this was partly because I had acquired a very detailed
grasp of the current knowledge of proteins. We could not at
all see what the answer was, but we considered it so important
that we were determined to think about it long and hard, from
any relevant point of view. Practically nobody else was
prepared to make such an intellectual investment, since it
involved not only studying genetics, biochemistry, chemistry
and physical chemistry (including x-ray diffraction — and who
was prepared to learn that?), but also sorting out the
essential alloy from the dross. If you didn't worry about
the tautomeric forms of the bases you weren't going to find
the structure and believe me, I did worry about them, even
though I was for some time misled by Hunter's ideas. Such
discussions, since they tend to go on interminably, are very
demanding and sometimes intellectually exhausting. Nobody
could sustain them without an overwhelming interest in the
problem.

And yet history of other theoretical discoveries often
shows exactly the same pattern. In the broad perspective of
the exact sciences we were not thinking very hard but we were
thinking a lot harder than most people in that corner of
biology, since in those days, with the exception of
geneticists and possibly the people in the phage group, most
of biology was not thought of as having a highly structured
logic to it.

Of course it is obvious now that nucleic acid is the main
if not the only genetic material, but in the late forties and
early fifties this was far from clear. Everybody knew of the
work on transforming principle by Avery and his colleagues.
Even such a conservative body as the Royal Society gave
Avery a prestigious medal for the discovery as early as 1945.
The citation shows clearly they understood its genetic
implications but not everybody else was convinced so easily.
Mirsky in particular thought for some time that the effect
was due to contaminating protein but I do not think that was
the main stumbling block. The real difficulty was to decide

whether transformation was of general significance or whether
it was a freak. Initially it had been found only in
pneumococcus — and it was not even known whether that
organism had genes in the ordinary Mendelian sense.
Moreover it appeared to affect only one character, the nature
of the coat. A little later the very careful work of Rollin
Hotchkiss showed that other characters could be transformed.
He also made the idea of a protein impurity highly unlikely.
But transformation still remained an almost isolated case.
Moreover one could always argue that the experiments fitted
equally well the idea that a gene contained two essential
and specific components, nucleic acid and protein.   The
importance of the Hershey-Chase experiment on phage T4 was
that it provided a second quite separate instance of the
genetic specificity of DNA, even though, by comparison, the
experiments were far dirtier than those of Avery and
Hotchkiss. Hershey's results made a deep impression on Jim
and myself, even though many people couldn't see what the
fuss was all about. From then on we had few reservations
that DNA was biologically important. Whether its structure
would tell us anything interesting we could only guess —
and hope for the best.

     Looking back I can see that it is also important not to
be too clever. Consider the following argument. DNA fibers
show a very good x-ray diffraction pattern, implying that
the microcrystalline structure which produced the many spots
is very regular. But for a genetic material to have any
interest it must necessarily be somewhat irregular.
Therefore nothing of interest is likely to come from studying
diffraction patterns of DNA. A similar argument could be
used about its base composition. Only an imperceptive
person could possibly spend time measuring the exact amounts
of the four bases in DNA since how could that possibly reveal
anything of genetic interest? Fortunately neither of these
arguments influenced us at the time. We can see now that
they are wrong because of the overwhelming importance of
base-pairing and because one base-pair looks very like
another in shape and, at that resolution, to the x-rays.
The important thing is not to be deflected too much by
negative arguments of this general type, even though they
may indeed turn out to be correct and one's labors to have
been in vain. The much stronger rule is that if something is
of great scientific importance one can hardly learn too much
about it, even by what, at first sight, may seem rather
pedestrian methods. Of course, not everybody may be
equipped to appreciate the significance of some rather
simple observation (why the stars come out at night, for
example) as the history of the double helix shows rather
clearly.

But enough of the remote past. What has happened since then? It is instructive to sketch, although only very briefly, our newer knowledge of nucleic acid. We now know that RNA, too, can form a double helix similar in its base-pairing rules to DNA. Both nucleic acids can form triple helices and some DNA molecules, having the purines and pyrimidines segregated on different chains, may even be able to form a four-stranded structure. Poly I also appears to be four-stranded. Double helices with parallel chains (rather than anti-parallel) can be formed by special molecules such as poly A though whether such a structure has any biological significance (at the tails of messenger RNA, for example) remains to be seen. In transfer RNA unusual base-pairs and base-triples abound. At least one of them, the pairing of guanine with uracil, appears to be used more widely. DNA molecules, both single- and double-stranded, can be circular. Intact circular double-stranded DNA is often found supercoiled. In 1953 we wondered whether the DNA in a single chromosome might consist of one very long molecule but, luckily for us, we never considered that a DNA molecule might be a circle.

DNA duplication, though apparently following the simple base-pairing mechanism we proposed has turned out to be far more complex than we imagined then. I understand that, at the last count, no less than 19 distinct proteins may be involved. We expected that there ought to be repair mechanisms but we never imagined there would be quite so many, though it is easy now to see why. Even at that time we suggested that the best way to unwind DNA was to nick the backbone occasionally and then repair it, so we half anticipated the nicking-closing enzyme but we never suggested there might be a gyrase as well.

The history of reverse transcriptase is slightly different. In those days Tobacco Mosaic Virus and its RNA were very much on our minds. We often wondered, since no RNA double helix had then been discovered, whether TMV replicated via a DNA intermediate. When it was shown that the replicative form was double-stranded RNA the earlier idea was discarded but, as far as I was concerned, it was never abandoned completely. It did not violate the Central Dogma, as I had originally defined it. Jim unfortunately used this term incorrectly in his influential and widely-read textbook. Chargaff was misled and no doubt others as well but no serious harm was done. In particular I reassured Howard Temin on the point (by way of Dick Matthews) some time before reverse transcriptase was eventually discovered.

We did make a first tentative suggestion about mutagenesis but we never foresaw phase-shift mutants until the experimental

facts drove us relentlessly in that direction. We saw that
DNA must, almost inevitably, code for proteins and we soon
realized that it was likely to be a triplet code, but it was
not till much later that the idea of suppressor mutations
surfaced. Nor did we seriously consider, in those early
days, that nucleic acid as well as being informational
could also fulfill a structural role. In fact ribosomal RNA,
when first identified, was thought to be messenger RNA and
transfer RNA was, for a brief period, thought to be too big
to be the postulated adaptor molecule. Thanks to the double
helix we saw the broad scheme of things in outline but we
were often confused over the details.

Clearly nucleic acids had many ramifications which were
too far ahead for us to glimpse. But even more striking has
been the blossoming of new experimental methods, especially
in the last few years, starting with density gradient
centrifugation, followed by nucleic acid hybridization (it
seems strange now to realize that at one time we had serious
doubts as to whether it would work), gel electrophoresis,
the visualization of nucleic acid in the electron microscope,
and many others, culminating in restriction enzyme technology,
DNA cloning and the new, extremely rapid methods for DNA
sequencing. All these are leading to an information
explosion of almost unprecedented magnitude. The recent
discovery of split genes in eucaryotes is likely to be only
one of the striking results which these new methods will
reveal.

The source of these new techniques seem to be two-fold:
the use of highly specific enzymes to carry out particular
chemical operations with a far higher accuracy than any
classical chemist could ever achieve, coupled with methods,
based on biological magnification (replication in intact
cells) for obtaining almost unlimited amounts of the required
nucleic acid molecule, starting from as little as one copy
of it. Restriction enzymes, rapid growth and rapid screening
between them have achieved results which only a few years ago
we would have thought impossible. Nor do I see any end in
sight to these developments, not only of ingenious tricks
based on present methods, but on radically new methods to
meet new demands, so powerful are the tools which are
potentially at our disposal. The Zen of Molecular Biology is
to prompt Nature to do the job for you.

But what was it like to live with the double helix? I
think we realized almost immediately that we had stumbled
onto something important. According to Jim, I went into the
Eagle, the pub across the road where we lunched every day,
and told everyone that we'd discovered the secret of life.
Of that I have no recollection, but I do recall going home

and telling Odile that we seemed to have made a big discovery.
Years later she told me that she hadn't believed a word of it.
"You were always coming home and saying things like that,"
she said, "so naturally I thought nothing of it." Bragg was
in bed with 'flu at the time but as soon as he saw the model
and grasped the basic idea he was immediately enthusiastic.
All past differences were forgiven and he became one of our
strongest supporters. We had a constant stream of visitors,
a contingent from Oxford which included Sydney Brenner was
one of them, so that Jim soon began to tire of my repetitious
enthusiasm. In fact at times he had cold feet, thinking
that perhaps it was all a pipe dream, but the experimental
data from King's College, when we finally saw it, was a great
encouragement. By the summer most of our doubts had vanished
and we were able to take a long cool look at the structure,
sorting out its accidental features (which were somewhat
inaccurate) from its really fundamental properties which time
has shown to be correct.

For a number of years after that things were fairly quiet.
I named our house in Portugal Place "The Golden Helix" and
eventually erected a simple brass helix on the front of it,
though it was a single helix rather than a double one. It
was supposed to symbolize not DNA but the basic idea of a
helix. I called it golden in the same way that Apuleius
called his story "The Golden Ass," meaning beautiful. People
have often asked me whether I intend to gild it. So far
we've got no further than painting it yellow.

Gradually DNA became better known. Paul Doty told me
that shortly after lapel buttons came in he was in New York
and to his astonishment saw one with "DNA" written on it.
Thinking it must refer to something else he asked the vendor
what it meant. "Get with it, bud," the man replied in a
strong New York accent, "dat's the gene." Nowadays most
people know what DNA is, or if they don't, they know it must
be a dirty word, like "chemical" or "synthetic." Fortunately
people who do recall that there are two characters called
Watson and Crick are often not sure which is which. Many's
the time I've been told by an enthusiastic admirer how much
they enjoyed my book — meaning, of course, Jim's. By now
I've learned that it's better not to try to explain. An
even odder incident happened when Jim came back to work at
Cambridge in 1955. I was going into the Cavendish one day
and found myself walking with Neville Mott, the new Cavendish
Professor (Bragg had gone on to the Royal Institution in
London). "I'd like to introduce you to Watson," I said,
"since he's working in your lab." He looked at me in
surprise. "Watson?" he said, "Watson? I thought your name
was Watson-Crick." I did arrange an appointment for Jim

to meet him, but rather typically, Jim forgot to turn up.

I must not finish without saying a few words about my own subsequent work on DNA. In the summer of 1953 Jim and I decided that we would not start to do experiments involving the x-ray diffraction of DNA but would leave that problem in the capable hands of Maurice Wilkins. Jim and Alex Rich tried to take x-ray pictures of RNA, without much success, and later we all worked together on the structure of poly A. In subsequent years my main interest became protein synthesis and the problem of the genetic code. In fact the most striking thing is how little, in all this time, I have actually worked on DNA. Naturally I followed attentively Meselson and Stahl's work and also Arthur Kornberg's pioneer studies on the synthesis of DNA in the test-tube but I took very little part in them. Leslie Orgel and I more than once told Kornberg that, for theoretical reasons concerning mutation rates, there should be an inherent error correcting mechanism in DNA synthesis but he was very reluctant to believe us until he found it for himself. By the late sixties I had become interested in developmental biology and was rather out of touch with most work on DNA itself. Eventually I was led back to it as I became involved in the puzzle of chromatin. This really blossomed at Cambridge because of Roger Kornberg's work. Aaron Klug and I wondered whether DNA on nucleosomes might be kinked. So far there is no evidence that kinks exist and recent calculations by Michael Levitt suggest that the energy required to bend DNA smoothly to the required diameter is rather less than we had guessed, so that kinks are unlikely.

To understand the coiling of DNA I had to master the mathematics of the subject. This involved understanding the linking number, the twist, and the writhing number. The fundamental relationship between them was first established by James White in 1968 but most molecular biologists learned of them from the Brock Fuller account. I found that in spite of Fuller's paper almost everybody working in the field was confused about twist and linking. I remember explaining patiently the difference to one of my more irrepressible visitors but a few months later, when I received a preprint of his paper, I found he had once more fallen into error. I wrote a long and careful letter to him, making the difference quite explicit, but he was quite unconvinced by these theoretical arguments until he had obtained a display of his proposed structure from his computer. To his astonishment he found that I was right! It was as if someone could not accept Pythagoras' theorem until he had actually measured a right-angled triangle.

These concepts have been especially useful in dealing with supercoiled DNA. The fact that such DNA forms a set of discrete bands on electrophoresis under suitable conditions and that any single band, if melted and reannealed, still runs as a single band in the same place on the gel, shows that adjacent bands must differ by a characteristic which is both quantized and also survives heating and cooling. This can only be the linking number, since this is a topological property of the two circular DNA backbones.

Starting from this it is possible to show, from existing experimental data (such as D-loops in mitochondia) that there must be roughly 10 base-pairs per turn for DNA in solution. Very recently James Wang has performed a very elegant experiment, also based on supercoiling, which shows that under standard conditions of salt, temperature, etc., the exact value is 10.4 ± 0.1. This effectively disposes of the recent models proposed for DNA in which the two backbones do not twine round one another but run effectively side-by-side. It is interesting to note that such a powerful counter-argument could only have been made in the last few years.

Finally let me say a few words about split genes and RNA splicing, the main topic of our symposium earlier today. Although new experimental data are coming in all the time what we have at the moment is very patchy. For example, we do not ever know whether messenger RNA is ever spliced in Drosophila, nor is there any published example of a spliced housekeeping gene, such as one from the Krebs cycle.

The most obvious next step is to get hold of the splicing enzymes and to discover their specificity. The enzyme which processes some of the tRNAs of yeast is unlikely to splice mRNA in vertebrates correctly since these introns in tRNA do not obey Chambon's rule. I notice that this is a problem which the Genetic Engineers would rather leave to the Enzymologists and vice versa. Naturally a good assay is essential but I do not believe this will prove impossible to obtain.

We do not yet know what happens if an intron is cleanly deleted. Even if such a gene will still function, deletion may be a rather slow step in evolution since it must be done very exactly if it is to produce the correct mRNA. Thus once introns are common enough it may be almost impossible to eliminate them all in evolution except possibly under extreme selective pressure. If one splicing enzyme splices in many places in the genome, as seems highly likely, then it will be difficult for it to change its specificity. For this reason its specificity may be much the same over a wide range of species. Notice that, whatever the exact signals are for splicing they must avoid incorrect splicing, from the

beginning of one intron to the end of another. For this
reason the signals cannot be too simple. Some contribution
from secondary or tertiary structure is an obvious way to
avoid this problem.

The most extensive data so far on introns in related proteins comes from the work, mainly by symposium speakers, on
the various globins. All globins examined so far have just
two introns. Relative to the amino acid sequence they are
always in the same place. Since ∝ and β globin diverged a
long time ago this strongly suggests that these two introns
are very old. It will be interesting to see if they are also
there in the myoglobin gene.

While the position of the introns has remained fixed,
their length and sequence appear to have changed rapidly in
evolution. Nevertheless they have not been deleted. Nor does
it look as if additional introns have frequently been introduced successfully. The data on the ќ and λ light chains
of the immunoglobulins point to similar conclusions.

We still have no idea when introns first emerged in evolution, whether they came in with eucaryotes or whether, as has
been already suggested, they started much earlier, even
before the existence of cells. Nor do we know by what
mechanism they originated. They could arise by multiple
mutations producing the splicing signals; by specific insertion of introns; or by the shuffling of exons together with
fragments of their flanking introns. More taxonomic data may
help us to see this problem more clearly.

What we know about the DNA in chromosomes suggest that,
in evolutionary terms, it is far from static. There appear to
be many mechanisms for DNA addition, translocation and multi-
plication. If these processes are at least partly random
then in an organism where most DNA is not exon DNA the
commonest sequences to be moved around are likely to be those
of introns and intergenic spacers. Thus intron size in evolution may be a dynamic balance between additions and deletions.

Another useful idea, suggested to me by Dr. Leslie Orgel,
is that of "selfish DNA" in the genome. If some sequences are
preferentially replicated in evolution they will spread over
the genome in evolution even if they have no particular function, provided they do not harm the "host" organism too much.
These two ideas, the Dynamic Genome and Selfish DNA, may be
necessary for us to understand non-exonic sequences in higher
organisms. It seems highly likely that some introns will have
some function (the control of gene expression, the reduction
of unequal crossing over, etc.) but it does not follow that
all introns have a use. Some may be simply waiting to be
deleted.

Fascinating and unexpected as introns are they should not deflect us too much from what is even more important, RNA transcription and its control. This necessity brings us back to the double helix. What a remarkable molecule it is! Modern man is perhaps 50,000 years old, civilization has existed for scarcely 10,000 years, and the United States for only just over 200 years, but DNA and RNA have been around for at least a billion years, if not longer. It is indeed remarkable that the double helix has been there, and active, all that time and yet we are the first creatures on earth to become aware of its existence and the beauty of its structure.

THE OVOMUCOID GENE
ORGANIZATION, STRUCTURE AND REGULATION

Bert W. O'Malley
Joseph P. Stein
Savio L.C. Woo
James F. Catterall
Ming-Jer Tsai
Anthony R. Means

Department of Cell Biology
Baylor College of Medicine
Houston, Texas

INTRODUCTION

The molecular mechanism by which steroid hormones regulate specific gene expression has been an area of acute
interest during the past several years. One particularly
attractive model system for studying this hormonal regulation
has been the hen oviduct (1). A number of laboratories, in
addition to our own, have utilized this model system for
investigations of eucaryotic molecular biology (2-8). Administration of estrogen to the newborn chick stimulates oviduct
growth and differentiation and results in the appearance of a
number of new specific intracellular proteins (1,6,9-14). The
synthesis of one of these proteins, ovalbumin, has been
studied extensively. Ovalbumin mRNA has been purified (15)
and a full-length dsDNA copy synthesized (16) and cloned in
a bacterial plasmid (17). More recently, ovalbumin genomic
DNA sequences have been isolated from restriction enzyme
digests of hen DNA and cloned (18). The other three major
proteins under estrogenic control in the oviduct tubular
gland cell, ovomucoid, conalbumin and lysozyme, have been less
extensively studied (19,6). In order to study the mechanisms
by which steroid hormones coordinately regulate the expression

[1]Supported by NIH grant HD-8188 (B.W.O.), American Cancer
Society Research Grant BC-101 and Robert A. Welch Foundation
Grant Q-611 (A.R.M.) and American Cancer Society Fellowship
PF-1211 (J.P.S.). S.L.C.W. is an Associate Investigator of
the Howard Hughes Medical Institute.

of different genes in one tissue, we attempted to isolate a
second estrogen-induced gene from the chick oviduct. We now
wish to describe the purification of ovomucoid mRNA, the synthesis and cloning of a dsDNA from this ovomucoid mRNA, and
the use of this ovomucoid structural gene clone to isolate
and amplify the natural ovomucoid gene from chick DNA, and to
determine the sequence organization of the native ovomucoid
gene.

CLONING OF THE OVOMUCOID STRUCTURAL GENE SEQUENCE

Ovomucoid mRNA Purification

Sequential chromatography of a total hen oviduct RNA
extract on oligo (dT)-cellulose, Sepharose 4B, which substantially separated the ovalbumin mRNA activity from ovomucoid and lysozyme mRNA activities, and a second oligo (dT)-
cellulose column, yielded a mRNA preparation which consisted
primarily of two different size classes of RNA (20). To
separate these two size classes, the mRNA preparation was
denatured and applied to 12.2 ml linear 5-20% sucrose
gradients. Although the RNA banded in a rather broad peak,
judicious collection of fractions from the leading edge of
the peak resulted in the removal of the bulk of the smaller
contaminating mRNA. The final purified mRNA (as recovered
from the sucrose gradient) migrated as a single band on a
2.5% agarose slab gel in 0.025 M sodium citrate, pH 3.5, containing 6 M urea. The size of the purified mRNA (mRNA$_{om}$) was
estimated to be 800 nucleotides.

The enrichment of mRNA$_{om}$ relative to other mRNA species
during the purification was assessed by translation of the
RNA preparations in a cell-free translation system derived
from wheat germ (15,20). Ovomucoid mRNA activity was determined by specific immunoprecipitation of ovomucoid peptides
using a partially purified goat antiovomucoid immunoglobin G.
The measurement of radioactivity incorporated into TCA-
insoluble material was used as an indication of total mRNA
activity.

The purification of ovomucoid mRNA as followed by translation is shown in Table 1. The combination of purification
steps listed resulted in a 270-fold increase in the specific
activity of ovomucoid mRNA compared with the total extract.
The percentage of immunoprecipitable ovomucoid in the wheat
germ postribosomal supernatant fraction compared with the
total peptides synthesized increased during purification from
3% to 52%.

TABLE I. Purification of Ovomucoid mRNA

	sp act.[a] (cpm/μg)	purification[b] (fold)	ovomucoid synthesized[c] total protein synthesized (%)
total extract	100	1	3
dT-cellulose bound	1,900	19	12
Sepharose peak	8,000	80	24
dT-cellulose bound	10,500	105	35
sucrose gradient peak	27,000	270	52

[a]Defined as the total amount of ovomucoid polypeptides synthesized in the wheat germ translation assay in response to 1 μg of the various RNA preparations.

[b]Purification of mRNA_om over the total RNA in the preparation.

[c]Determined as the ratio of immunoprecipitated ovomucoid in the wheat germ translation assay to the total Cl_3CCOOH-precipitated polypeptides.

Synthesis and Cloning of a Complementary DNA Copy

Since it was clear that $mRNA_{om}$ was not a pure species and conventional methods of RNA purification had been exhausted, we decided to complete the purification and isolate the coding ovomucoid DNA sequence at the same time by molecular cloning. To this end, dsDNA was prepared from the partially purified $mRNA_{om}$ and the hairpin loop cut with S_1 nuclease as described previously (16,17,20,21). To obtain only full-length dsDNAs for restriction analysis and cloning, a 2% preparative agarose gel was run and a 0.6 cm band containing dsDNA of about 750-900 base pairs was excised. The DNA was recovered from the agarose gel by diffusion (22).

For molecular cloning experiments, the dsDNA was "tailed" with $[\alpha-^{32}P]dATP$ to an average length of 51 dAs per 3' terminus by terminal transferase. Chimeric plasmids were formed by annealing the $[^{32}P]dsDNA$ to Pst 1-cut pBR322. Ligation of the hybrid molecule as well as repair of any gaps which might occur due to the difference in length of the poly(dA) and poly(dT) regions occurred in vivo after transformation (23). This hybrid molecule was used to transform E. coli strain X1776. Transformed colonies containing chick DNA inserts were selected by in situ hybridization with $[^{32}P]cDNA$ synthesized from $mRNA_{om}$ (24). In all, fifteen colonies gave a positive signal. However, since the $mRNA_{om}$ from which the probe was synthesized was not homogeneous, the probe should represent many sequences. Thus, it was still necessary to identify a clone unambiguously as containing an ovomucoid insert.

Identification of an Ovomucoid Structural Gene Clone

To select a clone containing an ovomucoid DNA insert, an assay based on mRNA hybridization and translation was developed. Recombinant plasmid DNA was isolated from several positive clones, as described by Katz (25). Plasmid DNA (100 µg) was then digested with the restriction endonuclease Hhal, heat denatured and bound to Millipore filters as described in Figure 1. Either pure ovalbumin mRNA ($mRNA_{ov}$) or the enriched $mRNA_{om}$ was then hybridized to the DNA filters for 12 hr. at 42° in 600 µl of 50% formamide buffer. An equal amount of poly(dA) was added to prevent hybridization of the poly (A) tract of noncomplementary mRNAs to the linkers present in the filter-bound plasmid DNA. After removing the buffer and washing the filters several times to remove non-specifically adsorbed RNA, any hybridized RNA was eluted by denaturing in 0.01 M Tris-HCl, pH 7.4, 3 mM EDTA for 2 min.

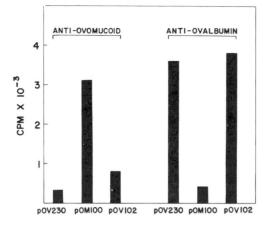

FIGURE 1. Hybridization-translation assays of cloned
DNAs. Plasmid DNA (100 μg: either pOV230, pOM100 or pOV102)
was digested with 30 units of Hhal at 37°C for 16 hr.
followed by phenol extraction. After denaturing each
restricted DNA in a small volume by heating at 100°C for 10
min and quick cooling, each was diluted into 6 mM Tris-HCl,
pH 8.0, 25 mM MgCl$_2$ in 4 X SSC to a final concentration of
10 μg/ml and passed twice through a Millipore filter at 4°C.
About 20% to 30% of the DNA bound to the filters. The filters were then baked at 70°C in a vacuum oven to fix the DNA.
One-half of each filter was then hybridized to pure mRNA$_{ov}$ and
the other half to the enriched mRNA$_{om}$ in 70 mM Tris-HCl, pH
7.7, 7.2 mM EDTA, 0.42 M NaCl, 50% in formamide. After denaturing the RNA/DNA hybrids, this hybridized nucleic acid was
recovered and translated in the wheat germ system. The polypeptides synthesized were precipitated with either anti-
ovomucoid or anti-ovalbumin. The radioactive peptides precipitated with anti-ovomucoid are shown for all three plasmid
DNAs on the left, and the precipitates with anti-ovalbumin on
the right. No correction factor for nonspecific trapping of
label was applied to the antibody precipitations.

at 90°C. Carrier tRNA was added (10 μg/ml) and the RNA precipitated in ethanol. The recovered hybridizable RNA was
translated in the wheat germ system, and the amounts of
ovalbumin- and ovomucoid-specific products were compared
(Figure 1). pOV230, a clone that contains the full-length
ovalbumin structural gene sequence, was used to test the
validity of the assay. The RNA that hybridized to this
plasmid DNA was translated into ovalbumin and not ovomucoid.
This indicated that filter-bound recombinant plasmid DNA

containing a known structural gene DNA sequence could
selectively hybridize to a complementary mRNA species, and
that this RNA could be removed from the filter and translated
into antibody-precipitable peptides in the wheat germ system.
The other two plasmids tested in Figure 1 both resulted from
the transformation with dsDNA$_{Om}$ reported here. Translation
of the mRNA that hybridized to the pOM100 DNA filter yielded
immunoprecipitable ovomucoid. The clone from which pOM100 DNA
was derived must therefore contain an inserted ovomucoid DNA
sequence. The other plasmid tested in Figure 1, pOV102,
hybridized only to mRNA$_{Ov}$. The fact that this plasmid contained an ovalbumin DNA sequence was not surprising since
about 48% of the dsDNA used in the transformation represented
non-ovomucoid sequences. Since the mRNA used to synthesize
the dsDNA was considerably shorter than native ovalbumin mRNA,
the mRNA preparation must contain fragments of about 800 base
pairs derived from the 3'-end of ovalbumin mRNA.

Confirmation of an Ovomucoid Insert by DNA Sequencing

In order to confirm the identity of the cloned DNA of
pOM100 as ovomucoid cDNA a partial DNA sequence was determined. Recombinant plasmid pOM100 was cleaved with Pst I and
digested with snake venum phosphodiesterase (26). This
treatment enabled labeling of the 5' termini with bacterial
alkaline phosphatase and T4 polynucleotide kinase (27). The
200-bp Pst I fragment (designated P1) was isolated in a 12%
polyacrylamide slab gel, then incubated with Hinf I, which
cleaves P1 into two fragments of 116 and 84 bp in length.
The DNA sequence of fragment P1 of pOM100, shown in Figure 2,
codes for portions of two glycosylated ovomucoid peptides
whose amino acid sequence was determined by Beeley (28). The
matched sequence included a thirteen amino acid peptide, as
shown in Diagram 1 (below).

^{72}SER TYR ALA ASN THR THR SER GLU ASP GLY LYS VAL MET^{84}
5'AGT TAT GCC AAC ACG ACA AGC GAG GAC GGA AAA GTG ATG3'

This sequence match is well beyond random probability and
proved that the inserted DNA of pOM100 contained ovomucoid
sequences.

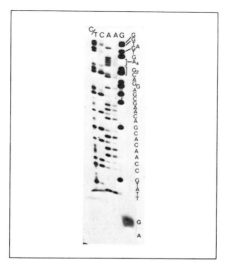

FIGURE 2. Autoradiograph of the nucleotide sequence
derived from Pst I fragment Pl of pOM100. Plasmid DNA was
cleaved with Pst I and the 5'-ends were dephosphorylated and
labeled with $[\gamma-^{32}P]$dATP. Fragment Pl was isolated and
cleaved with Hinf I. The sequence shown corresponds to the
coding (mRNA) strand near the left Pst I site toward the Hinf
I site in Pl (see map, Figure 3). Band doubling in the C + T
and C lanes resulted in some anomalous bands. This may be due
to incomplete β elimination.

A Partial Restriction Map of pOM100

A restriction map of the pOM100 clone is shown in Figure
3. The cloned DNA contains 646 base pairs of ovomucoid DNA,
including a 138 base pair sequence of non-coding DNA at the
3'-end. The cloned DNA is not a complete copy of the mRNA,
but rather is missing about 120 nucleotides at the 5'-end.
The cutting sites for the restriction endonucleases shown in
Figure 3 were determined from the nucleotide sequence; many
have been confirmed through restriction endonuclease digestions. The two Pst I sites, which define the Pl fragment
used for the sequence analysis described above, are located at
130 and 335 bp from the 5'-end. These are the only Pst I
sites in plasmid pOM100. Eco Rl and Bam Hl each cut the
cloned DNA only once, at sites in the 3' non-coding sequence
separated only by 12 base pairs. Hinc II also cuts only
once, in the middle of the cloned DNA. Hha I does not cut the
cloned ovomucoid sequence, but cuts the plasmid DNA at two

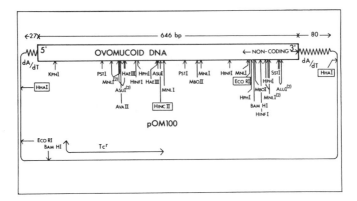

FIGURE 3.   A partial restriction map of pOM100.

sites close to either end of the insert.   This enzyme was
therefore used to cut out the entire ovomucoid DNA sequence
for the preparation of hybridization probes, as described
below.

REGULATION OF EXPRESSION OF THE OVOMUCOID GENE BY ESTROGEN

Preparation of a Pure Hybridization Probe

     A specific probe for ovomucoid mRNA sequences was pre-
pared from pOM100, as illustrated diagrammatically in Figure
4.   pOM100 was digested with the restriction endonuclease
Hha I.   The Hha I fragment containing the ovomucoid specific
sequences and flanking plasmid sequences was then purified
from the remaining plasmid fragments by agarose gel electrophoresis (29).   The purified fragment was labeled with [^{3}H]-
dCTP by nick translation to a specific activity of 6 X 10^6
cpm/μg.   A single-stranded ovomucoid probe was then prepared
by hybridizing these labeled fragments with a 100-fold excess
of ovomucoid mRNA to a R_0t of 2.5 X 10^{-2}.   Twenty micrograms
each of poly(U) and poly(A) was added to avoid self-reanneal-
ing of the Hha I fragment caused by the poly(A) and poly(T)
linker.   Under these conditions, the hybridization reaction
was complete and essentially no DNA-DNA reassociation
occurred (C_0t of 2.5 X 10^{-4}).   At the end of the incubation
period, the reaction mixture was treated with S_1 nuclease to
destroy the flanking plasmid sequences and anti-coding
sequences which remained single-stranded.   The hybridization

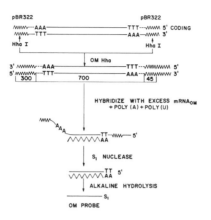

FIGURE 4. Schematic representation of the preparation of [3H]DNA$_{Om}$ probe. Fragments corresponding to DNA$_{Om}$ were labeled with [3H]dCTP by nick translation to a specific activity of 6 X 10^6 cpm/μg. Pure single-stranded probes were obtained as described in the text.

with excess ovomucoid mRNA and S$_1$ nuclease digestion were repeated once more to ensure complete removal of the plasmid and anticoding sequences. Finally, the single-stranded DNA probe ([3H]DNA$_{Om}$) was obtained by alkaline hydrolysis of the messenger RNA.

Ovomucoid is a Single Copy Gene

In order to determine the copy number of the ovomucoid gene in the chick genome, the kinetics of hybridization of the single-stranded [3H]DNA$_{Om}$ probe to an excess of chick oviduct and liver DNA was determined (Figure 5). The probe reacted similarly to both chick oviduct and liver DNA, with an apparent $C_0t_{\frac{1}{2}}$ value of 1.5 X 10^3. This agrees well with the value predicted for a single copy gene in the haploid chick genome (11,30-32). Therefore, the synthesis of ovomucoid mRNA resembles that observed for other proteins which are synthesized in large amounts (ovalbumin, globin, silk fibroin, etc.) and is accomplished by preferential transcription rather than by gene amplification (32,33).

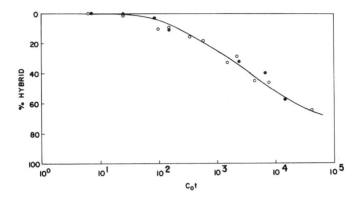

FIGURE 5. Hybridization of ^{3}H-labeled ovomucoid probe to
excess chick DNA. Chick DNA (1.3 mg) isolated from oviducts
(●) or from liver (0) was hybridized to 0.06 ng of $[^3H]DNA_{Om}$
in 200 µl.

The Effect of Estrogen on Ovomucoid mRNA Accumulation

The $[^3H]DNA_{Om}$ probe was also used to quantitate ovomucoid
RNA sequences in target and non-target tissues, and to
quantitate the induction of ovomucoid mRNA during secondary
estrogen stimulation (34). As shown in Figure 6, nuclear RNA

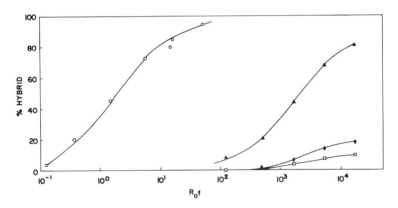

FIGURE 6. Hybridization of nuclear RNA to $[^3H]DNA_{Om}$.
$[^3H]DNA_{Om}$ (0.12 ng) was hybridized to 0.6 µg and 0.15 µg of
DES-stimulated chick oviduct RNA (0), 150 µg of chick liver
RNA (□), 150 µg of chick spleen RNA (◆), or 150 µg of
withdrawn chick oviduct RNA (▲).

isolated from chick oviducts stimulated chronically with estrogen reacted rapidly with the probe. The apparent $R_0t_{\frac{1}{2}}$ for the reaction is 1.7, indicating that 0.1% of the RNA sequences in stimulated oviduct nuclei are ovomucoid sequences. Nuclear RNA prepared from chicks first stimulated and then withdrawn from estrogen reacted 1000 times more slowly ($R_0t_{\frac{1}{2}}$ of 1.6 X 10^3) than nuclear RNA isolated from stimulated chicks. Nuclear RNA prepared from liver and spleen produced only 10-18% hybridization at a R_0t value of 2 X 10^4, indicating that ovomucoid is expressed only at extremely low levels in non-target tissue.

The marked diminution of ovomucoid sequences in the RNA preparation upon withdrawal of estrogen made it interesting to study the temporal effect of hormone on accumulation of $mRNA_{om}$ during acute stimulation. Chicks withdrawn from hormone for 14 days were subsequently given a single injection of estrogen, sacrificed at various time intervals, and the nuclear RNA isolated and hybridized to the $[^3H]DNA_{om}$ probe. Figure 7 shows the analyses of various RNA preparations. Within 2 hr after readministration of estrogen to withdrawn chicks, the rate of hybridization increased 4-fold. It gradually

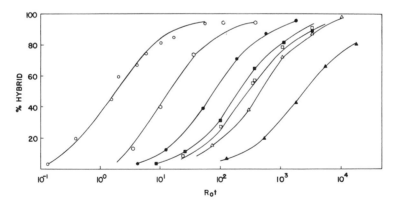

FIGURE 7. Analysis of ovomucoid sequences in nuclear RNA during secondary stimulation. Nuclear RNA was prepared from 14-day DES-withdrawn (▲, 150 µg), 2-h DES-treated (Δ, 90 µg), 4-h DES-treated (□ , 30 µg), 5-h DES-treated (■ , 30 µg), 16-h DES-treated (●, 15 µg), and 48-h DES-treated chick oviducts (O, 6 µg). Hybridization was carried out in a final volume of 30 µl. The kinetic curve for nuclear RNA isolated from 14-day DES-stimulated chick oviduct is included for comparison (O, 0.15 µg and 0.6 µg).

increased to 10-fold after readministration of estrogen for 8
to 16 hr, and finally reached a rate of hybridization at 48 hr
that was only 7-fold slower than that calculated for nuclear
RNA from oviducts of chronically stimulated chicks. Since
the proportion of tubular gland cells in the chick oviduct
during secondary stimulation has been determined (2,35) one
can estimate the number of molecules of ovomucoid RNA per
tubular gland cell nucleus from the $R_0t_{\frac{1}{2}}$ values in Figures 6
and 7. In this manner the number of ovomucoid RNA molecules
in chronically stimulated chick oviducts was calculated to be
about 1900. Upon withdrawal of hormone for 14 days, the
number of ovomucoid RNA molecules decreased to 3 per tubular
gland cell nucleus. Four hours after readministration of
hormone, 38 molecules per cell were found. The number of
molecules increased further to 600 per cell after 48 hrs,
which is 200 times higher than that observed for withdrawn
chicks. However, this increase is gradual and linear for a
long period of time and is significantly different from that
observed for the induction and accumularion of ovalbumin RNA.
As shown in Figure 8, ovalbumin and ovomucoid mRNA sequences
increase at similar rates during the first phase of induction
(0-4 hrs). Four hours after readministration of hormone, the
accumulation of ovalbumin mRNA sequences increases at a rate

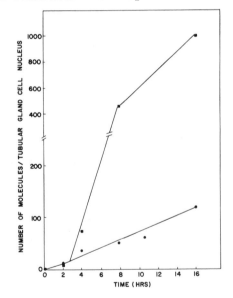

FIGURE 8.   Induction of ovomucoid and ovalbumin messenger
RNA sequences during secondary stimulation. Data for accumulation of ovalbumin mRNA sequences were taken from Roop et
al. (37).

ten-fold greater than that of the earlier phase. In contrast,
the concentration of ovomucoid mRNA sequences continues to
increase at a rate similar to that observed during the first
few hours. After 16 hr, 1000 molecules of ovalbumin mRNA as
compared with 120 molecules of ovomucoid mRNA are present per
tubular gland cell nucleus. At steady-state conditions,
ovalbumin mRNA accounts for about half of the total cellular
mRNA, while ovomucoid mRNA is approximately 6-9% of the total.
The difference in the rate and level of accumulation of
the two gene products may be due to two reasons. First,
the rate of transcription of the ovalbumin gene appears to be
greater than the ovomucoid gene after the initial phase of
induction. Second, ovalbumin mRNA sequences may be stabilized
by an unknown mechanism and thus degraded at a slower rate
than ovomucoid mRNA.

Ovomucoid mRNA Transcription in Isolated Nuclei

The transcription of ovomucoid sequences in isolated
nuclei from several tissues and hormonal states was also
investigated. [^{3}H]RNA was synthesized in nuclei isolated
from hormone-stimulated chick oviducts, withdrawn oviducts,
or chick liver or spleen and then assayed for the presence of
mRNA$_{om}$ sequences by hybridization to filters containing pOM100
DNA. As shown in Table II, radiolabeled ovomucoid mRNA
sequences were found in the in vitro transcripts from chronically stimulated chick oviducts but very little if any were
observed in withdrawn oviduct, liver or spleen. Thus,
transcription of the ovomucoid gene sequences is relatively
tissue specific, confirming the results of the isolated
nuclear RNA hybridization studies presented previously.

Of the total RNA synthesized in nuclei isolated from
stimulated oviducts, 0.018% corresponds to mRNA$_{om}$ sequences.
This value is approximately 200-fold greater than would be
expected if random transcription of the haploid chick genome
occurred. Since only 10% of the DNA is expressed in oviduct
cells (36), ovomucoid sequences are actually transcribed at a
frequency 20-fold greater than random transcription of the
available sequences. It is thus clear that the ovomucoid
gene is transcribed preferentially in stimulated oviduct
nuclei. Of interest, also, is the observation that 0.1% of
the RNA synthesized corresponds to mRNA$_{ov}$ sequences. This
suggests that the ovomucoid gene is transcribed at a rate
5.5 X slower than the ovalbumin gene on a mass basis. The
amount of mRNA$_{om}$ (nucleotides) present in the nuclear RNA
extracted from oviducts of chronically stimulated chicks is
0.11% while that for ovalbumin sequences is 0.4% (37). The

TABLE II. In Vitro Transcription of $mRNA_{om}$ and $mRNA_{ov}$ from Isolated Nuclei

Filter	$[^3H]RNA$[a]	Competitor	$[^3H]RNA$ Hybridized (cpm)	Hybridizable gene sequences (cpm)	% mRNA in total RNA
pOM100	$oviduct_{DES}$	—	232	1077	0.018
		$mRNA_{om}$	78		
		yeast RNA	248	1287	0.021
E. coli DNA	$oviduct_{DES}$	—	96	N.D.	
		$mRNA_{om}$	46		
rat liver DNA	$oviduct_{DES}$	—	84	N.D.	
		$mRNA_{om}$	98		
pHb1001	$oviduct_{DES}$	—	86	N.D.	
		$mRNA_{om}$	68		
pOM100	$oviduct_W$	—	74	N.D.	
		$mRNA_{om}$	86		
pOM100	liver	—	106	N.D.	
		$mRNA_{om}$	90		
pOM100	spleen	—	76	N.D.	
		$mRNA_{om}$	80		
pOM100	$oviduct_{DES}$ + α-amanitin	—	36	N.D.	
		$mRNA_{om}$	38		
pOV230	$oviduct_{DES}$	—	1130	6300	0.106
		$mRNA_{ov}$	166		

[a] Input $[^3H]RNA$: $oviduct_{DES}$, 5.9×10^6 cpm; $oviduct_W$, 4.4×10^6 cpm; liver, 4.9×10^6 cpm; $oviduct_{DES}$ + α-amanitin, 1.6×10^6 cpm.

four-fold difference in the relative amount of these two
nuclear mRNA sequences in vivo is very similar to the relative rate of transcription of the two genes in isolated nuclei.
Therefore, transcription rates might contribute significantly
to differential expression of the two genes. This by no means
rules out potential contributions of post-transcriptional factors as the differences in concentration of these two proteins
in the cytoplasm is greater than the difference in concentration with respect to their mRNAs in nuclei.

RESTRICTION MAPPING OF THE OVOMUCOID GENE IN CHICK DNA DIGESTS

     Recently, it was shown that the structural sequences of
the ovalbumin gene in native chick DNA are not contiguous (29,
38-39). Rather, multiple non-coding DNA sequences (intervening sequences) are interspersed within the structural ovalbumin gene. To determine if the ovomucoid structural gene is
similarly split, limited restriction mapping studies of hen
liver DNA were undertaken.

Preparation of Ovomucoid Probes

     To generate ovomucoid-specific probes for these studies,
pOM100 was digested with the restriction endonuclease Hha I.
This generated a 1.04 Kb fragment (OM_T) that contained the
entire ovomucoid DNA insert as shown in Figure 9. After
resolving this fragment from the other Hha I fragments of
pBR322 by electrophoresis in a 1% agarose gel, it was further

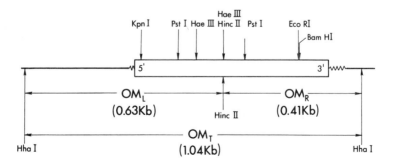

FIGURE 9. Diagrammatic illustration of the ovomucoid-
specific hybridization probes prepared from the recombinant
plasmid pOM100.

digested with <u>Hinc</u> II. This enzyme cuts only once, in the
middle of the ovomucoid sequence, to generate a 0.63 Kb fragment (OM_L) and a 0.41 Kb fragment (OM_R) corresponding to the
left and right halves respectively of the cloned ovomucoid
DNA sequence. These fragments were resolved in agarose gels,
excised separately and recovered by diffusion (22). The
purified probes were nick-translated with $[^{32}P]$dGTP to about
1×10^8 cpm/μg by the method of Maniatis <u>et</u> <u>al</u>. (40).

Does the Ovomucoid Gene Contain Intervening Sequences

Hen liver DNA was then digested with various restriction
endonucleases, and the restricted DNA fragments were separated
by agarose gel electrophoresis and transferred onto a nitrocellulose filter by the method of Southern (41). The filter
was treated according to the procedure of Denhardt (42) and
those DNA fragments on the filter containing structural ovomucoid sequences were identified by hybridization to $[^{32}P]OM_T$
followed by radioautography (43). The radioautograms of five
of these digests are shown in Figure 10. Only one of the
endonucleases used, <u>Hinc</u> II, cuts the ovomucoid structural
gene clone. It thus was expected to produce two hybridizable
fragments. Instead, three were observed. The other four
enzymes displayed here do not cut the cloned ovomucoid DNA,
and thus should have resulted in only one hybridizable band.
However, <u>Hind</u> III, <u>Taq</u> I and <u>Pvu</u> II resulted in three hybridizable fragments, and <u>Hpa</u> II digestion yielded four. Therefore, non-coding sequences which contain target sites for
these enzymes must be interspersed within the ovomucoid
natural gene.

Restriction Mapping of the Ovomucoid Gene

In order to determine the relative positions of these
intervening sequences in the native ovomucoid gene, hybridization of restricted DNA with the left (OM_L) and right (OM_R)
ovomucoid probes was performed (Figure 11). Panel A shows
the fragments from seven different digests that hybridized to
OM_L and Panel B the fragments that hybridized to OM_R. Only
one restriction endonuclease, <u>Bgl</u> II, yielded the same hybridization pattern with both OM_L and OM_R. This result indicated
that this 8.5 Kb <u>Bgl</u> II fragment contained the entire region
of the ovomucoid gene included in the structural gene clone.
The hybridization pattern observed with each of the other
restriction enzymes used differed with OM_L and OM_R, indicating that each of these enzymes cuts within the gene sequence

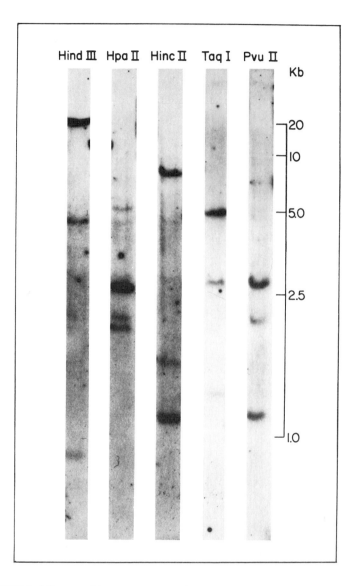

FIGURE 10.   Radioautogram of an agarose gel containing hen
liver DNA after digestion with various restriction endonucleases and Southern hybridization.   12.5 µg of hen liver DNA
was digested by the respective restriction endonuclease listed
above each lane in the radioautogram.   Electrophoresis was
carried out in 1.5% agarose gels and DNA was transferred to
nitrocellulose filters.   Hybridization was carried out with
$[^{32}P]$labeled OM_T.

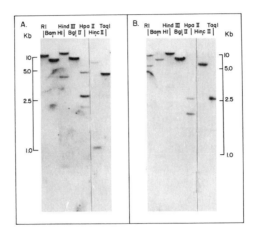

FIGURE 11.   Radioautogram of agarose gel containing hen
liver DNA after digestion with various restriction endonucleases and Southern hybridization.   The experiment was
carried out as in Figure 10, except the hybridization was
carried out with $[^{32}P]$labeled OM_L (Panel A) and $[^{32}P]$labeled
OM_R (Panel B), respectively.

at least once.   The sizes of these fragments, and of fragments derived from other, double digestions are summarized in
Table III.   These data were employed to construct a preliminary model for the sequence organization of the natural
ovomucoid gene as shown in Figure 12.   Hinc II digestion of
hen liver DNA resulted in two DNA fragments of 6.7 and 1.1 Kb
in length that hybridized with OM_L and a 6.0 Kb DNA fragment
that hybridized with OM_R but not with OM_L (Table III).   By
virtue of the method used to prepare the two probes the Hinc
II site at the 3'-end of either the 6.7 or 1.1 Kb fragment
must be the unique Hinc II cleavage site present in the structural ovomucoid gene (Figure 9).   Due to the large size of the
6.7 Kb fragment, it was ordered to the 5'-end of the 1.1 Kb
fragment, as shown in line 2 of Figure 12.   Thus, the Hinc II
site between the 1.1 Kb and 6.0 Kb Hinc II fragments, Hinc II_C
should be the Hinc II site present in the structural gene.

TABLE III. Sizes of Ovomucoid Sequence-Containing Chicken DNA Fragments After Restriction Endonuclease Digestion

	Size after digestion, Kb						Probe Used
	Eco RI	Bam HI	Hind III	Bgl II	Hinc II	Taq I	
	15.0	8.0	18.0 4.5 0.8	8.4	6.7 1.1	4.5	OM_L
+ Eco RI	15.0	8.0	4.5 3.9 0.8	5.8	6.7 1.1	4.5	OM_L
+Hinc II	6.7 1.1	6.3 1.1	3.7 1.1 0.8	4.1 1.1	6.7 1.1	4.5 1.1	OM_L
	15.0 7.0	8.0 1.2	18.0	8.4	6.0	2.9	OM_R
+ Eco RI	15.0 7.0	8.0 1.2	6.0 3.9	5.9 2.6	5.4 0.6	2.5 0.4	OM_R
+Hinc II	5.4 0.6	1.2 0.6	6.0	3.2	6.0	2.9	OM_R

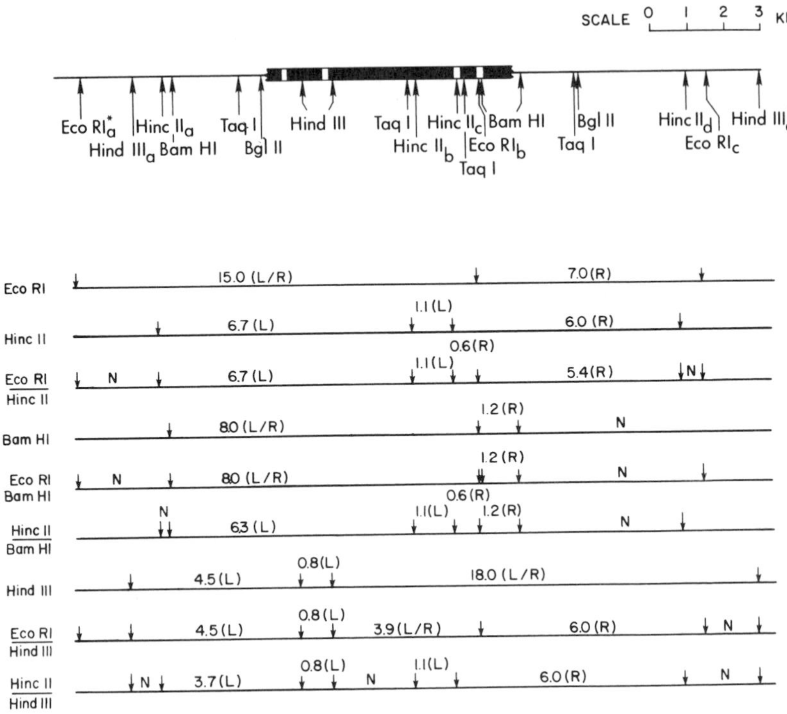

SCALE 0 1 2 3 Kb

Eco RI$_a^*$ | Hinc II$_a$ | Taq I | Hind III | Taq I | Hinc II$_c$ | Bam HI | Bgl II | Hinc II$_d$ | Hind III$_d^*$
Hind III$_a$ Bam HI | Bgl II | Hinc II$_b$ | Eco RI$_b$ | Taq I | Eco RI$_c$
Taq I

Eco RI — 15.0 (L/R) — 7.0 (R)

Hinc II — 6.7 (L) — 1.1 (L) — 6.0 (R)

Eco RI / Hinc II — N — 6.7 (L) — 0.6(R) 1.1(L) — 5.4(R) — N

Bam HI — 8.0 (L/R) — 1.2 (R) — N

Eco RI / Bam HI — N — 8.0 (L/R) — 1.2 (R) — N

Hinc II / Bam HI — N — 6.3 (L) — 0.6(R) 1.1(L) 1.2(R) — N

Hind III — 4.5 (L) — 0.8 (L) — 18.0 (L/R)

Eco RI / Hind III — 4.5 (L) — 0.8 (L) — 3.9 (L/R) — 6.0 (R) — N

Hinc II / Hind III — N — 3.7 (L) — 0.8(L) — N 1.1(L) — 6.0 (R) — N

FIGURE 12. Model for the organization of the natural ovomucoid gene in chicken DNA. DNA sequences present in cDNA$_{Om}$ are represented by □ , intervening sequences are represented by ■ , and flanking DNA sequences are represented by —. Various restriction sites on the DNA are shown by the arrows. Two exceptions to the scale are the EcoRI$_a^*$ site (14.4 Kb to the left of the Hinc II$_c$ site) and the Hind III$_d$ site (14.5 Kb to the right of the Hinc II$_c$ site). The generation of DNA fragments of different sizes by various restriction digestions of this DNA is shown below the model: numbers represent the sizes of DNA fragments in kilobases; (L), (R) and (L/R) indicate that the DNA fragment was detected by hybridization with OM$_L$, OM$_R$, or both OM$_L$ and OM$_R$, respectively. N represents DNA fragments that should have resulted from various restriction digestions but were not detected in the radioautograms because there was no DNA sequence complementary to the probes used or the lengths of the complementary sequences were insufficient to form stable hybrids under the conditions used.

Eco RI digested the hen DNA into a 15 Kb DNA fragment
hybridizable to both OM_L and OM_R, and a 7 Kb fragment
hybridizable only to OM_R (Table III). Furthermore, Eco RI
cuts the 7 Kb Hinc II fragment into 0.6 and 5.4 Kb fragments,
both hybridizable only to OM_R (line 3, Figure 12). Thus there
should be an Eco RI site (Eco RI_b) 0.6 Kb to the right of
Hinc II_c, and 6.4 Kb to the left of Hinc II_d. Since there is
an Eco RI site 0.22 Kb to the right of the Hinc II site in the
ovomucoid structural gene and no smaller fragments were found
in the Eco RI/Hinc II double digests, Eco RI_b must be this
structural Eco RI clevage site. However, since it is 0.6 Kb
from Hinc II_c, there must exist an intervening sequence of
several hundred base pairs between Hinc II_c and Eco RI_b.

Similar analyses were performed with the restriction
endonucleases Bam HI, Hind III, Taq I and Bgl II, and the
results were incorporated into the map of the ovomucoid gene
presented in Figure 12. The ovomucoid structural gene
sequences are not contiguous and are apparently separated
into at least four portions by multiple intervening sequences
within the chicken genome. Although the precise location
(and structure) of the ends of the ovomucoid gene have not
been determined, as indicated in Figure 12 by the broken
intervening sequence DNA, the ovomucoid gene within native
chicken DNA appears to be about 6 Kb in length.

CLONING OF THE NATURAL OVOMUCOID GENE

Although the restriction mapping of the ovomucoid gene in
digests of hen liver DNA did substantiate the existence of
intervening sequences within this gene, the number and location of these sequences can only be roughly approximated. In
order to precisely define the sequence organization of this
gene within genomic chick DNA, it was necessary to isolate
and clone natural gene fragments. Since one of the two Eco RI
ovomucoid fragments had been shown by total DNA mapping to
contain most of the structural gene sequences, as well as the
5' region, this 15 Kb fragment was enriched from total Eco RI-
digested chick DNA by preparative agarose gel electrophoresis
and employed for cloning of the ovomucoid gene.

Lambda phage Charon 4A contains 3 Eco RI cleavage sites
and yields 4 DNA fragments upon digestion with Eco RI. The
two upper phage DNA arms were separated from the middle Eco
RI DNA fragments by preparative agarose gel electrophoresis.
The isolated phage DNA arms were allowed to ligate in the
presence of the DNA fraction enriched for ovomucoid DNA using
T4 DNA ligase. The ligation mixture was employed for cloning

in a P3 facility by in vitro packaging according to the
method of Sternberg et al. (44) as modified by Faber,
Kiefer and Blattner (personal communication). Approximately
20,000 phage plaques were generated in this experiment.
These phages were screened for the recombinant phage containing the ovomucoid gene by hybridization with the OM_T probe
using an amplification procedure as described previously (18).
Three phage plaques gave positive signals on radioautograms
in this test. These phages were cultured individually and
their DNA prepared. Digestion of DNA from any of these
phages with Eco RI produced two DNA fragments corresponding
to the arms of Charon 4A and an additional 15 Kb DNA fragment
which hybridized with the ovomucoid structural sequence OM_T
upon Southern analysis. This 15 Kb DNA fragment of the
natural ovomucoid gene was subsequently recloned in pBR322 in
E. coli X1776 and was designated pOM15.

RESTRICTION MAPPING OF THE CLONED 15 KB OVOMUCOID DNA

The 15 Kb ovomucoid DNA (OM15) was excised from pBR322 DNA
by Eco RI digestion of the recombinant plasmid pOM15 followed
by preparative agarose gel electrophoresis. The excised ovomucoid DNA was further digested with a variety of other
restriction endonucleases and those DNA fragments containing
the left and right halves of the structural ovomucoid gene
sequence were identified by agarose gel electrophoresis
followed by Southern hybridization with the OM_L and OM_R
probes, respectively. Some of the data are shown in Figure
13 and summarized in Table IV. These data were employed to
construct a preliminary map of the natural ovomucoid gene as
shown in Figure 14. As expected the OM15 DNA hybridized with
both probes. Hinc II digestion of OM15 DNA resulted in two
DNA fragments of 8.0 and 1.1 Kb in length that hybridized
with OM_L and a unique 0.7 Kb DNA fragment that hybridized with
OM_R but not OM_L. Thus, there is a Hinc II site (Hinc II_C)
0.7 Kb to the left of Eco RI_b site (Figure 14). By virtue of
the method used to prepare the two probes the Hinc II_C site
must be the unique Hinc II cleavage site present in the
structural ovomucoid gene (Figure 9). Limited sequence analysis has also established that the Eco RI_b site is the one
present in the structural ovomucoid gene (data not shown).
The unique Hinc II and Eco RI sites in the structural gene,
however, are separated by only 0.2 Kb of DNA. Thus, it
appears that these two sites are separated by an intervening
sequence in genomic DNA, confirming the results observed with
total chick DNA mapping. Furthermore, there should be a

TABLE IV. Sizes of Ovomucoid Structural Sequence-Containing Fragments After Restriction Endonuclease Digestion of OM15

	Size after digestion (kb)								Probe Used
	EcoRI	Hind III	Bam HI	Taq I	Ava I	Kpn I	Hae III	Pst I	
-Hinc II	15.0	3.9 0.9	10.0	2.4	3.5 2.3	8.5 3.7	1.0 0.5	0.75 0.75	OM_L
+Hinc II	8.0 1.1	1.1 0.9	8.0 1.1	1.1	2.3 1.1	6.0 1.1	0.6 0.5	0.7 0.3	OM_L
-Hinc II	15.0	3.5	10.0 0.3	0.75	3.5	3.7	0.7	0.65	OM_R
+Hinc II	0.7	0.7	0.4 0.3	0.7	0.7	0.7	0.7	0.65	OM_R

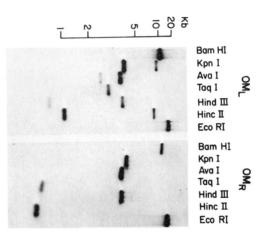

FIGURE 13. Radioautogram of 1% agarose gel containing
OM15 DNA after digestion with various restriction endonucleases
and Southern hybridization. OM15 DNA was extracted from Eco
RI-digested pOM15 by preparative agarose gel electrophoresis.
100 ng of purified OM15 DNA were digested by respective
restriction endonucleases as indicated above each lane in the
radioautogram. Electrophoresis was carried out in 1% agarose
gels and DNA was transferred to nitrocellulose filters.
Hybridization was carried out with [^{32}P]labeled OM$_L$ (left
panel) and [^{32}P]labeled OM$_R$ (right panel), respectively.

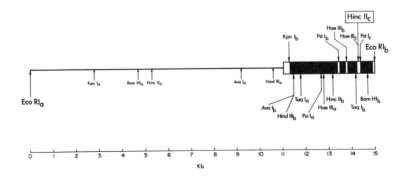

Hinc II site (Hinc II$_a$) 9.1 Kb to the left of Hinc II$_c$. The
Hinc II$_b$ site, however, could be either 8.0 or 1.1 Kb to the
left of Hinc II$_c$.

The presence of an intervening sequence in the genomic DNA
between Hinc II$_c$ and Eco RI$_b$ was also confirmed by Bam HI
digestion of OM15 DNA (Table IV). The results indicated the
presence of a Bam HI site (Bam HI$_b$), 0.3 Kb to the left of
Eco RI$_b$, which does not exist in the structural gene.

Hind III digested OM15 DNA into a 3.5 Kb DNA fragment
hybridizable with both OM$_L$ and OM$_R$ probes as well as a 0.9 Kb
DNA that hybridized with OM$_L$ only. Thus, there should be a
Hind III site (Hind III$_b$) 3.5 Kb to the left of Eco RI$_b$ and an
additional Hind III site (Hind III$_a$) 0.9 Kb to the left of
Hind III$_b$ (Figure 14). Indeed the 3.5 Kb DNA fragment is
cleaved by Hinc II into a 1.1 and a 0.7 Kb fragment hybridizable with OM$_L$ and OM$_R$, respectively, which are the same DNA
fragments observed by digestion of OM15 DNA with Hinc II alone
(Table IV). This experiment further suggests that the Hinc
II$_b$ site is 1.1 Kb to the left of Hinc II$_c$ and that there is
no structural gene sequence between Hind III$_b$ and Hinc II$_b$.
In addition, there should be some structural gene sequence
within the 0.9 Kb Hind III DNA fragment (Fig. 14). Since the
structural gene contains no Hind III cleavage site, it
suggests that there may be more intervening DNA sequences
within the natural ovomucoid gene.

Kpn I digestion of OM15 DNA confirmed both of the above
hypotheses. The Kpn I site present in the 5' region of the
ovomucoid structural gene (Fig. 9) was mapped to a site
(Kpn I$_b$) 3.7 Kb to the left of Eco RI$_b$, which is between
Hind III$_a$ and Hind III$_b$. Thus, there is a structural gene
sequence between these sites, and a large intervening
sequence must be present between Hind III$_b$ and Hinc II$_b$.

FIGURE 14. Structural organization of the natural ovomucoid gene (OM15). Ovomucoid structural sequences are
represented by ☐ , intervening sequences are represented by
■ , and flanking DNA sequences are represented by ——.
Various restriction endonuclease sites on the ovomucoid gene
are indicated by the arrows. Those above the schematic
drawing indicate restriction sites present in the structural
ovomucoid gene and those below are sites within the intervening or flanking DNA sequences. There are additional restriction sites present in the left hand portions of OM15 DNA, but
their positions have not been mapped and are therefore not
shown in this schematic representation of the ovomucoid gene.

The OM15 DNA was digested by Hae III into a unique 0.7 Kb
DNA fragment hybridizable to OM_R which is also resistant to
Hinc II digestion (Table IV). Hence there is a Hae III site
(Hae III_c) about 0.7 Kb to the left of Eco RI_b, which could
be one of the two Hae III sites located near the unique
Hinc II site in the structural gene (Fig. 9). Since there are
two Hae III fragments of 1.0 and 0.5 Kb that hybridized with
OM_L, there should be a Hae III site (Hae III_a) 1.5 Kb to the
left of Hae III_c. Furthermore the 0.5 Kb Hae III fragment
appears to be resistant to Hinc II while the 1.0 Kb fragment
is cleaved by this enzyme to a 0.6 Kb fragment hybridizable
to OM_L (Table IV). Thus, the Hae III_b site should be located
at 0.5 Kb to the left of Hae III_c. Otherwise, the 1.0 Kb Hae
III DNA fragment would be resistant to Hinc II while the 0.5
Kb fragment would be digested to a 0.4 and 0.1 fragment.
Furthermore, this Hae III_b site may be the second Hae III site
present in the structural gene (Fig. 9) since both the 0.5 and
1.0 Kb Hae III DNA fragments hybridized with OM_L. By
similar analysis a Taq I site (Taq I_b) has been located
between the Hae III_b and Hae III_c sites (Figure 14). Since
the Hae III sites in the structural gene are not separated by
a Taq I site, there may be a third intervening sequence within the ovomucoid gene.

Pst I digested the OM15 DNA into a unique 0.65 Kb fragment hybridizable to OM_R which is resistant to further digestion by Hinc II. Hence a Pst I site (Pst I_c) should be
located at 0.65 Kb to the left of Eco RI_b and must be the Pst
I site located at 0.05 Kb to the right of the unique Hinc II
site present in the structural gene (Fig. 9). Although Pst I
digestion yielded only one 0.75 Kb band when hybridized to OM_L,
there must be at least two DNA fragments of the same size
because one of the fragments was digested by Hinc II into a
0.3 Kb fragment while the other fragment remained resistant.
Hence there should be two Pst I sites to the left of Pst I_c
that are 0.75 Kb apart. Moreover, the Pst I_b site may be the
other Pst I site present in the structural gene (Figure 9)
because the 0.75 Kb Pst I fragment should have been digested
to 0.45 Kb instead of a 0.3 Kb fragment if Pst I_a is the site
present in the structural gene. Since the Pst I_b and Hae III_b
sites are separated by 0.3 Kb in genomic DNA but only by 0.07
Kb in the structural gene (Figure 9), it appears that there
may be a fourth intervening sequence within the ovomucoid gene.

These results therefore substantiate the preliminary
restriction studies carried out on total DNA digests of hen
liver. The structural ovomucoid gene sequences are not contiguous and are apparently separated into several portions by
multiple intervening sequences within the chick genome. The
existence of four intervening sequences as established by

limited restriction analyses of the cloned OM15 DNA should be a minimum estimation of the number of intervening sequences.

OVOMUCOID mRNA PRECURSORS IN CHICK OVIDUCT NUCLEI

Identification of High Molecular Weight Species
of Ovomucoid RNA in Chick Oviduct Nuclei

Recently several high molecular weight species of ovalbumin sequence-containing RNA were identified in chick oviduct nuclei (37). These species presumably represent a large nuclear precursor RNA in different stages of processing. Since we have demonstrated that the ovomucoid gene contains multiple intervening sequences, there might also exist in oviduct nuclei species of ovomucoid RNA larger than mature mRNA$_{Om}$. To search for these species, nuclear RNA from oviducts of estrogen-stimulated chicks was subjected to agarose gel electrophoresis in the presence of the denaturing agent, methylmercuryhydroxide. Following transfer of the RNA from the gel onto diazobenzyloxymethyl (DBM) paper and hybridization to a [^{32}P]labeled ovomucoid cDNA probe, multiple discrete bands containing ovomucoid structural sequences were observed (Figure 15, lane A). The band with the greatest mobility comigrated with mature mRNA$_{Om}$ (Figure 15, lane B). Five major bands were significantly larger than ovomucoid mRNA (800 nucleotides). In addition, two minor bands were detected in certain nuclear RNA preparations. As shown in Table V, the sizes of the seven species of high molecular weight ovomucoid RNA, which were labeled a-g, ranged from 1.7 to 6 times the size of ovomucoid mRNA. The mobilities of the prominent ovomucoid RNAs (bands a,b,d,e,g) were reproducibly observed in all 5 preparations of oviduct nuclear RNA that were analyzed. Based upon its intensity, band "g", which has 1650 nucleotides, appears to be the most abundant species of high molecular weight ovomucoid RNA. Based upon the radioactivity associated with each band, as determined by liquid scintillation counting, we estimate that 20-45% of the nuclear ovomucoid sequences reside in the forms higher in molecular weight than mRNA$_{Om}$.

High molecular weight species of RNA that contain ovomucoid sequences were not found in oviduct cytoplasm, where only mature mRNA$_{Om}$ was detected (Figure 15, lane C). In addition, the presence of nuclear high molecular weight ovomucoid RNA, as well as that of mature mRNA$_{Om}$, was tissue specific. No ovomucoid bands were detected in the nuclear RNA from spleen

TABLE V.   Sizes of the Ovomucoid Sequence-Containing
           RNA from Chick Oviduct Nuclei

RNA$_{Om}$ species	Relative mobility (to mRNA$_{Om}$)	Length (nucleotides)
a	0.25	5400+120
b	0.45	3100+ 60
c	0.48	2750+ 60
d	0.55	2500+ 60
e	0.59	2300+ 50
f	0.65	2000+ 30
g	0.76	1700+ 75
mRNA$_{Om}$	1.0	1100+110

Relative mobility was calculated as the average of 6
individual experiments.   Length was calculated by comparison with the electrophoretic mobility of RNA and
DNA standards.   RNA standards were chicken 27S rRNA
(4,530 nucleotides), chicken 18S rRNA (2,076 nucleotides), Escherichia coli 23S rRNA (3,129 nucleotides)
and E. coli 16S rRNA (1,550 nucleotides).   DNA standards were obtained by digestion of λ DNA with EcoRI
(21, 746, 7,524, 5,920, 5,523, 4,793 and 3,380
nucleotides) or with Hind III (23,222, 9,730, 6,460,
4,492, 2,301, 2,000 and 682 nucleotides).   In the conditions of electrophoresis (in the presence of methylmercury hydroxide), DNA becomes single stranded and
its mobility is similar to that of RNA.   Error limits
indicate the standard deviation.

or liver that was isolated from estrogen-stimulated chicks
(Fig. 15, lanes D, E).   Thus, high molecular weight ovomucoid
RNA, as well as mRNA$_{Om}$, is present in significant amounts
only in those tissues that actively synthesize egg-white
proteins.
     Although the conditions employed for electrophoresis were
rigorously denaturing (45,46), an attempt was made to generate
high molecular weight species of ovomucoid RNA by mixing liver nuclei
and oviduct cytoplasm (both prepared from chicks stimulated with DES)
and then extracting the RNA from the mixture.   However, mRNA$_{Om}$ was
the only species detected (Fig. 15, lane F).   Thus as was observed previously for ovalbumin RNA, the high molecular weight
species of ovomucoid RNA do not appear to be due to aggregation
of mRNA$_{Om}$ with itself or other RNA molecules.

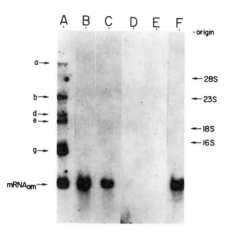

FIGURE 15.   Identification of ovomucoid mRNA sequences in
high molecular weight nuclear RNA.   Nuclear or cytoplasmic RNA
was isolated from the indicated tissues from White Leghorn
chicks stimulated chronically with diethylstilbestrol (DES).
RNA was subjected to agarose gel electrophoresis in the presence of 10 mM methylmercury hydroxide, then transferred by
blotting from the gel to diazobenzyloxymethyl (DBM) paper by
the method of Alwine et al. (46).   Following transfer the DBM
paper was hybridized to the $[^{32}P]$labeled ovomucoid probe (3 X
10^7 cpm, Spec. act. 5 X 10^8 cpm per μg), washed 3 times in 200
ml of 0.5 X SSC (75 mM NaCl, 7.5 mM Na citrate, pH 7.0) for 2
hours at 65°, dried and radioautographed for 16 hours using a
Dupont Cronex intensifying screen at -20°. A: oviduct nuclear
RNA, 20 μg.   (High molecular weight ovomucoid RNA species "c"
and "f" were not detected in this preparation.) B: ovomucoid
mRNA purified to approximately 40%, 15 ng.   C: oviduct cytoplasmic RNA, 20 μg. D: spleen nuclear RNA, 20 μg. E: liver
nuclear RNA, 20 μg. F: RNA isolated after mixing liver nuclei
with oviduct cytoplasm, 20 μg. The position of 28S, 23S, 18S
and 16S ribosomal RNA markers is indicated.

Identification of Ovomucoid Intervening Sequences in
High Molecular Weight Nuclear RNA

In order to confirm the hypothesis that these large nuclear
ovomucoid RNA species represent mRNA precursors in different
stages of processing, it was necessary to identify ovomucoid
intervening sequences in these species.   Therefore, a hybridization probe was prepared by nick-translation of the OM15
clone which contains all of the ovomucoid structural and

intervening sequence regions (as described above). When this
probe was hybridized to nuclear RNA from chronically stimulated chick oviducts, all of the ovomucoid nuclear RNA bands,
except "f", that hybridized to the ovomucoid cDNA probe were
again detected (Figure 16, lane -mRNA). In addition, two bands

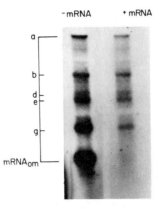

FIGURE 16.   Identification of ovomucoid intervening
sequences in high molecular weight ovomucoid nuclear RNA.
Oviduct nuclear RNA from chronically stimulated chicks, 20 µg,
was electrophoresed, transferred to DBM paper (2 x 14 cm) and
allowed to hybridize to the ^{32}P-labeled natural ovomucoid gene
probe (10^7 cpm, Spec. act. 6×10^8 cpm per µg) in the absence
(lane -mRNA) or presence of unlabeled ovomucoid mRNA as a
competitor (lane +mRNA). The conditions for competition were
as follows: Ovomucoid mRNA (20 µg, approximately 20% pure)
was preincubated with the ^{32}P-labeled probe in 1.0 ml of
hybridization buffer at 42° for 1 hr. The mixture was diluted
to 5.0 ml with hybridization buffer and then hybridized to the
DBM paper containing the bound nuclear RNA. Radioautography
of both samples was for the same length of time (about 1 week).

(located between bands "a" and "b") that were extremely faint
after hybridization to the ovomucoid cDNA probe were clearly
detected by the natural ovomucoid gene probe. In order to
detect only those bands that contained intervening sequences,
it was necessary to hybridize in the presence of excess
unlabeled ovomucoid mRNA to compete out the structural
sequences in the natural gene. Under these hybridization conditions, detection of the band corresponding to mature mRNA$_{om}$

was eliminated (Figure 16, lane +mRNA). In contrast, all of
the bands of high molecular weight ovomucoid RNA were still
detected. This indicates that all of the species of ovomucoid
RNA larger than mature mRNA$_{om}$ contain sequences complementary
to the intervening sequences of the ovomucoid gene.

Estrogen-Induced Accumulation of High Molecular Weight Species of Ovomucoid RNA

The accumulation in the oviduct nucleus of species of ovomucoid RNA larger than mature mRNA$_{om}$ was dependent upon estrogen. From 20 μg of oviduct nuclear RNA isolated from chicks
that had been withdrawn from DES treatment for 14 days,
neither high molecular weight ovomucoid RNA nor mRNA$_{om}$ was
detected (Figure 17, left, lane W). Thus, in agreement with
our studies described above using R_0t analysis, the amount of
ovomucoid RNA sequences in withdrawn oviduct nuclei is very
low. When 100 μg of withdrawn oviduct nuclear RNA was

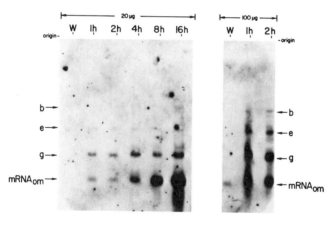

FIGURE 17. Induction of high molecular weight ovomucoid
RNA by secondary estrogen stimulation. Oviduct nuclear RNA
was isolated from chicks that had received daily subcutaneous
injections of 2.5 mg of DES for 14 days and subsequently withdrawn from hormone for 14 days (W). In addition, hormone
withdrawn chicks were injected with a single dose of 2.5 mg
of DES and the oviduct nuclear RNA was isolated after 1,2,4,8
and 16 hours. The RNA was electrophoresed, transferred to DBM
paper (15 X 15 cm) and hybridized to the [^{32}P] ovomucoid probe
(3.3 X 10^7 cpm, Spec. act. 8 X 10^8 cpm per μg). Left Panel:
20 μg of nuclear RNA was analyzed; Right Panel: 100 μg of
nuclear RNA was analyzed.

analyzed, a faint band of mRNA$_{Om}$ was observed (Figure 17
right, lane W). Thus, the few ovomucoid RNA sequences that
remain after hormone withdrawal appear to be present mainly in
the form of mature mRNA$_{Om}$.

As soon as one hour after administration of DES to hor-
mone-withdrawn chicks, both mature mRNA$_{Om}$ and high molecular
weight ovomucoid RNA species have accumulated. Analysis of
20 µg of nuclear RNA revealed the accumulation of the most
abundant high molecular weight species, band "g" (Figure 17,
left), while analysis of 100 µg of RNA revealed the accumulation of the three most predominant species ("b", "e" and "g")
larger than mRNA$_{Om}$ (Figure 17, right). Species "a", "c", "d"
and "f" were not detected, probably because these are the
least abundant of the high molecular weight ovomucoid RNA
molecules.

After short periods (1 to 2 hours) of secondary estrogen
stimulation, the relative intensity of the ovomucoid RNA bands
(Figure 17, right) was similar to that observed after chronic
estrogen stimulation (Figure 15, lane A). The bands corresponding to species "g" and mRNA$_{Om}$ were high and of roughly
equal intensity, while that of species "e" was of lesser
intensity and that of species "b" was lesser yet. After
longer periods (greater than 4 hours) of secondary stimulation,
this pattern shifted. The concentration of mRNA$_{Om}$ increased
at each time point so that 16 hours after the DES injection a
very intense mRNA$_{Om}$ band was observed. By contrast, the concentration of the most abundant high molecular weight ovomucoid RNA species, band "g", increased only slightly during
this period and no significant increases in the concentration
of ovomucoid species "e" were observed. Thus, from 4 to 16
hours after DES injection, mature mRNA$_{Om}$ was clearly more
abundant than of any of the high molecular weight ovomucoid
RNA species. Taken together, these results suggest that ovomucoid mRNA may be synthesized by the processing of a large
primary gene transcript and that the multiple bands of ovomucoid RNA larger than mRNA$_{Om}$ may represent transcripts at
different stages of processing.

ELECTRONMICROSCOPIC MAPPING OF THE OVOMUCOID GENE

Although intervening sequences within structural genes can
be detected by restriction enzyme mapping, a complete map
requires other methods independent of the chance occurance of
restriction sites. Short structural gene regions may be
missed due to the relative instability of the hybrids formed
during hybridization, resulting in weak signals upon Southern

filter analysis. In an attempt to more fully characterize the
structure of the ovomucoid gene, hybrid molecules formed
between ovomucoid mRNA and the cloned 15 Kb ovomucoid DNA were
examined by electronmicroscopy.

DNA was thermally denatured and incubated with ovomucoid
mRNA under conditions that permit only RNA/DNA hybridization
but not DNA/DNA reassociation. Thus, homologous regions
between the mRNA and the cloned DNA would appear as double-
stranded regions. DNA sequences non-homologous with the mRNA
would appear as single-stranded loops. Such a hybrid molecule formed between ovomucoid mRNA and the cloned OM15 DNA is
shown in Figure 18. A total of 6 intervening DNA loops of

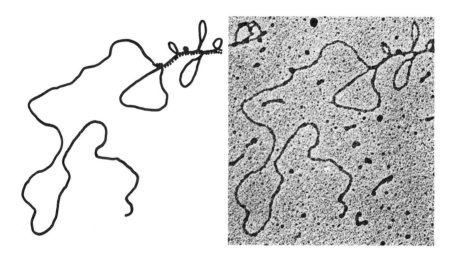

FIGURE 18. Electronmicrograph and line drawing of a hybrid molecule formed between the ovomucoid gene (OM15) and
ovomucoid mRNA. Hybridization was carried out using 10 µg/ml
of OM15 DNA and 20 µg/ml of ovomucoid mRNA in 70% deionized
formamide containing 100 mM Tris-HCl, pH 7.6, 10 mM Na_2EDTA
and 150 mM NaCl. The mixture was heated at 80°C for 5 minutes to denature the DNA and hybridization was carried out at
43°C for 2 hrs. The samples were immediately prepared for
electronmicroscopy as follows. The hybridization mixture containing 0.1 to 0.5 µg of nucleic acids was diluted into 100 µl
of a solution containing 70% formamide, 0.1 M Tris-HCl, pH 8.4
0.01 M Na_2EDTA and 100 µg/ml of cytochrome C. The mixture was
spread onto a hypophase of distilled water. Samples were collected on collodion coated-300 mesh copper grids, stained with
uranyl acetate, rotary shadowed with platinum-palladium and
examined at 80Kv on a Joel 100 C electron microscope. ——,
OM15; ---- $mRNA_{Om}$.

various sizes are present in this molecule. All 6 loops were
present in 10 individual hybrid molecules examined. The
relative positioning of these loops was established in a
histogram as shown in Figure 19. All of the loop structures
occurred at one DNA terminus and occupied about 1/3 of the
total length of the OM15 DNA molecule. Two additional short
intervening sequences, C and F, were detected in these studies,
but not by restriction mapping analysis. Although restriction
mapping is more quantitative in that fragment sizes and position can be correlated with known distances in the structural
sequence, in the absence of favorably placed restriction sites
some intervening sequences may be missed. Electronmicroscopy
permits more direct analysis of intervening sequences. Hence,
EM analysis has detected two additional intervening sequences,
yet the overall structure is in good qualitative agreement
with the limited restriction map. Thus, there are at least
six intervening sequences in the ovomucoid gene.

     This is, necessarily, a minimum number. Quite recently,
recent indications that there might be a seventh intervening
sequence in the ovomucoid gene were obtained (47). In an
mRNA/OM15 DNA hybridization experiment performed as in Figure
18, several molecules with 7 loops were observed. It appears

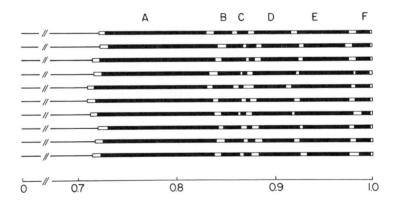

FIGURE 19. Histogram of 10 individual hybrid molecules
formed between OM15 DNA and ovomucoid mRNA under conditions
described in legend of Figure 18. The molecules were measured
using a map measure, and the sizes and positions of the loops
were plotted on a linear scale of 0 to 1, with 1.0 representing 15 Kb of DNA. Ovomucoid structural sequences are represented by ☐ , intervening sequences are represented by ■ ,
and flanking DNA sequences are represented by ──── .

that the very large intervening sequence, the one closest to
the 5'-end of the gene in Figure 18, might actually consist of
two large intervening sequences with a very short sequence of
structural DNA between them. This structural sequence is presumably so short that it does not form a stable hybrid with
the homologous mRNA sequence when surrounded by two large non-
homologous regions. Thus, if no additional intervening sequence is observed in the 3' non-coding region, a situation
not previously observed in the natural genes mapped so far
(7-8, 18, 48-57), the ovomucoid gene contains 7 intervening
sequences. Since the ovomucoid structural gene is only about
750 nucleotides pairs in length, the frequency of occurrence
of these intervening sequences is striking. At 1 intervening
sequence per 107 nucleotide pairs of structural gene sequence,
it represents the highest frequency of intervening sequences
observed among all eucaryotic genes examined to date.

We now have the entire ovomucoid gene, along with several
kilobases of flanking DNA on the 5'- and 3'-ends, in one clone
isolated from a library of oviduct DNA fragments, supplied to
us by Tom Maniatis and Richard Axel. Fragments of this clone
are now being labeled and used for hybridization to in vivo
nuclear RNA to further delineate the extent of the gene that
is transcribed. This clone is now being sequenced, and in the
near future comparative sequence data on the intron/exon
boundaries should be available. Finally, attempts to isolate
and study the high molecular weight nuclear RNA precursors of
ovomucoid mRNA are underway. We have already established
(unpublished observations) that the poly(A) tail at the 3'-end
of the mRNA is added very early, perhaps at the same time as
transcription of the primary gene product itself, as recently
determined for Adeno 2 viral mRNAs (58). Further studies
of those molecules should provide much information about the
synthesis and processing of ovomucoid mRNA.

REFERENCES

1. O'Malley, B.W., McGuire, W.L., Kohler, P.O., and
    Korenman, S.G. (1969) Recent Prog. Horm. Res. 25, 105-160.
2. Oka, T., and Schimke, R.T. (1969) J. Cell Biol. 43, 123.
3. Palmiter, R.D., and Schimke, R.T. (1973) J. Biol. Chem.
    248, 1502.
4. Palmiter, R.D., Moore, P.B., and Mulvihill, E.R. (1976)
    Cell 8, 557.
5. Cox, R.F. (1977) Biochem. 16, 3433.
6. Hynes, N.E., Groner, B., Sippel, A.E., Njuyen-Huu, M.C.,
    and Schutz, G. (1977) Cell 11, 923-932.

7. Garapin, A.C., Lepennec, J.P., Roskam, W., Perrin, F., Cami, B., Krust, A., Breathnach, R., Chambon, P., and Kouvilsky, P. (1978) Nature 273, 349.
8. Mandel, J.L., Breathnach, R., Gerlinger, P., LeMeur, M., Gannon, F., and Chambon, P. (1978) Cell 14, 641.
9. Palmiter, R.D. (1973) J. Biol. Chem. 248, 8260-8270.
10. Chan, L., Means, A.R., and O'Malley, B.W. (1973) Proc. Natl. Acad. Sci. U.S.A. 70, 1870-1874.
11. Harris, S.E., Means, A.R., Mitchell, W.M., and O'Malley, B.W. (1973) Proc. Natl. Acad. Sci. U.S.A. 70, 3776-3780.
12. Harris, S.E., Rosen, J.M., Means, A.R., and O'Malley, B.W. (1975) Biochem. 14, 2072-2080.
13. Sullivan, D., Palacios, R., Staunezer, J., Taylor, J.M., Taras, A.J., Kiely, M.L., Summers, N.M., Bishop, J.M., and Schimke, R.T. (1973) J. Biol. Chem. 248, 7530-7539.
14. O'Malley, B.W., and Means, A.R. (1974) Science 183, 610.
15. Rosen, J.M., Woo, S.L.C., Molder, J.W., Means, A.R., and O'Malley, B.W. (1975) Biochem. 14, 69-78.
16. Monahan, J.J., McReynolds, L.A., and O'Malley, B.W. (1976) J. Biol. Chem. 251, 7355-7362.
17. McReynolds, L.A., Catterall, J.R., and O'Malley, B.W. (1977) Gene 2, 217-231.
18. Woo, S.L.C., Dugaiczyk, A., Tsai, M.-J., Lai, E.C., Catterall, J.R., and O'Malley, B.W. (1978) Proc. Natl. Acad. Sci. U.S.A. 75, 3688-3692.
19. Palmiter, R.D. (1972) J. Biol. Chem. 247, 6450-6461.
20. Stein, J.P., Catterall, J.F., Woo, S.L.C., Means, A.R., and O'Malley, B.W. (1978) Biochem. 17, 5763-5772.
21. Woo, S.L.C., Chandra, T., Means, A.R., and O'Malley, B.W. (1977) Biochem. 16, 5670-5676.
22. Sharp, P.A., Gallimore, P.M., and Flint, S.T. (1974). Cold Spring Harbor Symp. Quant. Biol. 39, 457-474.
23. Chang, S., and Cohen, S.N. (1977) Proc. Natl. Acad. Sci. U.S.A. 74, 4811-4815.
24. Grunstein, M., and Hogness, D.S. (1975) Proc. Natl. Acad. Sci. U.S.A. 72, 3961-3965.
25. Katz, L., Williams, P.H., Sato, S., Leavitt, R.W., and Melinski, D.R. (1977) Biochem. 16, 1677-1683.
26. McReynolds, L.A., O'Malley, B.W., Nisbett, A., Fothergill, J.E., Fields, S., Givol, D., Robertson, M., and Brownlee, G.G. (1978) Nature (London) 273, 723-728.
27. Maxam, A., and Gilbert, W. (1977) Proc. Natl. Acad. Sci. U.S.A. 74, 560-564.
28. Beeley, A.G. (1976) Biochem. J. 159, 335-345.
29. Lai, E.C., Woo, S.L.C., Dugaiczyk, A., Catterall, J.F., and O'Malley, B.W. (1978) Proc. Natl. Acad. Sci. U.S.A. 75, 2205-2209.
30. Monahan, J.J., Harris, S.E., and O'Malley, B.W. (1976) J. Biol. Chem. 251, 3738-3748.

31. Harrison, P.R., Birnie, G.D., Hell, A., Humphries, S.,
    Young, B.D., and Paul, J. (1974) J. Mol. Biol. 84, 539-
    544.
32. Bishop, J.O., and Rosbach, M. (1973) Nature (London) New
    Biology 241, 204-207.
33. Suzuki, Y., Gage, L.P., and Brown, D.D. (1972) J. Mol.
    Biol. 70, 637-649.
34. Tsai, S.Y., Roop, D.R., Tsai, M.-J., Stein, J.P., Means,
    A.R., and O'Malley, B.W. (1978) Biochem. 17, 5773-5780.
35. Yu, J.Y.L., Campbell, L.D., and Marquardt, R.R. (1971)
    Can. J. Biochem. 49, 348.
36. Liarakos, C.D., Rosen, J.M., and O'Malley, B.W. (1973)
    Biochem. 12, 2309-2816.
37. Roop, D.R., Nordstrom, J.L., Tsai, S.Y., Tsai, M.-J., and
    O'Malley, B.W. (1978) Cell 15, 671-685.
38. Dugaiczyk, A., Woo, S.L.C., Lai, E.C., Mace, Jr., M.L.,
    McReynolds, L.A., and O'Malley, B.W. (1978) Nature 274,
    328-333.
39. Lai, E.C., Woo, S.L.C., Dugaiczyk, A., and O'Malley, B.W.
    (1979) Cell 16, 201-211.
40. Maniatis, T., Jeffrey, A., and Cleid, D.G. (1975) Proc.
    Natl. Acad. Sci. U.S.A., 1184-1189.
41. Southern, E.M. (1975) J. Mol. Biol. 98, 503-518.
42. Denhardt, D. (1966) Biochem. Biophys. Res. Commun. 23,
    641-646.
43. Laskey, R.A., and Mills, A.D. (1977) F.E.B.S. Letters
    82, 314-316.
44. Sternberg, N., Tiemeier, D., and Enquist, L. (1977) Gene
    1, 255-280.
45. Bailey, J.M., and Davidson, N. (1976) Anal. Biochem. 70
    75-85.
46. Alwine, J.C., Demp, D.J., and Stark, G.R. (1977) Proc.
    Natl. Acad. Sci. U.S.A. 74, 5350-5354.
47. Lai, E.C., Stein, J.R., Catterall, J.F., Woo, S.L.C.,
    Means, A.R., and O'Malley, B.W. (manuscript in prepara-
    tion).
48. Garapin, A.C., Cami, B., Roskam, W., Kourilsky, P.,
    LePennec, J.P., Perrin, F., Gerlinger, P., Cochet, M.,
    and Chambon, P. (1978) Cell 14, 629-639.
49. Weinstock, R., Sweet, R., Weiss, M., Cedar, H., and Axel,
    R. (1978) Proc. Natl. Acad. Sci. U.S.A. 75, 1299-1303
    (1978).
50. Tilghman, S.M., Tiemeier, D.C., Polsky, F., Edgell, M.H.,
    Seidman, J.G., Leder, A., Enquist, L., Norman, B., and
    Leder, P. (1977) Proc. Natl. Acad. Sci. U.S.A. 74,
    4406-4410.
51. Tiemeier, D., Tilghman, S. ., Polsky, F.I., Seidman, J.C.,
    Leder, A., Edgell, M.H., and Leder, P. (1978) Cell 14,
    237-245.

52. Maniatis, T., Hardison, R.C., Lacy, E., Lauer, J., O'Connell, C., Quon, D., Sim, G.K., and Efstratiatis, A. (1978) Cell 15, 687-701.
53. Tonegawa, S., Brack, C., Hozumi, N., and Schuller, R. (1977) Proc. Natl. Acad. Sci. U.S.A. 74, 3518-3522.
54. Brack, C., and Tonegawa, S. (1977) Proc. Natl. Acad. Sci. U.S.A. 74, 5652-5656.
55. Tonegawa, S., Maxam, A.M., Tizard, R., Bernard, O., and Gilbert, W. (1978) Proc. Natl. Acad. Sci. U.S.A. 75, 1485-1489.
56. Seidman, J.G., Leder, A., Edgell, M.H., Dolsky, F., Tilghman, S.M., Tiemeier, D.C., and Leder, P. (1978) Proc. Natl. Acad. Sci. U.S.A. 75, 3881-3885.
57. Seidman, J.G., Edgell, M.H., and Leder, P. (1978) Nature 271, 582-585.
58. Nevins, J.R., and Darnell, J.E., Jr. (1978) Cell 15, 1477-1493.

DISCUSSION

F.H.C. CRICK: Let me make two remarks. It is obviously very interesting to know what are the functions, if any, of the flanking sequences, and I would certainly agree with you that it looks as if they are not going to be transcribed. But one has to, I think, realize something which you implied, but did not say implicitly. It does depend on how fast a chain is degraded because if it is degraded very fast you would not see very much of the precursor. So until we know how fast the degradation is, it is difficult to be sure that the flanks are not being transcribed. Obviously it would be much more satisfactory if we could locate the actual signals for initiating and terminating transcription. I am sure that you would agree with me.

B.W. O'MALLEY: Absolutely. All we can say at present is that we are making an educated guess using the available evidence while keeping our options open. However, it should be realized that we are working at a sensitivity of about 1 molecule per cell under these conditions and really ought to be able to detect flanking sequence transcripts if they exist.

F.H.C. CRICK: I was going slightly further in saying that it might be degraded as it is transcribed. That would be the extreme form of the idea. Personally, I think that it is not true.

The other question is the matter of the number of intermediates. You stressed very rightly that until that can be done using a pulse-chase, it is very difficult to say what the processing is. But one has to remember one elementary fact, that a small change of activation energy can make a big difference to the rate of an enzymatic reaction. Consequently, if you look at all the loops that will have to be processed, it would be surprising if they were all processed at the same rate, since they are clearly not all exactly the same. We do not know yet what the precise signals are which make the splicing enzyme work. If you have a distribution of rates, some very fast and some very slow, then as you have implied because you see a limited number of precursors, the processing is not necessarily done in a non-random way. This is quite important because we really want to know whether there is a defined sequence in the processing, which means there must probably be some interaction between the loops when you do the processing, or whether it is random, in the sense that each one is done independently, though at different rates. I think Dr. O'Malley has shown that it is very unlikely that all are processed at random at the same rate.

B.W. O'MALLEY: I agree. I should also mention that our recent pulse-chase experiments also confirm the precursor-product relationship of the high molecular weight RNA and mature mRNA.

F.H.C. CRICK: Is it purely random, but with the same base composition? It will be very interesting to see, when we have all the other introns, for hemoglobin and so on, if they have anything in common. Have you found any longer repeated sequences? You did have some hexons didn't you?

B.W. O'MALLEY: Yes, hexons were present in all introns.

STRUCTURAL ORGANIZATION AND EXPRESSION OF OVALBUMIN
AND RELATED CHICKEN GENES

P. Chambon, C. Benoist, R. Breathnach, M. Cochet, F. Gannon,
P. Gerlinger, A. Krust, M. LeMeur, J.P. LePennec, J.L. Mandel,
K. O'Hare and F. Perrin

Laboratoire de Génétique Moléculaire des Eucaryotes du CNRS,
Unité 184 de l'INSERM et Institut de Chimie Biologique,
Faculté de Médecine
Universite Louis Pasteur
Strasbourg - France

INTRODUCTION

Chicken ovalbumin accounts for 50 to 65% of total protein synthesis in laying hen oviduct tubular gland cells (1). As for other egg white proteins such as conalbumin, ovomucoïd and lysozyme, the rate of transcription of its mRNA (which represents 50% of the total mRNA population) is under hormonal control (for references, see 2-5). Therefore the genes corresponding to these proteins provide a very useful model system for the study of the transcriptional regulation of gene expression by hormones. Moreover, administration of oestrogen to immature chicks results in cytodifferentiation of tubular gland cells (1), which comprise up to 90% of the cells in the magnum portion of laying hen oviduct, thereby providing a very interesting model for investigating whether gene rearrangement could be one of the mechanisms involved in cell differentiation. With the advent of *in vitro* Recombinant DNA technology it is possible to undertake studies aimed at understanding, in molecular terms, how these hormonal regulations operate. This is mainly why we initiated a detailed analysis of the organization of the ovalbumin gene. Unexpectedly, we found that, as for many other eukaryotic genes (for references see 6-9), the ovalbumin gene is split (7). Very rapid progress has been made since this first report, and the organization

55

of the complete ovalbumin gene is now elucidated (8, 10-18).
We will here first summarize our present knowledge of the
ovalbumin gene structure, discuss the possible length of the
ovalbumin transcription unit and provide evidence that the
ovalbumin primary RNA transcript is colinear with the gene. We
will next present some sequencing data which possibly provide
some insights into the mechanisms involved in the RNA splicing
events and the initiation of transcription. Finally, we will
present the results of our recent studies on the conalbumin
and ovomucoïd gene organization.

STRUCTURE OF THE SPLIT OVALBUMIN GENE

The 1872 nucleotides of ovalbumin mRNA (ov-mRNA) (see
Fig. 1c, and Ref. 19, 20) are encoded for by a chicken genomic
DNA region of about 7.7 kb which contains a leader-coding
region (17) and 7 mRNA coding regions (exons 1 to 7) separated
by 7 introns (intervening sequences) A to G. A schematic representation of this structure is given in Fig. 1b, which
corresponds to the restriction enzyme map of the ovalbumin
region inserted in a λ Charon 4A clone (λC4-ov5) (14). The
split organization of the ovalbumin gene is visualized in
Fig. 2 as a hybrid DNA-RNA molecule between ov-mRNA and the
DNA of λC4-ov5 clone. The leader-coding region and the 7
exons (1 to 7) (thicker DNA-RNA hybrid regions) are separated
by 7 single-stranded DNA loops (A to G) which correspond to
the introns. The sizes of the leader coding region (L) and of
the exons 1 to 7 which are known from sequencing studies (14,
17, 20) are 47, 185, 51, 129, 118, 143, 156 and 1043 nucleotides, respectively (see Fig. 1c). It is unknown at present
whether these exonic RNA domains could correspond to protein
domains (for a discussion of this problem, see Refs. 8, 46). From
electron microscopic measurements the sizes of the 7 introns
A to G are 1560 ± 147, 238 ± 45, 601 ± 35, 411 ± 58, 1029 ± 71,
323 ± 36, 1614 ± 88 bp, respectively. For introns B, C and D
which have been sequenced, these electron microscopic measurements are in very good agreement with the sequence results
which give 251, 582 and 401 bp, respectively (C. Benoist,
R. Breathnach and K. O'Hare, unpublished results). From all of
these results, the length of the ovalbumin gene from the region coding for the 5' end of the ov-mRNA to the region coding
for its 3' end is about 7.7 kb, approximately four times
longer than the size of the mature mRNA.

In no case have we found a different arrangement of the
ovalbumin split gene, when we compared the genomic DNA of
cells in which the gene is expressed (for example, hen ovi-

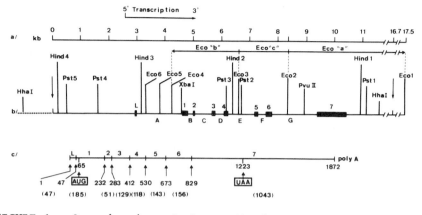

FIGURE 1. Organization of the ovalbumin gene.
a) Scale in kilobase pairs (kb) for Fig. 1b.
b) Localization of the exons (1 to 7, heavy lines) and introns (A to G) of the ovalbumin gene in λC4-ov5 (see Ref. 14)
within and upstream from the Eco RI fragments Eco "a", Eco
"b" and Eco "c" (8, 17, 18). The horizontal dotted line represents λ Charon 4A sequences. The arrows indicate the limits
of the 16.7 kb genomic DNA fragments inserted in λC4-ov5. L
corresponds to the location of the leader-coding sequences as
determined by electron microscopy and DNA sequencing (see Ref.
14). The Eco, Pst, Hind and Hha I sites correspond to Eco RI,
Pst I, Hind III and Hha I restriction enzyme sites (taken
from references 8, 12 and 14).
c) Schematic representation of ov-mRNA. The total length,
the position of the AUG and UAA codons are taken from references 17, 19 and 20. L refers to the leader sequence and 1
to 7 correspond to the 7 domains of ov-mRNA coded by exons 1
to 7. The arrows indicate the limits of these domains, assuming that the splicing events obey   the "GT-AG rule" (see
Ref. 17 and text). Numbers in parentheses indicate the length
(in nucleotides) of the domains.

duct) with that of cells in which the gene is not transcribed
(for instance, erythrocyte). This indicates that extensive
gene rearrangement, such as found during lymphocyte differentiation (21), is not involved in the mechanisms leading to the
expression of ovalbumin during differentiation. It is interesting that none of our studies (7, 8, 13, 14) have provided
evidence that the ovalbumin intronic sequences are repeated
in their entirety in the chicken genome. This may have some

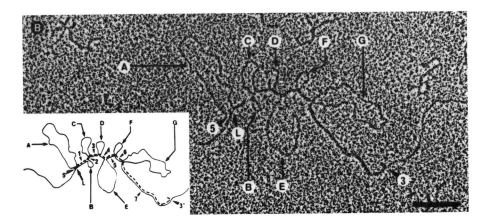

FIGURE 2. Electron microscopy of an RNA-DNA hybrid molecule
between ovalbumin mRNA and the DNA of clone λC4-ov5 (see text
and Ref. 14). In the line drawing, ov-mRNA is represented by
a dashed line, whereas the solid line corresponds to single-
stranded DNA. 5' and 3' arrow heads : 5' and 3' ends of ov-
mRNA, respectively. 1 to 7 (the thicker DNA-RNA hybrid regions)
correspond to the ovalbumin gene exons, whereas A to G correspond to the 7 introns. L is the hybrid region corresponding
to the leader-coding sequences. The RNA-DNA hybrid region is
1893 ± 114 bp long. The lengths of intronic loops A to G and
of the exonic mRNA-DNA hybrid segments 1 to 7 are (in nucleotides or base pairs) : 1560 ± 147 (A), 238 ± 45 (B), 601 ± 35
(C), 411 ± 58 (D), 1029 ± 71 (E), 323 ± 36 (F), 1614 ± 88 (G),
197 ± 34 (1), 85 ± 14 (2), 164 ± 25 (3), 143 ± 18 (4), 154 ±
25 (5), 201 ± 28 (6) and 1097 ± 51 (7), respectively. There
is about 7.7 kb between the 5' end of loop A and the 3' end
of exon 7. The bar represents 0.1 μm.

bearing on the mechanisms which have led to the split gene
organization. In contrast, we have found that some amino-
coding regions of the ovalbumin gene are repeated elsewhere
in the chicken genome (see Ref. 8). In fact some recent studies of A. Royal et al. (personal communication) suggest that
duplication of at least one part of the ovalbumin gene could
have occurred during the course of evolution.

THE OVALBUMIN TRANSCRIPTION UNIT : EVIDENCE FOR A COMPLETE
COLINEAR RNA TRANSCRIPT AND PROCESSING INTERMEDIATES.

Our studies have established that about 7.7 kb separate
the genome region coding for the 5' and 3' extremities of ov-
mRNA. Recent *in vivo* (22) and *in vitro* (23) studies on primary
transcripts from the major late transcription unit of Adeno-
virus-2 strongly suggest that the cap site and the transcription initiation site are coincident and that the initiating
residues of the primary transcript are conserved in mRNA as
the 5' capped terminus. If the situation is identical in the
case of ovalbumin, the minimal size of the ovalbumin transcription unit (Ov-TU) should be about 7.7 kb. A complete colinear
transcript should then be in the range of 7700 nucleotides.
The recent studies of Roop et al. (24) have provided evidence
for ovalbumin RNA transcripts of at least 7800 nucleotides in
length. The results shown in Fig. 3 confirm this result and
indicate that the length of the largest RNA found in total
hen oviduct RNA (lane 2) is in fact about 7900 nucleotides.
This is in very good agreement with our estimate of the Ov-TU,
when the existence of a poly A-tail is taken into account.
This 7900 nucleotide-long molecule is indeed polyadenylated,
since it is found in the RNA fraction which is retained on
oligodT-cellulose (dt-RNA, lanes 1 and 5). Evidence that this
7900 nucleotide long RNA molecule contains transcripts of
both exonic and intronic sequences is provided by its hybridization to both exonic (double-stranded cDNA, ds-cDNA, lanes
1, 2 and 5) and intronic (introns E or G, lanes 4 and 3,
respectively) probes. Other hybridization results (not shown)
indicate that this RNA molecule hybridizes also to a probe
specific for intron A. In addition, electron microscopic
studies have provided direct evidence that laying hen oviduct
RNA contains RNA molecules which hybridize to the totality
of the Ov-TU (F. Perrin, unpublished). There is therefore
little doubt that the Ov-TU (as defined above) is transcribed
in its totality yielding a complete colinear RNA transcript
which can be polyadenylated.
Oviduct oligodT RNA contains many discrete RNA species
which hybridize to the exonic ds-cDNA probe with molecular
weights intermediate between that of the largest 7900 nucleotide RNA and of ovabumin mRNA (the very heavy dark spot
centred at a position corresponding to 1900 nucleotides –
Fig. 3, lanes 1 and 5). As many as 9 individual bands can
be clearly discerned on the original autoradiograms, with 3
main bands of 5050, 3250 and 2860 nucleotides in length, in
addition to the 7900 nucleotide band. To demonstrate that
these RNA molecules do in fact correspond to stepwise exci-
sions of the intronic transcripts, we have hybridized them to

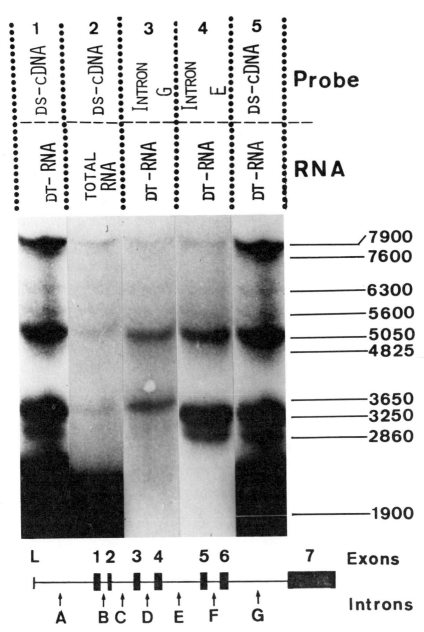

FIGURE 3. Identification of a complete colinear transcript of ovalbumin gene and of processing intermediate RNA species in total laying hen oviduct RNA (total RNA) and in oviduct RNA purified on oligodT-cellulose (dT-RNA). The purification of RNA, its fractionation by agrose gel electrophoresis in the presence of methyl mercuric hydroxide and its transfer to

specific intronic probes (Fig. 3, lanes 3 and 4). It is clear
that some RNA molecules hybridize to both intron E and G
probes (for instance the 5050 band), whereas other hybridize
specifically to the intron E probe (3250 and 2860 bands) or to
the intron G probe (the 3650 band). As expected ov-mRNA is not
revealed by these intronic probes. These results show that
one can find RNA molecules smaller than the complete colinear
transcript, which contain both intron E and G transcripts,
but also RNA molecules in which either the intron E or the
intron G transcript has been "spliced out". In addition, the
intronic transcripts do not appear to be necessarily removed
in the same order as they were transcribed, since the intron
E transcript is still present in the smaller 3250 and 2860
molecules, whereas the intron G transcript has already been
removed. In any case, these results strongly suggest that the
multiple hybridizing RNA bands which are seen on the autoradiogram correspond to intermediate species of RNA generated
during the processing of the initial poly(A)$^+$ complete
colinear RNA transcript.

To study in more detail these splicing steps, we have isolated the oviduct RNA molecules which contain specific intronic transcripts. Total laying hen oviduct RNA was hybridized
to either an intron E or an intron G DNA filter, eluted,
fractionated on agarose gel in the presence of methyl mercuric
hydroxide, transferred to DBM-paper and hybridized either to
a probe representative of all of the ovalbumin gene sequences
(Gene in Fig. 4) or to a probe specific for intron G. The
results are shown in Fig. 4. In agreement with the result
shown in Fig. 3 lane 4, 4 major RNA species (7900, 5050,
3250 and 2860 nucleotides in length) are specifically retained
on the intron E filter (Fig. 4, lane 2). However, and in
keeping with the results presented in Fig. 3 (lanes 3 and 4),
only two of these RNA species (7900 and 5050) hybridize to the
intron G probe (Fig. 4, lane 3), indicating that the intron G
transcript has been excised from the lower molecular weight
species, whereas the intron E transcript is still present.

DBM-paper (31) will be described elsewhere. The immobilized
RNA species were hybridized either to the [^{32}P]-labelled
"Hhaov" ovalbumin double-stranded DNA (ds-cDNA, see Ref. 7)
or to nick-translated [^{32}P]-labelled intron E and G DNAs
(cloned in pBR 322), as indicated in the figure. 20 µg of
total laying hen oviduct RNA (lane 2) or of oligodT-cellulose
purified RNA were electrophoresed (lanes 1, 3, 4 and 5) as
indicated in the figure. Further experimental details will be
published elsewhere. A schematic representation of the ovalbumin gene is shown below the autoradiogram.

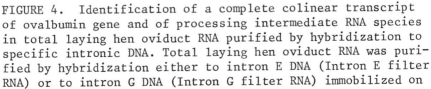

FIGURE 4. Identification of a complete colinear transcript of ovalbumin gene and of processing intermediate RNA species in total laying hen oviduct RNA purified by hybridization to specific intronic DNA. Total laying hen oviduct RNA was purified by hybridization either to intron E DNA (Intron E filter RNA) or to intron G DNA (Intron G filter RNA) immobilized on

When the laying hen oviduct RNA was fractionated by hybridization with an intron G filter and hybridized to a total ovalbumin gene probe (lane 4) at least 6 bands (7900, 6850, 6300, 5050, 4825, 3650) were revealed (the band at 1900 corresponds to a minor contamination by ov-mRNA). Not surprisingly the same RNA bands are found after hybridization to a probe specific for intron G (Fig. 4, lane 5), but a faint band at 1600 which was barely visible in lane 4 is now clearly seen : this RNA is an excellent candidate for the "spliced out" transcript of intron G, the length of which is precisely 1600 bp (see above).

To further support the conclusion that the multiple bands which are seen on the autoradiograms (below the 7900 complete colinear transcript) correspond to intermediate species in the processing of this initial transcript, we have analyzed by electron microscopy the RNA-DNA hybrid molecules which can be formed between the ovalbumin genomic DNA (clone λC4-ov5, see above) and the RNA species purified by hybridization to an intron DNA filter. The results are shown in Fig. 5. A control mature ov-mRNA-DNA hybrid molecule is presented in panel A. The single-stranded DNA loops B to G correspond to intronic regions whose transcripts are not anymore present in the mature ov-mRNA (loop A is not seen in this hybrid molecule as in other molecules presented in panel B and C, because the 47 bp DNA-RNA hybrid region between the mRNA and the leader-coding region (see above) is markedly unstable under the hybridization conditions used in this study). Panel B and C show two hybrid molecules between λC4-ov 5 DNA and RNA molecules containing sequences complementary to intron E. Examination and measurements of the hybrid molecule of panel B indicate that the transcripts of introns A, B, C, D and G have been removed, whereas the transcripts of introns E and F are still linked to the ov-mRNA sequences. A similar hybrid molecule, but from which the intron D transcript has not yet been excised, is shown in panel C. It is noteworthy that in both cases, and in agreement with the biochemical results presented in Fig. 3 and 4, one finds RNA molecules containing the intron E transcript, whereas the intron G transcript has already been "spliced out". The presence, at the 3' end of

DBM-paper (31). After elution the RNA was electrophoresed on agarose gels as described in legend to Fig. 3 and hybridized either to nick-translated [32P]-labelled ovalbumin gene (Gene probe, lanes 1, 2, 4 and 6) or intron G DNA (Intron G probe, lanes 3 and 5) probes. Further experimental details will be published elsewhere. A schematic representation of the ovalbumin gene is shown below the autoradiogram.

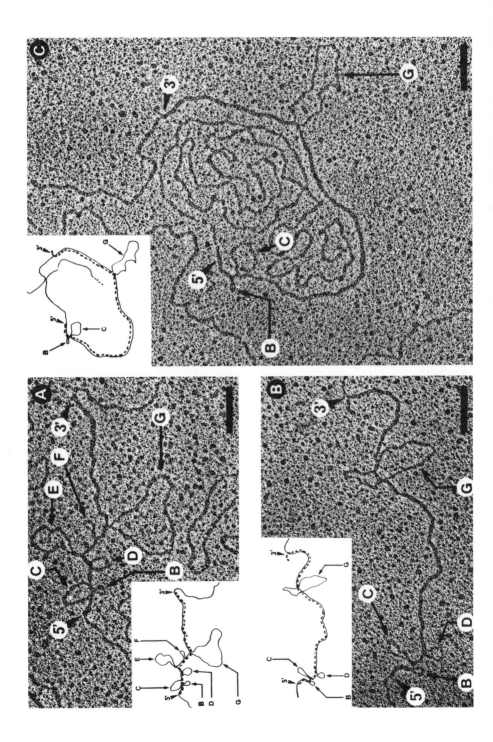

the molecule, of a poly A tail (100 to 200 nucleotides in length), much longer than that associated to mature ov-mRNA (panel A), is also remarkable. There is no doubt that one is here dealing with poly(A)$^+$ RNA species which are processing intermediates between the complete colinear ovalbumin gene transcript and the mature ov-mRNA.

From all of these results we conclude : 1) the ovalbumin gene is transcribed as a complete colinear RNA molecule which is very rapidly polyadenylated. A similar conclusion has been reached in the case of both Adenovirus-2 (25) and globin (26) initial RNA transcripts for which poly(A) addition precedes splicing. Whether, as it is the case for Adenovirus-2 late major RNA transcription (27), the RNA polymerase reads past the site where poly(A) is added and whether the 3' end which serves as the poly(A) addition site results from endonucleolytic cleavage of a longer primary transcript are at present unknown. It should however be pointed out that no discrete RNA transcripts longer than 7900 nucleotides were found among the RNA molecules which were selected by hybridizing total oviduct RNA to intron E or intron G filters (Fig. 4, lanes 2-5); 2) the processing of the complete colinear transcript is a stepwise mechanism involving the generation of a series of intermediates which differ in structure by an additional step of splicing. However, our data suggest that RNA splicing can occur through multiple pathways involving several combinations of intermediates, rather than being a direct RNA:RNA splicing of the ultimately conserved mRNA domains in a specific sequential temporal order in the 5' to 3' direction.

FIGURE 5. Electron micrographs of hybrid molecules between the ovalbumin gene integrated in λC4-ov5 DNA (see legend to Fig. 2 and text) and processing intermediate RNA species. (A): A control ov-mRNA-DNA hybrid molecule (see Fig. 2 for the symbols and line drawing); (B) and (C): RNA molecules were purified by hybridization to an intron E filter (see legend to Fig. 4). Length measurements on the hybrid molecule shown in panel B gave the following values : 230, 550, 351, 1579 for introns B, C, D and G, respectively; 204, 76, 128 and 1047 bp for exons 1, 2, 3 and 7, respectively (see line drawing on Fig. 2); 1850 bp for the hybrid region between intron D and G. The length measurements on the hybrid molecule shown in panel C were as follows : 244, 578 and 1488 bp for introns B, C and G, respectively; 220, 70 and 1100 bp for exons 1, 2 and 7, respectively (see line drawing on Fig. 2); 2400 bp for the hybrid region between introns C and G. The bar represents 0.1 μm. Further experimental details will be published elsewhere.

A similar situation may exist in the case of Adenovirus-2
(28), although evidence to the contrary has been reported
(29); 3) the finding of a discrete complete colinear transcript and of discrete processing intermediates supports very
strongly the concept that mature ov-mRNA, as other cellular
mRNAs, like globin mRNA (30), is derived from an initial
transcript which contains covalently linked intronic and
exonic transcripts; 4) the excision of a given intron could
be a single-step event, since at least in one case (intron G -
see above and Fig. 4, lane 5), we have found evidence that a
full-length intron transcript is "spliced out".

SEQUENCE AT INTRON-EXON BOUNDARIES

The finding of a complete colinear RNA transcript and of
RNA intermediates raise the obvious question : what is the
mechanism of splicing? At least one enzyme should be involved
and in the case of the tRNA from yeast an enzyme activity has
been partially characterized by two groups (32, 33). How does
then the enzyme(s) recognize where to splice with such great
precision? To tackle this problem, we have sequenced all of
the exon-intron junctions of the ovalbumin gene (Fig. 6 and
Refs. 14 and 17). Our main conclusions are : 1) there is no
evident way to form base-paired structures that would bring
into close proximity the ends of consecutive exon transcripts
and to loop out intron transcripts to allow excision and
splicing of the mRNA precursors, as was previously suggested
(34); 2) due to the existence of directly repeated sequences
(boxed in Fig. 6) at the beginning and at the end of a given
intron, not one exon-intron boundary of the ovalbumin gene can
be uniquely defined; 3) introns of the ovalbumin gene may be
divided into three types depending on the sequences present
at their extremities (Fig. 6 and Ref. 17). These types are
themselves closely related and the sequences at the extremities of introns A to G have been aligned in Fig. 7 to emphasize their similarities. All of the 3' extremities are related to the model sequence 5'-TCAGGTA-3' and similarly the 3'
intron extremities to the model sequence 5'-TXCAGG-3'. A similar conclusion has been reached by Catterall et al. (35).
When aligned as in Fig. 7, it becomes apparent that, in all
cases, common excision-ligation points could be defined
(broken lines in Fig. 7; see Fig. 6 for mRNA sequences). The
following rule emerges : the base sequence of an intron
transcript begins with GU(T) and it ends with AG; 4) tracts
rich in pyrimidine (particularly U) are always found in the
intronic transcript region preceding the beginning of a new
exonic transcript (17).

It is striking that the above conclusions hold true for all of the exon-intron junctions of viral or other higher eukaryotic genes sequenced to date (see Fig. 7, similar exon-intron junction features are present in the case of the Adeno-virus-2 major late gene - Ref. 37). In the case of Simian Virus 40 genes, there are remarkable similarities with the different types of ovalbumin intron extremities (Types 1 - 3). But even more striking is the observation that in all cases a unique common splicing point can be defined which obeys the GU(T) - AG rule, suggesting very strongly that these dinucleo-tides could be recognized by the splicing enzyme(s). It is unlikely, but not excluded, that the above dinucleotides and junction sequences related to the model sequences shown in Fig. 7 are sufficient in themselves to select the splicing positions. The pyrimidine-rich tract mentioned above could confer additional specificity, but the obvious hypothesis is

---

FIGURE 6.   DNA sequences at exon-intron boundaries. Numbering of the nucleotides is from the 5' end of ov-mRNA. The DNA of the non-coding strand is shown. The maximum possible extent of the exons is indicated by the horizontal arrows. Beyond these points, the DNA and RNA sequences diverge and are sepa-rated. The numbers define the messenger nucleotides that could be encoded for on either side of the introns. Sequences of nucleotides that are repeated directly at both ends of a given intron are boxed. Those sequences at the extremities of the different introns that are used to define the three types of ovalbumin introns (see text) are shown under an unbroken line. For further details, see Refs. 14 and 17.

FIGURE 7.   Comparison of DNA sequences at exon-intron bounda-ries of ovalbumin and other genes. Sequences have been aligned in order to stress their common features. Boxed nucleotides represent direct repeats (see text). The vertical broken line shows how the excision-ligation events could occur in all cases at unique positions with respect to the invariant di-nucleotides GU(T) and AG (see text). The frequency with which a given nucleotide appears in each position of the possible model (prototype) sequence is tabulated below these sequences. The ovalbumin sequences are taken from Refs. 14 and 17, where-as the SV40 sequences are from Ref. 36. The immunoglobulin sequences are from Ref. 48 and from O. Bernard, N. Hozumi and S. Tonegawa (personal communication). The globin sequences are from J. Van der Berg, A. Van Ooyen, N. Mantei, A. Scham-bock, R. Flavell and C. Weissmann (personal communication). For further details, see Refs. 11 and 17. The lines below the junction sequences are type-specific (see legend to Fig.6).

Intron A

```
                                                        4647
                                                        ↓
                                               UCAAA↓ACAAC
                                                        AGACAACUCAGAG...3'
5'...CAAGCUCAAAAG↓ACAAC
5'...CAAGCUCAAAAG GTCACTCTCTGGAATTAACCTTCTCTCTATATTAGC....CATCCTTACATTTTCACTGTTCTGCTGTTTGCTCAGACAACTCAGAG...3'
----- LEADER ---→                                                                    ---→ EXON 1 -----

         231  234                                              231  234
          ↓    ↓                                                ↓    ↓
  GAUAAAUAAGGU UGUUC                                   AAAUA AGGUUGUUCGCUU...3'
5'...GAUAAAUA AGGU UGUUC
5'...GATAAATA AGGT GAGCCTACAGTTAAAGATTAAAACCTTTGCCCTGCT....TAACCATTATTTCAGCTACTATATTTCAATTACAGGTT GTTCGCTT...3'
----- EXON 1 ---→                                                                    ---→ EXON 2 -----
                                    INTRON B

          280  283                                             280  283
           ↓    ↓                                               ↓    ↓
   AUUGAAGCUCAG UGUGG                                  GAAGC UCAGUGUGGCACA...3'
5'...AUUGAAGC UCAG UGUGG
5'...ATTGAAGC TCAG TACAGAGAAATAATTTCACCTCCTTCTCTATGTCCCT....AACTAGAATAACAACATCTTCTTCTTCTTGTATTCAG GTGGCACA...3'
----- EXON 2 ---→                                                                    ---→ EXON 3 -----
                                    INTRON C

          413                                                  413
           ↓                                                    ↓
   CAAUCCUGCCAG AAUAC                                  UCCUGCCA GAAUACUUGC...3'
5'...CAAUCCUGCCAG AAUAC
5'...CAATCCTGCCAG TAAGTTGA..................AAATTCGTATCTGAAAGCTGAATACTCTTGCTTTACAG AATACTTGC...3'
----- EXON 3 ---→                                                                    ---→ EXON 4 -----
                                    INTRON D

          530  551                                             530  551
           ↓    ↓                                               ↓    ↓
   UCAGACAAAUGG AAUUA                                  GACAAAU GGAAUUAUCAG...3'
5'...UCAGACAAAUGG AAUUA
5'...TCAGACAAATGG TAAGGTAGAACATGCTTTGTACATAGTGAGAGTTGG....GATATACGTAAACTCTCTTTTGTATTCATTTCAAGT AATTATCAG...3'
----- EXON 4 ---→                                                                    ---→ EXON 5 -----
                                    INTRON E

          672  673                                             672  673
           ↓    ↓                                               ↓    ↓
   AGAGUGACUGAG CAAGA                                  GUGACUG AGCAAGAAAGC...3'
5'...AGAGUGACUGAG CAAGA
5'...AGAGTGACTGAG TATGATGGGCATACCTTAGAG..............TTCTCTCTCTCTTTTTTTTTTTTTGGTTGCTCAG CAAGAAAGC...3'
----- EXON 5 ---→                                                                    ---→ EXON 6 -----
                                    INTRON F

          826  829                                             826  829
           ↓    ↓                                               ↓    ↓
   GGCCUUGAGCAG UUGAG                                  CUUGA GCAGCUUGAGAGU...3'
5'...GGCCUUGAGCAG UUGAG
5'...GGCCTTGAGCAG TATGGCCCTAGAAGTTGGCTTCAGAATATTAAAAA....TGTCGCCATTCCATGGATCCATTCATTTCCTTGCAG CTTGAGAGT...3'
----- EXON 6 ---→                                                                    ---→ EXON 7 -----
                                    INTRON G
```

5' →

```
LEADER  ... T C A A A [A G G] T C A C T C ... INTRON A ... G C T C T A [G G] A C A A C T ... EXON 1   TYPE 1

OVALBUMIN

EXON 1 ... A A A T A [A G G] G A G C C ... INTRON B ... A T T A C A [A G G] T G T T ... EXON 2   TYPE 3

EXON 2 ... A G C [T C A G] G T A C A G A ... INTRON C ... T A T [T C A G] T G T G G C ... EXON 3   TYPE 1

EXON 3 ... C C T G C C A [G G] T A A G T T ... INTRON D ... T T T A C A G [G] A A A T A C ... EXON 4   TYPE 2

EXON 4 ... A G A A A T [G G] T A A G G T ... INTRON E ... C T T A A A [A G G] A A T T A ... EXON 5   TYPE 2

EXON 5 ... G A C T G A [G G] T A T A T G ... INTRON F ... G C T C C [A G G] C A A G A A ... EXON 6   TYPE 1

EXON 6 ... T G A [G C A G] G T A T G G C ... INTRON G ... C T T [G C A G] C T T G A G ... EXON 7   TYPE 1


                  ... T C A G [A G G] T A C C T A ... INTRON ... T T T C C [A G G] C C A T ...      ⎫
SV40 LATE 19S (213-476)                                                                            ⎬  TYPE 3
                  ... C G T T A [A G G] T C G T A ... INTRON ... T T T C C [A G G] C C A T ...      ⎭
SV40 LATE 19S (291-476)

                  ... T T A A C T [G G] T A A G T T ... INTRON ... T T T C C A G [G] C C A T ...    ⎫  TYPE 2
SV40 LATE 19S (444-476)                                                                            ⎬
                  ... T T A A C T [G G] T A A G T T ... INTRON ... C T T C T A G [G] C C T G T ...  ⎭
SV40 LATE 16S (444-1381)

                  ... A A C T G A [G G] T A T T T G ... INTRON ... A T T T T A [A G G] A T T C C A ... ⎫
SV40 EARLY T (4837-4490)                                                                              ⎬ TYPE 1
                  ... C T A T A [A G G] T A A A T G ... INTRON ... A T T T T [A G G] A T T C C A ...   ⎭
SV40 EARLY t (4556-4490)

                  ... T G G G C [A G G] T T G G T A ... INTRON ... T T T T T A [G G] C T G C T ...
MOUSE β GLOBIN SMALL INTRON

                  ... T G G G G [C A G G] T T G G T A ... INTRON ... T T C T C A [G G] C T G C T ...
RABBIT β GLOBIN SMALL INTRON

                  ... C T T C A G [G] T G A G T C ... INTRON ... C C C A C A G [G] T C C T G ...
MOUSE β GLOBIN LARGE INTRON

                  ... C T T C A G [G] T G A G T T ... INTRON ... C C T A C A G [G] T C T C C T ...
RABBIT β GLOBIN LARGE INTRON

                  ... A G C T C A [G G] T C A G C A ... INTRON ... T T T G C A [G G] G G C C A ...
IG 99    SMALL INTRON  ⎫
IG 303 λI SMALL INTRON ⎭

                  ... T G C T C A [G G] T C A G C A ... INTRON ... T T T G C A [G G] A G C C A ...
IG 13 λII SMALL INTRON

                  ... G T C C T A [G G] T G A G T C ... INTRON ... T C C T G C [G G] C C A G C ...
IG 303 λI LARGE INTRON
```

POSSIBLE MODEL SEQUENCE ↙

```
T C A G G T A                T X C A G G
8 8 14 18 19 19 9 /19        14 11 16 17 10 /17
```

69

that some secondary or tertiary structure is formed. Such a
structure has been proposed in the case of SV40 to bring in
close proximity the ends of a given intronic transcript (36).
Finally, the possibility that recognition of modified bases
in mRNA precursors might be involved in the specificity of
the splicing mechanism should not be ignored (37). All of the
above similarities between intron-exon junctions of pre-mRNA
of various origins suggest that there could be only one
splicing enzyme for processing all mRNA precursor molecules.
However, there should be at least two splicing enzymes in
eukaryotic cells, since there is no relation whatsoever between the exon-intron junction sequences of mRNA- and tRNA-
coding genes (38, 39).

The existence of the sequences which at intron extremities are related to the model (prototype) sequences shown
at the bottom of Fig. 7 has suggested that they may have
evolved in the same way as the extremities of prokaryotic
translocatable (insertion) sequences (see Ref. 17 for references and a discussion of such a possibility). However,
since these sequences present at the ends of the introns are
clearly closely related (Fig. 7), this similarity with the
organization of the extremities of the prokaryotic insertion
sequences does not allow us to make predictions as to whether
it was the introns or the exons which were "inserted" in
evolution, if any were ever inserted (see Concluding Remarks).

POSSIBLE TRANSCRIPTIONAL CONTROL REGIONS.

A promoter is functionally defined in prokaryotes as the
region of the DNA which precedes a transcription unit and indicates the startpoint for RNA polymerase. Most sequenced
prokaryotic promoters have two common interacting sites, the
Pribnow box and the recognition site (for references, see 40).
The Pribnow box, which is of the form 5'-TATAATG-3', precedes
the transcription-initiation site by 5 to 7 nucleotides and
belongs to the model sequence 5'-ttggcGGTTATATTg-3' where
                                     ATA a
the upper and lower case letters correspond to bases present
at high and moderate frequency, respectively (see Ref. 40 for
further explanation). The recognition site is a region centred
at about 32 base pairs before the beginning of the mRNA and
has a structure clearly related to the sequence
5'-TGTTGACATTT-3' (see 40 for references). Assuming that the
ovalbumin mRNA cap site and the startpoint for transcription
are coincident (see above), sites analogous to the prokaryotic common interacting promoter sites could be contained within the region upstream from the leader-coding sequence. In a
search for such sequences, we have compared (14) the DNA

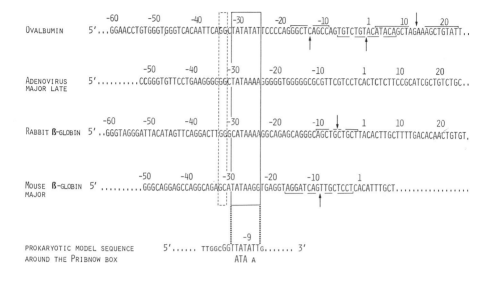

FIGURE 8. Comparison between DNA sequences of ovalbumin,
rabbit β globin (J. Van den Berg, A. Van Oogen, N. Mantei,
A. Schambock, R. Flavell and C. Weissmann; A. Efstratiadis,
T. Maniatis and E. Lacy; personal communications), mouse β
globin major (41) and Adenovirus-2 major late (22) genes upstream from the 5' end of the mRNAs. The sequences are aligned to show their similarities. Numbering is in bp and negative numbers are given to nucleotides upstream from the 5'
end of the mRNAs (position 1). The AT-rich region of homology
is boxed with a continuous line. The GC-rich region of homology is boxed with a dashed line. Lines above or below the
sequences delineate regions of partial two-fold rotational
symmetry. Centers of symmetry are indicated by arrows. The
prokaryotic promoter model sequence (40) which has the Pribnow box at one end is given for comparison (The upper and
lower case letters correspond to bases present at high and
moderate frequency, respectively).

sequences upstream from the cap site for ovalbumin, rabbit β
globin (J. Vand der Berg, A. Van Ooyen, N. Mantei, S. Scham-
bock, R. Flavell and C. Weissmann; A. Efstratiadis, T. Maniatis and E. Lacy; personal communications), mouse β globin
major (41) and Adenovirus-2 major late (22) genes. When these

sequences are aligned so as to emphasize sequence homologies
as shown in Fig. 8, there is clearly a 7 bp AT-rich region of
approximate homology centred at about 28 bp from the initial
nucleotide of the mRNAs. This region of homology which may be
a common interacting promoter site is very similar to the
"Hogness box" (D. Hogness, personal communication) which is
of the form 5'-TATAAATA-3' and has been found in an analogous
location upstream from the startpoint of transcription of
several genes. No similar sequence exists in analogous positions upstream from the 5' end of all of the ovalbumin exons
(17, and our unpublished results). It is noteworthy that in
all four cases compared, the AT-rich region of homology is
flanked by GC-rich segments (Fig. 8), suggesting that the
"Hogness box" could be extended by 3 or 4 residues. There is
some resemblance between this eukaryotic region of homology
and the prokaryotic model sequence which has the Pribnow sequence at one end (Fig. 8), suggesting that prokaryotic and
eukaryotic promoters could share some common features. However, the "Hogness box" is in a position resembling more to
that of the prokaryotic recognition site than to that of the
Pribnow box. In this respect, it is interesting that the recognition site of the E.coli lac promoter centred at about
−30 is bounded on both sides by GC-rich sequences (42).

Regions of partial (hyphenated) two-fold symmetry are
found around the probable start of transcription of the ovalbumin gene downstream from the possible RNA polymerase interacting site (Fig. 8). Regions of hyphenated two-fold symmetry
are also present in a similar location in the rabbit and
mouse β-globin genes, but are absent in the adenovirus case.
Elements of complete or partial two-fold rotational symmetry
are found in prokaryotic operators and at other sites controlling prokaryotic transcription (for references, see 14 and
40). Further studies will reveal whether specific regulatory
proteins, possibly related to the steroid hormone induction
system, could bind to these symmetrical sequences which
surround the probable ovalbumin transcription initiation site.

STRUCTURAL ORGANIZATION OF THE OVOMUCOÏD AND CONALBUMIN GENES.

We have recently cloned genomic DNA fragments containing
sequences coding for the ovomucoïd mRNA (ovom-mRNA) and the
conalbumin mRNA (con-mRNA). These two mRNAs are about 850
and 2500 nucleotides long (our unpublished results). Fig. 9
shows an electron micrograph of an RNA-DNA hybrid molecule
between ovom-mRNA and the ovomucoïd genomic DNA cloned in
λ Charon 4A (P. Gerlinger et al., in preparation). The ovomucoïd gene is obviously a very frequently split gene, since

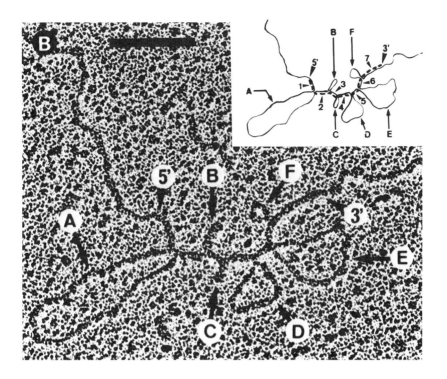

FIGURE 9. Electron micrograph of a hybrid molecule between
ovomucoïd mRNA and DNA from an ovomucoïd genomic clone. Exons
are numbered from 1 to 7 in the line drawing. Introns are
designated by letters A to F in the line drawing and the
electron micrograph. 5' and 3' arrow heads indicate the 5'
and 3' ends of ovom-mRNA, respectively. The experimental conditions will be described elsewhere. Length measurements are
as follows :   128 ± 50, 169 ± 35, 87 ± 22, 153 ± 28, 96 ± 30,
130 ± 49 and 201 ± 44 bp for exons 1 to 7, respectively;
1859 ± 200, 443 ± 104, 253 ± 100, 757 ± 98, 1131 ± 145 and
382 ± 93 bp for introns A to F, respectively. The total
length of the hybrid DNA-RNA region is 850 ± 90 bp. The bar
represents 0.1 μm.

it is composed of at least 7 exons (1 to 7) separated by 6
introns (A to F). Although electron microscopy suggests that
the mRNA is hybridized to its very 5' end (the length of the
ovom-mRNA-DNA hybrid regions is 850 ± 90 bp), we cannot exclude at present the existence of a short leader sequence similar to the ovalbumin one (see above). The genomic DNA frag-

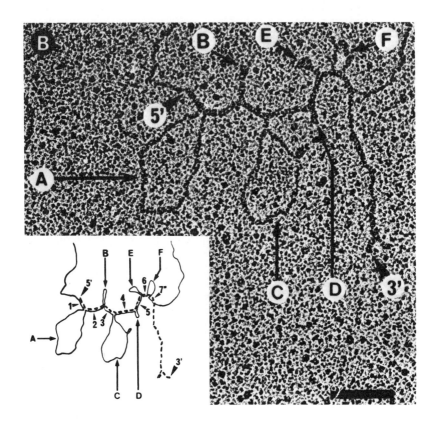

FIGURE 10. Electron micrograph of a hybrid molecule between
conalbumin mRNA and DNA from a conalbumin genomic clone. The
experimental conditions will be described elsewhere. Numbers
and letters in the line drawing and in the micrograph correspond to the different exons and introns, respectively.
5' and 3' arrow heads indicate the 5' and 3' ends of con-mRNA,
respectively. Length measurements are as follows :   113 ± 22,
174 ± 20, 133 ± 31, 197 ± 26, 145 ± 21, 84 ± 19 and 107 ± 12
bp for exons 1 to 7 respectively (see text); 1306 ± 125,
266 ± 61, 1108 ± 128, 134 ± 35, 333 ± 47 and 457 ± 44 bp for
introns A to F, respectively. The total length of the RNA-
DNA hybrid region is 942 ± 55 bp. The bar represents 0.1 μm.

ment which codes for ovom-mRNA is about 5800 bp long, about
seven times the length of the mature ovom-mRNA (we have evidence — P. Gerlinger, unpublished results — for a colinear
transcript of at least this size). It is striking that all
of the ovom-mRNA coding sequences are very short, none being

longer than 200 bp.

We have not yet isolated a complete conalbumin genomic clone. However, we have cloned a genomic DNA fragment which contains the sequences coding for about the first 900 nucleotides (from the 5' end) of this 2500 long mRNA (M. Cochet et al., in preparation). An electron micrograph of a hybrid molecule between this cloned DNA and con-mRNA is shown in Fig. 10. The conalbumin gene is also very frequently split, since sequences coding for the first 900 nucleotides of con-mRNA (the length of the DNA hybrid-region in Fig. 10 is 942 ± 55 bp) are contained in a genomic DNA segment of about 4600 bp, again about 5 times longer than the corresponding length of the con-mRNA sequence. There are 7 exons (1 to 7 – the genomic clone is interrupted within the sequence coding for exon 7, M. Cochet et al., in preparation) separated by 6 introns (A to F). As in the case of ovomucoïd, we cannot exclude at present the possible existence of an additional leader-coding region. Again, it is striking that all of the con-mRNA-coding sequences are very short with lengths between 70 and 200 bp. Finally, it should be stressed that we have not found any difference between the structure of the conalbumin gene in oviduct and liver DNA, even though the synthesis of transferrin (which is identical in its amino acid sequence to conalbumin, see Ref. 43) is not under hormonal control in chicken liver.

CONCLUDING REMARKS

Although we have not yet learnt very much about the molecular mechanisms involved in the hormonal control of transcription of the ovalbumin, conalbumin and ovomucoïd genes, our studies have contributed to the revelation of some unexpected features of the structure and function of the eukaryotic genome. It seems that we now understand better the paradox of the high DNA content of higher eukaryotic cells and the significance of hn-RNA. However, the basic and fascinating questions of why eukaryotic genes are split and whether eukaryotic split genes have evolved from previously uninterrupted genes, are still unanswered. Many speculative ideas have been put forward (for a review, see 44). Although some of them are particularly attractive, like shuffling of sequences coding for protein domains (45, 46), much more data are required before a general unifying evolutionary and functional picture of the eukaryotic genome can be drawn. For instance, protein domains should be very small to account for the structure of the ovalbumin, ovomucoïd and conalbumin genes. At present, we favor the idea that most eukaryotic genes

were never composed of "contiguous" mRNA-coding sequences,
but that it is the existence of a splicing enzyme(s) which
has allowed the emergence of functional genes from information scattered in a given region of the genome. Obviously,
this does not exclude the possibility that under some
circumstances shuffling of sequences coding for protein
domains could occur.   In our present view (see also
Ref. 47) prokaryotes would not be ancestors of eukaryotes,
but cells, which thanks to their very short generation time,
have successfully eliminated their original intronic sequences. Of course, this does not necessarily mean that some
other basic features of the bacterial genome structure and
function, particularly those related to the regulation of
gene expression at the transcriptional level, are intrinsi-
cally different in prokaryotic and eukaryotic cells.

ACKNOWLEDGEMENTS

     We are greatly indebted to Drs. J. Dodgson, R. Axel, D.
Engel and T. Maniatis who provided the "chicken library"
which was used to isolate a complete ovalbumin gene (14). We
thank all of those who sent us their results before publication. The technical assistance of C. Wasylyk, J.M. Garnier,
M.C. Gesnel, E. Sittler, E. Taubert, A. Landmann and B. Boulay
is greatly acknowledged. This work was supported by grants to
P. Chambon from the INSERM (CRT 76.5.468 and 76.5.462), the
CNRS (ATP 2117) and the Fondation pour la Recherche Medicale
Francaise. J.P. LePennec and K. O'Hare were supported by fel-
lowships from the University Louis Pasteur, Strasbourg.

REFERENCES

 1. Oka, T., and Schimke, R.T., J. cell. Biol. 43, 123-132
     (1969).
 2. Palmiter, R.D., Cell 4, 189-197 (1975).
 3. McKnight, G.S., Cell 14, 403-413 (1978).
 4. Bellard, M., Gannon, F., and Chambon, P., Cold Spring
     Harbor Symp. Quant. Biol. XLII, 779-791 (1977).
 5. Schütz, G., Nguyen-Huu, C., Giesecke, K., Hynes, N.Y.,
     Groner, B., Wurtz, T., and Sippel, A.E., Cold Spring
     Harbor Symp. Quant. Biol. XLII, 617-624 (1977).
 6. Chambon, P., Cold Spring Harbor Symp. Quant. Biol. XLII,
     1209-1234 (1977).
 7. Breathnach, R., Mandel, J.L., and Chambon, P., Nature
     270, 314-319 (1977).
 8. Mandel, J.L., Breathnach, R., Gerlinger, P., LeMeur, M.,
     Gannon, F., and Chambon, P., Cell 14, 641-653 (1978).

9. Dawid, I.B., and Wahli, W., Develop. Biol., in press.
10. Garapin, A.C., LePennec, J.P., Roskam, W., Perrin, F., Cami, B., Krust, A., Breathnach, R., Chambon, P., and Kourilsky, P., Nature 273, 349-354 (1978).
11. Perrin, F., Garapin, A.C., Cami, B., LePennec, J.P., Royal, A., Roskam, W., and Kourilsky, P., *in* Proceedings of the Internat. Symp. on Genetic Engineering, pp. 89-98. Elsevier North Holland, (1978).
12. Garapin, A.C., Cami, B., Roskam, W., Kourilsky, P., LePennec, J.P., Perrin, F., Gerlinger, P., Cochet, M., and Chambon, P., Cell 14, 629-639 (1978).
13. LePennec, J.P., Baldacci, P., Perrin, F., Cami, B., Gerlinger, P., Krust, A., Kourilsky, A., and Chambon, P., Nucl. Acids Res., in press.
14. Gannon, F., O'Hare, K., Perrin, F., LePennec, J.P., Benoist, C., Cochet, M., Breathnach, R., Royal, A., Garapin, A., Cami, B., and Chambon, P., in press.
15. Dugaiczyk, A., Woo, S.L.C., Lai, E.C., Mace, M.L. Jr., McReynolds, L., and O'Malley, B.W., Nature 274, 328-333 (1978).
16. Breathnach, R., Mandel, J.L., Gerlinger, P., Krust, A., LeMeur, M., Gannon, F., and Chambon, P., Cell 14, 641-653 (1978).
17. Breathnach, R., Benoist, C., O'Hare, K., Gannon, F., and Chambon, P., Proc. Natl. Acad. Sci. USA 75, 4853-4857 (1978).
18. Kourilsky, P., and Chambon, P., Trends in Biochemical Sciences 3, 244-247 (1978).
19. McReynolds, L., O'Malley B.W., Nisbet, A.D., Fothergill, J.E., Givol, D., Fields, S., Robertson, M., and Brownlee, G.G., Nature 273, 723-728 (1978).
20. O'Hare, K., Breathnach, R., Benoist, C., and Chambon, P., in preparation.
21. Hozumi, N., and Tonegawa, S., Proc. Natl. Acad. Sci. USA 73, 3628-3632 (1976).
22. Ziff, E.B., and Evans, R.M., Cell, in press.
23. Manley, J.L., Sharp, P.A., and Gefter, M.L., Proc. Natl. Acad. Sci. USA, in press.
24. Roop, D.R., Nordstrom, J.L., Tsai, S.Y., Tsai, M.J., and O'Malley, B.W., Cell 15, 671-685 (1978).
25. Nevins, J.R., and Darnell, J.E., Cell 15, in press (1978).
26. Ross, J., J. Mol. Biol. 106, 403-420 (1976).
27. Fraser, N.W., Nevins, J.R., Ziff, E., and Darnell, J.E., J. Mol. Biol., in press.
28. Chow, L.T., and Broker, T., Cell 15, 497-510 (1978).
29. Berget, S.M. and Sharp, P.A., J. Mol. Biol., in press.

30. Tilghman, S., Curtis, P., Tiemeier, D., Leder, P., and Weissmann, C., Proc. Natl. Acad. Sci. USA 75, 1309-1313 (1978).

31. Alwine, J.C., Kemp, D.J., and Stark, G., Proc. Natl. Acad. Sci. USA, 74, 5330-5354 (1977).

32. O'Farrell, P.Z., Cordell, B., Valenzuela, W., Rutter, W., and Goodman, H.M., Nature 274, 438-445 (1978).

33. Knapp, G., Beckman, J.S., Johnson, P.F., Fuhrman, S.A., and Abelson, J., Cell, in press.

34. Klessig, O.F., Cell 12, 9-21 (1977).

35. Catterall, J.F., O'Malley, B.W., Robertson, M.A., Staden, R., Tanaka, Y., and Brownlee, G.G., Nature 275, 510-513 (1978).

36. Ghosh, P.K., Reddy, V.B., Swinscone, J., Lebowitz, P., and Weissman, S.M., J. Mol. Biol., in press;

37. Akusjärvi, G., and Pettersson, U., Cell, in press.

38. Goodman, H.M., Olson, M.V., and Hall, B.D., Proc. Natl. Acad. Sci. USA 74, 5453-5457 (1977).

39. Valenzuela, P., Venegas, A., Weinberg, F., Bishop, R., and Rutter, W.J., Proc. Natl. Acad. Sci. USA 75, 190-194 (1978).

40. Scherrer, G.E.F., Walkinshaw, M.D., and Arnott, S., Nucl. Acids Res. 5, 3759-3773 (1978).

41. Kunkel, D., Tilghman, S., and Leder, P., Cell, in press.

42. Dickson, R.C., Abelson, J., Barnes, W., and Reznikoff, W.S., Science 187, 27-35 (1975).

43. Thibodeau, S.N., Lee, D.C., and Palmiter, R.D., J. Biol. Chem. 253, 3771-3774 (1978).

44. Crick, F., Science, in press.

45. Gilbert, W., Nature 271, 501 (1978).

46. Blake, C.C.F., Nature 273, 267 (1978).

47. Doolittle, W.F., Nature 272, 581-582 (1978).

48. Tonegawa, S., Maxam, A.M., Tizard, R. Bernard, O., and Gilbert, W., Proc. Natl. Acad. Sci. USA 75, 1485-1489 (1978).

DISCUSSION

R. HAMILTON: You were showing previously that removal of the intervening sequences may be in a sequential order and it appeared that sequences A,B, C were removed before D from one end of the transcript and that G had been removed from the other end. Is it a possibility that the sequential order in which they are being removed was from the ends going in toward the middle and does that have any significance?

P. CHAMBON: My answer is probably no – because what we have done here is to select the molecules which still have Intron E transcripts not excised. We are presently doing the same study with filters containing the other intronic DNA and then we will be able to answer your question. But it seems from preliminary results that the intron A transcript may be removed first and that any of the other intron transcripts could then be removed without having the preceding transcript either from the 5' end or from the 3' end necessarily removed beforehand.

F. CRICK: You have to do a lot of work to show that it is not random in the broad sense because you have got to show the rate constants are such that you cannot in fact account for all the fragments and there must be some defined order.

P. CHAMBON: I just want to add a further comment to your remark. When you see wide bands as those you can see when you analyze the total RNA they could, in fact, correspond to ten RNA species which could be resolved only when using specific probes. So I would not guess at present how many intermediate RNA species are present in the oviduct. It could be as many as 30 or even more.

A. SHATKIN: Could I ask Dr. O'Malley (who implied that the nuclear transcripts upon completion are translatable) if he has evidence to support that? More specifically, is it feasible to collect, for example, on cDNA-cellulose, nuclear transcripts in amounts sufficient to check translation?

B. O'MALLEY: I could not quite hear your questions, but I believe you were asking whether the nuclear precursor is translatable. Most likely not. But we actually have not done that study at the present time. We plan to do it, but it is a question of mass amounts of the material. I think it may have been done in some other labs, so someone else might

have an answer. It is likely though that you would not have
transcript all the way through because there are many protein
synthesis stop signals in the message so whereas you may
initiate transcription at the AUG, it is unlikely to
transcribe high molecular weight peptides.

A. SHATKIN: My impression of your statement was that the
finished messenger still present in the nucleus could be
translated.

B. O'MALLEY: Yes, there is some finished message. I think
there is a finished message pool in the nucleus and I do for
two reasons. An obvious possibility is some contamination
from the cytoplasm, but there are some studies that militate
against that. One is the size is slightly larger, perhaps
because the poly(A) is slightly larger in the nucleus, and
the size does not exactly correlate if you do fine studies
with the cytoplasm. Then, probably even more impressive, if
one looks at the cell during turn-on of the gene and build-up
of message in the cell, the message increases in the nuclear
pool and then saturates. It saturates while the concentration is low in the cytoplasm. As you drive the concentration
in the cytoplasm up at very late times to as many as 70,000
molecules a cell, you do not significantly change the nuclear
pool. So, I think if you make the nuclei right it is not
overwhelming contamination you are looking at, but that is
still a dangerous study. There could be some contamination,
and we have not utilized these studies to answer a precursor-
product relationship for the cytoplasm.

A. PETERSON: Could somebody please tell me, what is the role
of the intron in evolution?

F. CRICK: No we cannot tell you at this stage because we
agreed that we would not discuss that until we had at least
the evidence from hemoglobin. It is not reasonable to
discuss a broad question before you have heard all the facts.

D. APIRION: Could the size of the transcription unit be
identified by the UV inactivation mapping technique, in order
to see whether there is a correlation between the size of the
largest transcript identified and the size of the transcription unit as determined by the UV mapping technique.

P. CHAMBON: Unfortunately the chicken oviduct is not amenable to this kind of study because we cannot obtain cells in
culture which will still make ovalbumin very efficiently. So
the UV inactivation studies which were done in other systems

cannot be done in that case. But there are other ways to answer this question. Dr. O'Malley has alluded to some of them, but my personal feeling is that we need more sensitive methods.

O. SMITHIES: As I understand both of our speakers today, it has not been possible to show the presence of accumulated amounts of the full transcript in cells that are not induced for the production of ovalbumin. If this is the case, doesn't this tell us that the processing of this transcript cannot be the means of controlling the synthesis of ovalbumin?

P. CHAMBON: You are right. There is no evidence that the rate of processing could be a control mechanism.

O. SMITHIES: That is the point I want to make, namely that there is not any evidence at this point that the processing of the full length transcript can be the control mechanism for these genes.

J. BAG: When you prepare the cDNA to ovalbumin mRNA it does not contain the introns and one can make a double stranded DNA from the cDNA - can this DNA which does not contain introns be transcribed in vitro to give ovalbumin mRNA?

P. CHAMBON: We have not tried that with the bulk. On principle it should be possible provided you can initiate at one end.

AN ENDONUCLEASE-SENSITIVE REGION
IN SIMIAN VIRUS 40 CHROMATIN

Walter A. Scott, Dianne J. Wigmore
and Richard W. Eaton
Departments of Biochemistry and Microbiology
University of Miami School of Medicine
P. O. Box 016129
Miami, Florida

Abstract

Simian virus 40 chromatin obtained from nuclei of infected monkey kidney cells (BSC) is cleaved in a nonrandom manner by endonucleases. A short region (about 300 base pairs in length) extending from the origin of viral DNA replication toward the late portion of the viral genome is preferentially cleaved by deoxyribonuclease I and by an endonuclease endogenous to BSC cells. Staphylococcal nuclease cleaves at sites scattered around the genome. Viral mutants containing duplicated DNA segments including the origin of viral DNA replication and contiguous regions have been used to define cis genetic elements responsible for the endonuclease-sensitive structure adjacent to the replicative origin. More than one cis genetic element must be postulated to account for the results.

Introduction

When simian virus 40 (SV 40) infects a cell, it forms a nucleoprotein structure shown to have biochemical and morphological features similar to cellular chromatin (1-6). The resulting viral chromatin provides an attractive experimental system because it can readily be isolated free of contamination by cellular chromatin.

In recent years endonucleases have provided considerable insight into the structure of chromatin. In particular, active genes are more susceptible to degradation by certain endonucleases (notably deoxyribonuclease I) than are inactive genes (7-9). We started this work on simian virus 40 chromatin with the hypothesis that, since endonucleases detect a feature of cellular chromatin that is biologically interesting, they may provide a sensitive probe for inhomogeneity in the nucleoprotein architecture related to biologically meaningful structural features.

A Nuclease-Sensitive Region in Wild Type SV40 Chromatin

In these experiments, only the single most accessible site in any given chromatin molecule is cleaved by endonuclease. This is feasible since linear DNA is readily separated from circular DNA by electrophoresis on agarose gel slabs. The ends of the linear DNA molecules can be subsequently mapped using restriction enzymes. This is illustrated in Figure 1.

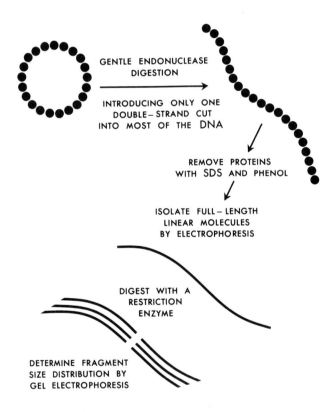

Figure 1. *Procedure for mapping nuclease-sensitive sites in SV40 chromatin.*

A restriction enzyme which cuts SV40 at only one site will produce two fragments when it acts on a full-length linear molecule which has been previously cleaved by another endonuclease. The length of these two fragments is determined by the distance between the initial endonucleolytic cleavage site and the restriction enzyme site. If the initial cut was randomly distributed, the fragments produced by subsequent cleavage with a one-cut restriction enzyme will show a random size distribution. If the initial cut was at a unique site, even in a substantial minority of the molecules, subsequent cleavage by a one-cut restriction enzyme will produce two bands on electrophoresis (corresponding to the distances in each direction around the circular genome from the restriction site to the initial cleavage site) which will stand out over a random background.

Figure 2 shows fragment patterns produced by initial cleavage of SV40 chromatin by one of three endonucleases followed by isolation of full-length linear DNA and subsequent cleavage with either Eco RI or Taq I. An endonuclease extracted from BSC cells (either SV40-infected or uninfected) shows strong preference for a limited region of the genome as indicated by the appearance of a cluster of fragments 67 to 73% of SV40 genome length if the second cleavage was introduced by Eco RI and a cluster 84 to 90% of SV40 genome length if the second cleavage was introduced by Taq I. DNAase I shows preference for the same region of the genome that is cleaved by the endogenous nuclease except that cleavage is less selective. Staphylococcal nuclease also cleaves SV40 chromatin nonrandomly, but sensitive sites are located throughout the viral genome. These patterns are not obtained if the initial cleavage is introduced into isolated SV40 DNA rather than into SV40 chromatin. Endogenous nuclease and DNAase I cleavage of isolated DNA is essentially random. Staphylococcal nuclease shows some site specificity even on isolated DNA, although the extent of this is variable in our hands.

From the size of DNA fragment clusters seen after initial cleavage by DNAase I or by endogenous nuclease and subsequent cleavage by Eco RI or by Taq I, it is possible to map the most nuclease-sensitive region of SV40 chromatin in a short region of the genetic map, (Figure 3). This location (map position 0.67 to 0.73) has been substantiated with a number of restriction enzymes (10). Using different methods for chromatin isolation or analysis, Waldeck et al. (11) and Varshavsky et al. (12) have detected a nuclease-sensitive region in SV40 chromatin from infected cells which is located in approximately the same region of the genome that we have seen.

CHROMATIN CHROMATIN
Eco R I   Taq I
E   I   M   E   I   M

FRACTION
SV 40
LENGTH

1.00 →

0.83 →

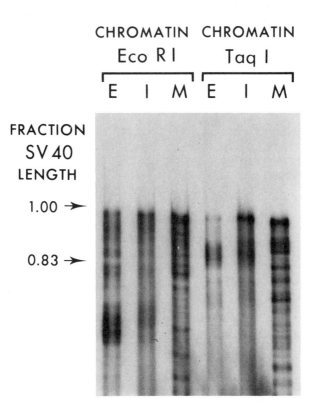

Figure 2. *Eco RI or Taq I digestion of SV40 DNA after a double-strand cleavage has been introduced in SV40 chromatin by various endonucleases.* 32P-Labeled SV40 chromatin isolated by sucrose density gradient fractionation, was cleaved by endogenous endonuclease (E), DNAase I (I), or staphylococcal nuclease (M). Incubation conditions are given in reference (10). Endogenous nuclease was provided by adding unlabeled SV40-infected BSC cell nuclear extract (1/5 of incubation volume) to the labeled isolated SV40 chromatin. Following incubation with each endonuclease, linear SV40 DNA was isolated, and digested to completion with Eco RI or Taq I. The resulting fragments were fractionated by electrophoresis on 1.4% agarose and detected by autoradiography.

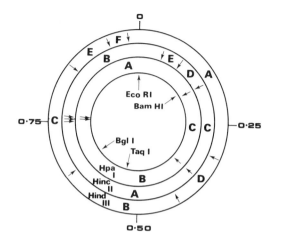

Figure 3. *Sites of cleavage of SV40 DNA by restriction enzymes.* Arrows in each concentric circle indicate sites of cleavage by the restriction endonuclease designated in the lower left part of that circle. Arrows in the inner circle show sites of cleavage by single-cut enzymes. This figure is reproduced from reference 10 by permission of the publisher (MIT Press). Copyright c MIT.

It is clear that the unique nucleoprotein configuration detected in these experiments spans an interesting segment of the viral genome (Figure 4). At one edge of the nuclease-sensitive region is the origin for viral DNA replication; at the other is a major 5'-terminus for late gene transcripts (14,15). 5'-Termini for early gene transcripts are also located in this region. The relationship of this unique nucleoprotein feature to viral genetic signals will be considered below.

Nuclease-sensitive regions in SV40 variants

Because the nuclease-sensitive feature depends on the protein component of SV40 chromatin and maps uniquely on the viral genome, it is reasonable to propose that specific DNA-protein interactions are responsible. A protein attached

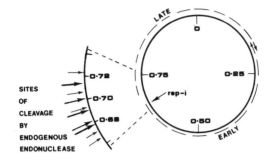

Figure 4. *The genome of SV40 showing sites of preferential cleavage of SV40 chromatin by endogenous nuclease.* Sequences transcribed early or late after viral infection (13) are indicated by the dashed lines. Arrows indicate the approximate positions for most frequent initial cleavage of SV40 chromatin by endogenous nuclease determined with Hind III or Hpa I (10).

firmly to a unique region of the genome (such as those reported by Griffith et al., (16) and by Kasamatsu and Wu (17) could provide this specificity. In addition, SV40 tumor antigen recognizes binding sites in this region of the genome (18,19). We have attempted to look at the DNA component of this specific interaction by use of mutants containing duplication of part or all of the region that is nuclease-sensitive in wt SV40 chromatin.

Maps of two of these mutants are shown in Figure 5. Each mutant contains a duplicated segment including the origin of replication and extending for varying distances in either the late direction (ev-1114) or in the early direction (ev-1119). These mutants provide a series of deletions approaching the origin from either direction (illustrated in Figure 6).

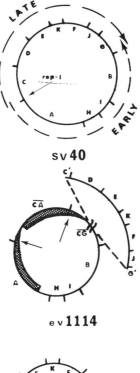

sv **40**

ev **1114**

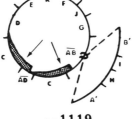

ev **1119**

Figure 5. *Hind II/III maps of mutant genomes in relation to the map of wt SV40 DNA (from reference 20).* Duplicated regions are indicated by stippled areas and deleted regions by wedge-shaped extensions from the circles. Letters designate segments corresponding to wt SV40 Hind II/III fragments (21) or portions thereof. Arrows indicate origins of replication. This figure is reproduced from reference (22) by permission of the publisher. Copyright c Academic Press.

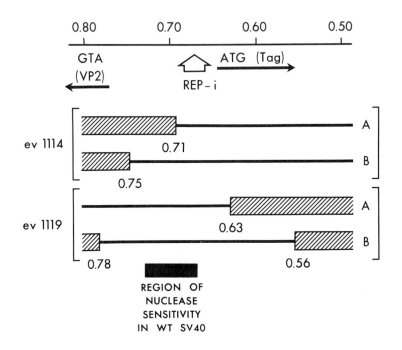

Figure 6. *Segments of DNA containing the origin or replication and contiguous sequences from two SV40 variants.* The
scale at the top indicates map units on the wt SV40 map (the
distance in fractional units of the SV40 genome clockwise
from the Eco RI site.) Translational start signals for proximal late (VP2) and early (T antigen) proteins are given to
delineate the boundaries of the untranslated region around
the origin of replication (23). Each variant contains a
duplication (designated by segments A and B) of part of the
region of the genome shown. The presence of sequences
corresponding to the wt SV40 map is indicated by a solid
line; the absence of sequences, by a hatched segment (//////).
The number below the line indicates the map position for each
recombination joint (20 and M. Gutai, personal communication).
The letters, A and B, are needed to distinguish the two
segments of each mutant genome in subsequent discussion.

Analysis of these mutants follows the same plan used for analysis of wild type SV40 chromatin. A single cleavage is introduced into a substantial fraction of the mutant genomes by action of endogenous nuclease. Linear mutant viral DNA is separated from other species by electrophoresis on 1.4% agarose and the site of initial cleavage is mapped by subsequent cleavage with a restriction enzyme which introduces one cut in the variant genome. If nuclease-sensitive nucleoprotein configurations are organized near each replication origin, then four clusters of fragments should be observed (Figure 7).

Results for the experiment illustrated in figure 7 are shown in figure 8. The expected fragment size clusters are seen (the smallest cluster runs off the bottom of both gel types). This means that both segments contain sufficient cis genetic information to organize the nuclease-sensitive nucleoprotein configuration. An analogous experiment with ev-1119 (Figure 9) also shows that nuclease-sensitivity occurs in the genome segments adjacent to each origin of replication.

The simplest interpretation of these results is that the DNA segment between 0.63 and 0.71 map units on the viral genome (the segment that is common to all four nuclease-sensitive segments in ev-1114 and ev-1119, Figure 6) contains cis genetic information responsible for organizing the nuclease-sensitive chromatin structure. Closer inspection of figures 8 and 9 suggests that, while both segments A and B in ev-1119 are cut with about equal frequency, there is a difference in cutting frequency at the two origins in ev-1114. Segment A is cleaved less frequently than segment B. This suggests that there is additional cis genetic information between map position 0.71 and 0.75 (that portion of the genome present in ev-1114 segment B but missing in A) which plays a role in the nuclease-sensitive structure of this portion of the genome.

Results reported by Lai et al. (14) and by Reddy et al. (15) may be relevant to this latter observation. Late nuclear viral-specific RNA species have been mapped on the SV40 genome. 5'-Termini map at several discrete positions with the major species mapping at about map position 0.72. Other species contain 5'-ends mapping at 0.71, 0.70, 0.67, 0.64, and 0.59. This may indicate that there are multiple sites for initiation of late viral RNA synthesis, and these sites may also be involved in the organization of the nuclease-sensitive structure detected in these experiments.

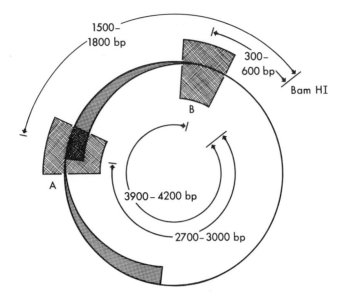

Figure 7. *Mapping procedure for determining preferential cleavage sites in ev-1114 chromatin using restriction enzyme Bam HI.* The stippled segments indicate duplicated portions of the genome of ev-1114 and allow reference to the more detailed map of this variant shown in figure 5. The cross-hatched wedges indicate genome sections which correspond to the nuclease-sensitive region in wt SV40 chromatin (extending approximately from the origin of replication for 300 bp in the late direction, reference 10). Letters A and B refer to the two duplicated segments depicted in figure 6. The distance of each predicted nuclease-sensitive region from the single Bam HI site in this mutant is computed from the known restriction map of ev-1114 (20) and indicated as a range of fragment sizes which would result from an initial cleavage within the cross-hatched area followed by a second cleavage at the Bam HI site.

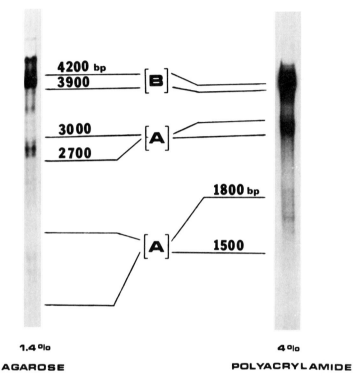

Figure 8. *Endogenous nuclease cleavage of ev-1114 chromatin mapped with Bam HI.* Chromatin was extracted from ^{32}P-labeled BSC nuclei infected by SV40 variant ev-1114. MnCl$_2$ was added to 2.0 mM and the extract was incubated 30 min. at 37°C. Variant-length linear DNA was isolated by electrophoresis on 1.4% agarose, digested with Bam HI and the resulting fragments were fractionated by electrophoresis on 1.4% agarose or on 4% polyacrylamide and detected by autoradiography. Segments indicated A or B correspond to the range of fragment sizes predicted for each possible nuclease-sensitive region of variant chromatin (Figure 7) and were determined by DNA size markers.

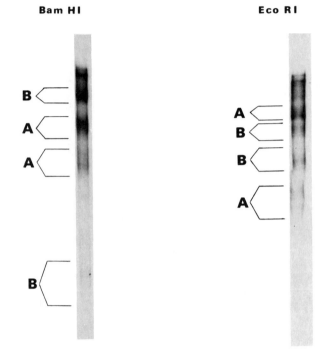

Figure 9. *Endogenous nuclease cleavage of ev-1119 chromatin mapped with Bam HI and with Eco RI.* ^{32}P-Labeled chromatin from ev-1119-infected BSC cells was prepared and cleaved as described in the legend to Figure 8. Variant-length linear DNA was recovered, digested with either Bam HI or Eco RI, and fractionated by electrophoresis on 4% polyacrylamide. A and B correspond approximately to the range of fragments predicted for initial cleavage in the segment A and B of ev-1119 (Figure 6) and subsequently cleaved by restriction enzyme Bam HI or Eco RI.

Possible biological significance of the nuclease-sensitive chromatin structure

Assuming that this unusual chromatin segment has biological significance -- an assumption encouraged by its location on the genetic map -- what possible viral functions may be involved?

a) Viral late gene transcription is the most obvious, both from the location on the viral genome (extending in the late direction from the origin of replication) and from the fact that viral chromatin was isolated during a period of maximal late viral gene expression.

b) Early viral gene transcription is also a possible candidate. In particular, early genes may be organized in an active configuration to be packaged in virus particles so that they will be active on release in a newly infected cell. From this argument, it would be predicted that chromatin isolated from viral particles would retain the specifically located nuclease-sensitive region.

c) Viral DNA replication seems a less likely candidate. The nuclease-sensitive region detected in these experiments is located assymetrically with respect to the origin of replication, yet DNA replication proceeds in both directions from the origin at about the same rate (24,25). In addition, ev-1114 utilizes both origins for replication with about equal frequency (20), yet nuclease-sensitivity of the two segments is different.

d) Virion assembly must be considered, but no details of this process are available which will allow us to evaluate the significance of nuclease-sensitivity of a specific region of the viral chromatin with respect to this process.

In conclusion, we have detected a nuclease-sensitive segment of SV40 chromatin from infected cells extending in the late direction from the origin of viral DNA replication. More than one cis genetic element appears to be involved in the organization of this structure. Its role in viral processes remains to be explored.

Acknowledgments

SV40 variants were isolated by Brockman and Nathans and generously supplied by D. Nathans.

This work was supported by NIH grant AI-12852 and NSF grant PCM78-06847. W.A.S. is a recipient of a Research Career Development Award (AM-00549).

REFERENCES

1. White, M. and Eason, R. (1971). J. Virol. 8, 363–371.
2. Hall, M. R., Meinke, W. and Goldstein, D. A. (1973). J. Virol. 12, 901–908.
3. Griffith, J. D. (1975). Science 187, 1202–1203.
4. Meinke, W., Hall, M. R. and Goldstein, D. A. (1975). J. Virol. 15, 439–448.
5. Cremisi, C. Pignatti, P. F. and Yaniv, M. (1976). Biochem. Biophys. Res. Commun. 73, 548–554.
6. Bellard, M., Oudet, P., Germond, J. E. and Chambon, P. (1976). Eur. J. Biochem. 70, 543–553.
7. Gottesfeld, J. M., Garrard, W. T., Bagi, G., Wilson, R. F. and Bonner, J. (1974). Proc. Nat. Acad. Sci. USA 71, 2193–2197.
8. Garel, A. and Axel, R. (1976). Proc. Nat. Acad. Sci. USA 73, 3966–3970.
9. Weintraub, H. and Groudine, M. (1976). Science 193, 848–856.
10. Scott, W. A. and Wigmore, D. J. (1978). Cell 15, 1511–5118.
11. Waldeck, W., Fohring, B., Chowdhury, K., Gruss, P., and Sauer, G. (1978). Proc. Nat. Acad. Sci. USA 75, 5964–5968.
12. Varshavsky, A. J., Sundin, O., and Bohn, M. (1979). Cell 16, in press.
13. Khoury, G., Carter, B. J., Gerdinand, F. J., Howley, P. M., Brown, M. and Martin, M. A. (1976). J. Virol. 17, 832–840.
14. Lai, C.-J., Dhar, R. and Khoury, G. (1978). Cell 14, 971–982.
15. Reddy, V. B., Ghosh, P. K., Lebowitz, P. and Weissman, S. M. (1978). Nucl. Acids Res. 5, 4195–4213.
16. Griffith, J., Dieckmann, M. and Berg, P. (1975). J. Virol. 15, 167–172.
17. Kasamatsu, H. and Wu, M. (1976). Biochem. Biophys. Res. Commun. 68, 927–936.
18. Jessel, D., Landau, T., Hudson, J., Lalor, T., Tenen, D. and Livingston, D. M., (1976). Cell 8, 535–545.
19. Tjian, R. (1978). Cell 13, 165–179.
20. Brockman, W. W., Gutai, M. W. and Nathans, D. (1975). Virology 66, 36–52.
21. Danna, K. J., Sack, G. H., Jr. and Nathans, D. (1973). J. Mol. Biol. 78, 363–376.
22. Scott, W. A., Brockman, W. W. and Nathans, D. (1976). Virology 75, 319–334.

23. Reddy, V. B., Thimmappaya, B., Dhar, R., Subramanian,
     K. N., Zain, B. S., Pan, J., Gosh, P. K., Celma, M. L.
     and Weissman, S. M. (1978). Science 200, 494-502.
24. Danna, K. J. and Nathans, D. (1972). Proc. Nat. Acad.
     Sci. USA 69, 3097-3100.
25. Fareed, G. C., Garon, C. F. and Salzman, N. P. (1972).
     J. Virol. 10, 484-491.

DISCUSSION

J. SYGUSCH: I was wondering if you have considered the
transcriptional   complexes   among   the   SV40   nucleoprotein
complexes.   Is nuclease-sensitive SV40 chromatin enriched in
transcriptional or potential transcriptional activity?

W.A. SCOTT: I do not know the answer to that.   Most of the
studies with transcriptional complexes have used conditions
that take off the histones.   I expect that would make the
structure more nuclease-sensitive.   It should be possible to
isolate transcriptionally active chromatin complexes with all
of their proteins, and then look for nuclease sensitivity, but
we haven't done that.

J. SYGUSCH: In Science (201, 406, 1978), Keller et al. isolated
the entire nucleoprotein complex and showed a small fraction of
the population was transcriptionally active and also apparently
contained a full complement of histones.

W.A. SCOTT: We have not studied complexes isolated in that
manner.

P. CHAMBON: P. Garaglio in my lab, has isolated the
transcriptional complex in the form of a mini-chromosome. One
relevant point is that the transcriptional complex represents
only 1.5% of the total SV40 DNA which is found late in
infection. 1.5% should not be detectable in your study and I am
wondering   whether   what   you   are   seeing   is   related   to
transcription, or rather to replication. Of course we may argue
that the template is ready for transcription, but that the
amount of RNA polymerase is limiting.   I have one question:

there are SV40 mutants where you have triplication of the origin
of replication. It would be interesting to look at the
chromatin of these mutants to see what is the nuclease-
sensitivity in that case.

W.A. SCOTT: Do you know what fraction of the complexes at this
time in the infection would actually be involved in
replication? I believe that is low, too. We have estimated the
nuclease-sensitivity of this region to be a property of at least
30% of the total labeled viral DNA. Alexander Varsharsky's
experiments seem to indicate that it is 80% or more. So you are
right, there is a problem in explaining this phenomenon in terms
of physiological function. The results from Weintraub's and
Palmiter's laboratories seem to say that a gene can be nuclease-
sensitive and therefore in an active configuration, even if it
is not actively involved in transcription. Maybe what we are
detecting is the potential for transcription.

J. ALWINE: I was wondering whether you had done any experiments
with tsA mutants. The question arises from the fact that T
antigen binds to the origin of replication and is involved in
early transcription and initiation of DNA replication.
Therefore, it seems that T antigen could induce the nuclease-
sensitive site. A very simple test of this would be to do
temperature shift experiments with a tsA mutant, and find out if
your nuclease-sensitive site is lost in conjunction with the
loss of active T antigen. Have you done any such experiments?

W.A. SCOTT: That is an interesting experiment which I would
like to do and which we are going to do, but haven't yet done.

EXPRESSION OF CLONED VIRAL AND CHROMOSOMAL
PLASMID-LINKED DNA IN COGNATE HOST CELLS

Charles WEISSMANN, Ned MANTEI, Werner BOLL,
Robert F. WEAVER, Neil WILKIE*, Barklie CLEMENTS*,
Tadatsugu TANIGUCHI, Albert VAN OOYEN, Johan VAN
DEN BERG, Mike FRIED[+] and Kenneth MURRAY[‡]

Institut für Molekularbiologie I
Universität Zürich
8093 Zürich, Switzerland

There has been a good deal of speculation on
the possible hazards inherent to hybrid DNA technology. An important question in this regard is
whether a viral genome, linked to a prokaryotic
vector, can be expressed in a host cell. This is
shown to be the case both for prokaryotic viral
DNA in a prokaryotic host, and for eukaryotic viral
DNA in a eukaryotic host. We find that an intact
copy of Qβ DNA joined to pCRI elicits formation
of Qβ in E.coli, that a hybrid consisting of a
head-to-tail polyoma DNA dimer and pBR322 causes
polyoma formation in cultured mouse fibroblasts,
and that the cloned thymidine kinase (TK) gene of
herpes simplex type I linked to pBR322 can transform TK⁻ mouse L cells. Moreover we show that chromosomal β globin DNA, linked to a cloned thymidine
kinase gene, can be introduced into TK⁻ L cells
and is expressed therein.

I. Qβ phage formation by E.coli transformed with
   a Qβ DNA-pCRI hybrid plasmid (1).
   Preparation of a hybrid plasmid containing a
   complete DNA copy of Qβ RNA.
       Phage Qβ contains a single-stranded RNA
   of about 4500 nucleotides which serves both as
   genome and as mRNA. It codes for 4 proteins;
   three of these are part of the capsid, the
   fourth is a subunit of Qβ replicase, the virus-

---

* Institute of Virology, University of Glasgow
[+] Imperial Cancer Research Fund Laboratories, London
[‡] Dept. Molecular Biology, University of Edinburgh

specific RNA polymerase. Purified Qβ replicase
produces Qβ RNA in vitro, and synthesis of infectious viral RNA can be elicited in vitro by
Qβ RNA plus strands or their complement, Qβ minus strands (2,3).

    In designing approaches for the preparation
of a hybrid plasmid containing a complete copy
of the Qβ genome, we first considered the potential lethality of the hybrid for the host cell.
We anticipated that the in vivo transcription
of a viral RNA copy from a Qβ DNA plasmid would
be inefficient in at least one of the two possible orientations of the insert, and that, moreover, the processing of the primary transcript
to yield an RNA molecule capable of being translated and replicated would be a rare event within the span of the bacterial generation time.
We therefore decided to use wild-type Qβ RNA
as our starting material, rather than a temperature-sensitive variant.

    Previous experience had shown that double-
stranded cDNA copies of mRNA, prepared by conventional means, lack at least a few nucleotides at the end corresponding to the 5' terminus of the RNA (4). A modified approach, outlined in Fig. 1, was devised to overcome this
difficulty. Qβ plus and minus strands, synthesized in vitro with Qβ replicase (5), were
elongated with Ap residues using poly(A) polymerase (6), and separately used as templates
for reverse transcriptase in an oligo(dT)-primer
reaction. The cDNA copies were freed of RNA
by very mild alkaline digestion followed by
sucrose gradient centrifugation. The chain
length of the minus and plus strand Qβ DNA
was estimated by alkaline agarose gel electrophoresis to range approximately over 3500-4500
and 1600-4500 nucleotides, respectively (data
not shown). About equimolar amounts of the
complementary strands were hybridized to yield
incomplete double strands. Repair synthesis
with DNA polymerase I yielded a double-stranded
DNA which on agarose gel electrophoresis moved
as a discrete band with a mobility similar
to that of linear pBR322 DNA, (4,361 base
pairs; G. Sutcliffe and W. Gilbert, personal
communication).

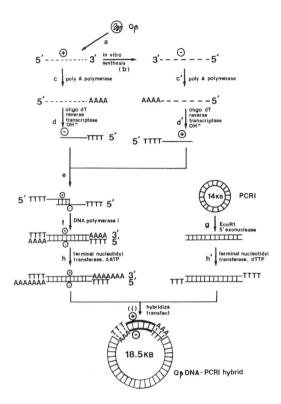

Fig. 1.   Synthesis of a hybrid plasmid containing a complete DNA copy of the Qβ genome. Qβ RNA
extracted from plaque-purified phage (a) is used as
template for the synthesis of Qβ minus strands by
Qβ replicase (b). Qβ RNA and purified Qβ minus
strands are elongated with AMP residues by poly(A)
polymerase (c and c', respectively) and used as
templates for reverse transcriptase, with oligo(dT)
as primer (d and d', respectively). The minus and
plus cDNA copies are hybridized (e) and the incomplete duplex is repaired with DNA polymerase I (f).
Plasmid pCRI is cleaved with EcoRI and the ends are
trimmed back with 5' exonuclease (g). The double-
stranded Qβ DNA and the pCRI DNA are elongated with
dAMP and dTMP residues, respectively (h and h'),
using terminal nucleotidyl transferase. The two
components are hybridized and transfected into E.
coli HB101 (i) (from ref. 1).

The double-stranded Qβ DNA was elongated
with dA residues, hybridized to poly(dT)-pCRI
DNA and transfected into Escherichia coli
HB101, a strain resistant to infection by
Qβ phage. 96 kanamycin-resistant clones were
gridded and assayed by the Grunstein-Hogness
in situ hybridization assay (7), using Qβ
^{32}P RNA as probe; 72 gave a positive response.
Two sets of the colonies were reassayed by the
same procedure, using as probes plus strand
and minus strand segments, respectively, ^{32}P-
labeled at the ɣ position of the 5' terminal
pppG residue. 32 colonies were labeled by plus,
and 22 colonies by minus strand termini; four
colonies, ZpQβ-32, ZpQβ-33, ZpQβ-74 and ZpQβ-87,
were labeled by both probes and were presumed
to contain a complete copy of Qβ DNA.

Properties of E.coli HB101 containing Qβ DNA
plasmids.
    To test whether any of the clones containing
Qβ DNA hybrid plasmids were producing Qβ phage,
colonies of the four clones considered to contain complete Qβ DNA copies, as well as clones
with Qβ DNA which apparently lacked either one
or the other terminus, were replica-plated onto
a lawn of indicator bacteria (E.coli Q_{13}) sensitive to Qβ phage. Three of the four strains containing the complete Qβ genome (ZpQβ-32, ZpQβ-33
and ZpQβ-87), but none of the others, gave an
area of lysis (Fig. 2). Cultures of several bacterial strains were prepared. The bacteria were
collected at a cell density of about 10^8 ml^{-1}
and resuspended in the original volume of medium.
The bacterial suspension and the supernatant
were titrated on indicator bacteria. The supernatants of the cultures containing the plasmids
ZpQβ-32, ZpQβ-33 and ZpQβ-87 had about 5 x 10^8
plaque-forming units (PFU) per ml, whereas the
value for the bacterial suspensions was about
5 x 10^7 PFU per ml. E.coli HB101 containing
pCRI or plasmid ZβQβ-74 produced no plaque-forming agents. The plaque-forming activity in

the supernatant was identified as phage Qβ by
neutralization with Qβ antiserum.

The generation times of the three Qβ-producing strains were all greater (40-46 min) than
those of E.coli HB101 containing pCRI or a non-
producer pCRI-Qβ DNA hybrid (36 and 35 min, respectively), showing that Qβ production reduced
the growth rate of the bacterial population.
The release of phage could be due to the occa-
sional lysis of a plasmid-containing bacterium

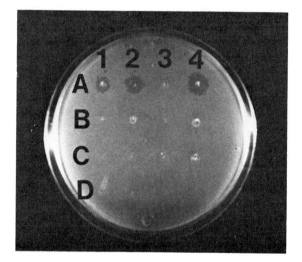

Fig. 2.   Identification of phage-producing colonies of E.coli HB101 transfected with Qβ DNA
hybrid plasmids. The four bacterial clones believed
to contain complete Qβ DNA as well as 12 clones
with incomplete copies were inoculated onto agar
plates. The colonies were replica-plated onto an
agar plate freshly covered with soft agar containing about 5×10^7 E.coli Q_{13}, a strain susceptible
to infection by phage Qβ. Positions A-1 to A-4 are
occupied by colonies containing ZpQβ-32, ZpQβ-33,
ZpQβ-74 and ZpQβ-87, respectively, positions B-1
to D-4 by colonies containing incomplete Qβ DNA
plasmids (from ref. 1).

TABLE 1

Analysis of phage production and population-forming capacity of single E.coli HB101 containing the Qβ-inducing hybrid plasmid ZpQβ-32

Bacteria (colony-forming units per tube)	Phage (PFU per tube)							
	0	10^0-10^1	10^1-10^2	10^2-10^3	10^3-10^4	10^4-10^5	10^5-10^6	$>10^6$
0	45							
10^0-10^3		4	14	9		1	1	1
10^3-5×10^3								12
5×10^3						1		8

E.coli HB101 containing plasmid ZpQβ-32 were grown in tryptone broth to a cell density of 7×10^7 cells per ml and diluted in tryptone broth to 5 cells per ml. 100-µl aliquots were distributed into 96 wells of a microtitre plate and incubated for about 20 h at 37°C. Phage titers and bacterial counts were determined. In a control experiment, E.coli HB101 was distributed into 192 wells (on average, 0.5 cells per well) in a similar fashion: 71 wells (37%) showed bacterial growth (expected value, 39%).

as shown for E.coli chronically infected with
f_2 phage (8), or to the constant leakage of a
small number of phage from most or all bacteria,
as in the case of phage fd-infected E.coli (8,9).
To distinguish between these possibilities, E.coli
HB101 (ZpQβ-32) were washed to remove free phage,
and distributed into 96 tubes at a dilution such
that each tube contained on average about 0.5
bacteria. As calculated from the Poisson distribution, 60.6% of the tubes should contain no
bacteria, and 30, 7.6 and 1.8% of the tubes 1,
2 or 3 and more bacteria, respectively. After
incubation, the tubes were scored for bacterial
growth and phage content. In a typical experiment, 45 tubes (47%) contained neither phage
nor bacteria, 28 tubes contained phage only,
and 23 tubes both bacteria and phage; no tubes
contained bacteria only (Table 1). It is interesting that the number of phage per tube varied
from 10 to 10,000 in the tubes containing phage
but no bacteria, and was in general $> 10^6$ when
bacteria were also present.

We interpret these findings as follows. If,
on placing a single bacterium in a tube, lysis
occurs before the first division, the tube will
contain the progeny of a single burst, on average $\sim 1,300$ phage particles (10) and no bacteria. If the bacterium divides, both progeny
may lyse to yield a tube with two bursts and
no bacteria, one bacterium may lyse and the
other divide, or both bacteria may divide. With
successive generations the majority of the tubes
will be either devoid of bacteria (but contain
varying numbers of phage) or they will contain
a large population of bacteria (and phage).

Physical and biological characterization of a
hybrid plasmid containing Qβ DNA.
The orientation of the inserts was determined by restriction analysis and is given in
Fig. 3.
The specific infectivity of the Qβ DNA plasmid 32 (that is, its capacity for eliciting Qβ
production on transfection) was compared to that
of Qβ RNA in an infectious center assay, using
E.coli K12W6 spheroplasts. As shown in Table
2, the specific infectivity of the plasmid was

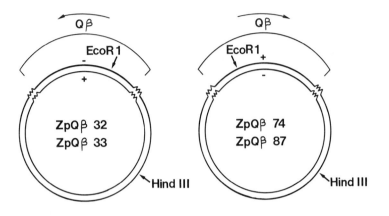

Fig. 3.   Orientation of the Qβ DNA insert in pCRI (from ref. 1).

almost equal to that of the RNA on a weight basis, and seven times greater on a molar basis. Treatment with restriction enzyme BspI (an isoschizomer of HaeIII (11) abolished the Qβ-producing activity of the plasmid, but not that of Qβ RNA, whereas the opposite result was found with RNase A. No plaques were obtained when 1 μg of plasmid DNA was plated on E.coli Q_{13}. These experiments show that the plasmid DNA, and not contaminating Qβ RNA or Qβ phage, is responsible for eliciting Qβ phage formation. In a further experiment, E.coli HB101 was infected with the hybrid plasmid ZpQβ-32 and kanamycin-resistant clones were screened for Qβ phage production. All of 100 clones examined produced phage as determined by replica-plating the colonies onto a lawn of indicator bacteria (data not shown). Thus, within about 40 bacterial generations (the number required to expand a single bacterium to the culture from which the DNA was prepared), less than 6×10^{-8} mutations lethal to Qβ occurred per nucleotide and doub-

TABLE 2

Specific infectivity (Qβ formation) of ZpQβ-32 DNA or Qβ RNA by the infectious center assay.

Nucleic acid added, ng	Buffer	Treatment		Plaques found	PFU µg^{-1}
		BspI	RNase A		
Qβ RNA					
10	0.5 mM EDTA	-	-	250	2.5 x 10^4
400	Y-50	-	-	1062	2.7 x 10^3
400	Y-50	+	-	1186	3.0 x 10^3
400	Y-50	-	+	6	15
ZpQβ-32 DNA					
10	20 mM Tris-HCl	-	-	208	2.1 x 10^4
10	Y-50	-	-	165	1.7 x 10^4
10	Y-50	+	-	0	0
10	Y-50	-	+	233	2.3 x 10^4

Spheroplasts of E.coli K12W6 were prepared and infected as described (10), and plated on lawns of E.coli Q13. 0.5 µg DNA or 20 µg RNA were treated with 5 units of BspI restriction enzyme or with 1 ng pancreatic RNase for 30 min at 37°C. Self-digested Pronase and 0.2% sodium dodecyl sulphate were added and incubation was carried out for 30 min at 37°C. 5 µl were withdrawn and mixed with 20 µl of 20 mM Tris-HCl (pH 7.5). Of this, 5 µl (corresponding to 1/50 of original mate-rial) were diluted with 0.2 ml 0.5 mM EDTA, mixed with spheroplasts and plated on E.coli Q13. Y-50 buffer is 10 mM Tris-HCl (pH 7.5), 6 mM MgCl$_2$, 50 mM NaCl, 6 mM mercaptoethanol, 200 µg ml^{-1} bovine serum albumin.

ling. It remains to be seen whether on prolonged
culturing there is a selection for bacteria containing a reduced number of Qβ hybrid plasmids
capable of yielding infectious Qβ RNA.

Characterization of the Qβ RNA formed following
infection with Qβ DNA-containing plasmids.
    Multiply-passaged Qβ phage population con-
tain a high proportion of viable variants (12).
Although the Qβ RNA used to generate the Qβ DNA
plasmids was prepared from plaque-purified virus, it was of interest to compare the T_1 fingerprints of authentic wild-type Qβ RNA and
the RNA of phage recovered from Qβ DNA plasmid--
infected cells. ^{32}P-labeled RNA of phage derived from cells infected with the hybrid plasmids ZpQβ-32, ZpQβ-33 and ZpQβ-87 as well as
wild-type Qβ ^{32}P RNA was treated with RNase T_1
and the oligonucleotides were separated by two-
dimensional polyacrylamide gel electrophoresis.
All autoradiograms were indistinguishable, and
identical with that shown in Plate 3(a) of ref.
10, confirming the identification of the virus
as "wild type" Qβ. In particular, the presence
of the 3' terminal T_1-resistant oligonucleotide
T-18 showed that the 3' end of the plasmid-in-
duced phage RNA was identical with that of the
wild-type RNA and had no supernumerary residues.

Biological implications.
    The synthesis of a DNA copy of the Qβ genome
involved the use of Qβ replicase, AMV reverse
transcriptase and E.coli DNA polymerase I, the
first two of which are reported to have a high
error rate, of about 10^{-4} errors, respectively,
per nucleotide and doubling (13,14). Nonetheless, three out of four hybrid plasmids which
contained a complete copy of the Qβ genome
proved capable of eliciting Qβ phage formation
in E.coli. This proportion may be a low estimate,
as the colony-forming capacity of Qβ-producing
transformants is only about 45% (see Table 1).
    E.coli propagating a Qβ DNA genome and pro-
ducing Qβ phage constitute a novel biological
system, although superficially they resemble
a population chronically infected with Qβ phage,
where the bulk of the bacteria replicate normally, a certain fraction lyses, releasing phage

particles, and the population is largely resistant to infection from outside, that is, does not permit plaque formation when used as lawn (8,9,15-17). The exact mechanism of this so-called "carrier-state" is not known; however, it is probably based on horizontal (8) and/or vertical (17) transmission of the phage RNA genome in conditions which (1) preclude efficient viral expression and (2) reduce susceptibility to infection. There is no evidence that DNA of E.coli acutely infected with an RNA phage contains phage-specific sequences (18); however, the DNA of a carrier population has to our knowledge not yet been examined.

How is an infectious cycle initiated in a cell containing a Qβ DNA hybrid plasmid? The first event must be the synthesis of a plus strand transcript which then serves as messenger for the synthesis of the virus-directed replicase β subunit (for a review, see ref. 3). As it is unlikely that transcription by the DNA-dependent RNA polymerase of the host would initiate precisely at the beginning of the Qβ sequence and terminate at its end, we must assume that either the primary transcript is trimmed back, at least at the 3' end, or that Qβ replicase initiates RNA synthesis at the internal site corresponding to the viral 3' terminus; perhaps both processes occur. If the newly synthesized minus strand is not terminated at the correct site, it in turn may be trimmed back, or here too replicase may initiate at the site corresponding to the natural 3' terminus of the minus strand. Some such combination of events would lead to Qβ RNA of the correct length.

We have previously shown that Qβ RNA elongated in vitro with about 30 AMP residues lost its capacity to serve as template for Qβ replicase (19). Nonetheless, the specific infectivity of poly(A)-Qβ RNA in the spheroplast assay was the same as that of natural Qβ RNA and the progeny RNA was devoid of supernumerary AMP residues (6,19). Phosphorolysis in vitro of poly(A)-Qβ RNA with polynucleotide phosphorylase removed all but a few AMP residues and restored the template activity of the RNA (19). It is

of interest that the same treatment did not diminish the infectivity of natural Qβ RNA, presumably because the secondary and/or tertiary
structure of the RNA protects it against exo-
nucleolytic degradation by this enzyme (19).
From these observations we conclude that an
oversized transcript of the Qβ plus strand could
be processed in vivo to a size where it can be
used as template by Qβ replicase, allowing initiation of progeny RNA synthesis at the site
corresponding to the true 3' terminus.
    Where on the hybrid plasmid does transcrip-
tion of the Qβ DNA-containing sequence origi-
nate? We have not yet characterized the primary
transcript, so that we cannot tell whether a
natural promotor of pCRI is being used, or whether a low level of random initiation, possibly
in the AT-linker region, is occurring. The fact
that both orientations of the Qβ insert lead
to Qβ production at about equal efficiency
(Table 2) seems to argue against the use of a
natural promotor of the plasmid.
    The synthesis of a Qβ DNA-containing plasmid
which leads to the formation of viable Qβ particles demonstrates the fidelity of the cloning
procedure. Moreover, it opens a new approach
to genetic experimentation with phage Qβ and
will facilitate the determination of the nucleotide sequence of its genome.

II. Lytic infection caused by polyoma head-to-tail
    dimer DNA integrated in pBR322 (20).
        As part of a risk-testing program sponsored
    by EMBO we undertook to determine if and under
    which conditions polyoma DNA, joined to a pro-
    karyotic vector, would lead to virus formation
    in a permissive eukaryotic host cell.

    Preparation and characterization of hybrids be-
    tween polyoma and pBR322 DNA.
        To allow identification of progeny DNA re-
    covered at later stages of the experiment, we
    used a variant polyoma DNA which lacks an HhaI
    restriction site at 26.4 map units from the
    EcoRI site (21) and thus yields only two rather
    than three fragments on cleavage with HhaI.
        Polyoma DNA was digested with EcoRI or BamHI,
    which cleave at a single site (Fig. 4), and

joined with DNA ligase to appropriately cleaved
pBR322 DNA (Fig. 4). The molar ratios of viral
to plasmid DNA used were between 2:1 and 1:1.
E.coli HB101 was transformed with the recombinant DNA, and colonies containing polyoma DNA
were identified by in situ hybridization (7)
to ^{32}P-labeled polyoma RNA. The DNA from 28
positive colonies obtained by transformation
with recombinants made from BamHI-cleaved DNA
was purified and analyzed by agarose gel electrophoresis. Two of the hybrid plasmids had a
mobility relative to polyoma DNA form I of about
0.5 and 0.52, respectively, the others had mobilities of 0.68 to 0.70. The smaller recombinants
(with the higher mobility) were shown by restriction analysis to contain one unit each
of polyoma and pBR322 DNA with both possible
orientations being represented (Fig. 4c and d);
these we call "monomeric" hybrids. One of the
larger hybrids (D3) consisted of one molecule
of pBR322 and two of polyoma DNA, linked head-
to-tail, in the orientation shown in Fig. 4e,
and is called "dimeric" hybrid. Another large
hybrid, D5, consisted of one molecule of pBR322
linked to a head-to-tail polyoma dimer with
a deletion of about 800 bp   in the region
shown in Fig. 4e.
      None of the 48 polyoma-pBR322 hybrids in
which the components were joined at their EcoRI
sites contained more than one unit of polyoma,
but two contained two units of pBR322. Both
orientations of polyoma with respect to pBR322
DNA were found amongst the 46 monomeric recombinants.

Infectivity of pBR322-polyoma DNA hybrids.
      The infectivity of the polyoma-pBR322 hybrid
DNAs before and after cleavage with the restriction enzyme used to generate the joined fragments was compared to that of polyoma virus DNA,
using the DEAE method on mouse embryo cells
(22). In Table 3 the specific infectivities
of some of the polyoma-pBR322 hybrids joined
via the BamHI site are presented. Hybrids containing one unit length of polyoma virus DNA
gave less than 1 PFU/μg of hybrid DNA (less than
1 PFU/10^{11} DNA molecules), however, cleavage

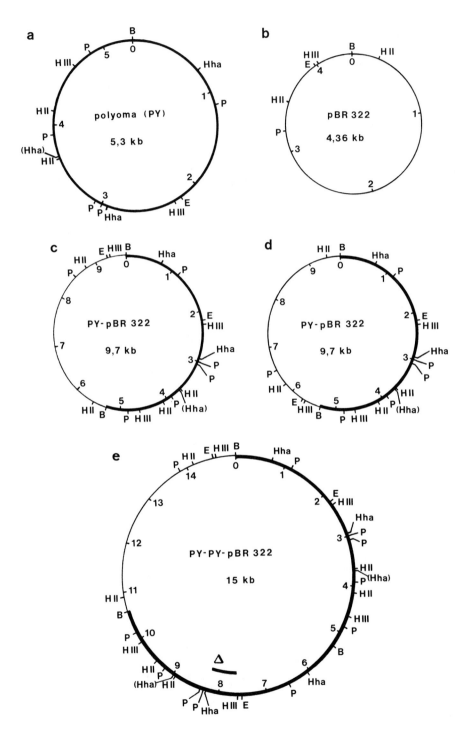

of these hybrid DNAs with BamHI resulted in a
specific infectivity (in terms of unit lengths
of polyoma DNA) similar to that of viral DNA
cleaved with BamHI. Similar results were obtained for 5 out of 6 hybrids that contained
monomers of polyoma virus DNA joined to pBR322
via the EcoRI cleavage site (data not shown).
In no case did the uncleaved DNA show infectivity; 5 of the hybrid DNAs were infectious
after cleavage with EcoRI, which liberated a
unit length linear viral genome.

In contrast to the uncleaved hybrids containing one unit of polyoma DNA, the intact hybrids containing 2 or 1.9 tandem head-to-tail
genomes of polyoma virus DNA were infectious.
The hybrid containing a dimer of polyoma virus
DNA was one tenth as infectious as intact form
I viral DNA (in terms of PFUs per polyoma DNA
unit), whereas the hybrid containing 1.9 viral
genomes was 1/100th as infectious as viral DNA.
The specific infectivity of the BamHI-cleaved
DNA of the two hybrids containing more than one
unit of polyoma virus DNA was similar to that
of BamHI-cleaved viral DNA in terms of PFUs
per unit length polyoma DNA molecules. Polyoma
virus was isolated from a number of cultures
infected with "monomer" and "dimer" hybrid DNA;

Fig. 4. Restriction maps of pBR322 (G. Sutcliffe, personal communication), polyoma DNA (36)
and pBR322 polyoma DNA hybrids (20). (a) Polyoma
DNA. The variant used in this work lacks the HhaI
site (in parentheses) at 26.4 map units from the
EcoRI site; (b) pBR322; (c) "monomeric" polyoma
pBR322 hybrid, joined at the BamHI sites so that
the reading direction of the "early" genes of polyoma and the tetracycline gene of pBR322 coincide
(defined as head-to-tail). (d) As (c), however the
reading directions are opposed (defined as head-to-
head). (e) Dimeric polyoma pBR322 hybrid, joined
at BamHI sites so that all units are head-to-tail,
as defined above. B, BamHI; P, PstI; E, EcoRI; HII,
HindII; HIII, HindIII; Hh, HhaI. $\triangle$, site of deletion in hybrid D_5.

TABLE 3

Infectivity of polyoma hybrids*.

Preparation	Units of Py DNA/hybrid	Specific infectivity of DNA†	
		uncleaved (PFU per molecule $\times 10^7$)	BamHI-cleaved‡ (PFU per complete polyoma molecule $\times 10^7$)
A2	1	0 ($<10^{-4}$)	1.9
D6	1	0 ($<10^{-4}$)	1.6
E8	1	0 ($<10^{-4}$)	1.5
G3	1	0 ($<10^{-4}$)	2.2
D3	2	12	1.4
D5	1.9	1.4	1.3
Py DNA	-	130	4.6

* From Fried et al. (20).

† Appropriate dilutions of unrestricted and restricted DNA, respectively, were assayed on mouse embryo cells for plaque formation using the DEAE method (22). Py, polyoma DNA; PFU, plaque-forming units.

‡ The DNA was cleaved to completion with BamHI endonuclease and the enzyme was inactivated by heating at 60°C for 15 min.

in all cases, the progeny had only the two HhaI
sites characteristic for the mutant DNA used
to produce the recombinant DNAs (data not
shown). To eliminate the possibility that complete polyoma DNA (linear or circular) was present in the hybrid DNA preparations prior to
transfection, 1 μg-samples of the uncleaved
polyoma dimer hybrids were subjected to agarose
gel electrophoresis, the DNA was transferred
to cellulose acetate membranes (23) and poly-
oma-specific DNA visualized by hybridization
with ^{32}P-labeled polyoma virus DNA followed by
autoradiography. No radioactive band was detected in the positions of form I or form II polyoma DNA; 0.1 ng of polyoma DNA could be detected in parallel lanes. This experiment shows
that the infectivity was due to the polyoma
pBR322 hybrid molecules and not to free viral
genomes in the preparations, since a possible
contamination could not exceed about 0.01%.

The fact that cloned polyoma DNA, when excised at the site of joining, has an infectivity very similar to that of form III derived
from natural polyoma DNA is once again testi-
mony to the remarkable fidelity of the cloning
procedure and the stability of eukaryotic sequences replicated in a prokaryotic host. The
finding that hybrid DNA containing only one
unit of polyoma DNA is non-infectious to cells
permissive for polyoma shows that events leading
to the intracellular excision of full-length,
circular or linear polyoma DNA from the hybrid
do not occur or are very rare. In contrast, excision of polyoma DNA from a hybrid containing
a head-to-tail dimer is a relatively efficient
process in mouse fibroblasts, and is most likely
due to legitimate recombination. A hybrid containing a head-to-tail polyoma dimer with an
800 bp deletion in which, due to the location
of the deletion, only 1.3 unit of polyoma DNA
are available for excision of a unit length viral DNA, is 10% as infectious as the dimer-con-
taining plasmid. This suggests that the frequency of excision may be related to the target
size of possible recombinational events. The
finding that hybrid DNA containing oligomers

of polyoma DNA is infectious is in agreement
with the results obtained by Jänisch and Levine
(24), and Israel et al. (25) who found that linear and circular oligomeric SV40 DNA is infectious to permissive cells and gives rise to SV40
particles containing monomeric DNA.

It is clear that risk-testing experiments
designed to answer the question whether recombinant DNA can be transferred from within E.coli
to an animal cell or to an animal are only meaningful if the recombinant DNA is potentially
infectious. The demonstration that hybrid DNA
containing head-to-tail polyoma dimer is highly
infectious to mouse fibroblasts sets the stage
for the next step in the risk-assessment program, namely the administration of bacteria containing such hybrid plasmids to cultured mouse
cells and mice and scoring for polyoma infection.

III. Transformation of thymidine kinase negative
mouse cells with hybrid DNA containing the thymidine kinase gene of herpes simplex type I
(26).

Herpes simplex type I DNA was (a) cleaved
with BamHI and joined to BamHI-cleaved pBR322,
or (b) cleaved with KpnI, elongated with (dA)-
residues and joined to EcoRI cleaved, dT-elon-
gated pBR322. The DNA was cloned in E.coli X1776
and ampicillin-resistant colonies were screened
by the Grunstein-Hogness assay (7) using as
probe the purified 4800 bp KpnI fragment known
to contain the thymidine kinase (TK) gene (N.W.
and B.C., unpublished results). Several hybrid
plasmids were purified,and characterized in regard to their hybridization and restriction
properties. Fig. 5 shows the map of Z-pBR322/TK·
HSV-M4 (TK-M4 for short), a hybrid containing
the TK-specific KpnI fragment.

It has been shown earlier that herpes simplex DNA as well as certain restriction fragments derived therefrom are able to transform
TK$^-$ cells in TK$^+$ (27-29). The capacity of TK-
specific hybrid plasmids to transform TK$^-$ cells
was tested by treating the cells with BamHI-
or SalI-cleaved hybrid plasmids by the calcium
phosphate method (30) and determining the number

of colonies appearing after incubation in HAT
medium (31) for 3 weeks. As shown in Table 4,
all hybrids tested transformed TK⁻ L cells efficiently; for example, Z-pBR322/HSV·TK-P1 gave
rise to 40 HAT-resistant colonies per 0.25 µg
plasmid DNA per 8×10^5 cells. This may be compared with the values obtained with authentic

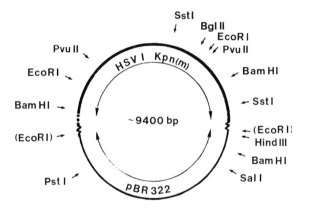

Fig. 5.   Restriction map of a hybrid of pBR322
and the thymidine kinase gene-containing KpnI (m)
fragment of herpes simplex virus type I DNA. Plasmid Z-pBR322/HSV·TK-M4 consists of the KpnI (m)
fragment of HSVI DNA joined to EcoRI-cleaved pBR322
by the AT-tail method. The site of linkage is indicated by squiggles; the dashed line indicates an
area where the precise length is not known. The map
of the pBR322 moiety is from Sutcliffe (personal
communication); the sites of the KpnI fragment were
measured relative to the BglII site. Only the 2
PvuII sites closest to the BglII site are indicated. The EcoRI sites in parentheses are obliterated.

TABLE 4

Transformation of TK⁻ L cells by HSV thymidine kinase gene-containing hybrid plasmids.

TK hybrid plasmid	μg DNA/dish	Colonies/number dishes*		Transformants per μg hybrid DNA per dish†
		Expt. 1(a)	Expt. 2(b)	
P1	0.02	6/1		
	0.07	5/1		
	0.25	35/1	81/2	160
	0.30			
M2	0.02	2/1		
	0.07	2/1		
	0.25		83/2	95
	0.30	4/1		
M4	0.02	2/1		
	0.07	10/1		
	0.25		78/2	122
	0.30	17/1		
Mock trans-fected		0/2		
Untreated		0/1	0/6	–

herpes simplex DNA (1 per µg) (27) or with the
purified BamHI fragment (2500 per µg) (30). No
colonies were recovered after mock transformation.

IV. Introduction into mouse L cells of a cloned chromosomal rabbit β globin gene linked to a thymidine kinase gene-containing hybrid (32).

The hybrid plasmid Z-pCRI/RchrβG-1 consists
of a DNA fragment containing a chromosomal rabbit
β globin gene linked to pCRI (Fig. 6a, ref. 33);
the nucleotide sequence of the globin gene and
the surrounding regions has recently been completed (Fig. 6b, A. van Ooyen, unpublished results). The globin gene-containing plasmid and
a thymidine kinase gene-containing hybrid (TK-
M4, TK-M2 or TK-P1) were digested with SalI
(which cleaves both hybrids once in the vector
moiety) and joined by DNA ligase at high DNA
concentration. As determined by agarose gel electrophoresis, all of the monomeric HSV plasmid
DNA disappeared and a broad band of high molecular weight DNA was found (data not shown). The
concatenated DNA was used to transform TK$^-$ mouse
L cells by the calcium phosphate method. After
19 days in HAT medium 1 - 18 colonies per µg total DNA per 8×10^5 cells had appeared. 21 colonies were isolated, each was expanded to about
6×10^7 cells by multiple passaging in HAT medium, and DNA and RNA were prepared from each
sample. DNA was cleaved by PstI and EcoRI, run
on agarose gels, transferred to cellulose nitrate
membranes (23) and hybridized to the ^{32}P-labeled,
recloned β globin-specific segment of PβG (4),
which represents the almost complete cDNA copy
of rabbit β globin mRNA (38). As shown in Fig. 7,

Legend to Table 4.

[*]8×10^5 TK$^-$ L cells mouse/dish, cultured in HAT
medium were transformed as described (30).

[(a)]Plasmids digested with BamHI.

[(b)]Plasmids digested with SalI.

[‡]Average of all experiments.

several DNA preparations gave 2 bands with a mobility corresponding to about 900 and 1200 bp
respectively, the values expected from the map
shown in Fig. 6. The relative intensities of
these bands are similar to those found with the
β globin chromosomal DNA plasmid digested in the
same fashion and run in parallel. Altogether, 19
of 21 preparations from TK$^+$ transformants showed
the two bands, at varying degrees of intensity,
while DNA from cells transformed with the TK
plasmid alone did not.

In vivo transcription of the extraneous rabbit globin gene was monitored by a modification
of the Berk-Sharp procedure (34) developed by

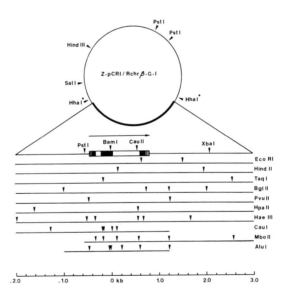

Fig. 6. a) Restriction map of hybrid of pCRI
and a chromosomal β globin gene-containing fragment
of rabbit DNA. Plasmid Z-pCRI/RchrβG-1 (33) consists
of a KpnI fragment of rabbit DNA linked to EcoRI-
cleaved pCRI by the AT-tail method. Within the pCRI
moiety, only the HhaI sites closest to the insert
are indicated. Black bars, exons; white bars, introns; hatched bars, leader and trailer of mRNA.
b) Nucleotide sequence of the β globin gene-containing region (A. van Ooyen, unpublished results).

Nucleotide sequence of a chromosomal rabbit DNA segment containing a β globin gene.

```
TAGCAATTAGTACTGCTGGTATGGGTCTGGGAGATACATAGAAGGAAGGCTGAGTCTGTCAGACTCCTAAGCCATTGCCATAACTGCCAA
  -220       -210       -200       -190       -180       -170       -160       -150       -140

                                              PstI                    BspI
GGACAGGGGTGCTGTCATCACCCAGACCTCACCCTGCAGAGCCACACCCTGGTGTTGGCCAATCTACACACGGGGTAGGGATTACATAGT
  -130       -120       -110       -100        -90        -80        -70        -60        -50

                                PvuII       Cap
TCAGGACTTGGGCATAAAAGGCAGAGCAGGGCAGCTGCTGCTTACACTTGCTTTTGACACAACTGTGTTTACTTGCAATCCCCCAAAACA
  -40        -30        -20        -10         0         10         20         30         40

                                                                     MboII              BspI
GACAGAATGGTGCATCTGTCCAGTGAGGAGAAGTCTGCGGTCACTGCCCTGTGGGGCAAGGTGAATGTGGAAGAAGTTGGTGGTGAGGCC
                     Met Val His Leu Ser Ser Glu Glu Lys Ser Ala Val Thr Ala Leu Trp Gly Lys Val Asn Val Glu Glu Val Gly Glu Ala
  50         60         70         80         90         100        110        120        130

CTGGGCAGGTTGGTATCCTTTTTACAGCACAACTTAATGAGACAGATAGAAACTGGTCTTGTAGAAACAGAGTAGTCGCCTGCTTTTCTG
Leu Gly Arg
              140        150        160        170        180        190        200        210        220

                                                                               TaqI
                                                                       MboII  ┌──┐
CCAGGTGCTGACTTCTCTCCCCTGGGCTGTTTTCATTTTCTCAGGCTGCTGGTTGTCTACCCATGGACCCAGAGGTTCTTCGAGTCCTTT
                                                    Leu Leu Val Val Tyr Pro Trp Thr Gln Arg Phe Phe Glu Ser Phe
  230        240        250        260        270        280        290        300        310

GGGGACCTGTCCTCTGCAAATGCTGTTATGAACAATCCTAAGGTGAAGGCTCATGGCAAGAAGGTGCTGGCTGCCTTCAGTGAGGGTCTG
Gly Asp Leu Ser Ser Ala Asn Ala Val Met Asn Asn Pro Lys Val Lys Ala His Gly Lys Lys Val Leu Ala Ala Phe Ser Glu Gly Leu
  320        330        340        350        360        370        380        390        400

                              AluI                        AluI        BamHI
AGTCACCTGGACAACCTCAAAGGCACCTTTGCTAAGCTGAGTGAACTGCACTGTGACAAGCTGCACGTGGATCCTGAGAACTTCAGGGTG
Ser His Leu Asp Asn Leu Lys Gly Thr Phe Ala Lys Leu Ser Glu Leu His Cys Asp Lys Leu His Val Asp Pro Glu Asn Phe Arg
  410        420        430        440        450        460        470        480        490

AGTTTGGGGACCCTTGATTGTTCTTTCTTTTTCGCTATTGTAAAATTCATGTTATATGGAGGGGGCAAAGTTTTCAGGGTGTTGTTTAGA
  500        510        520        530        540        550        560        570        580

  MboII
ATGGGAAGATGTCCCTTGTATCACCATGGACCCTCATGATAATTTTGTTTCTTTCACTTTCTACTCTGTTGACAACCATTGTCTCCTCTT
  590        600        610        620        630        640        650        660        670

                                  AluI
ATTTTCTTTTCATTTTCTGTAACTTTTTCGTTAAACTTTAGCTTGCATTTGTAACGAATTTTTAAATTCACTTTTGTTTATTTGTCAGAT
  680        690        700        710        720        730        740        750        760

TGTAAGTACTTTCTCTAATCACTTTTTTTTTCAAGGCAATCAGGGTATATTATATTGTACTTCAGCACAGTTTTAGAGAACAATTGTTATA
  770        780        790        800        810        820        830        840        850

ATTAAATGATAAGGTAGAATATTTCTGCATATAAATTCTGGCTGGCGTGGAAATATTCTTATTGGTAGAAACAACTACATCCTGGTCATC
  860        870        880        890        900        910        920        930        940

                                                                               HpaII  BspI
ATCCTGCCTTTCTCTTTATGGTTACAATGATATACACTGTTTGAGATGAGGATAAAATACTCTGAGTCCAAACCGGGCCCCTCTGCTAAC
  950        960        970        980        990        1000       1010       1020       1030

         MboII       AluI                                                      EcoRI
CATGTTCATGCCTTCTTCTTTTTCCTACAGCTCCTGGGCAACGTGCTGGTTATTGTGCTGTCTCATCATTTTGGCAAAGAATTCACTCCT
                          Leu Leu Gly Asn Val Leu Val Ile Val Leu Ser His His Phe Gly Lys Glu Phe Thr Pro
  1040       1050       1060       1070       1080       1090       1100       1110       1120

                                BspI                              BglII
CAGGTGCAGGCTGCCTATCAGAAGGTGGTGGCTGGTGTGGCCAATGCCCTGGCTCACAAATACCACTGAGATCTTTTTCCCTCTGCCAAA
Gln Val Gln Ala Ala Tyr Gln Lys Val Val Ala Gly Val Ala Asn Ala Leu Ala His Lys Tyr His
  1130       1140       1150       1160       1170       1180       1190       1200       1210

                                                                     pA
AATTATGGGGACATCATGAAGCCCCTTGAGCATCTGACTTCTGGCTAATAAAGGAAATTTATTTTCATTGCAATAGTGTGTTGGAATTTT
  1220       1230       1240       1250       1260       1270       1280       1290       1300

TTGTGTCTCTCACTCGGAAGGACATATGGGAGGGCAAATCATTTAAAACATCAGAATGAGTATTTGGTTTAGAGTTTGGCAACATATGCC
  1310       1320       1330       1340       1350       1360       1370       1380       1390
```

Fig. 6b

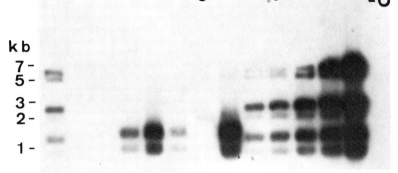

Fig. 7.   Detection of rabbit β globin sequences
in EcoRI-PstI-digested DNA of transformed mouse
L cell lines.
DNA was digested with EcoRI and PstI, denatured,
electrophoresed, transferred to cellulose nitrate
membranes, and hybridized as described (37). The
^{32}P-labeled, nick-translated probe was the rabbit
β globin-specific cDNA moiety of PβG which had been
excised by the S_1-formamide procedure, joined to
HindIII linkers and recloned in the HindIII site
of pBR322 (38). (I) DNA from cell lines transformed
with the rabbit β globin DNA-containing hybrid
Z-pCRI/RchrβG-1 linked to a thymidine kinase (TK)
gene-containing hybrid (TK-P1 or TK-M2). Lanes d-f:
cell lines P1/R-3 to P1/R-5, lane h: cell line M2/-
R-1. (II) DNA from cell lines transformed only with
a SalI-cleaved TK gene-containing plasmid. Lanes
b and c: cell lines P1/S-1 and P1/S-2, lane g: cell
line M2/S-2. The marker is a mixture of PβG digested
with (i) EcoRI, (ii) BglI plus AvaI, and (iii)
Z-pCRI/RchrβG-1 digested with PstI plus EcoRI.
Lanes a and i-m contain about 0.018, 0.018, 0.036,
0.072, 0.144 and 0.36 fmol of (i), (ii) and (iii).
The chain length scale was determined from a denatured EcoRI digest of unlabeled Ø29 DNA corun
in the marker lane, and from marker (iii) above.

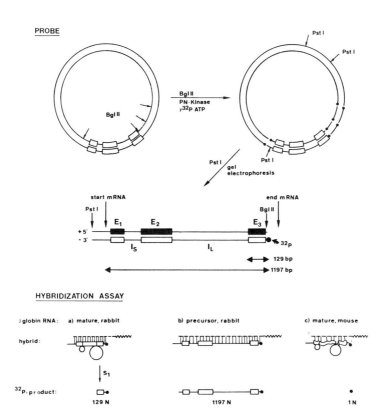

Fig. 8. Method for distinguishing mature and precursor rabbit β globin mRNA in the presence of excess mouse β globin mRNA.
The procedure is a modification of the method of Berk and Sharp (34). The probe is a DNA restriction fragment ^{32}P-labeled at one 5' terminus (rather than being uniformly labeled by nick translation). The denatured DNA is hybridized to the RNA sample in 80% formamide, digested with S_1 nuclease, denatured and analyzed by polyacrylamide gel electrophoresis in 7 M urea. Only if the 5' terminus of the probe was hybridized to RNA is a labeled DNA fragment recovered. The Pst-BglII (1293) fragment derived from the rabbit chromosomal β globin plasmid (cf. Fig. 6), labeled at the BglII site, is diagnostic for plus strand β globin sequences and allows the distinction between rabbit precursor RNA, rabbit mature RNA and mouse RNA. E, exon; I_S and I_L , small and large intron, respectively.

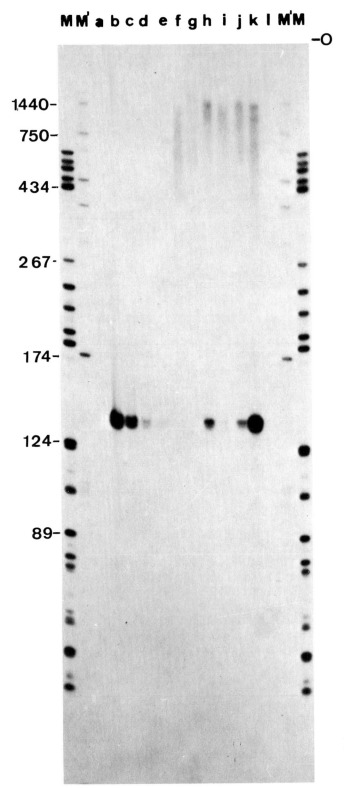

Fig. 9

Weaver et al. (35), the principle of which is outlined in Fig. 8. The hybrid plasmid PβG contains a single BglII site immediately following the 3' end of the β globin coding sequence. The hybrid was cleaved with BglII and labeled to a high specific activity with ^{32}P at the 5' termini. After digestion with PstI, the PstI-BglII (1293) fragment, labeled at the BglII site, was isolated. The labeled DNA was hybridized to cell RNA in 80% formamide, treated with S_1 nuclease, denatured, and analyzed by polyacrylamide gel electrophoresis in 7 M urea. Only if the 5' terminus of the DNA is protected by complementary RNA is a labeled fragment recovered. Since the probe is labeled on the minus strand, only plus strand RNA is detected. Mature rabbit β globin mRNA protects a fragment of 129 nucleotides length (from the BglII site to the 3' end of the large intron), whereas the precursor is expected to protect a 1197-nucleotide fragment (cf.

Fig. 9.   Detection of rabbit β globin RNA in transformed mouse L cell lines.
The assay was carried out as outlined in Fig. 8, using as probe the 1293 bp PstI-BglII fragment of Z-pCRI/RchrβG-1 labeled with ^{32}P at the BglII end (cf. Fig. 8). The denatured samples were run on a 5% acrylamide gel in 7 M urea. The samples were: (I) total RNA from cell lines transformed with the rabbit β globin DNA-containing hybrid Z-pCRI/RchrβG-1 linked to a thymidine kinase (TK) gene-con-containing hybrid (TK-P1 or TK-M2). Lanes h to j: cell lines P1/R-3 to P1/R-5, lane k: cell line M2/R-1. (II) Total RNA from cell lines transformed only with the SalI-cleaved TK plasmid P1. Lanes f and g: cell lines P1/S-1 and P1/S-2. Lane a: no RNA, lanes b to e: 3, 1, 0.3 and 0.1 ng rabbit globin RNA, respectively, lane 1: 50 ng mouse globin RNA. M and M': size markers.

Flavell, this volume). It is possible to discriminate between mouse and rabbit globin mRNA because the rabbit mRNA has a BglII recognition
sequence while mouse RNA does not and therefore
does not protect the labeled 5' terminus of the
probe (35). Fig. 9 shows that as little as 0.1
ng rabbit β globin mRNA (lane e) gave the expected 129 nucleotide band, while 50 ng of mouse
globin mRNA (lane l) gave no such band. The
transformed cell lines P1/R-3, P1/R-4, P1/R-5
and M2/R-1 (lanes h, i, j and k) yielded as major
product a fragment of about 150 nucleotides. Some
weak bands between 750 and 1400 nucleotides in
lanes h, j and k may be due to unprocessed precursor molecules. Most of the rabbit β globin--
specific RNA was of the mature type. Mouse cells
transformed with only a TK plasmid (lines P1/S-1
and P1/S-2, lanes f and g) gave no detectable
bands. We estimate that the cell line with the
most abundant rabbit globin transcripts (M2/R-1)
contained about 400 RNA molecules per cell.

    We have yet to characterize the state and
copy number of the rabbit β globin DNA in the
transformed cells, to determine the size of the
primary and processed transcripts, the origin
of transcription, and to ascertain whether rabbit globin is being produced. However, it is
already clear that a number of problems will
be susceptible to analysis by reversed genetics
(39) with the system as it is available now.

ACKNOWLEDGMENTS

    We thank Dr. R.J.C. Harris for making available
to us the C_3 and C_4 containment facilities of the
MRE in Porton, and the staff of that institution,
particularly Dr. P. Greenaway, for the hospitality
and generous help extended to C.W., W.B., N.W. and
B.C. The containment used for the cloning of viral
DNA and propagation of the hybrids was that recommended by the GMAG (Great Britain). The travel involved in this work (cf. Fig. 10) was supported in
part by short-term EMBO grants. Financial support
was provided by the Schweizerische Nationalfonds
and the Kanton of Zürich.

Fig. 10

REFERENCES

(1) T. Taniguchi, M. Palmieri and C. Weissmann, Nature 274 (1978) 223-228.

(2) C. Weissmann, M.A. Billeter, H.M. Goodman, J. Hindley and H. Weber, Annu. Rev. Bioch. 42 (1973) 303-328.

(3) C. Weissmann, FEBS Lett. 40 (1974) S10-S18.

(4) T. Maniatis, K.G. Sim, A. Efstratiadis and F.C. Kafatos, Cell 8 (1976) 163-182.

(5) R.A. Flavell, D.L. Sabo, E.F. Bandle and C. Weissmann, J. Mol. Biol. 89 (1974) 255-272.

(6) C. Gilvarg, F.J. Bollum and C. Weissmann, Proc. Natl. Acad. Sci. USA 72 (1975) 428-432.

(7) M. Grunstein and D.S. Hogness, Proc. Natl. Acad. Sci. USA 72 (1975) 3961-3965.

(8) T.J. Lerner and N.D. Zinder, Virology 79 (1977) 236-238.

(9) H. Hoffmann-Berling and R. Mazé, Virology 22, (1964) 305-313.

(10) E. Domingo, R.A. Flavell and C. Weissmann, Gene 1 (1976) 3-25.

(11) A. Kiss, B. Sain, E. Csordas-Toth and P. Vene-tianer, Gene 1 (1977) 323-329.

(12) E. Domingo, D. Sabo, T. Taniguchi and C. Weissmann, Cell 13 (1978) 735-744.

(13) E. Batschelet, E. Domingo and C. Weissmann, Gene 1 (1976) 27-32.

(14) M.A. Sirover and L.A. Loeb, J. Biol. Chem. 252 (1977) 3605-3610.

(15) C.I. Davern, Aust. J. Biol. Sci. 17 (1964) 719-725.

(16) N. Tsuchida, M. Nonoyama and Y. Ikeda, J. Mol. Biol. 20 (1966) 575-583.

(17) V. Zgaga, Nature 267 (1977) 860-862.

(18) R.H. Doi and S. Spiegelman, Science 138 (1962) 1270-1272.

(19) C. Gilvarg, H. Jockusch and C. Weissmann, Biochim. Biophys. Acta 454 (1976) 587-591.

(20) M. Fried, K. Murray, B. Klein, P. Greenaway, J. Tooze, W. Boll and C. Weissmann, submitted for publication.

(21) L.K. Miller, B.E., Cooke and M. Fried. Proc. Natl. Acad. Sci. USA 73 (1976) 3073-3077.

(22) M. Fried, J. Virol. 13 (1974) 939-946.

(23) E.M. Southern, J. Mol. Biol. 98 (1975) 503-517.

(24) R. Jaenisch and A. Levine, Virology 44 (1971) 480-493.

(25) M.A. Israel, J.C. Byrne and M.A. Martin, Virology 87 (1978) 239-246.

(26) W. Boll, N. Mantei, N. Wilkie, B. Clements, P. Greenaway and C. Weissmann, Experientia, in press.

(27) S. Bacchette and F.L. Graham, Proc. Natl. Acad. Sci. USA 74 (1977) 1590-1594.

(28) A. Pellicer, M. Wigler, R. Axel and S. Silverstein, Cell 14 (1978) 133-142.

(29) N.J. Maitland and J.K. McDougall, Cell 11 (1977) 233-241.

(30) M. Wigler, A. Pellicer, S. Silverstein and R. Axel, Cell 14 (1978) 725-731.

(31) J. Littlefield, Proc. Natl. Acad. Sci. USA 50 (1963) 568-573.

(32) N. Mantei, W. Boll and C. Weissmann, Experientia (1979) in press.

(33) J. van den Berg, A. van Ooyen, N. Mantei, A. Schamböck, G. Grosveld, R.A. Flavell and C. Weissmann, Nature 276 (1978) 37-44.

(34) A.J. Berk and P.A. Sharp, Cell 12 (1977) 721-732.

(35) R. Weaver, W. Boll and C. Weissmann, Experientia (1979) in press.

(36) M. Fried and B.E. Griffin, In "Advances in Cancer Research", vol. 24 (1977), Academic Press, pp. 63-113.

(37) A.J. Jeffreys and R.A. Flavell, Cell 12 (1977) 1097-1108.

(38) F. Meyer, H. Heijneker, H. Weber and C. Weissmann, Experientia (1979) in press.

(39) C. Weissmann, TIBS 3 (1978) N109-N111.

DISCUSSION

W. MARZLUFF: Do you think the rabbit genes are under control in the mouse cells?

C. WEISSMANN: That is a very interesting question, but I cannot answer it now. These cells have probably received multiple doses of the rabbit gene, and it may be just transcribed by brute force. We were a little bit surprised that we did find transcription, actually. In fact one of the cell lines has as many as about 200 copies of mouse globin messenger per cell. That is really quite a lot, not just sort of a symbolic amount, but I cannot answer your question.

P. REDDY: Did you formally exclude the possibility that the RNA band which you see on gels is an incomplete transcript of rabbit precursor mRNA rather than a processed transcript. Did you find any unprocessed precursor mRNA in these cells?

C. WEISSMANN: No. All I can say is we could not detect any precursor at the level of sensitivity we are working at. In erythroleukemic mouse cells the ratio of precursor to messenger RNA is about 1:1000. We certainly would not have been able to pick up anything at that low level, so I think the question is still open. One thing which is clear is that although this is a rabbit precursor, it is processed perfectly well in the mouse system. I think that much can be said. I would also add that the introns are very different in mouse and rabbit.

J. LEBOWITZ: I am just curious, since you had the other
probe for the negative strand, whether there is any
expression of the opposite strand. I ask this question
because in SV40 there is symmetric synthesis and degradation
and I have not seen an answer for whether this occurs in
globin or ovalbumin.

C. WEISSMANN: We have not done the experiments yet. I cannot
answer the question. It is something which we want to do.

P. CHAMBON: I wanted to ask you whether you had any evidence
that the thymidine kinase gene is still linked to the globin
gene or vice versa; whether the globin gene is still linked
to the thymidine kinase in the genome? Secondly, whether you
have evidence that they are in fact integrated into the
genome? And lastly, whether the globin gene was inserted in
pBR322; was it inserted by the A-T tail technique or not?

C. WEISSMANN: We do not know whether the two are linked within the cell. We also do not know whether they are integrated
or not. One thing I can answer is that the globin gene was
inserted by the A-T tail method, inserted into the pCRl
plasmid.

P. CHAMBON: The reason I asked the last question is that it
could introduce some difficulty with respect to recognizing
promotor sequence because you have an A-T rich region there.

C. WEISSMANN: Yes, and of course we have absolutely no idea
whether the transcription we are observing has anything to do
with the natural promotor. Incidently, this will be quite a
difficult problem to resolve.

B. O'MALLEY: Actually my question goes a little further than
that of Chambon. Must you actually link covalently your Hb
DNA to the viral DNA in order to effect a transfer. This may
sound initially like a strange question, but the TK transfer
is not a dose dependent phenomenon and there is some reason
to believe that in animal cells that there may be a phenomenon of competence for accepting DNA. Thus, within a
population of cells, certain cells will take up the DNA and
other cells will not take up the DNA no matter what level to
which you push the concentration. For this reason you may
simply monitor transfer by mixing the 2 DNAs using TK as a
marker for the cells that will pick up any DNA molecule.

C. WEISSMANN: I understand that Axelrod has done experiments of this kind. The reason that we prefer to do it this way anyway is because we want to maintain the globin gene, and we have some hopes that if they were linked, physically linked, then selection by HAT medium would also tend to select for cells which are retaining the globin gene. But I understand from actual experiments that in fact what you can do is this: (I think the experiment was done with ϕX174.) You just mix thymidine kinase DNA with ϕX174 DNA, transfect cells and the ones which are transformed will have a relatively high proportion of ϕX174 DNA as well. This is from hearsay.

F. CRICK: Following what Dr. Weissman said in his introduction about how successful these new techniques have been I would like to remark how extremely competent, as well as interesting, this work is and to complement the speakers on the very high professional standard with which they have presented their results. Thank you very much.

THE ARRANGEMENT, REARRANGEMENT AND ORIGIN OF IMMUNOGLOBULIN GENES

Philip Leder, Jonathan G. Seidman, Edward Max,
Yutaka Nishioka, Aya Leder, Barbara Norman and Marion Nau

Laboratory of Molecular Genetics
National Institute of Child Health and Human Development
National Institutes of Health
Bethesda, Maryland

INTRODUCTION

As a number of the talks preceding and following ours concern the organization of globin genes, I shall refer you to two papers (Konkel et al., 1978; Leder et al., 1978) that summarize our most recent globin gene findings and pass on to the consideration of another interesting, but rather unique genetic system that we have been studying for some years: the immune system.

The immune system requires sufficient genetic information to encode approximately a million different antibody molecules, each exhibiting remarkable specificity with respect to the antigens that it binds. These antibody molecules, comprised of pairs of heavy and light chains, display an interesting structural feature wherein their 100 or so N-terminal amino acids differ from antibody chain to antibody chain, while their C-terminal portions remain identical in primary amino acid sequence.

The problems that this family of proteins presents are as follows: How is this unusual half constant-half variable structure encoded in the genome? How does the diversity displayed by the variable region arise--either during evolution or somatic mutation or both--and what mechanisms might serve to assure that only one of a given set of variable region genes is expressed?

A simple stringent germline model was proposed to answer these questions by envisioning a thousand contiguous variable-constant regions arrayed along a chromosome. This model, however, was ruled out by hybridization kinetic analyses that clearly indicated that the constant region was a relatively unique sequence in the mammalian genome (see refs. in Seidman et al., 1978b). More likely was the model proposed by Dryer and Bennett (1965) that held that there was but one copy of the constant region gene that somehow was joined to a variable region gene (or their products were joined) during somatic development. This model was supported by hybridization studies referred to above and by recent in situ hybridization studies (Hozumi and Tonegawa, 1976; Rabbits and Forster, 1978; Seidman et al., 1978a) as well as by direct cloning experiments involving the lambda and kappa light chain genes of the mouse (Tonegawa et al., 1977; Seidman et al., 1978a).

In what follows we would like to advance four arguments concerning the questions raised by the immune system. First, we would like to suggest that immunoglobulin kappa variable region genes constitute a multiple gene family that has arisen through evolution and now consists of many small subfamilies of variable region genes that are closely related both in their structural and flanking sequences. Second, we would like to suggest that the kappa light chain is encoded in at least four discrete coding segments, two of which join by a DNA recombination event occurring at the chromosomal level, the remainder joining by elimination of two intervening sequences from a large mRNA precursor. Third, we would like to suggest that there is an array of several small genetic segments encoding 10 or so amino acids any of which can join variable and constant region genes: the J segments. These have two critical faces, one for DNA and one for RNA recombination or splicing. Fourth, and finally, we would like to suggest a simple recombination model that follows from the distinctive structure of the immunoglobulin variable region genes that may be responsible for generating their diversity both in sequence and numbers during evolutionary development.

These points are supported by evidence obtained by cloning relevant gene segments and by analyses of their structures using the electron microscope and by direct sequence determination. The recombination model we refer to is supported by molecular genetic analysis of various inbred strains of the laboratory mouse and their comparison to an interesting assortment of Asian mice species.

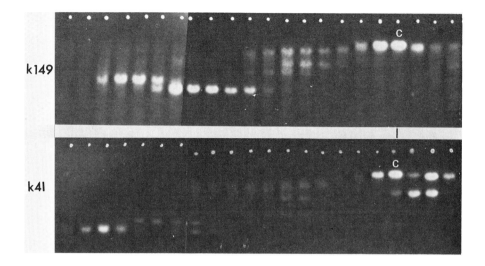

FIGURE 1. Two-dimensional analysis of EcoRl restriction
fragments of MOPC-149 DNA hybridized to cloned MOPC-149
(K149) and MOPC-41 (K41) variable and constant region cDNA
sequences. The two-dimensional separation and blot were
carried out as described (Seidman et al., 1978a). The probes
were cloned MOPC-149 (K149) and MOPC-41 (K41) variable plus
constant region cDNA sequences. Each band represents a DNA
fragment carrying either K149 or K41 variable region sequences. The C-region encoding fragment is indicated. The
figure is reproduced from Seidman et al. (1978a).

Kappa Variable Regions Constitute a Multigene Family

The best evidence for the multiplicity of kappa variable
region genes comes from in situ hybridization and cloning
studies that used hybridization probes directed against the
mRNAs corresponding to two different myeloma-produced light
chains, MOPC-149 and MOPC-41 (Seidman et al., 1978a; Seidman
et al., 1978b). A two-dimensional, fingerprint analysis of
the EcoRl-DNA fragments corresponding to these genes revealed
8-10 fragments corresponding to each probe (Figure 1). Each
set consisted of a different (i.e. non-overlapping) set of
fragments. Three of the variable region genes detected by
the MOPC-149 probe have been cloned and their sequence determined (Seidman et al., 1978a; Y.N., unpublished). Several
interesting points were made from this sequence comparison.

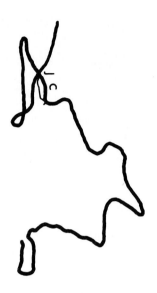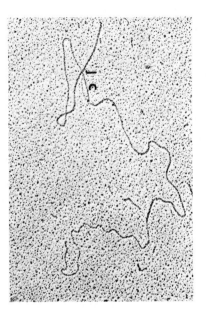

FIGURE 2. Visualization of a κ light chain constant and J region sequence encoded by a cloned EcoRl fragment derived from embryonic DNA. The 16.0-kilobase EcoRl fragment isolated from the phage λCH4A-EC-5 was annealed to MOPC-41 mRNA and spread for electron microscopic analysis. The constant region R loop (C) is 475 + 78 base pairs long and the variable region R loop (V) is 299 + 68 base pairs long. The analysis of 35 molecules suggested that the constant region R-loop is 10.4 + 0.24 kilobases, from one end and 5.1 + 0.20 kilobases from the other end of the fragment. The location of the J region gene was determined by the distance, 1.3 + 0.19 kilobases, from the end of the fragment to the point where the DNA falls across the end of the R loop. The figure is reproduced from Seidman and Leder (1978).

First, the sequences <u>very closely resembled but were not identical to one another</u> (314 of 336 bases were identical). The differences were largely, but not exclusively, clustered in the so-called hypervariable or complementarity determining regions. Thus, the MOPC-149 probe was detecting a closely related family of variable region genes that existed in both indifferent and committed (myeloma) cells. Small sequence differences were, therefore, encoded in multiple germline genes. Furthermore, the variable region coding sequence

stopped at amino acid codon 95-97 indicating that the remaining ten or so amino acids generally associated with the variable region were encoded elsewhere. We shall see below that
these important segments are encoded close to the constant
region gene.

Since each probe detected a distinct subgroup of 10 or
so light chain variable region genes, it is likely that probes
corresponding to the 30 kappa light chain families already
discovered will correspond to similar sets. Thus the variable
region gene repertoire is likely to consist of at least 300
germline variable region genes. More recent saturation hybridization experiments have led to similar conclusions
(Valbuena et al., 1978).

The Tetrasegmental Structure of Kappa Light Chain Genes

Our sequence studies in the kappa gene (Seidman et al.,
1978a) and those of Tonegawa et al. (1978) on the lambda gene
demonstrated a variable region sequence ending approximately
at amino acid positions 95-98. Subsequent cloning experiments
involving cloned λ (Brack et al., 1978) and kappa genes
(Seidman and Leder, 1978) located a short segment of DNA
several thousand bases from the constant region gene that
encoded the adjacent joining nucleotide sequences (the J
region). A DNA fragment encoding the kappa constant region
and the MOPC-41 J segment was cloned from embryonic DNA
(Figure 2). The J segment was separated by 3700 bases from
the constant region gene.

The fragment cloned from embryonic DNA represents the
gene segment prior to the recombination event that converts
it into a complete (and presumably active) gene. The fragment containing the active gene was subsequently cloned from
a MOPC-41 myeloma cell and yielded a fragment in which a recombination event had occurred linking the MOPC-41 variable
region sequence to the J segment 3.7 Kb from the constant
region sequence (Figure 3). Subsequent sequence studies showing further detail (J.S., unpublished results) indicate that
the variable region is further divided into a hydrophobic
leader segment separated by an approximately 100 bp intervening sequence from the body of the variable region gene. Thus
the entire gene consists of four segments, two of which are
joined at the DNA level (Figure 4).

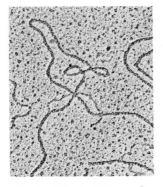

FIGURE 3. The R loop formed between a fragment cloned from
MOPC-41 DNA and MOPC-41 mRNA. The variable and constant
region R loops formed are separated by 3.7 ± 0.187 kilobases
of DNA. The constant region R-loop can be identified to the
reader's right by the single stranded tail of polyA hanging
from the end of the R-loop. The figure is reproduced from
Seidman and Leder (1978).

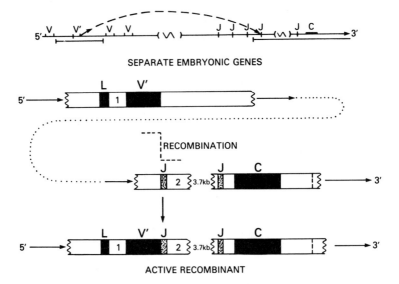

FIGURE 4. The arrangement of kappa light chain genes in
indifferent (embryonic) and light chain producing (myeloma
MOPC-41) cells. The map is derived from electron microscopic
and sequence studies of cloned fragments of mouse DNA
(Seidman and Leder, 1978). The fragments are oriented from
an EcoRl site at their 5' ends and the letters V, J and C
refer to the variable, J and constant regions, respectively.
The hatched area represents the fragment of DNA carrying the
variable region gene that has joined to the J segment. The
second (5' to 3') J segment was discovered by sequence
analysis (E.M., unpublished).

Obviously the J segment is of particular interest in
that it can undergo two very different types of splicing at
each of its two faces--one involving DNA, the other, RNA
ligation. Further evidence (E.M., unpublished) indicates that
other J-like sequences are encoded on the 5' side of the C
region gene. Amino acid sequence comparisons are also consistent with several distinct J region coding segments
(Kabah et al., 1978; Weigert et al., 1978). Thus, a picture
emerges in which there is an array of J segments to which
can be joined an even larger array of variable region genes
(Fig. 4). This recombination event appears to be required to
activate an immunoglobulin gene and suggests an attractive
mechanism to explain lymphocytic allelic exclusion, the fact
that only a single light chain gene is expressed in each
lymphocyte. Apparently, only one member of an allelic pair
may undergo accurate recombination (Seidman and Leder, 1978).

The Possible Role of Unequal-Crossing in Generating
Immunoglobulin Gene Diversity: The Model and
an Evolutionary Test

The model illustrated in Figure 4, though still not
proven in every detail, is likely to be a fairly accurate
picture of the kappa light chain genes before and after
somatic recombination, a step apparently required to activate gene expression. Still remaining, however, is the question of how this diverse family of variable region genes
arose and how it is maintained in the genome, and further,
whether it is adequate to provide all the necessary diversity
required of the immune system. That somatic mutation, i.e.
the alteration of a few bases in a germline gene to produce
a slightly different sequence in a myeloma tumor can and does
occur seems entirely reasonable and is supported by direct
gene sequence data (S. Tonegawa, personal communication).
But even where evidence is best for somatic mutation, in the
mouse lambda myeloma light chains (Cohn et al., 1974), by far
the most frequently expressed sequence is the germline sequence. Thus, the unaltered germline gene must be used, and
used most often. We have already shown that the mouse germline encodes many kappa variable region genes, certainly
several hundred and possibly close to a thousand (Seidman et
al., 1978b). It is possible that this repertoire is adequate
and that minor deviations from the germline sequences are
idiosyncrasies of myeloma tumors. Putting aside this question for the moment, we should like to turn our attention to
this large repertoire of variable region genes to ask how it
might have arisen during evolution and how it is maintained

and adapts its diversity to suit the selective purposes of
the organism.

That all the immunoglobulin genes arose by duplication,
amplification and evolutionary drift from a common ancestral
gene has long been accepted by immunochemists (Hood et al.,
1970). We have shown above that for the mouse kappa system
these genes have apparently amplified and drifted in such a
way as to consist of many small subfamilies of ten or so
members whose structural gene sequence is closely homologous.
When we compared the structures of several of these cloned
genes from a single subfamily, we found that they shared
sequence homology not only in their coding sequences, but
also in the thousands of bases that flanked their coding
sequences (Seidman et al., 1978b). This structure was in
sharp constrast to what we had observed when we compared the
structures of two very closely related mouse globin genes,
beta major and beta minor (Tiemeier et al., 1978). These
genetically stable globin genes shared only their coding
sequences and a few hundred nucleotides of homology at their
ends and at the borders of their intervening sequences. We
argued that this arrangement of small, dispersed regions of
homology separated by intervening sequences presented a small
target size for recombination between closely linked
homologous genes. Obviously recombination between non-
allelic β -globin genes could result in disadvantageous gene
loss or amplification. The arrangement of the genes with
minimal homology tended to reduce the probability of an
unequal crossover event.

On the other hand, recombination between immunoglobulin
variable regions might be advantageous to the organism. Indeed, over ten years ago, Edelman and Gally (1967) had described how unequal crossing over might serve to produce
sequence diversity somatically through gene conversion (i.e.
mismatch repair). Unaware of their proposal, we saw the
large targets of homologous sequence present among members of
variable region gene subfamilies as logical candidates for
just the unequal crossing over and mismatch repair that Edelman and Gally had previously envisioned. We further noted
that extensive flanking sequence homology extended only to
members of a subfamily. This would tend to prevent a catas-
trophic gene loss in which an entire family or class of variable region genes would be deleted by a fortuitous recombination event between variable region genes at the extreme ends
of the presumably clustered genes.

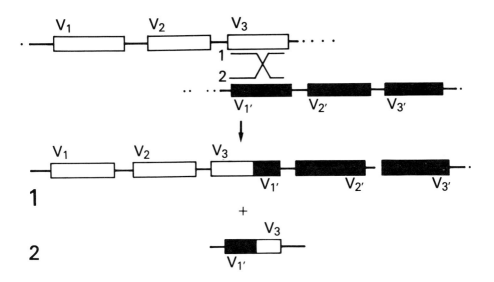

FIGURE 5. Gene expression and loss through unequal crossing-
over. The diagram represents an unequal crossover event
between two homologous, but nonallelic, hypothetical variable
region sequences. The reciprocal products to be inherited by
separate daughter cells represent an expanded set of genes
(three genes to five) and a diminished set of genes (three
genes to one).

    The evidence for this difference in flanking sequence
homology between globin and immunoglobulin gene has been published elsewhere (Seidman et al., 1978b). A diagrammatic
representation of how unequal crossing-over can amplify or
diminish members of a multigene family and how mismatch
repair can introduce new base sequence diversity is reproduced in Figures 5 and 6.

    This model suggests how much diversity, particularly
but not necessarily exclusively, in the germline could arise
by relatively frequent recombination events occurring within
subfamilies. Such a model has several testable predictions, among them that closely related mouse species will
occasionally differ in the number and sequence of variable

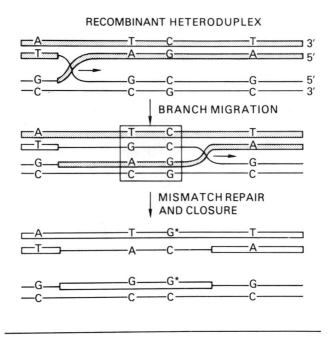

RECOMBINANT HETERODUPLEX

FIGURE 6. Generation of diversity by recombination and mis-match repair. The top strands represent a recombinant hetero-duplex structure in which single strands of DNA join two aligned DNA duplexes at a region of close, but not complete homology. Points at which base sequences differ in the two strands are indicated. In the second set of strands, branch migration has occurred, matching extensive homologous regions from each duplex but also creating small regions of mis-matched base pairs (that is T-G, A-C, C-C, and G-G, as indi-cated). These mismatched base pairs are converted into authentic Watson-Crick base pairs by appropriate repair enzymes. There are two ways of correcting the mismatch. In one, the original sequence contained in that duplex will be restored. (The first set of mismatches is restored in the last set of strands). In the other, a new sequence is cre-ated (the new sequence in the second set of mismatches is shown as G*).

region genes corresponding to a given subfamily. The pattern of restriction fragments encoding variable region genes can be visualized using in situ hybridization wherein bands may differ in length and intensity (reflecting gene

CONSTANT AND VARIABLE REGION GENES IN SEVERAL SPECIES OF MUS

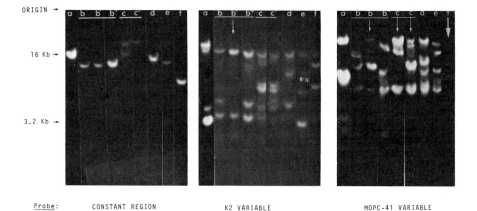

FIGURE 7. Comparison of EcoRl fragments of DNA encoding
variable region genes among several species of mice. Fragments encoding K149 (K2) and K41 variable region genes
were detected using appropriate probes (Seidman et al.,
1978a) against nitrocellulose filter blots containing EcoRl
digests of DNA separated by agarose gel electrophoresis. The
DNA was derived from the following species of mice: a. Mus
musculus (Balb/c), b. Mus cervicolor cervicolor, b'. Mus
cervicolor popaeus (Chantaburi Provence), b". Mus cervicolor
popaeus (Tak Provence), c. Mus cookii (Loei Provence),
c'. Mus cookii (Tak Provence), d. Mus caroli, e. Mus dunni,
f. Mus pahari.

dosage). Such judgments are difficult to make, though we
have seen suggestive changes in several inbred lines of Mus
musculus (unpublished observation). The prediction that is
least ambiguous to assess is the gene loss that might occur
between distinct members of a subfamily, eliminating most
members in a mini-catastrophic gene loss. Such a gene loss
can, in fact, be observed by comparing the subfamilies corresponding to the MOPC-149 (K2) and MOPC-41 variable region
genes in Mus musculus (the Balb/c laboratory mouse) and

several Asian species of Mus (Figure 7). While all species contain several members of the K2 class there is a clear change in band intensity between two isolated populations of Mus cervicolor popaeus (Fig. 6; K2, b' and b") as well as a possible band loss between two isolated populations of Mus cookii (Fig. 7; K2, c and c'). But what is impressive is the loss of the MOPC-41 subfamily from Mus pahari (Fig. 6; MOPC-41, 8). A single faintly hybridizing band can be visualized on the original radioautograph. This must represent the recombinant formed between distant members of the MOPC-41 subfamily whose recombination deleted the bulk of the subfamily. This result is consistent with the model and likely to be found repeatedly when other species and subgroups are examined.

CONCLUSIONS

We have shown that the kappa light chain genes of the mouse are encoded as four separate genetic elements. The first two comprise a large set of split variable region genes encoding a major portion of antibody diversity. One of these variable genes must be joined to one of the J segments that encode the next 10-12 amino acids. The J segments must encode signals for DNA and RNA splicing at their 5' and 3' ends, respectively. The final element is the single constant region gene that is separated from the VJ segment by a few thousand bases of intervening sequence that must be eliminated to make an intact light chain mRNA.

The structure of the variable region gene subfamilies further suggests that they are suitable substrates for intra-subfamily recombination that could serve to generate additional diversity. Evidence from the representation of two variable region gene subfamilies in several species of mice are consistent with this notion.

ACKNOWLEDGMENTS

We are grateful to Dr. Michael Potter for providing us with the original myeloma tumors used in these studies and to Dr. Robert Callahan for providing us with and educating us about the Asian mice. These species were originally trapped by Dr. Joe T. Marshall, Jr.. We wish to thank Mrs. Margery Sullivan for her advice and assistance in the analysis of the electron micrographs. We are particularly grateful to Ms. Terri Broderick for her expert assistance in the preparation of this manuscript.

REFERENCES

Brack, C., Hirama, M., Lenhard-Schuller, R., and Tonegawa, S.
    (1978). Cell 15, 1.
Cohn, M., Blomberg, B., Geckeler, W., Raschke, W., Riblet, R.,
    and Weigert, M. (1974). In "The Immune System" (E. E.
    Sercarz, A.R. Williamson, and C.F., Fox, eds.), p. 89.
    Academic Press, New York and London.
Dreyer, W.J., and Bennett, J.C. (1965). Proc. Natl. Acad.
    Sci. USA 54, 864.
Edelman, G.M., and Gally, J.A. (1967). Proc. Natl. Acad. Sci.
    USA 57, 353.
Hood, L.E., Potter, M., and McKean, D.J. (1970). Science
    170, 1207.
Hozumi, N., and Tonegawa, S. (1976). Proc. Natl. Acad. Sci.
    USA 73, 3628.
Kabat, E.A., Wu, T.T., and Bilofsky, H. (1978). Proc. Natl.
    Acad. Sci. USA 75, 2429.
Konkel, D., Tilghman, S., and Leder, P. (1978). Cell,
    in press.
Leder, A., Miller, H., Hamer, D., Seidman, J.G., Norman, B.,
    Sullivan, M., and Leder, P. (1978). Proc. Natl. Acad.
    Sci. USA, in press.
Loh, E., Schilling, J., and Hood, L. (1978). Nature 276, 785.
Rabbitts, T.H., and Forster, A. (1978). Cell 13, 319.
Seidman, J.G., Leder, A., Edgell, M.H., Polsky, F., Tilghman,
    S.M., Tiemeier, D.C., and Leder, P. (1978a). Proc. Natl.
    Acad. Sci. USA 75, 3881.
Seidman, J.G., Leder, A., Nau, M., Norman, B., and Leder, P.
    (1978b). Science 202, 11.
Seidman, J.G., and Leder, P. (1978c). Nature 276, 790.
Tiemeier, D.C., Tilghman, S.M., Polsky, F., Seidman, J.G.,
    Leder, A., Edgell, M.H., and Leder, P. (1978). Cell 14,
    237.
Tonegawa, S., Brack, C., Hozumi, M., and Schuller, R. (1977).
    Proc. Natl. Acad. Sci. USA 74, 3518.
Tonegawa, S., Maxam, A.M., Tizard, R., Bernard, O., and
    Gilbert, W. (1978). Proc. Natl. Acad. Sci. USA 75, 1485.
Valbuena, O., Marcu, K.B., Weigert, M., and Perry, R. (1978).
    Nature 276, 780.
Weigert, M., Gatmaitan, L., Loh, E., Shilling, J. and Hood, L.,
    (1978). Nature, 276, 785.

DISCUSSION

C. WEISSMANN: Your explanation of the allelic exclusion puts it down on a different level does it not? My question is then why do you only get a recombinant on one of the genes? Why not on both?

P. LEDER: You can just take the simplest explanation and say that recombination is a random event that occurs with a certain frequency and that frequency is sufficiently rare so it does not occur on two genes in one cell. I would like to emphasize, though, that the conclusion of a single gene moving has to be tempered by the fact that we cannot exclude the possibility that our embryonic genes arise in undifferentiated tissues in our cells.

P. REDDY: You suggest that all the V genes in the embryonic DNA are contiguous. Am I correct in assuming that there are no intervening sequences between V genes?

P. LEDER: The kappa V region genes are thought to be on the same chromosome. I assume that they are not contiguous, that is, that coding sequences are not cheek by jowl with coding sequence. There are thousands of bases that separate them.

P. REDDY: Between V and C genes you said there are about six or seven J sequences. Are these sequences identical to one another?

P. LEDER: We have direct evidence for two, and other recombinants that suggest that there are more.

P. REDDY: Are the sequences of the J chains identical?

P. LEDER: No, it turns out that when you look at the two that we have, they differ slightly from one another. That explains of course why people thought all along that these ten amino acids were part of the variable region sequence. They turned out to have slightly different sequences when compared among different light chains. In fact, Marty Wigert and Lee Hood, whose work I should have mentioned, and Elvin Kabat have analyzed the protein structures of these genes and have found in fact that the ten terminal amino acids segregate independently from other regions of the V sequence. This is obviously consistent with several different J region segments.

F.W. PUTNAM: We have in press evidence that the C regions of
some human alpha chain allotypes represent recombinants of
individual domains (Tsuzukida, Wang and Putnam, PNAS, 1979),
and I wonder if you have any evidence for, or if you are willing
to speculate, on the question of whether the exons of heavy
chains have a correlation with the individual domains of heavy
chains and likewise whether the J regions have a correlation
with the hinge sequence or inter-domain sequences of heavy
chains.

P. LEDER: We do not have any heavy chain clones at all so I
cannot comment on that.   I know that Lee Hood has analysed his
heavy chain clones and that Tonegawa has analysed his heavy
chain clones and the results that I have seen suggest that they
fall into domains that could be correlated with the domains of
the heavy chains.

THE STRUCTURE AND EXPRESSION OF NORMAL AND
ABNORMAL GLOBIN GENES

R.A.Flavell, G.C.Grosveld, F.G.Grosveld, R.Bernards,
J.M.Kooter and E.De Boer

Section for Medical Enzymology and Molecular
Biology, Laboratory of Biochemistry, University
of Amsterdam, Amsterdam, The Netherlands

P.F.R.Little

Department of Biochemistry, St. Mary's Hospital
Medical School, University of London, London,
Great Britain

I. Introduction

Most mammalian genes are present as a single
copy per haploid genome and hence comprise only
about one part in 10^6 of the total nuclear DNA.
This fact impeded work on single-copy genes, but
recently recombinant DNA technology (e.g. ref. 1)
and sensitive gene mapping [2-4] has led to the
elucidation of the structure of a number of
eukaryotic genes [5-10].

II. The Structure of the Rabbit β-Globin Gene

Our initial approach was to construct a
detailed physical map of the rabbit β-globin gene
using the blotting-filter hybridization approach
[4,5]. This showed that the β-globin gene consists
of at least two non-contiguous blocks of coding
sequences separated by an intervening sequence
(IVS) of 600 base pairs (bp). The rabbit β-globin
gene has now been cloned and we have analysed its

149

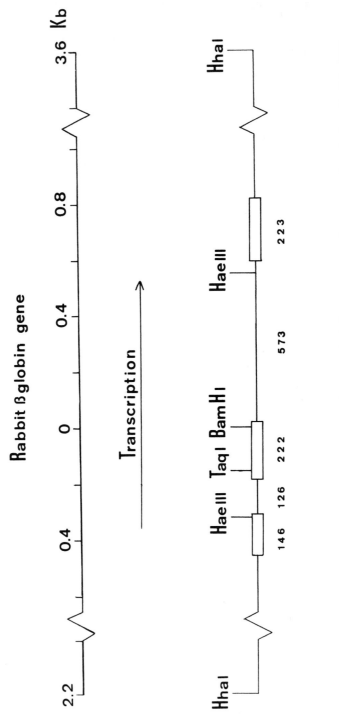

Fig. 1. A restriction map of the rabbit β-globin gene. The map depicts RβG1 DNA (a 5 kb segment of rabbit chromosomal DNA inserted into plasmid pCRI). The blocks signify the coding segments of the β-globin gene.

structure in more detail including the determination
of the DNA sequence (in collaboration with Charles
Weissmann and his colleagues [11]). These experiments show that two intervening sequences are
present in the rabbit ß-globin gene (Fig. 1), one
of 126 bp in length present between the DNA
sequences for amino acid 30 and 31 and the one
already mentioned above of 573 bp long, present
between the sequences coding for amino acids 104
and 105 [11]. For a discussion of the DNA sequence
of the rabbit and mouse ß-globin genes, see the
article by Weissmann et al. [this volume].

III. Transcription of the Rabbit ß-Globin Gene

     Of the many models to explain the way in which
mature mRNA is obtained from split genes (see e.g.
refs 12,13,5), RNA splicing [12-15] is by far the
most probable, although formal proof for many of
the steps is lacking. Splicing was proposed to
explain the primary observation that adenovirus
messenger RNAs (mRNAs) were transcripts of non-
contiguous segments of the viral genome. Two
subsequent observations on cellular gene transcripts
have supported this model.
     First, it has been known for some time that
in mouse a ß-globin pre-mRNA of 15S (about 1200-1500
nucleotides (NT) long) exists. The size of this
precursor is approximately that expected of a
transcript of the coding segments plus intervening
sequences. R-loop mapping of the pre-mRNA on the
cloned mouse ß-major globin gene showed a continuous
R-loop in the same region as the R-loops formed
with ß-globin mRNA which establishes that the
pre-mRNA is a transcript of the coding regions
plus at least the large intervening sequence [16].
     Second, Rutter's and Abelson's groups have
shown that an enzyme activity is present in yeast
which can remove the intervening sequence present
in yeast pre-tRNA transcripts to generate mature
tRNA (see e.g. ref. 17).

IV. Transcription Mapping of Rabbit ß-Globin pre-mRNA

     Does the 15S RNA contain RNA complementary to
the extragenic regions 5' and 3' of the coding
regions or is it simply a transcript of the coding
sequences and intervening sequences? The electron
microscopical data cannot answer this definitively

and we have, therefore, taken another approach.

Berk and Sharp [18] have shown that it is possible to map transcripts on a DNA accurately by forming DNA-RNA hybrids, degrading the unhybridized DNA with S_1 nuclease and analysing the size of the protected DNA on alkaline agarose gels. Spliced mRNAs form hybrids with single-stranded DNA loops which are destroyed on digestion with S_1 nuclease to give DNA fragments of the same length as the coding segments. A pre-mRNA containing all the intervening sequences forms a single S_1 nuclease resistant hybrid of the same length as the pre-mRNA.

We have analysed the transcripts of the rabbit β-globin gene in this way. Bone-marrow RNA was prepared from an anaemic rabbit, size-fractionated on a sucrose gradient and then hybridized to various fragments of the cloned rabbit β-globin gene. The single-stranded nucleic acids were degraded with S_1 nuclease, the resistant DNA fractionated by alkaline agarose gel electrophoresis and detected by Southern blotting and hybridization to probes for the β-globin structural gene (pβG1 DNA [19]) or the cloned chromosomal DNA fragment (RβG1 DNA [11]). Fig. 1 gives a schematic map of the RβG1 DNA in which the HhaI, BamHI and TaqYI sites discussed here are shown.

Hybridization of RNA of approximately 15S (12-18S) to RβG1 DNA cleaved with HhaI (which cuts in the plasmid regions of RβG1 to give an approximately 6 kb fragment containing all of the rabbit DNA), yields a series of fragments in the alkaline gel (Fig. 2). The largest fragment is 1250 NT long, a size consistent with a transcript of the coding sequences plus intervening sequences (1295 NT). In addition, fragments of 223 and 146 NT are present which are exactly the size expected for the DNA pieces generated by S_1 nuclease treatment of hybrids formed between β-globin mRNA and the cloned DNA. The IVS loops will be cut in such hybrids to give segments of 146, 222 and 223 NT in length. Finally, there is evidence for a pre-mRNA in which only the small IVS has been removed by processing. A prominent 1000 NT fragment (Fig. 2) is found in hybrids formed with RβG1 x HhaI (from this pre-mRNA we should expect a 1018 NT fragment) and a 735 NT fragment is found with RβG1 x HaeIII hybrids (expected 753 NT). Other faint bands are present which may be derived from still other splicing intermediates. These have not yet been

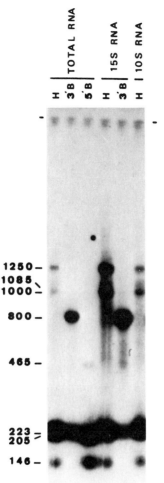

Fig. 2. Detection of the β-globin pre-mRNA from
rabbit bone-marrow DNA by S_1 nuclease mapping.
Total bone-marrow DNA, a 12-18S RNA fraction and
an 8-12S fraction were hybridized with RβG1 DNA
cleaved with HhaI (H), or the two separated globin
gene-containing fragments of RβG1 cleaved with
HhaI + BamHI (3' B and 5' B: see Fig. 1). Hybridization was performed under conditions where only
DNA-RNA hybrids form. The sample was digested with
S_1 nuclease and electrophoresed on a 2% alkaline
agarose gel [18]. Phage ØX DNA cleaved with TaqYI,
HaeIII and HapII (5' end-labelled with ^{32}P) was
used as a molecular weight marker. The DNA was
transferred to a nitrocellulose filter and hybridized with ^{32}P-labelled RβG1 DNA as described
[4,5,23]. The figure shows an autoradiogram of the
nitrocellulose filter.

mapped.

To map the termini of the 1250 NT transcript
we hybridized 15S RNA with fragments of RßG1 where
the gene is cut at various points. Cleavage of
RßG1 with HhaI and BamHI gives two large fragments;
a fragment of 2.2 kb containing the 5' extragenic
region, the first coding segment, small intervening
sequence and second coding segment up to the
sequence coding for amino acid 99, and a fragment
of 3.6 kb containing the remainder of the second
coding segment, the large intervening sequence,
the final coding segment and the 3' extragenic
region. We separated these two fragments by gel
electrophoresis and hybridized them to the '15S'
RNA pool. Since the entire DNA sequence of the
ß-globin gene (coding and intervening sequences)
has been determined, we can predict precisely the
positions of the 5'- and 3'-termini of the ß-globin
mRNA on the map in Fig. 1. If the 1250 NT pre-mRNA
is a transcript reading from the first coding
segment through to the last, we should obtain
hybrids of 477 and 815 bp in length. We find 465
bp and 800 bp (Fig. 2). In a similar experiment
with the separated TaqY1 + HhaI fragments we
should expect 307 bp and 989 bp; we find 320 bp
and 960 bp. These experiments show that the 1250
NT transcript contains the coding regions of the
ß-globin gene together with both intervening
sequences. Within the error of the measurements we
can conclude that the 5' and 3' termini of the
mRNA and this pre-mRNA are the same. We have no
evidence for a pre-mRNA larger than the 1250 NT
component. The question whether the termini of
this transcript represent the initiation and
termination points for transcription, however,
still remains open. This is likely to remain
unanswered until in vitro transcription systems
have been set up or until cloned DNA segments have
been brought to expression in cell systems [20].

V. The Human Globin Genes

Two aspects of the human globin gene complex
prompted our initial attention:

1. Genetic evidence suggested 'close' linkage
of the non-α-globin genes in one of two arrangements,
either Gγ-Aγ-δ-β or Aγ-δ-β-Gγ (the two non-allelic
foetal Υ-globins differ by a single amino acid at
position 136, Gγ has glycine and Aγ alanine at

this position; the δ- and β-globin genes are also closely related but differ at 10 amino acid sites). It was, however, unclear what the physical distance is between two linked genes on a mammalian chromosome. The human globin system seemed, therefore, appropriate to determine this physical linkage directly.

2. Numerous well-defined cases of defects in the functioning of the human globin genes have been described. They fall into two classes:

a) Abnormal globin proteins are produced usually because of amino acid substitutions which result from point mutations in the structural gene; less commonly, deletions or gene fusions cause other types of abnormal haemoglobins.

b) The level of - otherwise normal - haemoglobin can be reduced or even be zero in the diseases known as the thalassaemias. Here we shall restrict the discussion to the β-thalassaemias and related diseases. In some cases of β°-thalassaemia no β-globin mRNA can be detected in the cell [21]; in others nuclear β-globin RNA is found but no cytoplasmic globin mRNA [22]; in still other cases cytoplasmic globin mRNA is found which is apparently inactive in translation. In β⁺-thalassaemia reduced levels of β-globin mRNA are found with a parallel decrease in the levels of β-globin in the cell.

Our approach has been to construct physical maps of the normal human globin gene using the blotting procedure [2,4] and then to compare these with the corresponding maps of the DNA of patients with the various thalassaemias. In this way, large deletions in DNA regions within or around the globin genes can be detected directly in total genomic DNA.

VI. The Structure of the Human δβ-Globin Linkage Group

We have constructed a map of the two linked δ- and β-globin genes [23] (see Fig. 3). This map shows:

a) That the δ-globin gene is located about 7000 bp to the 5'-side of the β-globin gene. Both genes are transcribed from the same DNA strand.

b) The structures of the human δ- and β-globin genes strongly resemble the structures of the rabbit and mouse globin genes described earlier [5,6]. The blotting experiments show the presence

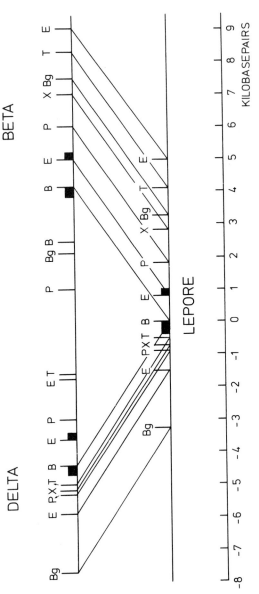

Fig. 3. A physical map of the β- and δ-globin genes in normal and Hb Lepore DNA. The probable positions of the coding regions of the two globin genes are shown as filled boxes. It should be stressed that the only extragenic cleavage sites which can be detected for a given enzyme by this analysis are those closest to the gene examined. Although the β- and δ-genes are presented as being composed of two coding segments in each case, the possibility that these segments are further split cannot be excluded from these data. Bg, BglII; E, EcoRI; P, PstI; X, XbaI; T, TaqYI; B, BamHI. Reproduced from ref. 23 with permission.

of the large intervening sequence in both human genes within the same region as in the rabbit and mouse genes. The human intervening sequences are 800-1000 bp long instead of 600 bp in the case of rabbit and mouse.

Both Mears <u>et al</u>. [24] and Lawn <u>et al</u>. [25] have also analysed the structure of the δ-β gene region and the results of the three groups do not show significant differences. In the latter case [25], the linkage of the δ- and β-globin genes has also been demonstrated.

VII. The Structure of the Human γ-Globin Genes

More recently we have also mapped the human γ-globin genes (Fig. 4) [26]. The map shows two linked γ-genes and the general organization of the genes resembles the β-δ locus.

a) Both genes are transcribed from the same DNA strand.

b) The Gγ and Aγ genes are distinguished from one another by the presence of a PstI site in the latter gene. This is the result of the single amino acid difference at position 136. By this criterium, the Aγ gene is at the 3'-side of the Gγ gene; the gene order in this region must, therefore, be 5'-GγAγδβ-3', which is in agreement with one of the two models proposed on the basis of the fusion proteins Hb Lepore and Hb Kenya.

c) The distance between the two γ genes is 3500 bp.

d) Both γ genes contain at least the large intervening sequence within the same region as in β-globin genes (coding for amino acids 101-121; but presumably also 104-105 as in the other β-globin genes).

VIII. Analysis of Abnormal Globin Genes: Hb Lepore

The structure of the Hb Lepore globin gene has been analysed by performing a blotting analysis on DNA of patients with homozygous Hb Lepore [23]. This investigation has proven that the Hb Lepore gene is the result of a fusion of the δ- and β-globin gene as originally proposed by Baglioni [27] and not a δ-RNA-β-RNA splicing event [5].

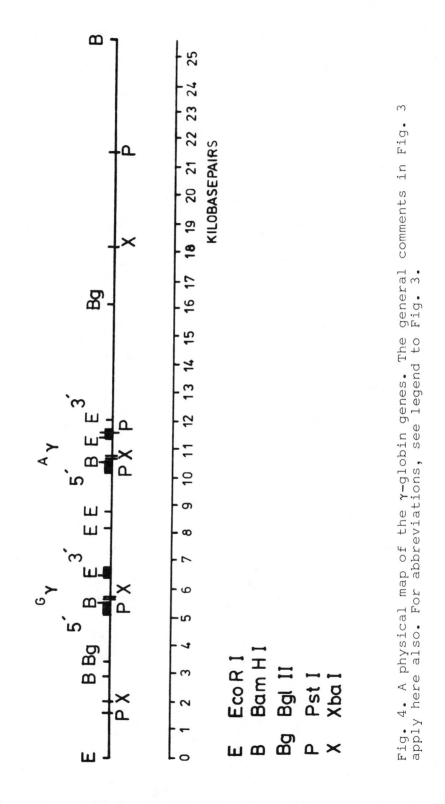

Fig. 4. A physical map of the γ-globin genes. The general comments in Fig. 3 apply here also. For abbreviations, see legend to Fig. 3.

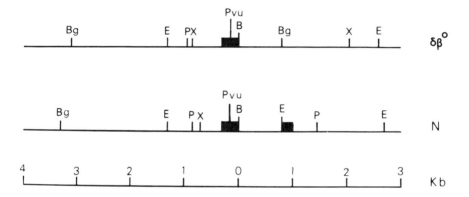

Fig. 5. A physical map of the δ+β-globin gene
region in DNA from a patient homozygous for δβ°-
thalassaemia. The DNA was isolated from a lymphocyte cell line, digested with the restriction
enzymes indicated in single and in various double
digests. The map is based upon the following
fragments: BglII (Bg), 3.8 kb; BglII + BamHI (B),
3.1 kb; BamHI + PstI (P), 0.95 kb; BamHI + XbaI
(X), 0.85 kb; BglII + EcoRI (E), 2.1 kb; BglII +
PstI, 1.7 kb; PvuII (Pvu) + XbaI, 0.8 kb; BglII +
XbaI, 1.65 kb; XbaI, 2.9 kb; EcoRI, 3.9 kb. N is
the normal δ-globin gene.

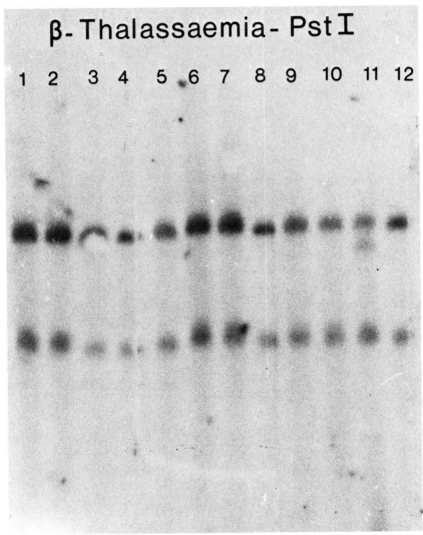

Fig. 6. β- and δ-globin gene fragments in digests of DNA from patients homozygous for β-thalassaemia. DNA was isolated from β-thalassaemics as described [4] and digested to completion with PstI. The samples were analysed for the β-δ-globin genes as described [23]. Patient 5 is described by Comi et al. [22]: Patient 11 in ref. 21. Patients 2, 3 and 10 have been diagnosed as β°-thalassaemia (Southern Italian; Ottolenghi, S., personal communication) and the remainder as homozygous β+-thalassaemics. 1, 4, 8 and 12 show control normal DNA (derived from a placenta from a Dutch individual).

IX. The β-Thalassaemias

δβ°-Thalassaemia is a rare condition that is characterized by the complete absence of β- and δ-globin chains. Complementary DNA (cDNA) titration hybridizations have shown that at least partial deletion of the β- or δ-globin genes has occurred although the extent of the deletion could not be defined. Recently we have constructed a map of cleavage sites in δβ°-thalassaemia. We have analysed DNA from two sibling patients homozygous for δβ°-thalassaemia and also from peripheral blood from two patients heterozygous for the condition. We find evidence for two components, one of which is the 5' regions of the δ-globin gene (Fig. 5) and a second component, which gives faint bands, that has not yet been identified. The deletion in δβ°-thalassaemia starts somewhere to the 3'-side of the intragenic BamHI site. We are not yet sure where the DNA sequences which are fused to the 3'-side of the δ-gene are derived from, although the most logical possibility is a DNA region to the 3'-side of the β-globin gene. Curiously, no transcripts of the δ-globin gene can be detected in δβ°-thalassaemia. This suggests either that the mere presence of the 5' regions of the δ-globin gene are not sufficient to ensure transcription of this region, or that transcription of this half-gene does occur but the transcripts are rapidly degraded.

In contrast to δβ°-thalassaemia, the structure of the β-globin gene in β°- and β⁺-thalassaemia is in most cases indistinguishable from the normal β-globin gene in our blotting hybridization experiments. Fig. 6 shows PstI digests of DNA from a normal placental DNA preparation and from patients with various forms of β-thalassaemia. In only one case can a difference be seen (no. 11). Here a second β-globin DNA fragment is found which is 600 bp shorter than the normal PstI β-gene fragment. Mapping experiments have shown that this results from a 600 bp deletion, which includes the 3' coding regions of the β-globin gene (coding for amino acids 104-146) and probably the 3' untranslated region (Fig. 7).

In the majority of β-thalassaemia cases, however, we can detect no alterations in the β-globin gene. To determine the molecular nature of the lesion, it will probably be necessary to

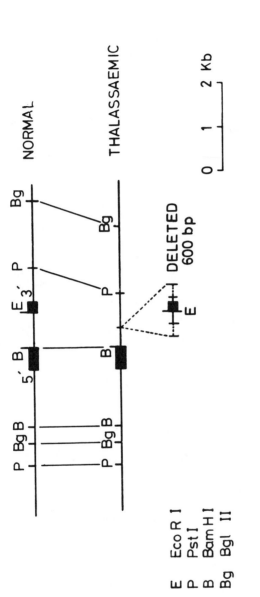

Fig. 7. A physical map of the β-globin gene region in a normal individual and a patient with β°-thalassaemia. The map shows a 600 bp deletion centered round the β-globin intragenic EcoRI site which is missing in this β°-thalassaemia gene. The 600 bp deletion is arbitrarily drawn centered on this EcoRI site. Also shown are the possible limits of the deletion which stretch 600 bp either side of this site.

clone the thalassaemic genes and to compare these
genes with the normal β-globin gene. The problems
involved with this type of approach have been
discussed elsewhere [20].

X. Acknowledgements

We are grateful to Prof. P.Borst and Dr.
H.H.M.Dahl for helpful discussions. RAF also
thanks Dr. R.I.Kamen (ICRF, London, UK) for his
hospitality and help with the S_1 nuclease mapping
procedure. Our work described here has been supported (in part) by a grant to RAF from The Netherlands Foundation for Chemical Research (SON) with
financial aid from The Netherlands Organization
for the Advancement of Pure Research (ZWO). Work
by PFRL was supported by a grant to R.Williamson
from the British Medical Research Council.

XI. References

1   Tilghman, S.M., Tiemeier, D.C., Polsky, F.,
     Edgell, M.H., Seidman, J.G., Leder, A., Enquist,
     L.W., Norman, B. and Leder, P. (1977) Proc.
     Natl.Acad.Sci.U.S. 74, 4406-4410.
2   Southern, E.M. (1975) J.Mol.Biol. 98, 503-517.
3   Botchan, M., Topp, W. and Sambrook, J. (1976)
     Cell 9, 269-287.
4   Jeffreys, A.J. and Flavell, R.A. (1977) Cell
     12, 429-439.
5   Jeffreys, A.J. and Flavell, R.A. (1977) Cell
     12, 1097-1108.
6   Tilghman, S.M., Tiemeier, D.C., Seidman, J.G.,
     Peterlin, B.M., Sullivan, M., Maizel, J.V.
     and Leder, P. (1978) Proc.Natl.Acad.Sci.U.S.
     75, 725-729.
7   Breathnach, R., Mandel, J.L. and Chambon, P.
     (1977) Nature 270, 314-319.
8   Tonegawa, S., Brack, C., Hozumi, N. and
     Schuller, R. (1977) Proc.Natl.Acad.Sci.U.S.
     74, 3518-3522.
9   Goodman, H.M., Olson, M.V. and Hall, B.D.
     (1977) Proc.Natl.Acad.Sci.U.S. 74, 5453-5457.
10   Valenzuela, P., Venegas, A., Weinberg, F.,
     Bishop, R. and Rutter, W.J. (1978) Proc.Natl.
     Acad.Sci.U.S. 75, 190-194.
11   Van den Berg, J., Van Ooyen, A., Mantei, N.,
     Schambock, A., Grosveld, G., Flavell, R.A. and
     Weissmann, C. (1978) Nature 276, 37-44.

12   Berget, S., Moore, C. and Sharp, P.A. (1977)
     Proc.Natl.Acad.Sci.U.S. 74, 3171-3175.
13   Chow, L.T., Gelinas, R.E., Broker, T.R. and
     Roberts, R.J. (1977) Cell 12, 1-8.
14   Klessig, D.F. (1977) Cell 12, 9-21.
15   Dunn, A.R. and Hassell, J.A. (1977) Cell 12,
     23-36.
16   Tilghman, S.M., Curtis, P., Tiemeier, D.C.,
     Leder, P. and Weissmann, C. (1978) Proc.
     Natl.Acad.Sci.U.S. 75, 1309-1313.
17   Knapp, G., Beckmann, J.S., Johnson, P.F., Fuhr-
     man, S.A. and Abelson, J. (1978) Cell 14, 221-
     235.
18   Berk, A.J. and Sharp, P.A. (1977) Cell 12,
     721-732.
19   Maniatis, T., Kee, S.G., Efstratiadis, A. and
     Kafatos, F.C. (1976) Cell 8, 163-182.
20   Flavell, R.A., Little, P.F.R., Kooter, J.M.
     and De Boer, E. (1979) in Models for the Study
     of Inborn Errors of Metabolism (Hommes, F.A.,
     Ed.), North-Holland, Amsterdam, in press.
21   Tolstoshev, P., Mitchell, J., Lanyon, G.,
     Williamson, R., Ottolenghi, S., Comi, P.,
     Giglioni, B., Masera, G., Modell, B., Weather-
     all, D.J. and Clegg, J.B. (1976) Nature 259,
     95-98.
22   Comi, P., Giglioni, B., Barbarano, L., Otto-
     lenghi, S., Williamson, R., Novakova, M. and
     Masera, G. (1977) Europ.J.Biochem. 79, 617-622.
23   Flavell, R.A., Kooter, J.M., De Boer, E.,
     Little, P.F.R. and Williamson, R. (1978) Cell
     15, 25-41.
24   Mears, J.G., Ramirez, F., Leibowitz, D. and
     Bank, A. (1978) Cell 15, 15-23.
25   Lawn, R.W., Fritsch, E.F., Parker, R.C.,
     Blake, G. and Maniatis, T. (1978) Cell, in
     press.
26   Little, P.F.R., Flavell, R.A., Kooter, J.M.,
     Annison, G. and Williamson, R. (1979) Nature,
     in press.
27   Baglioni, C. (1962) Proc.Natl.Acad.Sci.U.S.
     48, 1880-1886 .

DISCUSSION

P. LEDER: Richard, I wonder if you would mind recapitulating the situation with your usual β°-thalassaemia.   This individual has a deletion that has removed the back coding half of the β-globin gene and has a second gene in which no defect can be seen.   Is that the situation, and is it then correct that no stable β-globin message is made, but that δ-globin message is made?

R. FLAVELL: Yes, that is correct.   In other words he has two β-globin genes, both are β°, with one showing a deletion and the other one no apparent defect.   Obviously, there must be a defect in the latter gene, which has not yet been detected. Regarding the mRNA- the experiments which were done at the time measured the steady state RNA levels.   They could not distinguish δ and β because it was just done by cDNA excess hybridization using β probes, and of course they measure against a background of δ in normal individuals because β probes cross hybridize to δ.   I think what you see there is a slight amount of hybridization, but again the experiments were done with non-cloned β cDNA probes, so you could have contamination in the probe.   I think the question of whether there is δ RNA is not established, but it is very likely that the δ RNA is there.

GENOMIC CLONES FROM UNFRACTIONATED DNA

O. Smithies[1]
J. L. Slightom
P. W. Tucker[2]
F. R. Blattner[3]

Laboratory of Genetics
University of Wisconsin
Madison, Wisconsin

Many problems of gene control in higher organisms are
illustrated by the changes which occur in the hemoglobins
of mammals at different stages of their development. For
various reasons, this series of changes is best documented in
humans, who have at least seven different globin structural
genes active at specific periods of time. There are two
alpha-type globin genes, the embryonic ζ and adult α, and
five beta-type globin genes, the embryonic ε, fetal $^A\gamma$ and
$^G\gamma$, adult major β, and adult minor δ. A considerable amount
of genetic data is available suggesting that the β-type genes
are closely linked in humans with the most probable order
being $-^G\gamma-^A\gamma-\delta-\beta-$. Judging from data on the embryonic Y
chain gene in mice it is possible that the human embryonic ε
gene will prove to be in the same chromosomal region but this
is not established. The alpha type genes are on a different
chromosome. We heard yesterday from Dr. Flavell that the
order of the $^G\gamma$ and $^A\gamma$ genes deduced from genetic data is
now confirmed by molecular studies at the DNA level (1).
Recent work by Maniatis and his colleagues (2) shows that
the deduced order of the adult δ and β globin genes is also
confirmed at the DNA level. Thus humans have a series of
closely linked genes which are differentially expressed
during development.
These globin structural genes and the DNA intervening
between and within them are excellent targets for DNA
cloning with the aim of investigating gene control. Particularly favourable are the fetal γ and adult β genes. The
amino acid sequences of all the protein products are known,
and the base sequences of much of the protein coding regions

[1] Supported by NIH grants GM20069 and AM20120.
[2] Supported by NIH grants GM06526 and GM07131.
[3] Supported by NIH grant GM21812.

167

have been determined. If the genomic DNA corresponding to
these genes can be isolated then one can look for structural
clues to the mechanism which permits the fetal genes to be
active when the adult genes are inactive, and vice versa.
But I do not want to give you too naive a view of the total
problem. The "fetal to adult switch" - as it is often
referred to - is not likely to be a localized mechanism. My
reason for saying this is that concommittant with the changes
in the activity of the globin genes are other more complex
changes. For example, in mice the embryonic-type globins
present at around 10 days of gestation occur in relatively
large nucleated erythrocytes that are the progeny of cells
arising in the blood islands of the yolk sac, while the
adult-type globins present after about 12 days of gestation
occur in smaller non-nucleated erythrocytes which are
produced in the liver. Comparable but more subtle differences are found if one looks carefully at the human fetal
to adult change. There are differences in the new born
human in the membrane proteins and isozymes of red cells
which contain mainly fetal hemoglobin compared to those
which contain mainly adult hemoglobin. The "switch" is not
a simple "one gene on/one gene off" switch. Nevertheless we
hope to make a contribution to understanding the anatomical
differences in the DNA of fetal and adult globin genes by
studying the structural gene for the human fetal globin gene,
$^A\gamma$, which we have recently cloned.

The work I shall describe is the result of a happy
collaboration between the laboratories of Dr. Fred Blattner
and myself at the University of Wisconsin. Details of the
work are published in three papers: Blattner et al. (3),
Blattner et al. (4) and Smithies et al. (5). Dr. Jerry L.
Slightom, who with Dr. Philip W. Tucker has done the
sequencing work, will present some experimental details in
the poster session on Wednesday.

The first major problem encountered in the course of
this work was the need for a vector able to meet the
requirements of the N.I.H. guidelines for recombinant DNA.
The required biological containment is EK2. Simply stated,
an EK2 vector must be unable to propagate itself (or its
cloned foreign DNA) under non-laboratory conditions with
an efficiency no greater than 10^{-8} of the efficiency obtained
under special laboratory conditions. Current EK2 vectors
are of two categories, plasmids or bacteriophages, and each
category has its merits and disadvantages.

Plasmids have the advantages that, in general, their
DNA molecules are considerably smaller than the phage DNAs
and there is no pre-determined size limit to the foreign DNA
which can be cloned in them. Consequently the cloned DNA

can be a high proportion of the total recombinant molecule.
Yields of the foreign DNA obtainable from a healthy bacterial
host can accordingly be high.   The EK2 plasmids have, however,
some disadvantages that are serious for cloning mammalian
genomes.   First, in order to meet EK2 specifications, they
must be grown in the Curtiss host $\chi 1776$ which is debilitated
to the extent that high yields are difficult to obtain.
Secondly, it is difficult to screen very large numbers of
bacterial colonies in searching for rare clones, even when
using a colony replica technique.

Bacteriophages have the advantage that screening large
numbers of plaques (in the hundreds of thousands) is
relatively easy using the Benton-Davis (6) replica procedure.
In addition it has proved possible to develop phage strains
which can meet the EK2 requirement in conjunction with
bacterial hosts that are minimally handicapped under laboratory conditions.   For example, phage titers with our EK2
bacteriophage Charon 3A average about 5 X 10^{10} when propagated in liter quantities in the Pereira-Curtis-Leder host
DP50SupF.   These excellent titers compensate for the fact
that the cloned DNA is a relatively small percentage of the
recombinant molecule.   With a phage containing a 5000 base
pair cloned fragment, the yield can approach 50 μg of fragment per liter.   A disadvantage of the phage systems is that
the size of cloned fragments is limited by the size of the
bacteriophage head.

Table I shows the major characteristics of the six
Charon phages which have been certified at the EK2 level.
Clearly Charon 3A, 16A, 21A and 23A are suitable for cloning
smaller DNA fragments.   Charon 4A and 24A are useful for
larger fragments.   The results I will be considering today
were obtained with Charon 3A.

TABLE I.   Characteristics of EK2 Charon Phages

Phage	Cloning Sites	*Eco* RI Fragment Size Range (kbp)
3A (*23A)	*Eco* RI, *Xho* I	0 - 10
4A (*24A)	*Eco* RI, *Xba* I	8 - 20
16A	*Eco* RI, *Sal* I, *Sst* I	0 - 8
21A	*Xho* I, *Eco* II, *Hin*d III, *Sal* I, *Xho* I	0 - 10

*Improved versions of 3A and 4A

Considerable success has been achieved by other investigators in experiments in which clones have been constructed using DNA enriched for the genes of primary interest. We heard yesterday of the beautiful results obtained by Dr. Leder's group using this approach (7). He also referred to the successes of Dr. Tonegawa using the same general type of strategy (8). In Wisconsin we decided to attempt the cloning of the human genome by the simplest possible means, using complete Eco RI digests of unfractionated DNA. This approach was chosen because we wanted to construct a collection of phages containing a random sample of the human genome - a "shotgun" collection. A random shotgun collection can be used in searches for many different gene fragments, which need not be related in either size or DNA sequence. In practice there are deviations from randomness imposed by the choice of the restriction enzyme and by the range of fragment size which the vector can accept when a shotgun is made in a phage vector.

The overall basic procedures used in our shotgun experiments have four steps.
  i) Constructing the recombinant DNA molecules.
  ii) Obtaining viable phages from these molecules.
  iii) Screening the phages for the cloned fragments of
       interest.
  iv) Purifying and characterizing the cloned fragments.

Constructing the recombinant DNA molecules presented no particular problems. The vector DNA was prepared by Eco RI digestion of DNA prepared from the phage Charon 3AΔlac. Ligation is straightforward, except that it is essential to keep the concentration of DNA high. This favours the formation of large concatenates and reduces the tendency of the target DNA fragments to form circles by self-closing reactions.

Three ways were considered for obtaining viable phages from this ligated recombinant DNA: calcium shock transfection, spheroplast transformation and in vitro packaging. The efficiencies of the first two techniques allow the practical production of a few thousand phages, but only the in vitro packaging method allowed us to produce several millions. We used the in vitro packaging method after showing that it does not compromise the EK2 status of the system.

Screening the phages for cloned fragments of interest in a random shotgun collection of phages made with complete Eco RI digested target DNA requires the screening of millions of phage plaques. A conventional bacterial plate can be used to grow about 15,000 phage plaques without

excessive crowding. Screening 500 phage plates presents
considerable logistic problems, and we chose instead to
increase the size of the plates. We use cafeteria trays
lined with aluminum foil as the bacterial plates, and scale
all the Benton-Davis filter replica procedures to the size of
medical X-ray films. In this way we can screen about 400,000
phage plaques on a single tray - or megaplate, as we call
them.

The choice of probe to hybridize to the filter replica
in the cloning experiments is clearly critical. We found
that in our globin experiments we could use cDNA made
directly from purified globin mRNA. This choice is clearly
only available when the mRNA is relatively easy to obtain
and purify. In other experiments we have used plasmids
containing cloned cDNA copies of the mRNA of interest. This
is a more generally applicable approach. Both approaches
have their problems.

In the early stages of this work we attempted to screen
our "shotguns" with a mixture of probes. [We now do this in
a different way by taking several filter replicas of the
megaplates, and hybridizing each replica to a single probe.]
We wished to take advantage of the near random nature of the
shotgun phage collection to look for members of several
different gene families. We expected to be able to sort out
the hybridization positive plaques at a later stage of
purification by testing them against single probes. To our
surprise, it did not work that way - we found that most of
the candidate phages hybridized to all of the probes!

In one of these early experiments with a mouse shotgun
we screened four megaplates (Nos. 28 to 31) with a mixed
probe made from cDNA for mouse α and β globin, and from nick
translated plasmid DNAs containing cDNA copies of rat insulin
and rat growth hormone mRNAs. Forty two positive hybridizing
plaques were picked, including some very faint ones. Figure
1A illustrates a quarter of the autoradiograph from megaplate
30. Twenty-one of the forty-two candidate phages were still
positive after replating on small plates. [The negative ones
were, in general, the very faint candidates.] But twenty of
the still positive candidates hybridized to all the probes!
One of the candidates Mm (Mus musculus) 30.5 hybridized only
to the globin cDNA probes on retest. (This is labeled
5 in Figure 1A.) The only element in common to the cDNA
and plasmid probes which we could think of was the stretches
of poly(A/T) in the "tails" of the cDNA plasmids and the
poly(T) region of the cDNA. We therefore tested a few of
these cross-reacting phages for hybridization in the presence
of poly(rA). The hybridization of Benton-Davis phage

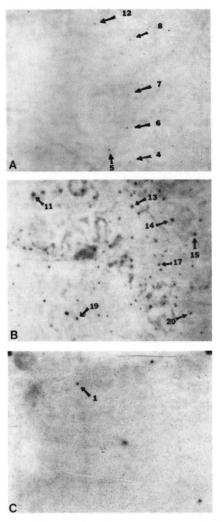

Fig. 1. Autoradiographs from approximately one quarter of three megaplates.

A. Megaplate 30 from a mouse shotgun screened with ^{32}P labeled mouse globin cDNA without poly(rA) in the hybridization mix. A plaque under the signal labeled 5 yielded Mm30.5.

B. Megaplate 39 from a human shotgun screened for histone clones with a ^{32}P labeled plasmid containing the sea urchin histone genes H2B, H4 and H1.

C. Megaplate 51 from a human shotgun screened with ^{32}P labeled human globin cDNA with poly(rA) in the hybridization mix. A plaque under the signal labeled 1 yielded Hs51.1.

replicas, and of Southern transfers of DNA electrophoresis
gels, was inhibited by poly(rA). The hybridization could
also be greatly decreased by increasing the stringency of
hybridization. It appears very likely that we had isolated
a whole series of clones of A/T-rich DNA - possibly simple
sequence satellite DNA.

The problem of simple sequence DNA clones is not
confined to poly(A/T) regions. We also encountered what
appears to be a similar situation when seeking histone clones
in the mouse and human shotguns. We were using as a probe
a plasmid containing sea urchin histone genes kindly
supplied by Drs. A.C.Y. Chang and Stanley N. Cohen. The
particular plasmid is pSp202 and contains the sea urchin
genes for histones H2B, H4 and H1. When we screened megaplates with this probe we obtained an incredible number of
hybridization positive plaques - about 1 in every 1000 clone-
bearing phages. Figure 1B illustrates the autoradiograph
corresponding to a quarter of the human shotgun megaplate
No. 39. On further analysis we found that the cloned
fragments hybridized at high levels of stringency to a
portion of pSp202 that contains only the histone gene H1 and
some adjacent DNA sequences. Since histone H1 is the least
conserved of all the histones during evolution we became
very suspicious. We thought we might have isolated some
simple sequence G/C-rich DNA clones. However the hybrid
was only partially inhibitable with poly(rC). It now seems
much more likely that the clones contain simple sequence DNA
related to the stretch of poly(CT) recently described by
Sures, Lowry and Kedes (9) as occurring in the fragment of
sea urchin histone genes which we used as a probe.

My purpose in telling you all this is to warn you that
even well-defined probes made from cloned genomic DNA or
cloned cDNA can lead you into pitfalls. Of course we now
have a whole set of clones from humans and mice which
probably contain related simple sequence DNAs. They are
interesting in their own right - but they are not what we
set out to find. The unwanted isolation of this general
category of clone is probably best avoided by using probes
that are known not to contain any simple sequences.

Once we had learned how to avoid isolating A/T-rich
clones, by including poly(rA) in our hybridization mixtures,
we had fewer difficulties. In the series of megaplates used
to isolate the human fetal globin gene fragment we used
9 megaplates to screen approximately 3 X 10^6 phages, of which
1 X 10^6 had inserts. Five megaplates were tested without
poly(rA). 107 candidates were obtained which hybridized to
globin cDNA, but none gave hybridization when retested in

the presence of poly(rA). The other four megaplates were
screened with poly(rA) in the hybridization mix. Ten candidates were obtained. One (labeled 1 in Figure 1C, which
illustrates a quarter of megaplate 51) gave us the clone
Hs (Homo sapiens) 51.1 which we will see contains most of
the structural gene for the fetal $^A\gamma$ globin chain.

Before going on to describe more about this clone, I
should comment on the isolation of single phages from the
large megaplates. The large size (a standard sized medical
X-ray film) of the autoradiograph and of the megaplates
(cafeteria trays) caused us some problems. The problems are
of aligning the autoradiographs with the megaplates while
maintaining sterility and at the same time taking care of the
few percent shrinkage of the nitrocellulose replicas which
occurs during processing.

The first part of the solution to the problems is to use
alignment marks that are fairly closely spaced. At the time
the nitrocellulose filter is on the megaplate we punch a grid
of squares with approximately 3" sides, using a syringe
needle dipped in India ink. At the time that the autoradiograph is to be made from the processed and dried nitrocellulose filter we further mark the filter with colored radioactive ink. After the X-ray film is developed, the radio-
actve marks enable the non-radioactive grid pattern of
syringe needle marks to be transferred to the film. Dr.
Philip W. Tucker devised the solution to the final problem –
aligning the autoradiographic signals with the plaques without violating sterility. Dr. Tucker had done research at an
earlier date in X-ray crystallography. He was therefore
familiar with the Richards box used to construct 3-dimension-
al atomic models from a stack of electron density maps. He
devised what we call the "Tucker Box" to solve our sterility
problem. (Figure 2) The half-silvered mirror of the Tucker
Box allows the viewer to see images of the autoradiographic
signals in the same plane as the top agar of the megaplate.
By using an adjustable intensity yellow incadescent light to
illuminate the film, the images can readily be seen. No
parallax difficulties occur and the grid marks enable very
precise alignment over any part of the plate. Picks can then
be made of the candidate phages. Since the plaques on the
megaplates are frequently touching, the picks are not of a
single phage plaque. However it is a simple matter to
separate the phages giving the hybridization signal from the
few other contaminating phages. Usually two rounds of plat-
ing, filter imprinting, hybridization, autoradiography and
picking are sufficient, but the process is always repeated
until all plaques are hybridization positive.

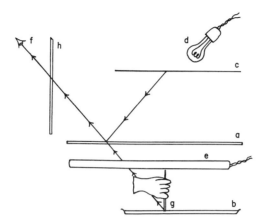

Fig. 2. <u>Tucker Box</u>. A half-silvered mirror (a) is
placed midway between the top surface of the megaplate (b)
and the autoradiograph (c). Variable intensity incandescent
yellow light bulbs (d) and fluorescent tubes (e) illuminate
the film and plate in such a way that the observer (f) can
see the images of the autoradiographic spots in the same
plane as the phage plaques. The plaques can be picked
sterilely (g) from outside the hood (h) without any parallax
problems.

The two globin candidates Mm30.5 and Hs51.1 isolated
from the mouse and human shotgun experiments were first
characterized by testing the purified phage DNAs for the
globin probes. We use a convenient small scale DNA purifica-
tions for this sort of preliminary work, which we call
"minilysates". (See note 29 in reference 4 for details.)
This method enables a single person to screen 20 or more
phages a day for their restriction enzyme patterns.

Figure 3 shows the result of testing Eco RI digests of
DNA from minilysates of Mm30.5 and Hs51.1. The autoradiograph was obtained using ^{32}P labeled human globin cDNA.
[The autoradiograph incidentally illustrates that mouse
globin DNA can be detected by human globin cDNA. This is
true also in plaque hybridization, and in the reverse direction.] The autoradiograph shows that Mm30.5 contains a 4.7
kbp globin-related fragment, and that the human clone,
Hs51.1, contains a fragment of length 2.7 kbp. In other
experiments we have shown that Hs51.1 contains a second
Eco RI fragment of human DNA of length 465 bp. This second
fragment is not globin-related. The method of preparing
the shotguns was expected to yield an appreciable proportion
of phages containing two unrelated DNA fragments, so we are
not surprised at finding two fragments in 51.1

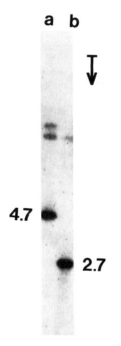

Fig. 3. Autoradiograph of a Southern transfer of an
agarose electrophoresis gel of Eco RI restriction enzyme
digests of DNA from minilysates of a) Mm30.5 and b) Hs51.1.
The probe was ^{32}P labeled human globin cDNA. The 4.7 and
2.7 kbp cloned fragments hybridizing to the globin cDNA are
indicated.

Let me now take up the question of the further characterization of Hs51.1, which is the clone of most interest to
this meeting on the transfer of information from genes to
proteins. Since the probe for isolating Hs51.1 was made
from a mixture of α and β globin mRNAs, we did not at first
know whether it was related to an α or β type globin. To
determine this we partially purified the 2.7 kbp fragment
of Hs51.1 and made a probe from it by nick translation. We
then hybridized this probe to Southern transfers of a gel on
which had been run restriction enzyme digests of DNA from
three phages: a) a phage carrying a clone of α-globin cDNA,
Hb117α; b) the phage, Charon 3AΔlac, used to construct the
clones of cDNA; c) a phage carrying a clone of mouse β-globin
cDNA, Hb4β. The probe made from Hs51.1 hybridized most
strongly to the band from phage c) labeled β in Figure 4.
This band is the only one which contains β-globin cDNA. [The
weaker positive reaction with other bands on the filter is
caused by the presence of phage DNA in the partially purified
probe which reacts with phage DNA in all the bands of the
digests.] This experiment told us that Hs51.1 contains DNA
related to a β-type globin.

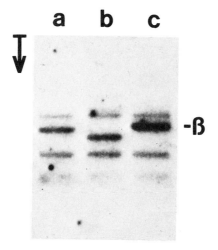

Fig. 4. Autoradiographs of filter transfers of an
agarose electrophoresis gel hybridized to a probe made by
nick-translating the partially purified 2.7 kbp insert from
Hs51.1. The gel contained Hpa I digests of DNA from a) a
mouse α globin cDNA phage, Hb117α, b) the parent phage,
Charon 3AΔlac, c) a mouse β globin cDNA phage, Hb4β. The
probe from Hs51.1 hybridizes preferentially to the band
labeled β which is the only one containing β globin cDNA.

```
DNA  5' ATCTTATTGTCTCCTTTCATTCAACAGCTCCTGGCAAATGTGCTGGTGACCGTTTGGCAATCCATTTCGGCAAAGAATTC 3'
Translation SerTyrCysLeuLeuSerSerGlnGlnLeuLeuLeuGlyAsnValLeuValThrValLeuAlaIleHisPheGlyLysGluPhe
                                                                               105                  122
G_γ, A_γ    LeuHisValAspProGluAsnPheLysLeuLeuGlyAsnValLeuValThrValLeuAlaIleHisPheGlyLysGluPhe
δ                                                                  Arg              Cys      ArgAsn
β                                                                  Arg              Cys      His
```

```
DNA  5' GGATCCTGAGAACTTCAAGGTGAGTCAGGAGATGTTTCAGCCCTGT 3'
Translation AspProGluAsnPheLysSerGlnGluMetPheGlnProTyr  (ValSerGlnGluMetPheGlnProTyr)
                              100                104
G_γ, A_γ    AspProGluAsnPheLysLeuLeuGlyAsnValLeuValThrVal
δ, β                                  Arg              Cys
```

```
DNA  5' ACCCTGGAAGGTAGGCTCTGCTGACCAGGACAAGGGAG 3'
Translation ThrLeuGlyArg * AlaLeuValThrArgThrArgGlu
                     27   30
G_γ, A_γ    ThrLeuGlyArgLeuLeuValValValTyrProTrpThrGln
δ, β        Ala
```

```
DNA  5' GCTGTCAACTGTTCTTGTCACCACAGGCTCTTGGTTGTC 3'
Translation AlaValAlaAsnCysSerCysHisHisArgLeuLeuValVal
                                          31       34
G_γ, A_γ    AspAlaGlyGlyGluThrLeuGlyArgLeuLeuValVal
δ           Ala           Ala
β           GluVal
```

Fig. 5. Comparisons of a translation of the DNA sequence of Hs51.1 with the amino acid sequences of the human β-type globins: G_γ, A_γ, δ and β. Normal underlined type shows regions of agreement of the DNA translation with the amino acid sequence of G_γ and A_γ globins. Italics show the regions of disagreement where intervening sequences occur. The asterisk is a termination codon.

We then went on to determine some DNA sequence from the
2.7 kbp fragment, using the Maxam-Gilbert procedure (10). The
top quarter of Figure 5 shows DNA sequence from one of the
Eco RI ends of the 2.7 kbp fragment and compares a translation
of this sequence with the known amino acid sequences of the
human β-type globins: $^{G}\gamma$, $^{A}\gamma$, δ and β. Clearly 51.1 contains
a portion of one of the fetal γ globin genes, $^{G}\gamma$ or $^{A}\gamma$. [I
shall present data later showing that it is $^{A}\gamma$.] Equally
clearly the genomic DNA before the amino acid residue leucine
at position 105 in the protein sequence does not agree with
any of the β-type globins. Having in mind Dr. Leder's mouse
globin data this will be no surprise to you. It appears that
the human $^{A}\gamma$ gene contains an intervening sequence between
residues 104 and 105, which is at the same position as in
the adult mouse globin genes.

The second quarter of Figure 5 shows DNA sequence data
obtained from the single Bam HI restriction enzyme site in
Hs51.1, and compares it with the protein sequence of the
human β-type globins. Once again only the amino acid
sequences of the fetal globins are in agreement with the
translated DNA sequence and, in this case, the sequence does
not agree <u>after</u> residue 104. This provides confirmation for
the location of the intervening sequence.

The lower half of Figure 5 shows DNA sequence data
obtained from an MboII restriction enzyme site in Hs51.1.
Again the translated amino acid sequences agree with the fetal
globin sequences. These sequences establish that there is a
second intervening sequence in the $^{A}\gamma$ fetal globin gene
beginning after amino acid residue 30. This position is at
the same position as the small intervening sequence in the
adult β globin genes from mouse and rabbit (7, 11).

These data consequently establish that this human fetal
globin gene has intervening sequences at the same positions
as those in the adult globin genes.

The length of the smaller intervening sequence IVS I,
between codons 30 and 31, is indicated by further DNA
sequence data (not given here). It is probably 118 base pairs,
but we still need to repeat some of the sequences to be sure
that we have not missed any bases before we can say that this
length is established.

The length of IVS 2, between codons 104 and 105, has
been determined in two ways - by electron micrographs of
heteroduplexes between Hs51.1 and the mouse cDNA clone Hb4β,
and by restriction enzyme mapping. The heteroduplex measurements (see reference 5) provide an estimate of IVS 2 of
849 ± 57 base pairs. The restriction enzyme data is presented

in Figure 6 which summarizes the relevant data from Hs51.1.
The lengths of the two pieces of DNA formed from the 2.7 kbp
fragment after digestion with Bam HI are 1740 $\pm$ 50 base pairs
and 960 $\pm$ 15 base pairs. The smaller of the two starts at
the Bam HI site formed by the codons for amino acid residues
98 - 100 and ends at the Eco RI site between codons 121 and
122. The length of this smaller piece permits a second
estimate of the length of IVS 2 as 894 + 15 base pairs [that
is 960 $\pm$ 15 - 3 (121 - 99)]. This agrees well with the
estimate, 849 $\pm$ 57 base pairs, from the heteroduplex.

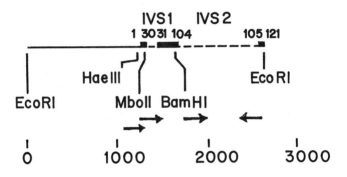

Fig. 6. A summary of restriction enzyme data used in the
structural analysis of the 2.7 kbp $^A\gamma$ globin gene fragment
from Hs51.1. The scale at the bottom is in base pairs.
IVS 1 and IVS 2 are the two intervening sequences (shown by
a broken line) which occur in the coding DNA (shown by heavy
black bars). The numbers above the black bars show the positions of IVS 1 and IVS 2. The arrows show the locations of
the DNA sequences presented in Figures 5 and 7.

There remains the question of identifying Hs51.1 as $^G\gamma$
or $^A\gamma$. This identification is now possible from the data
presented by Dr. Flavell yesterday (1). He showed that the
$^G\gamma$ globin gene yielded a 7.6kbp Eco RI fragment, while the
$^A\gamma$ globin gene yields a 2.7 kbp fragment. Hs51.1 obviously
is derived from the $^A\gamma$ gene.

Recently Chang et al. (12) have published the nucleotide
sequences of the 5' untranslated region of mixed $^G\gamma$ and $^A\gamma$
globin mRNA. A single sequence was obtained except for one
position 39 bases before the codon for the N-terminal amino

acid, where both G and C were detected. The authors suggested that this might represent a difference between the two
fetal γ globin genes. We find that our ^{A}γ gene has a G at
this position, (Figure 7). It will be interesting to see if
^{G}γ proves to have the expected C.

Thus we have shown that a single copy gene, for the fetal
globin chain ^{A}γ, can be isolated from an unfractionated
complete Eco RI digest of human DNA. We have also shown that
this fetal gene contains two intervening sequences, IVS 1 and
IVS 2, at the same positions as they occur in the adult globin
genes of man (2), mouse (7), and the rabbit (11).

What does this tell us about the differential control
of fetal versus adult genes? Well, it allows us unambiguously
to eliminate one very real possibility. We now know that the
differential control cannot be the consequence of the lack of
intervening sequences in fetal globin genes, or of the occurrence of IVS I and IVS 2 at sites different from those of the
adult genes.

Detailed comparison of fetal versus adult globin DNA
sequences to search for possible differential controls must
await the completion of sequences from both types of gene in
the same species and comparable pairs in other species. At
the present time, the only other human globin DNA sequence
data available to us is that in press in the paper by Lawn
et al. (2). These authors present 88 base pairs of the
sequences of the adult β and δ globin genes starting at 28
base pairs before the 3' end of IVS 2 and going to the Eco RI
site at codons 121 and 122. When we compare these sequences
with our data from the 3' end of IVS 2 we find that the human
fetal IVS 2 sequence differs from that of human adult β in
17/28 positions and from human adult δ in 15/28. However,
the differences between the human adult β and δ are comparable, being 19/28 positions.

Dr. Walter M. Fitch kindly compared the human globin
sequences for us, and comparable sequences from the adult
mouse β and rabbit β, by more sophisticated means. He used
computer programs designed to detect sequence homologies and
to detect possible secondary structures that might be formed
from the sequences. The sequence comparisons showed that the
3' ends of IVS 2 in all five globin genes are related to each
other, but no sequences appear to be "adult-specific".
Similarly the comparisons of potential secondary structures
in the 3' end of IVS 2 and in the neighboring coding region
showed no indications of specific structures related to
adult versus fetal globins. Clearly more data are needed
before any fetal specific features can be unambiguously
recognized.

A_γ

ACACTCGCTTCTGGAACGTCTGAGGTTATCAATAAGCTCCTAGTCCAGACGCCATGGTCATTTC

                    -50   -40   -30   -20   -10   -1

                                                                MetGlyHisPhe

γ mRNAs (m⁷Gpp)ACACUCGCUUCUGGAAC(G/C)UCUGAGGUUAUCAAUAAGCUCCUAGUCCAGACGCCAUGGGUCAUUUC
                                 ↑

Fig. 7. A comparison of the nucleotide sequence of the A_γ gene with that of the 5' untranslated region of mixed G_γ and A_γ-globin mRNAs (12). The arrow points to a possible difference between the two fetal globin genes (see text).

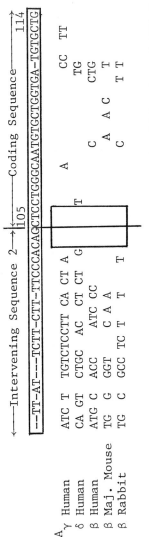

←——Intervening Sequence 2——→ | ←——Coding Sequence——→ 114
                              105

--TT-AT---TCTT-CTT-TTCCCACAG CTCCTGGGCAATGTGCTGGTGA-TGTGCTG

A_γ Human	ATC T	TGTCTCCTT	CA CT A	A	CC	TT
δ Human	CA GT	CTGC	AC CT CT G		TG	
β Human	ATG C	ACC	ATC CC	C	CTG	
β Maj. Mouse	TG G	GGT	C A A	A A C	T	
β Rabbit	TG C	GCC TC T	T T	C	T	T

Fig. 8. A comparison of the base sequences at the 3' border of IVS 2 in the five globin genes so far studied (references 5, 2, 2, 7, 11). The boxed sequence is that common to at least three of the genes. Bases differing from this sequence are given. The open box shows a region of conservation of six consecutive base pairs in all five genes.

There is one remarkably well conserved region at the 3'
end of IVS 2 which deserves special comment (and permits me
to correct an error in our paper, reference 5, caused by a
problem in transmitting DNA sequence data by telephone).
Figure 8 shows that three non-coding and three coding base
pairs at the 3' border of IVS 2 are completely conserved in
all five globin genes so far studied. There is a marked
although incomplete conservation of the base sequences in the
adjacent coding region; the degree of conservation exceeds
that necessary to conserve the amino acid sequence. In the
intervening sequence itself the degree of sequence conservation is less. If a fixed base sequence is required at the
border of IVS 2 for mRNA processing, it cannot exceed six base
pairs (not four, as stated in reference 5).

In summary, we have shown that two intervening sequences
occur in the human $^A\gamma$ fetal globin chain coding region at the
same positions as they occur in adult globin genes. We can
make the unequivocal statement that the differential expression of the $^A\gamma$ fetal globin gene during development
cannot be the result of the fetal genes lacking these intervening sequences or of having them at different positions.
We must wait for more data before being able to tell if the
specific base sequences of the intervening sequences or of
the coding regions are important in this context.

ACKNOWLEDGMENTS

This is paper number 2322 from the Laboratory of Genetics,
University of Wisconsin-Madison.

REFERENCES

1. Flavell, R. A., Grosveld, G. C., Grosveld, F. G.,
   De Boer, E., and Kooter, J. M., This Volume. (1979).
2. Lawn, R. M., Fritsch, E. F., Parker, R. C., Blake, G.,
   and Maniatis, T., *Cell*, In Press. (1978).
3. Blattner, F. R., Williams, B. G., Blechl, A. E.,
   Denniston-Thompson, K., Faber, H. E., Furlong, L. A.,
   Grunwald, D. J., Kiefer, D. O., Moore, D. D., Schumm,J.W.,
   Sheldon, E. L., and Smithies, O., *Science 196*, 161 (1977).
4. Blattner, F. R., Blechl, A. E., Denniston-Thompson, K.,
   Faber, H. E., Richards, J. E., Slightom, J. L.,
   Tucker, P. W., and Smithies, O., *Science 202*, 1279 (1978).
5. Smithies, O., Blechl, A. E., Denniston-Thompson, K.,
   Newell, N., Richards, J. E., Slightom, J. L., Tucker,P.W.,
   and Blattner, F. R., *Science 196,*1284 (1977).

6. Benton, W. D. and Davis, R. W., *Science 196*, 180 (1977).
7. Leder, P., Konkel, D., Nishioka, Y., Hamer, D., Leder,A., and Seidman, J., This Volume. (1979).
8. Brack, C., Hirama, M., Lenhard-Schuller, R., and Tonegawa, S., *Cell 15*, 1 (1978).
9. Sures, I., Loury, J., and Kedes, L. H., *Cell 15*, 1033 (1978).
10. Maxam, A. M. and Gilbert W., *Proc. Nat. Acad. Sci. USA 74*, 560 (1977).
11. Van den Berg, J., Van Ooyen, A., Mantei, N., Chambock,A., Grosveld, G., Flavell, R. A., and Weissmann, C., *Nature 276*, 37 (1978).
12. Chang, J. C., Poon, R., Neumann, K. H. and Kan, Y. W., *Nucleic Acids Research 5*, 3515 (1978).

DISCUSSION

P. LEDER: I would like to offer one comment and ask two questions. The comment is that in a more extensive comparison of the coding sequences that have been published for globin mRNAs and the sequence that we know for the mouse gene, we have also seen this conservation of sequence in the coding segment of the gene that borders the intervening sequence. This tends to break down as you reach portions of the coding sequence that are more distant from the intervening sequence. This suggests that there is a cold spot surrounding the intervening sequence which preserves this structural gene sequence.

I wonder, in the next to the last slide, where you showed the sequence of the chromosomal gene of the capping structure, if you have additional sequence data. Beyond that last base on that slide is a box of approximately 10 nucleotides, a capping box or whatever you might want to call it, which has a preserved sequence among globin genes, a preservation I think that was referred to by Pierre Chambon in his ovalbumin sequence. In addition to that, about 30 nucleotides down, is another sequence that Hogness has found in histone, a presumed promotor site, TATAA sequence, or something of that sort, very much like a Pribinow box. Do the human fetal genes demonstrate this?

O. SMITHIES: We have looked at the sequence just in front of the capping region, and preliminary sequence data do indeed show conservation in that region. I do not remember if we have looked for the area further in front. Jerry, can you tell us whether or not you have seen the Pribinow box?

J. SLIGHTOM: I do not really think we have seen that Pribinow box, but I have not really looked very closely. As far as the capping structure goes, we have seen that.

O. SMITHIES: It does agree, does it not?

J. SLIGHTOM: Not quite as well as Phil Leder's capping structure, but it does agree about 7 out of 12 base pairs.

O. SMITHIES: Concerning your comment, Phil, I agree about the conservation of the sequence in the coding area after IVS 2; it is much more than is needed to preserve the amino acid sequence. But it is very difficult to say just why this region is conserved in the sense that one does not know whether or not it is related to the use of specific transfer RNAs. What do you feel about that?

P. LEDER: I certainly agree that it is impossible to be certain. It is probably a matter that has not occurred by chance, and I doubt that it is related to the selection of code words that have to be used throughout the message, presumably from the same population of tRNAs. I think it is likely to be related to the splicing reaction.

My last question was a technical one. In the million hybrid clones that you screened you found ten positives and told us about one at this talk. How many of those were true positives and is that about the incidence that you would have expected by chance?

O. SMITHIES: In part, these numbers are determined by the fact that we were learning how to do the experiments at the time, and I would not like to say whether or not we would do better if we were to use our current technique again in the same way. However, the numbers are not far from what one might expect to obtain by chance. One would expect a million fragments to yield about one clone of a given piece of DNA. But since we were looking for any of the globin genes (not one in particular) we are probably lower, perhaps by a factor of five, than one would have expected if we had been fully efficient.

M. PHILIPP: We have found that the structure surrounding this intron in mouse and human beta globin seems to show some self complementarity. I wonder if you have found a lower mutation rate on the other side of the intron as well as on the side you illustrated.

O. SMITHIES: No we have not yet looked a great deal on the other side. We have not had the data long enough to do all the analyses we would like to do.

P. CHAMBON: One comment, there is clearly no relationship between the phasing of the amino acid codons and the position of the splicing points. This is true for the ovalbumin exon-intron junctions, as well as for other genes.

THE PROCESSING OF MESSENGER RNA AND THE
DETERMINATION OF ITS RELATIVE ABUNDANCE

R. P. Perry[1]
U. Schibler[2]
O. Meyuhas

The Institute for Cancer Research
Fox Chase Center
Philadelphia, Pennsylvania

I. INTRODUCTION

The ubiquity of mRNA processing as a fundamental aspect
of gene expression in metazoan cells is now firmly established, and the existence of complex processing pathways is
rapidly being revealed in a variety of systems. The expan-
sion of such knowledge has intensified our interest in questions dealing with the regulation of mRNA production. In a
mammalian cell the number of copies of individual mRNA
species may vary by more than four orders of magnitude (1).
To what extent is this wide variation in abundance determined
by differences in rates of transcription, in processing
efficiency, or in mRNA stability? Such questions can be
approached by studying individual species of mRNA or by
studying the properties of mRNA populations as a whole. For
individual mRNA species metabolic parameters can be accurately determined; yet, there is no assurance of their generality without examining a large number of examples. On the
other hand, generality is an inherent quality of population
studies, although such measurements may not necessarily reveal the extent of deviation from the norm. Thus, the information gained from one type of study should complement
that of the other. In this paper we illustrate examples of
both approaches: first, a study of the production of

_____

[1]Supported by NSF grant PCM77-06539; Institutional NIH grants
CA06927, RR05539; and an appropriation from the Commonwealth
of Pennsylvania
[2]Present address: Institut Suisse de Recherches Experimen-
tales sur le Cancer, Lausanne, Switzerland

immunoglobulin mRNAs in mouse myeloma cells, which indicates
that the high cellular content of these mRNAs is the result
of an exceptionally rapid transcription rate, efficient
processing and relatively high metabolic stability (2); and
second, studies of the total mRNA and hnRNA of mouse L cells,
which point to a general correlation between size, stability
and abundancy of mRNA (3) and relatively efficient processing
for the prevalent species of mRNA (4-6).

II. THE PRODUCTION OF IMMUNOGLOBULIN mRNAs

As described in an earlier presentation by P. Leder,
the DNA segments comprising the immunoglobulin (Ig) genes
undergo rearrangement during plasma cell differentiation to
form a transcriptional unit (Fig. 1) in which elements coding
for the variable and constant regions and for the signal peptide are separated by intervening sequences or introns (7-9).

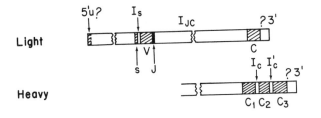

Expressed Immunoglobulin Genes

FIGURE 1.   Organization of expressed immunoglobulin
genes. 5'u:5' untranslated sequences of the mRNA; s:sequence
coding for NH_2-terminal portion of signal peptide; V:sequences
coding for variable region; J:joining segment; C:sequences
coding for constant region. The various introns are designated I_s, I_{JC}, I_c, etc.. The locations of the 5' untranslated portions of the mRNA and the sites of transcriptional
initiation and termination are still unknown.

In the case of the heavy chain gene there are also two small
introns that apparently demarcate the three constant region
domains (10). Moreover, the untranslated portions of the Ig
mRNAs may also be interrupted by introns. Given such complex gene arrangements it is reasonable to expect that the Ig
mRNAs would be produced by sequential processing of large
RNA transcripts (vide infra).

In MPC-11 myeloma cells the γ_{2b} heavy chain and κ light
chain mRNAs are sufficiently abundant so that they are
readily discerned as discrete peaks in a profile of poly A^+
mRNA (Fig. 2). The 1.8 kb H-mRNA and the 1.2 kb L-mRNA
components each comprise about 15% of the poly A^+mRNA in the

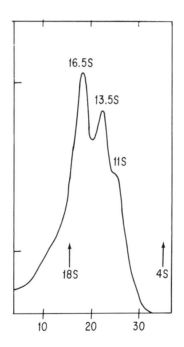

FIGURE 2. Sucrose gradient profile of membrane-bound
polyribosomal poly A^+RNA from MPC-11 cells. The positions
of 18S RNA and 4S RNA markers, run in a parallel gradient,
are indicated by arrows. The heavy chain mRNA (H-mRNA),
light chain mRNA (L-mRNA) and mRNA coding for the light chain
constant region fragment (L_f-mRNA) sediment at approximately 16.5S, 13.5S and 11S, respectively.

membrane bound polyribosomes. In addition these cells produce an mRNA (L$_f$-mRNA) of about 0.8 kb that codes for the
constant region portion of the L chain. Preparations of H-
and L-mRNA that are 55-80% pure as judged by T1 fingerprint
analysis and the relative homogeneity of 5'-terminal cap
structures (11) were obtained by successive sucrose gradient
fractionation or by a combination of sucrose gradient separation and formamide-polyacrylamide gel electrophoresis.
The cap-containing 5'-terminal sequences of the MPC-11 H-
and L-mRNAs are m^{7}Gppp(m^6)A^{m}pA^{m}pCp and m^{7}GpppG^{m}pA^{m}pAp, respectively. Similar to all other abundant mRNAs thus far
examined, the initial 5' nucleotide is a purine. As is
characteristic of other stable mRNAs in higher eukaryotes,
the penultimate nucleotide is methylated in a high proportion of the molecules. The H- and L-mRNAs have a mean lifetime of at least one cell generation (~24 h), as indicated
by labeling experiments (11), in which, after lengthy chase
intervals, the Ig-mRNAs exhibited relatively constant specific
activities and increased prominence with respect to other
mRNA species.

The content of the Ig-mRNA sequences in MPC-11 cells
was estimated by hybridization kinetic analysis of total
poly A$^+$mRNA or hnRNA, using ^{3}H labeled cDNA probes made from
the highly purified preparations of mRNA (2,12). For comparison, ^{32}P labeled cDNA representing the sequences of total
poly A$^+$mRNA was included in the hybridization reactions
(Fig. 3). From the Rot$_{\frac{1}{2}}$ values we estimate that there are
about 30 and 40 thousand copies, respectively, of the H and
L-mRNAs in the cytoplasm of each myeloma cell. In contrast,
a great many other mRNA species in these cells are present
in only a few copies. The nuclear content of Ig-mRNA sequences, estimated by hybridization kinetic analysis and by
saturation analysis with excess cloned cDNA (2), is equivalent to only about 100-150 molecules per cell. Although
the Ig sequences are among the most abundant mRNA species
in the nucleus, their numerical superiority over other mRNA
sequences is not nearly as great as it is in the cytoplasm
(Fig. 3). This is attributable to the high cytoplasmic
stability of the Ig-mRNAs relative to other mRNA species
and possibly also to their being more rapidly and efficiently
processed (vide infra).

The cDNAs representing the H- and L-mRNAs were purified
by recombinant DNA cloning, and used to identify the nuclear
molecules containing Ig sequences (2). When the bulk nuclear
RNA was fractionated on methylmercury-agarose gels, covalently affixed to cellulose paper (13), and probed with

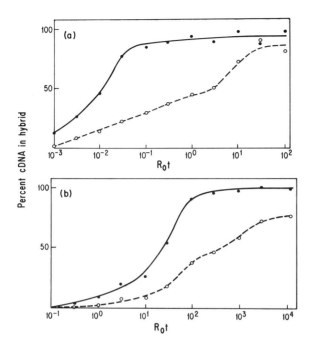

FIGURE 3. The relative abundance of L-mRNA sequences in poly A+mRNA and nuclear RNA. Purified (^{3}H) L-cDNA (●) and (^{32}P) cDNA made from poly A+mRNA (o) were simultaneously annealed in a solution containing 0.6 M NaCl with an excess of (a) poly A+mRNA (0.2-80 μg/ml) or (b) total nuclear RNA (2-4000 μg/ml). The hybrids were assayed after S1 nuclease digestion and plotted as percent of input cpm vs. Rot (in- itial RNA concentration in moles/liter x incubation time in sec.). The Rot values have been multiplied by a factor of five to correct to standard salt concentration. The data have been corrected for self-annealing (2).

nick-translated probes, we observe discrete bands corres- ponding to 11 kb, 3.7 kb and 1.8 kb for H-mRNA sequences and 5.4 kb, 3.3 kb, 1.2 kb and 0.8 kb for L-mRNA sequences (Fig. 4). A much more intense signal was discerned when the poly A+ nuclear RNA was used, indicating that most of the precursors contain poly A. In order to demonstrate that the large nuclear components are in fact precursors of the Ig

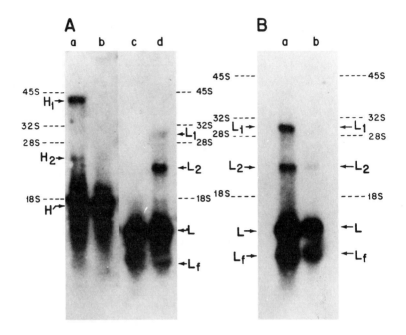

FIGURE 4. Analysis of the nuclear RNA molecules containing H- and L-mRNA sequences. Nuclear and cytoplasmic
RNAs were separated on 1% agarose slab gels containing 7.5
mM methylmercury hydroxide, transferred to DBM paper and hybridized with Ig-recombinant DNA probes labeled with ^{32}P by
nick translation. The paper was washed, dried and exposed
at -70°C to x-ray film backed by an intensifier screen. The
positions of pre-rRNA and rRNA markers [45S(13.7 kb), 32S
(6.4 kb), 28S(5.1 kb) and 18S(2.0 kb)] were determined on
parallel gels after staining with 2 μg/ml ethidium bromide
in 0.5 M ammonium acetate. The specific activity of the
probe and the autoradiographic exposure are different in
series A and B. (A) Tracks a and b are, respectively,
nuclear RNA (35 μg) and cytoplasmic RNA (5 μg) hybridized
with the H probe [pγ$_{2b}$(11)^7]. Tracks c and d are, respectively, cytoplasmic RNA (5 μg) and nuclear RNA (35 μg) hybridized with the L probe [pk(11)^{24}]. (B) Tracks a and b
are, respectively, poly A$^+$ nuclear RNA (12 μg) and total
nuclear RNA (50 μg) hybridized with pk(11)^{24}.

mRNAs we labeled the MPC-11 cells for various intervals with
^{3}H uridine and detected the labeled RNA species containing
Ig sequences by hybridization to excess cloned cDNA that was
immobilized on nitrocellulose filters (2). The 11 kb component appears to be the primary transcript of the H-mRNA

since it is the largest molecule labeled at very short
pulses (Fig. 5). As the labeling time increases, we observe

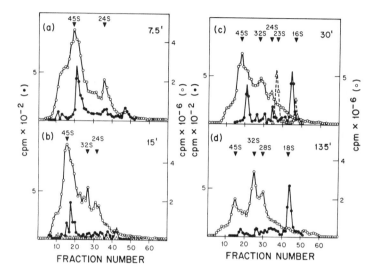

FIGURE 5. Kinetics of labeling of the nuclear molecules containing H-mRNA sequences. For each of the indicated
time periods 5×10^8 cells were concentrated by centrifugation and resuspension in prewarmed medium, and labeled with
10 mC of ^{3}H uridine. Nuclear RNA was isolated and 100-200 μg
portions electrophoresed in cylindrical methyl mercury-agarœe
gels. The gels were sliced into 1 mm fractions, which were
dissolved and hybridized to 5 μg of pγ2b(11)[7] DNA (●). A
1/50 aliquot of each sample was counted for analysis of the
total nuclear RNA (o). In the experiment of panel (b) an
additional set of filters containing 5 μg of the plasmid
vector (pMB9) was also included in the hybridization incubation (△). For size markers 50 μg portions of nuclear RNA
containing 45S and 32S pre-rRNA and 28S and 18S rRNA were
run on parallel gels and visualized by ethidium bromide
staining. In the experiments of panels (c) and (d) ^{32}P
labeled 23S and 16S rRNA from E. coli (--△--) was included
with the nuclear RNA for additional size markers. (a) cells
concentrated 15-fold and labeled for 7.5 min; (b) cells concentrated 10-fold and labeled for 15 min; (c) cells concentrated 10-fold and labeled for 30 min; (d) cells concentrated
2.5-fold and labeled for 135 min.

the 3.7 kb intermediate component and eventually the mature
1.8 kb H-mRNA. The patterns for the L-mRNA precursors are
somewhat more complicated because of the presence of both
L-mRNA and L_f-mRNA related molecules (Fig. 6). Most likely
the 5.4 kb component is the primary precursor of the L-mRNA
and the 3.3 kb component, the initial precursor of the L_f-
mRNA. A pair of intermediate size components was also reproducibly observed. Transfer of label from the large to
small Ig-components could be demonstrated by suitable pulse-
chase experiments (2), thus establishing unequivocally that
the complex Ig genes are read into large transcripts which
are subsequently cut and spliced into the mRNA components.
It was further observed that all of the nuclear Ig components
are polyadenylated to the same extent even at brief labeling
times. This is clear from a comparison of the relative
amounts of the large and mature Ig-components in total 30
min. labeled nuclear RNA (Figs. 5c,6c) and polyadenylated
nuclear RNA (Fig. 7). Thus polyadenylation must occur prior
to splicing of most Ig-mRNA precursors.

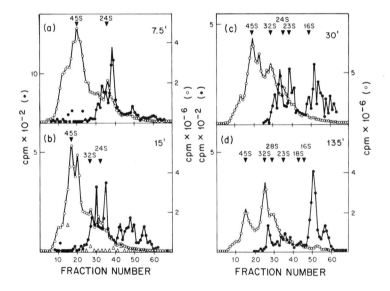

FIGURE 6. Kinetics of labeling of the nuclear molecules
containing L-mRNA sequences. The experiment is the same as
that described in Fig. 5 except that hybridization was to
filter bound pk(11)[24] DNA.

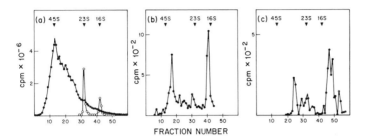

FIGURE 7.   Analyses of poly A$^+$ nuclear RNA. A portion
of the 30 min. labeled nuclear RNA shown in Figs. 5c and 6c
was applied to an oligo dT cellulose column in 0.15 M NaCl.
The bound fraction was run on a methylmercury-agarose gel,
and analyzed as described in Figs. 5 and 6. (a) total
poly A$^+$hnRNA (•), (^{32}P) E. coli rRNA size markers (o); (b)
RNA hybridized to pγ2b(11)[7]; (c) RNA hybridized to pk(11)[24].

   In similar studies Wall and colleagues (14,15) reported
that the primary precursor to the κ-light chain mRNA of
MOPC-21 cells is about 10 kb, almost twice the size of the
MPC-11 precursor. Their analyses suggest that processing
proceeds by the removal of a large segment on the 5' half of
the precursor (possibly splicing in part of the 5' untranslated portion of the mRNA) to produce an intermediate containing the J-C intron, which would subsequently be spliced
out in a later processing step. The substantial difference
in size of the presumptive primary transcripts of the κ-
mRNA genes in MOPC-21 and MPC-11 cells is not necessarily
unexpected because the variable elements of these genes are
evolutionarily very distantly related. One might then ask
whether transcription units, corresponding to more closely
related V_k genes, are more similar in size. To investigate
this point we examined the mRNA precursors in a series of
myeloma cells belonging to the V_k21 group. The members of
this group of related light chains have been divided into
subgroups that appear to be derived from the combination of

about six different V gene elements and several J elements
(16). Within the V_k21 group we have observed some tumors
with presumptive primary transcripts as large as 11 kb and
others with no detectable components larger than 6.5 kb.
Intermediate-size components ranging from 4.5 to 5.8 kb are
also observed. Thus the arrangement of the transcriptional
unit, and possibly also the mode of processing, may vary
considerably, even among closely related κ light chain genes.

The fact that an MPC-11 cell is able to produce 30 to
40,000 Ig-mRNA molecules (or more, given some mRNA turnover)
within a cell generation of about a day means that at least
20 molecules per min. are transcribed from the Ig gene.
Even at maximum estimated rates of RNA chain elongation it
would take several minutes to transcribe an RNA of 11 kb.
It follows, therefore, that such Ig genes are simultaneously
transcribed by at least 50 polymerase molecules. Thus,
highly efficient transcription as well as high stability
contributes to the high abundance of Ig-mRNAs. Given a
transcription rate of at least 20 molecules/min., the small
pool of nuclear precursors (100-150 molecules) indicates that
processing rates are rapid, e.g., a halftime of about 5 min.
for the entire processing pathway. Since the primary transcripts are present at about 1/10 the concentration of total
Ig sequence (cf. Figs. 5d,6d), their processing halftime
should be only about 0.5 min.. These processing rates are
faster than is apparent from the labeling studies, presumably because the latter include delays due to equilibration
of radioactive precursor pools and possibly also to metabolic perturbations induced by the labeling procedures. If
unprocessed molecules are continually vulnerable to degradative activity, the relatively rapid processing of the Ig-
mRNAs could conceivably insure a relatively high processing
efficiency.

III. STUDIES WITH TOTAL mRNA POPULATIONS

Studies of the 5' termini of hnRNA and mRNA in mouse L
cells have led to several conclusions about mRNA processing
and its possible role in the regulation of mRNA production.
For example, at least one quarter of the cap structures
($m^7GpppX^mp\cdots$) on the 5' ends of mRNA and hnRNA have a pyrimidine in the "X" position (5). In contrast, the nuclear
RNA components with recognizable characteristics of primary
transcripts, i.e., those bearing polyphosphate 5' termini
($pppXp\cdot\cdot$ or $ppXp\cdot\cdot$), exclusively have purines at the X
position (6). Thus, one might envisage that the pyrimidine-
terminated mRNAs are produced by a cleavage of their

precursor that results in the removal of one or more nucleotides from the 5' end. Curiously, all of the highly abundant mRNA species for which cap sequences have been determined
contain a purine at position X; in these cases the X nucleotide may indeed correspond to the site of transcriptional
initiation. It will be interesting to see whether pyrim-
idine-containing caps occur on any abundant mRNA species,
or whether they are confined solely to the so-called sparse
mRNAs.

The striking similarity of the cap-containing sequences
of hnRNA and mRNA (5) and the relatively high degree of conservation of the cap methyl groups in pulse-chase experiments
(4) indicated that caps are largely conserved in the processing of hnRNA to mRNA. The hnRNA molecules smaller than
about 16S were not included in these measurements because of
the presence in this region of cap-containing low moleoular
weight nuclear RNAs that are unrelated to mRNA. Taking this
uncertainty into account, we estimate that at least 2/3 of
the caps are conserved (4). Since about 50% of the hnRNA
molecules have capped 5' termini, and another 20% have polyphosphate 5' termini, which may eventually be capped (6),
it seems reasonable to conclude that a substantial fraction
(at least half) of the hnRNA molecules are processed into
mRNA. Given that hnRNA is on average about 3 to 4 times
larger than mRNA the conservation of nucleotide mass would
be about 15%. This is in reasonable agreement with estimates based on hybridization measurements (17). Although
the overall mRNA processing efficiency seems to be relatively
high in these rapidly growing cells, the efficiencies for
individual mRNA species might still vary widely. For example, an overall processing efficiency of 50% would be observed if about half of the mRNA consisted of species that
are processed at nearly 100% efficiency and half of species
that are processed at less than 10% efficiency. Such differences in processing efficiency are believed to contribute to
the changes in the mRNA population which occur in developing
sea urchin embryos (18) and in resting → growing cell transitions (19).

As demonstrated for several individual species of mRNA
including the immunoglobulin mRNAs discussed above, mRNA
stability is also a factor contributing to mRNA abundance
(20-22). Evidence for a general correlation between stability and high abundance was obtained in a series of experiments in which the species diversity and stability of different size classes of mRNA were compared (3). A measurement

of the Rot$\frac{1}{2}$ values for the hybridization of different size
fractions of L cell poly A+mRNA to their respective cDNAs
revealed striking differences (Fig. 8). Correcting for the
average molecular weight and relative mass of each size class
we estimated that the large, medium and small mRNA fractions,
respectively, consist of approximately 5000 species present
at about five copies each, 1000 species present at about 100
copies each and 300 species present at about 1000 copies each.
This distribution is illustrated schematically in Fig. 9.

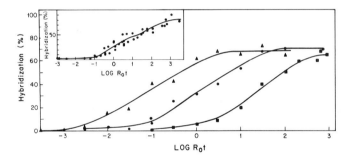

FIGURE 8. Kinetics of hybridization of cDNAs, transcribed from poly A+mRNA of different size classes, to their
respective template RNAs. L cell poly A+mRNA was sedimented through a sucrose gradient and fractionated into three
size classes: small (<15S), medium (15S-24S) and large (>24S).
Each mRNA fraction was used as template for the synthesis of
labeled cDNA and then hybridized in excess to the corresponding cDNA probe. The percent of hybridized cDNA was determined by S1 nuclease digestion. ▲ - ▲, small cDNA with
small poly A+mRNA (0.19-190 μg/ml); ● - ●, medium cDNA with
medium poly A+mRNA (0.58-580 μg/ml); ■ - ■, large cDNA with
large poly A+mRNA (0.5-500 μg/ml). Appropriate background
corrections were subtracted for each point (3). The inset
shows similar data for unfractionated poly A+mRNA and its
corresponding cDNA.

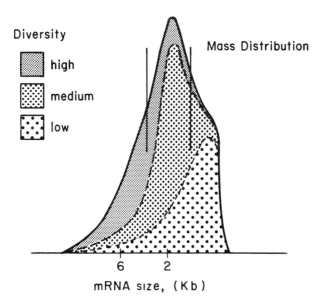

FIGURE 9. Schematic representation of the diversity of species comprising various size classes of the poly A^+mRNA of mouse L cells. The vertical lines indicate the fractionation into large, medium and small classes that was made for Fig. 8.

A comparison of the size distribution of pulse labeled mRNA, which emphasizes the rapidly turning over components, and mRNA from cells given a long term label and long term chase, which emphasizes the stable components, indicated that the stable mRNAs were on average considerably smaller than the mRNAs with shorter average lifetimes (Fig. 10). The fact that small mRNAs tend to be on the average both more abundant and more stable than large mRNAs suggests that stability and abundance might be directly correlated. This was indeed shown to be the case when excess cDNA prepared from steady state mRNA was simultaneously hybridized with predominately unstable and stable mRNAs (Fig. 11).

The foregoing considerations suggest that transcription rate, processing efficiency and mRNA stability all contribute to the determination of mRNA frequency distributions. In the case of very abundant mRNAs, like those coding for the immunoglobulins, transcription rate appears to be the major

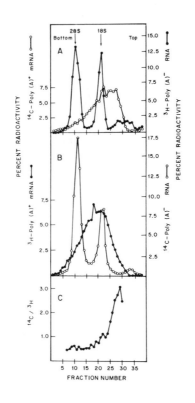

FIGURE 10. Size distributions of ^{3}H pulse-labeled and
[14]C stable poly A[+]mRNA. Mixtures of A: [14]C poly A[+]mRNA and
[3]H poly A[-]RNA or B: [3]H poly A[+]mRNA and ^{14}C poly A[-]RNA were
sedimented through 5-25% (w/w) sucrose gradients in aqueous
buffer containing 0.5% sodium dodecyl sulfate. The ^{14}C RNA
was derived from cells labeled for 24 h and chased for 18 h,
and the ^{3}H RNA from cells labeled for 1 h. Data plotted as
percent **total** radioactivity on gradient. C: percent ^{14}C/
percent ^{3}H plotted as a function of fraction number.

determinant. For sparse mRNAs, which are transcribed much
less frequently, processing efficiency and cytoplasmic turnover may play more crucial regulatory roles. These concepts are summarized for a hypothetical mRNA frequency distribution in Table I.

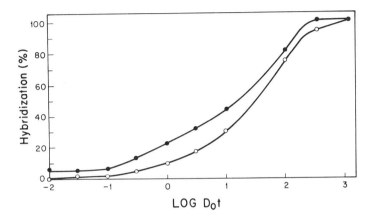

FIGURE 11. Kinetics of hybridization of excess cDNA
complementary to steady state poly A+mRNA to ^{14}C stable and
^3H pulse-labeled poly A+mRNA. ^{14}C poly A+mRNA ($\bullet$) and ^3H
poly A+mRNA (o) from cells labeled as in Fig. 10 were simultaneously hybridized to excess unlabeled cDNA made from
bulk poly A+mRNA using random oligodeoxynucleotide primers.
The percentage of mRNA hybridized was determined after RNAse
T1 and A digestion. The concentration of cDNA was 6-1250
μg/ml and the concentrations of ^{14}C poly A+mRNA plus ^3H
poly A+mRNA were 2.5-63 μg/ml (cDNA/RNA = 2.4-20). The percent hybridization values were corrected by subtracting for
the amount of ribonuclease resistant radioactivity found
after equivalent incubations in the absence of cDNA (3).

TABLE I. Determination of mRNA frequency distribution (hypothetical).

mRNA species	transcription rate (molecules/min.)	processing efficiency	mRNA lifetime (as multiple of cell generation time)	mRNA abundance (molecules/cell)
A	15	1	2	10^4
B	1.5	1	2	10^3
C	0.15	1	2	10^2
D	0.15	1/3	1/4	10
E	0.15	1/10	1/14	1

The abundance, A, is equivalent to $k_O \cdot E/(k_g + k_t)$ where k_O is the transcription rate in molecules/min., E is the efficiency of processing (number of mRNA molecules produced/number of pre-mRNA molecules synthesized) and k_g and k_t are, respectively, the exponential growth constant and the rate constant for mRNA turnover in min^{-1}. Given that $k_g = 1/\tau$, where τ is the mean generation time, and $k_t = 1/f\tau$ where f is the mRNA mean lifetime expressed as a multiple of the generation time, then $A = k_O E \tau/(1 + 1/f)$. The above values were calculated for a mean generation time of 1000 min.

REFERENCES

1. Lewin, B. Cell 4, 77 (1975).
2. Schibler, U., Marcu, K. B., and Perry, R. P. Cell 15, 1495 (1978).
3. Meyuhas, O., and Perry, R. P. Cell 16, 139 (1979).
4. Perry, R. P , and Kelley, D. E. Cell 8, 433 (1976).
5. Schibler, U., Kelley, D. E., and Perry, R. P. J. Mol. Biol. 115, 695 (1977).
6. Schibler, U., and Perry, R. P Nucleic Acids Res. 4, 4133 (1977).
7. Bernard, O., Hozumi, N., and Tonegawa, S Cell 15, 1133 (1978).
8. Rabbitts, T. H., and Forster, A. Cell 13, 319 (1978).
9. Seidman, J. G , and Leder, Philip Nature 276, 790 (1978).
10. Early, P. M , Davis, M. M., Kaback, D. B., Davidson, N , and Hood, L. Proc. Natl Acad. Sci USA (in press).
11. Marcu, K B., Schibler, U., and Perry, R P. J. Mol Biol. 120, 381 (1978).
12. Marcu, K. B., Valbuena, O., and Perry, R. P. Biochemistry 17, 1723 (1978).
13. Alwine, J C., Kemp, D. J., and Stark, G. R. Proc Natl. Acad. Sci USA 74, 5350 (1977).
14. Gilmore-Hebert, M., and Wall, R. Proc Natl Acad Sci. USA 75, 342 (1978).
15. Gilmore-Hebert, M , Hercules, K., Komaromy, M., and Wall, R. Proc Natl Acad. Sci. USA 75, 6044 (1978).
16. Weigert, M., Gatmaitan, L., Loh, E., Schilling, J., and Hood, L. Nature 276, 785 (1978).
17. Hames, B. D., and Perry, R P. J. Mol. Biol. 109, 437 (1977).
18. Wold, B. J., Klein, W H., Hough-Evans, B R., Britten, R. J., and Davidson, E. H. Cell 14, 941 (1978).
19. Benecke, B.-J., Ben-Zeev, A., and Penman, S. Cell 14, 931 (1978).
20. Aviv, H., Voloch, Z , Bastos, R. and Levy, S. Cell 8, 495 (1976).
21. Palmiter, R. D., and Schimke, R. T. J. Biol. Chem. 248, 1502 (1973).
22. Wahli, W., Wyler, T., Weber, R , and Ryffel, G. U. Eur. J. Biochem 66, 457 (1976).

DISCUSSION

A. TRAVERS: I think we should not exclude the possibility
that the capped pyrimidines derive from transcription initiation.   There is now a good precedent in E. coli.   Cashel,
Steitz and Nomura and their co-workers have all shown that
the distal promotor of several rRNA cistrons initiates in
vitro efficiently and often exclusively with CTP.

R. PERRY: Of course we never find pyrimidine terminated tri-
or di-phosphates in cells, although we find abundant purine
tri- and di-phosphates.   So it is hard to know whether such
initiations actually occur in vivo.   One would have to argue
that the capping efficiency for pyrimidines was far greater
than the efficiency for purines in order to explain our
observations.

A. TRAVERS: It is also true to say you do not find evidence
of pyrimidine tri-phosphate termini in E. coli either.   It
may just be that they are efficiently processed as you say.

P. CHAMBON: When you say that there are 35 to 50 polymerase
molecules per gene, is this for the 11 kb segment?   If yes,
we are still rather far from the very tight packing of RNA
polymerase molecules for DNA which is found on the actively
transcribed ribosomal transcription units.

R. PERRY: Yes, my calculation of 35 to 50 polymerase molecules was made for the 11 kb component that we believe to be
the primary transcript of the mRNA for the heavy chain
immunoglobulin.   That is a minimum estimate, assuming that
the messenger RNA is completely stable and that there is no
loss in processing.   Anything that would increase our estimate of the transcription rate, for instance, a significant
mRNA turnover or a low processing efficiency would increase
our estimate of the packing of polymerases."

P. LEDER: When you discussed the variable length of the
light chain precursors, you suggested that the distance
between the variable region gene and the promotor of the site
of transcriptional initiation might differ amongst these
various light chain genes.   I do not think you meant to
exclude the possibility that those region genes recombined
with J segments that were at various distances from the C
region.   One would expect ultimately that the large intervening sequence that separates the J and C region will be
variable in length and that accordingly the precursor length
will vary as well.

R. PERRY: The reason that I made that distinction was
because I have also examined the sizes of the processing
intermediates. Basically, there are three readily identifiable nuclear components: the one we think to be the primary
transcript, one very distinct intermediate-size component,
which is probably the intermediate that contains the intervening sequence between J and C, although I have not proven
that, and the mature size mRNA. By comparing the sizes of
the intermediates as well as the sizes of the putative
primary transcripts I arrived at the conclusion that there
was variability in the transcription unit size that could not
be accounted for solely by variation in the J—C intron size.
We have been studying the pre-mRNAs in myelomas of the V_k21
group. Martin Weigert and Lee Hood have sequenced many of
the kappa chains in this group, and find several distinct J
sequences, which implies that there are several distinct J
genes. Our data indicate that there is not a strict correlation between the size of the intermediates and the type of J
sequence. This result was unexpected, and so far I do not
have a reason for it. Maybe the gene arrangements or the
processing modes are more complicated than we imagine.

K. GUPTA: You mentioned the capping mechanism for pyrimidine
caps. Does this mechanism exclude the possibility of purine
caps by the same mechanism?

R. PERRY: Unless transcription can be initiated with pyrimidines, the presence of pyrimidines at the X position of caps
indicates that such messenger RNAs arise by cleavage and
removal of a portion of the RNA on the 5' side. If that is
the case, the cleavage would probably leave a monophosphate
at the 5' terminus of the mRNA. The capping enzyme that has
been described for vesicular stomatitis virus seems to be
able to deal with a monophosphate terminus, but so far, the
enzyme that has been found in mammalian cells requires a diphosphate terminus. Therefore, either one would have to have
a kinase to convert the monophosphate terminus to a diphosphate terminus, or one would have to have a VSV-like enyzme
which could cope with the monophosphate. That is all I meant
to imply about the capping mechanism for that class of
messenger RNAs which have pyrimidines at their 5' terminus.
I do not believe that there is any intrinsic difference
between the capping of termini-containing pyrimidines and
purines.

C.M. MCGRATH: I wonder about the model itself. Perhaps these
plasmocytomas have evolved unusual regulatory mechanisms. To
what extent do you think that any of the specific aspects of

processing that you are seeing in plasmacytoma lines are also occurring in normal tissues?

R. PERRY: I do not have any direct evidence that processing in plasmacytomas is a true reflection of the processing that occurs in normal plasma cells. However, other evidence, such as the distribution of immunoglobulin species, indicates that plasmacytomas are good models for normal plasma cells. However, until one can obtain comparable data on normal plasma cells, it is impossible to exclude the possibility that some of the results might be peculiar to the plasmacytoma cells themselves.

C. WEISSMANN: I might just add that in the case of the β -globin RNA, the 5' end of the 15S precursor RNA coincides precisely with the 5' end of the messenger so that unless there is a distinct species which is substantially larger than the 15S RNA then initiation is at the same point where the messenger is capped. Some of your arguments are based on the fact that the intermediate size messenger species are in fact intermediates in processing. Is there any evidence for that other than that they have intermediate sizes? Can you show that there is any chase from the intermediate sized RNA to the mature mRNA?

R. PERRY: For the most part the intermediates seem to be relatively short-lived in the sense that they are difficult to pick as kinetically distinct components. In MPC-11 cells the label appears more or less simultaneously in mature and intermediate size components. We interpreted this result to mean that the conversion from intermediate to mature mRNA is a very fast step. There is no detectable build-up of intermediate after a long chase, so if it were some abortive processing product, it would have to be degraded at a rate comparable to the processing rate.

P. CHAMBON: I have a question for Charlie Weissmann. Have you any evidence that in the globin case you have a very fast processing from the 15S RNA to a mature globin mRNA without intermediate steps?

C. WEISSMANN: Our resolution was not good enough to pick something up of intermediate size.

STEPS IN PROCESSING OF mRNA:
IMPLICATIONS FOR GENE REGULATION

J. E. Darnell, Jr.

The Rockefeller University
New York, New York

INTRODUCTION: The Problem of the Levels
of Gene Regulation in Eukaryotic Cells

One point of view has colored the work in our
laboratory over the past fifteen years. It has to
us seemed necessary that a detailed understanding
of the steps in mRNA formation in eukaryotic cells
(mammalian cells, in fact) must precede any logical
attempt to understand the level of regulation of
eukaryotic gene expression (1,2). This view may
have seemed parochial, or trivial, or unnecessary
in turn in the face of the brilliant demonstrations
of the universality of the genetic code, especially
including initiation and termination codons; the
broad similarity in all cells of protein factors
for initiation, elongation and termination during
polypeptide assembly (3); the presence of RNA
polymerases of high molecular weight and complicated structure that share many generally similar
properties in all cells (4,5); and even the possible similarity between histone and non-histone
chromosomal proteins to the recently discovered
generalized double-stranded DNA binding proteins
that join the already known bacterial regulatory
proteins that bind to DNA (5,6). My suspicion is
that all along the workers who contributed to the
very important areas listed above expected and
still expect pro- and eukaryotic gene regulation to
follow the same general rules with perhaps slight
adornments, and when the facts are all in their
general thesis may be borne out.

207

However, two types of experimental findings
have persuaded us that, until proven otherwise, we
should entertain the possibility that gene regulation in eukaryotes might not be accomplished <u>at</u>
<u>least</u> <u>entirely</u> by the same general logical rules of
positive or negative regulation of transcriptional
regulation that operate in bacteria. The two lines
of evidence referred to are 1) a basic structural
difference between prokaryotic and eukaryotic cells
with respect to transcriptional-translational coupling and 2) a class of molecular reactions in mRNA
formation that are almost completely or completely
confined to eukaryotes.

In bacterial cells transcription and translation are physically intimately associated and perhaps even physiologically coupled (7-9). A dia-
gramatic reminder of this fact is the electron
micrographs of <u>E</u>. <u>coli</u> by Miller et al. (7) showing
polyribosomes forming on the 5' portion of a new
mRNA molecule whose 3' portion has not yet been completed on the DNA template. Additional evidence of
the interaction between the translation and transcriptional apparatus comes from the studies of
attenuation (9,10), a phenomenon where premature
termination occurs in the transcription of certain
operons if, for example, amino acids synthesized by
enzymes encoded by the operon are plentiful. In
eukaryotic cells, on the other hand, protein synthesis is structurally separated from the nuclear
site of mRNA formation.

A second major difference between pro- and
eukaryotes, the enzymatic steps necessary to form
an mRNA, might necessitate the structural difference mentioned above. The mRNA in eukaryotes is
derived from primary nuclear RNA transcripts only
after an elaborate set of modifications (2). While
evidence for this major difference between the two
types of cells began to accumulate many years ago,
it is only within the last two years that a complete picture of how different the formation of
mRNA was between pro- and eukaryotes. In vertebrate cells (reyiewed in Ref. 2) and some insect
cells at least (11; but possibly all eukaryotes) there
are not only chemical modifications at both ends of
the final mRNA product prior to its engagement with
ribosomes in the cytoplasm (12,13) but cleavage of
the primary transcript and reunion of selected polynucleotide stretches--RNA:RNA splicing--is

necessary to generate a functional mRNA (14-17).

The question is now squarely posed--do cells
whose mRNA generating mechanisms differ so greatly
regulate the level of specific mRNAs in an identical fashion? Phrased another way, do eukaryotic
cells exercise regulatory decisions at levels
other than transcription?

The purpose of this article will be to review
breifly some recent experiments that attack this
general problem. While the questions of where and
how in the cell the selection is made of which
mRNAs are to be synthesized and in what quantities
have not by any means received a final answer, the
present results favor two levels of regulation in
eukaryotic systems. 1) Several kinds of transcriptional regulation: a) the off-on variety similar
to prokaryotes,   b) premature terminations exist
in animal cells and hypothetically could be regulatory c) a decision at the outset of transcription that controls success or failure in processing
the RNA transcript. In this latter case a decision
might be effected operationally at the level of
processing, but the issue might be settled at the
time of transcriptional initiation by the presence
or absence in the transcription complex of some
key factor. This possibility will become clearer
after the discussion of steps in the manufacture
of Ad-2 mRNA. 2) The stabilization or destabili-
zation of mRNA in the cytoplasm may be widely used
by eukaryotic cells especially in development.

What is missing from the above list of possible
levels of regulation is regulation after transcription but prior to cytoplasmic appearance, i.e. a
strictly post-transcriptional nuclear decision that
might be made after a transcript has been completed
and removed to a hypothetical processing site. The
details of mRNA processing that have been uncovered
recently seem to put this possibility in disfavor,
at least at the present time.

Initiation and Termination
in Transcription Units for mRNA

As emphasized above, the work to be summarized
from our group was based on the expectation that
steps in mRNA manufacture, specifically whether
processing of large nuclear precursor molecules
was a step in mRNA biosynthesis, had to be settled

before questions about regulation could be addressed. From this vantage point the problem reduced to determining whether nuclear RNA sequences
that eventually appeared in mRNA were only made
from regions of DNA longer than the length of the
mRNA or from regions of the DNA equal to the mRNA.
Obviously, transcription of both size nuclear
molecules might occur with mRNA production occur-
ing from only the smaller mRNA-sized transcripts.

A transcriptional unit for an mRNA is the segment of DNA between the presumed specific initiation site and the termination site(s) for RNA
transcription. The primary transcript is the RNA
produced by RNA polymerase copying such a region.
The definition of a transcriptional unit within
which a structural gene lies allows the determination of whether any or all of the transcriptional
unit is transcribed into RNA and later what happens
to that RNA.

Neither of the two classic techniques used in
the definition of prokaryotic transcription units,
genetic identification of start or stop sites (19,
20) or specific and correct transcription in vitro
by RNA polymerases and termination factors (4,21,22)
(RNA polymerase and rho factor from E. coli, for
example) are available to animal cell biologists.
Instead, analysis of very briefly labeled RNA,
"nascent chain analysis", and UV transcriptional
analysis have been used to identify the approximate
boundaries of transcription units (reviewed in
Ref. 2). The strength of these latter approaches
is that the products of a transcription unit as it
functions inside the cell can be studied to predict
the beginning and end of transcription; the weak-
ness is that until an enzyme or complex of enzymes
can be shown to produce the product beginning and
ending at the predicted sites it is nearly impossible to rule out almost instantaneous synthesis
and processing at the end of a transcriptional unit.
Consider the problem at the 5' end of the presumed
transcriptional unit for mammalian mRNAs. If the
start site is the cap site for the mRNA and capping
occurs quickly then no pppXp derivative will ever
be detected to mark the start site of a transcriptional unit for mRNA.

For example, the most extensively studied presumed start site for mRNA synthesis is the proposed initiation site for the major late

adenovirus-2 (Ad-2) transcriptional unit (Fig. 1).
From this transcription unit, 13 or 14 individual
mRNAs are synthesized (23,24). These mRNAs are
divisable into five 3' coterminal groups, i.e. five
different regions of the primary RNA transcript can
have poly(A) added (24-27). Within each group from
two to probably as many as five different splicing
arrangements can occur. Analysis of pulse-labeled
virus-specific RNA and UV transcription mapping of
the Ad-2 genome located the presumed promoter site
between 14 and 18 on the physical map of the genome (28-30). {The map is divided into 100 units,
350 bases each and 0 is at the left end of the
rightward reading strand (31).} No evidence for
short nascent labeled RNA complementary to sites
near the beginning of each late mRNA were found
(30), arguing that there are no specific initiation
sites before each mRNA. Likewise, the UV sensitivity of mRNA formation was equal to or greater
than that of the large nuclear RNA synthesis (32,
29). Furthermore, no RNA sequences could be detected that corresponded to transcripts upstream
from a site at 16.4 where the 5' cap structure that
is present in all the mRNAs from this transcriptional unit is encoded (15,16,33). Similar though
less complete evidence on nuclear RNA sequences
upstream from cap sites exists for two of the early
Ad-2 transcriptional units (Ziff, unpublished observations) and for the late SV40 transcriptional
unit (34,35). These results therefore tentatively
identify the initiation site with the cap site for
all of the virus transcriptional units so far
examined.

The identification of the presumed initiation
site for RNA synthesis late in Ad-2 infection has
allowed a determination of whether every polymerase
which initiates transcription in the region of
16.45 continues toward the end of the genome where
termination is thought to occur. The most direct
measurement of the distance traveled by each
polymerase is hybridization of nascent labeled RNA
to a series of DNA fragments from 16 to 100. After
an instantaneous pulse where no processing (or as
little as possible) has occurred, all equally transcribed genomic DNA segments will hybridize an
amount of labeled RNA proportional to the length of
the DNA segment (28,36,37). This experiment has
been performed by several different investigators

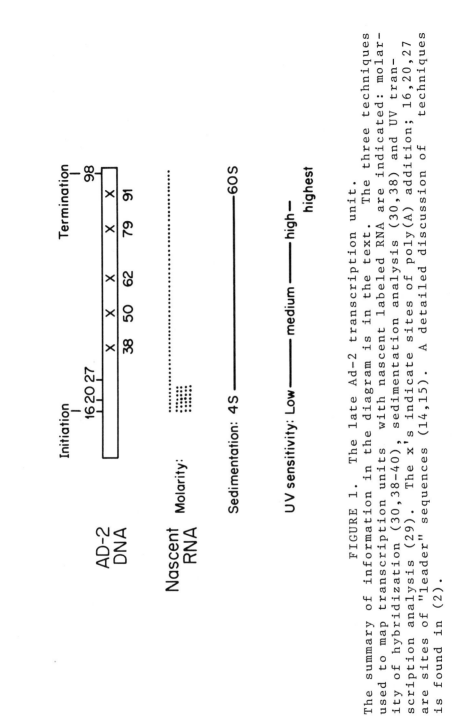

FIGURE 1. The late Ad-2 transcription unit. The three techniques used to map transcription units with nascent labeled RNA are indicated: molarity of hybridization (30,38-40), sedimentation analysis (30,38) and UV transcription analysis (29). The x's indicate sites of poly(A) addition; 16,20,27 are sites of "leader" sequences (14,15). A detailed discussion of techniques is found in (2).

in our group over the past several years (28,38-40)
and the collected results are given in Table 1.
From 25.7-27.9 all the way through 98 there is
equal transcription. The first part of the transcriptional unit however produces approximately
four times more RNA. Since some of the distinctive
oligonucleotides from this region of the genome can
not be found in fingerprints of the cytoplasmic
mRNA, it appears that these short nuclear molecules
from 16 to ∿21 or 22 are prematurely terminated RNA
chains. Autoradiography of RNA selected from nuclei of labeled, infected cells by hybridization
to the 11.1-18.2 DNA fragment that contains the
major late RNA start site reveals many discrete
bands that may mark frequent termination sites
(Sehgal, P. B., Fraser, N. W. and Darnell, J. E.,
unpublished observations). These short chains
appear to have caps at their 5' ends and seem to
start at the same site as those transcripts that
give rise to mRNA molecules downstream. Premature
termination also occurs in cellular transcription
where methylated cap structures are first labeled
primarily, if not exclusively, as part of short
molecules (Salditt-Georgieff, M., Harpold, M. and
Darnell, J. E., manuscript in preparation). After
a long label about one-half of the caps have shifted to molecules of mRNA length or longer suggesting
that about one-half of the chains containing caps
grow in size to approximate the distributions of
hnRNA but fully one-half of the chains bearing caps
remain less than 1000 nucleotides in length. The
short chains may be prematurely terminated hnRNA
molecules. Tamm (41) has observed the labeling of
short chains in vitro that are RNA polymerase II
products and that can be distinguished by their
resistance to the inhibitor DRB. The Ad-2 premature transcripts are also resistant to DRB (38)
suggesting a means of identifying premature hnRNA
transcripts.

     No evidence of a regultory role for premature
transcription has been obtained. For example, in
the early and middle hours (5 to 10 hours) of Ad-2
infection before the large late transcript has become the dominant functioning transcriptional unit,
there is no increase in the ratio for premature
transcripts complementary to the 11-18 region (42).

     If we now assume that the late Ad-2 transcrip-
tional unit begins at 16.45 and a polymerase has

TABLE 1.   Molarity of Late Transcription of
           Ad-2 Genome

Coordinates of DNA Fragment	Data from Reference A				
	(28)	(30)	(38)	(40)	(39)
11.1-18.2	2.2	6.1	4.5		7.5
18.8-36.7	1.4	1.64	2.04		1.33
25.7-27.9				.87	
36.7-40.5		.71		1.0	
40.5-52.6	.9	1.0	.85	.86	.97
52.6-56.9				.81	
58.5-70.7	1.0		1.0	1.0	1.0
70.7-75.9	.75		1.2	.61	
75.9-83.4	1.4		.76	.67	
83.4-89.7	.83		.76	.60	1.3
91.9-98.2				.70	
89.7-100	.77		.32		.95

Nuclear RNA from cells labeled 18 hours after Ad-2 infection by a 1 or 2 minute pulse of ^3H uridine (38,39) or RNA labeled in isolated nuclei (28,30) was hybridized to an excess of Ad-2 DNA with the indicated coordinates. The CPM hybridized (ranging from hundreds to thousands of CPM) was divided by the length of each fragment and normalized to one fragment which is given as 1.0 in the Table.

passed the region where premature termination can
occur, how far to the right does the enzyme go before termination? We already mentioned that equimolar transcription appears to include all of the
genome to 98. By assaying the size of nascent RNA
complementary to an ordered set of DNA segments and
by determining the relative UV sensitivity to
successive DNA fragments going from 16.45 toward
100, we conclude that the transcription extends all
the way to a site or sites between 98 and 99 (43).
If we conclude that the termination site is past 98
based on the molarity studies and the UV sensitivity of RNA complementary to that region, an important point emerges that we will return to for consideration later: the termination site for the
production of all the late Ad-2 mRNAs is downstream
to all of the five possible poly(A) sites in the
transcriptional unit.

Similar but less complete evidence exists for
two of the early regions of Ad-2 transcription and
for the late SV40 transcriptional unit where only
one poly(A) site exists per transcriptional unit.
In these cases also, the pulse-labeled nuclear RNA
molecules contain sequences past the poly(A) site,
documented in two of the cases to be initially part
of molecules larger than the poly(A) terminated
nuclear RNA (Nevins, J., Blanchard, J-M., and
Darnell, J. E., manuscript in preparation; Weber,
J., Blanchard, J-M. and Darnell, J. E., manuscript
in preparation). Thus we tentatively conclude
(Figs. 1 and 2) that transcription units for mRNAs
in mammalian cells and viruses can be thought of
as having an initiation site where the sequences
to which the cap is added are encoded and a termination site or sites. Premature termination, that
is, termination prior to reaching any mRNA sequences appears frequent for both cellular and virus transcriptional units. Within the primary RNA
transcript there is one or more sites where endonucleolytic cleavage results in the appearance of
a free 3' end to which poly(A) can be added. No
transcription unit has yet been identified where
termination occurs at the poly(A) site and no
transcription upstream from the cap site occurs in
the virus transcription units so far examined.

mRNA Biosynthesis

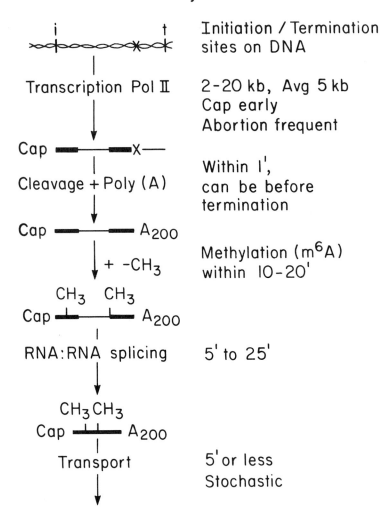

FIGURE 2. Summary of mRNA biosynthesis based on work with Ad-2 transcription units (28,30,38-40, 46,47) and Chinese hamster cells (see text; Salditt-Georgieff et al. and Harpold et al., in preparation.

The Steps in mRNA Manufacture

There are four common modifications of nuclear
RNA that occur during mRNA biogenesis: 1) capping
2) methylation of the 6 position of internal
adenylate residues 3) poly(A) addition and
4) RNA:RNA splicing (reviewed in Ref. 2). The
fourth of this list is of course the most recently
discovered and perhaps therefore the most surprising. The conclusion that splicing must occur was
reached by Chow et al. (15), Klessig (16), and
Berget et al. (14) because they discovered that
late Ad-2 mRNAs contained sequences that were non-
contiguous in the Ad-2 DNA. Because of the evidence that the only pulse-labeled nuclear molecules
that contain all the sequences necessary to make
an mRNA were the very long molecules described
earlier (28-30), they concluded that RNA:RNA splicing must occur. Many cases of sequences in mRNAs
that are non-contiguous in DNA have now been recorded as well as interrupted sequences for structural RNAs (reviewed in Ref. 2). Two cases of
in vitro splicing, for yeast pre-tRNA (44) and an
early Ad-2 mRNA (17), have been reported. It will
therefore be assumed that RNA·RNA splicing is a
common step in mRNA manufacture.

Again in line with the long-range goal of determining which steps in mRNA production might
serve a regulatory role, we have performed experiments to order the series of modification steps
and to determine which of the various modified
structures are conserved in transit to the cytoplasm. In the majority of cases the material used
in these studies was Ad-2 infected HeLa cells. The
results, summarized in Fig. 2, were that capping is
almost instantaneous and occurs on prematurely
terminated transcripts; poly(A) addition occurs before other modifications and sequences next to
poly(A) are conserved {this is very strongly indicated for Ad-2 mRNAs (40) and quite possibly true
also for cellular sequences (45)}; m^6A is conserved
in Ad-2 mRNA formation (46) but the evidence is not
yet clear on this point for cell mRNAs; splicing
is the final modification before exit from the
nucleus (40), and, at least for Ad-2 mRNAs, splicing occurs exclusively in the nucleus (24,47).

Capping Is Early

As described above, cellular hnRNA products
have been found to be capped fairly early in the
course of their synthesis. An experiment designed
to investigate this question for the large late
Ad-2 transcriptional unit has also been conducted
(Babich, A., Nevins, J. R. and Darnell, J. E., unpublished results). Cells were labeled for 20 minutes with ^{3}H adenosine which, of course, will label
the internal adenosine residues in an RNA chain and
also the cap structure which begins in m^{7}GpppACU---.
The adenosine-labeled RNA was hybridized to the
11-18 region of DNA which includes approximately
the first 600 nucleotides of the transcriptional
unit. From the direct sequence analysis, we know
that 100 of these 600 bases is A and therefore one
A out of each 100 from a transcript covering this
region could appear in a cap structure. If all of
the extra short prematurely terminated RNA molecules complementary to this region were capped then
1% or more of the total ^{3}H adenosine labeled RNA
should be found in a cap structure. This experiment has been performed twice and about 2% of the
total incorporated adenosine from molecules shorter
than 500 nucleotides has, in fact, appeared in the
cap structure. This indicates that all of the premature transcripts which make up the majority of
short RNAs as well as the forerunners to the larger
messenger precursor molecules (33) are capped.

Poly(A) Is the First Modification
of the Primary Transcript

With a knowledge of the boundaries of the transcription unit late in Ad-2 infection an ordering
of the modifications of the primary transcript was
investigated. The assumption in these experiments
was that either poly(A) or splicing might occur
first. If poly(A) addition occurred prior to splicing then the only molecules to which poly(A) would
be added would be larger than the mRNA. If splicing occurred first then poly(A) would be added to
molecules the size of mRNA. In addition to an-
swering this particular question, the experiments
have informed us about the time course of the addition of poly(A) relative to the transcriptional
process itself. The accumulation of ^{3}H uridine

was measured in poly(A) terminated nuclear molecules and revealed that ^{3}H uridine entered equally
rapidly into poly(A) terminated nuclear RNA complementary to each of the five possible poly(A) sites
(40). It appears that a polymerase crosses a
poly(A) site which is going to be used for that
particular transcriptional event and within a minute or so (equivalent to an additional 2000 to 3000
nucleotides if the polymerase continued moving)
        an endonucleolytic scission occurs and poly(A)
is added. The polymerase then continues to the end
of the transcriptional unit, otherwise equimolar
transcription from ∿25 to 100 would not occur
(Table 1). Finally, conservation of nuclear RNA
from the Ad-2 late transcriptional unit was studied.
The total nuclear RNA complementary to a region
containing an mRNA was labeled approximately four
times faster than the labeling of the same sequences in the cytoplasm, indicating approximately
a 20 to 25% conservation. However, it appeared
that conservation was close to 100% for nuclear
poly(A) terminated Ad-2 RNA. Only one mRNA is conserved from each transcriptional event that crosses
the late transcription unit and this mRNA is next
to the one poly(A) site which is used on that particular occasion. This poly(A) site is chosen
shortly after the polymerase has passed the site
(i.e. before transcription has been completed) and
the poly(A) addition is only to large molecules
that stretch apparently from 16 to the poly(A) site.
Therefore, as a general rule, splicing must not
occur until after poly(A) addition. Recent evidence suggests that the steps in splicing also
might be ordered because distinct intermediates can
be seen for the formation of mRNA which terminates
at 91 (47). The intermediates suggest the removal
first of a large piece of RNA indicating perhaps
that the 16.4-19.7-26.7 splicing occurs after the
splicing of the major body of the message to the
26 region.
        Finally, it is important to point out that also
in transcriptional units other than the large late
transcription unit the addition of poly(A) occurs
to nuclear molecules larger than mRNA, prior to
splicing. Early in Ad-2 infection, where at least
six independent transcriptional units function (18,
19,30), evidence has been obtained for two of these
transcriptional units that poly(A) is only added

to molecules substantially longer than the mRNA
derived from these transcriptional units. Similar
evidence exists for the late SV40 transcriptional
unit, although direct experiments determining the
site of addition of poly(A) have not been carried
out in that system. Thus, again, we reach a tentative general conclusion. Transcription goes past
the poly(A) site; the first event in the manufacture of mRNA is the recognition of the poly(A) site
followed by cleavage and poly(A) addition prior to
splicing (Fig. 2). What little is known about
cellular mRNA formation seems to fit this tentative
model as well. For example, in the case of hemoglobin precursors among the nuclear RNA, the 15S
nuclear precursor molecule is almost entirely polyadenylated (see Leder, this volume). Likewise in
the case of myeloma cells there are larger nuclear
polyadenylated immunoglobulin-specific RNA molecules than are found in the cell cytoplasm (Wall,
R., personal communication). Finally, a species
of nuclear RNA complementary to cloned DNA specific
for rat growth hormone is larger than the cytoplasmic growth hormone mRNA and contains poly(A)
(Harpold, M., unpublished results). In these instances of specific mRNA it is not sure whether
some splicing might not have occurred that reduced
the size of the primary transcript to a poly(A)
terminated molecule larger than the cytoplasmic
one. But at least all splicing has not occurred
prior to poly(A) addition in these cases as well as
in the case of the viral molecules.

Nothing is yet known in detail about how splice
points are chosen. Sequence analysis in several
transcription units suggests some structural features that may appear around many splice sites
(see O'Malley et al. and Chambon et al. in this
volume). But it seems highly unlikely that these
similar but non-identical oligonucleotide stretches
of 2-10 nucleotides are the only recognized features at the splice sites. Apart from the probable
three-dimensional RNA arrangement that allows the
enzymatic accomplishment of splicing there is in
the cases of virtually all the Ad-2, SV40 and
retrovirus transcription units, the necessity to
choose for any given nuclear molecule one of a
series of possible splicing choices. As pointed
out above, this choice occurs after poly(A) has
been added but no role for poly(A) in the process,

if any, has yet been described. One possible chemical mechanism that operates in the choice of
splice sites is the addition of methyl groups at
the 6 position of adenylate residues. Late Ad-2
mRNA has one m ^{6}A/400 nucleotides (50) and the m^{6}A
residues are not in the terminal several hundred
nucleotides of the mRNA. Since there are a total
of ~20,000 bases of the late transcription unit
represented in one or another mRNA molecule, there
could be as many as 40 to 50 different m^{6}A sites.
If the methyl groups are somehow related to splicing we reasoned that they might occur only to that
section of the primary transcript that was conserved to make an mRNA. When infected cells were
double-labeled with ^{3}H methyl methionine and ^{14}C
uridine, we found that the conservation of mehtyl
label from nucleus to cytoplasm was 3 to 4 times
better than that of ^{14}C uridine (46). A quantitative mode of label flow was developed and the data
from the accumulation of methyl and uridine label
analyzed. It appeared that the methyl groups were
entirely conserved even within regions where 3'
coterminal mRNAs exist. Thus the methylase appears
to know which splices are going to be made. However, we also found that methylation occurs mostly,
if not entirely, to larger, at least incompletely
spliced, molecules. Thus it seems possible that
methylation might have some role in choosing splice
points.

CONCLUSION

     How does all this new evidence affect our
thinking about the control of mRNA production? It
is still very likely that transcription must be
controlled to some extent at the level of initiation. For example, in Ad-2 infection the late large
transcription unit doesn't function early (42) and
preliminary evidence suggests the rate of synthesis
from early transcription units declines late
(Nevins, unpublished results). Potentially, regulation may also exist at the level of early termination or antitermination since prematurely terminated transcripts have been observed.
     Regulation of mRNA levels by differential pro-
cessing has been an intriguing possibility since
hnRNA was discovered. With the new evidence on
mRNA biosynthesis we can suggest the points where

such regulation might conceivably occur much more
precisely than in the past. Take the case of late
Ad-2 mRNS. Since poly(A) addition seems crucial
for sequences to emerge to the cytoplasm and is
accomplished at five different possible sites only
once per transcriptional event within 1 to 2 minutes
of transcription, it seems possible that this choice
might be made for a given transcriptional event before transcription begins. For example, a single
random choice of poly(A) sites made during transcription would presumably favor a polar distribution of usage of poly(A) sites whereas the actual
ratios of usage (from 5' to 3') are 10, 23, 28, 21,
18 (40). While it is still possible to arrange
binding affinities to meet this distribution, recall
that every (±25%) transcriptional event does give
rise to one mRNA. If choice of poly(A) sites were
random but occurred within a minute of transcription the fifth site in transcripts not yet polyadenylated might frequently fail to be utilized if
statistical choice were the basis for poly(A) site
selection. It seems to us somewhat more likely
that either 1) each polymerase (or processing complex) becomes equipped with a subunit that specifies
a type of poly(A) site or 2) in each template one
poly(A) site contains a recognition protein such
that when a polymerase passes that site, recognition
is made and poly(A) added.

The results on the immediacy of the decision to
add poly(A) are a potentially powerful aid in understanding cell mRNA biosynthesis. In mammalian
cells it has been clear for some time that more
nuclear RNA is formed than ever exits to the cytoplasm (1,2). Recent estimates of the size of primary transcripts compared to mRNAs shows them to be
3 to 5 times longer (2, Harpold, M., Evans, R.,
Salditt-Georgieff, M., and Darnell, J. in preparation). Thus, as much as 20 to 30% of each transcript would be expected to exit. However, estimates of what actually does exit are more like 5%
or less (51,52). In addition the "complexity" in
nucleic acid hybridization experiments is perhaps
20 times greater for hnRNA than for mRNA (53,54).
All these results indicate that perhaps 5 times as
many different kinds of transcripts are made as in
fact are ever processed into mRNA.

We have recently measured the rate of labeling
of caps compared to poly(A) segments in large
nuclear RNA molecules (not prematurely terminated

chains) and found about 5 times more caps than
poly(A)(Harpold, M., Salditt-Georgeiff, M. and
Darnell, J. E., unpublished observations). Thus
it may be that the selection of which transcripts
to process is signaled by the recognition of a
poly(A) site and addition of poly(A) to that site.
This choice might in a limited number of cases be
between one or more poly(A) sites within the same
transcription unit as in the case of late Ad-2, or
more frequently it could be a choice of add poly(A)
or not add poly(A). It is important to again point
out that while a choice of this type would be
effected on an hnRNA molecule, the mechanism of the
choice could lie in correctly equipping a tran-
scription-processing complex or the DNA template
itself at the time of transcriptional initiation.
This possibility gives a distinctly different slant
to the earlier discussions of "post-transcriptional" regulation (13,55-57).

Finally, it remains an open possibility that
differential splicing might occur between a series
of 3' coterminal mRNAs contained within a transcriptional unit (2). Cells apparently face this
possibility with integrated virus genomes such as
SV40, adeno, and retroviruses. What is not yet
established is that the proportion of different
mRNAs from such a transcriptional unit ever varies.
If the choice invariably results in a constant
ratio of the different 3' coterminal mRNAs then no
regulation occurs at this level even though such a
choice would theoretically be possible.

Again, even if regulation is accomplished by
varying splicing choices within a transcription
unit, it might be a decision that was exercised by
instructing a transcribing-processing complex.
Such decisions would not have to involve a transcript "floating off" the template awaiting a regulatory choice in solution.

Thus the framework for projecting the possibilities of how gene regulation is accomplished
in eukaryotes has been greatly improved and
strengthened by the knowledge of the steps required
for mRNA synthesis. The task still remains, but
now seems relatively much easier, to determine the
levels(s) of control and finally what proteins or
other factors are involved in mediating the control.

ACKNOWLEDGEMENTS

This research was supported by grants from
the National Cancer Institute and the American
Cancer Society. The author wishes to thank his
many colleagues for numerous discussions that have
helped shape the ideas put forth here but any blame
for overextended speculation resides entirely with
the author.

REFERENCES

1. Darnell, J. E. (1968). Bact. Rev. 32, 262.
2. Darnell, J. E. (1978). Prog. Nuc. Acid Res.
    Mol. Biol. 22, 327.
3. Lengyel, P. (1974) in "Ribosomes" (Nomura,
    Tissieres and Lengyel, eds.), p. 13, CSH
    Laboratory, New York.
4. Losick, R. (1972). Ann. Rev. Biochem. 41, 409.
5. Chambon, P. (1975). Ann. Rev. Biochem. 44,
    613.
6. Rouviere-Yaniv, J. (1977). CSH Symp. Quant.
    Biol. 42, 439.
7. Miller, O. L., Jr., Beatty, B. R., Hamkalo,
    B. A., Thomas, C. A., Jr. (1970). CSH Symp.
    Quant. Biol. 35, 505.
8. Jacobs, K. A., Shen, V., and Schlessinger, D.
    (1978). PNAS 75, 158.
9. Zurawski, G., Brown, K., Killingly, D., and
    Yanofsky, C. (1978). PNAS 75, 4271.
10. Bertrand, K., Corn, C., Lee, F., Platt, T.,
    Squires, G. L., Squires, C. and Yanofsky, C.
    (1975). Science 189, 22.
11. Suzuki, Y. (1979). Cell, in press.
12. Shatkin, A. J. (1976). Cell 9, 645.
13. Darnell, J. E., Jelinek, W., and Molloy, G.
    (1973). Science 181, 1215.
14. Berget, S., Moore, C., and Sharp, P. A. (1977).
    PNAS 74, 3171.
15. Chow, L. T., Gelinas, R. E., Broker, T. R.,
    and Roberts, R. J. (1977). Cell 12, 1.
16. Klessig, D. (1977). Cell 12, 9.
17. Blanchard, J-M., Weber, J., Jelinek, W., and
    Darnell, J. E. (1978). PNAS 75, 5344.
18. Aviv, H., Voloch, Z., Bastos, R., and Levy, S.
    (1976). Cell 8, 495.
19. Jacob, F., and Monod, F. (1961). J. Mol. Biol.
    3, 318.
20. Englesberg, E., Sheppard, D., Squires, C., and
    Meronik, F., Jr. (1969). J. Mol. Biol. 43,
    281.
21. Roberts, J. (1969). Nature 224, 1168.
22. Adhya, S., and Gottesman, M. (1978). Ann.
    Rev. Biochem. 47, 967.
23. Chow, L. T., Roberts, J.-M., Lewis, J. B., and
    Broker, T. R. (1977). Cell 11, 919.
24. Nevins, J. R., and Darnell, J. E. (1978).
    J. Virol. 25, 811.

25. McGrogan, M., and Raskas, H. (1978). PNAS
    75, 625.
26. Ziff, E., and Fraser, N. (1978). J. Virol.
    25, 897.
27. Fraser, N., and Ziff, E. (1978). J. Mol. Biol.
    124, 27.
28. Weber, J., Jelinek, W., and Darnell, J. E.
    (1977). Cell 10, 612.
29. Goldberg, S., Weber, J., and Darnell, J. E.
    (1977). Cell 10, 617.
30. Evans, R., Fraser, N. W., Ziff, E., Weber, J.,
    Wilson, M., and Darnell, J. E. (1977). Cell
    12, 133.
31. Flint, J. (1977). Cell 10, 153.
32. Goldberg, S., Nevins, J., and Darnell, J. E.
    (1978). J. Virol. 25, 806.
33. Ziff, E., and Evans, R. (1978). Cell 15.
34. Haegeman, G., and Fiers, W. (1978). Nuc.
    Acids Res. 5, 2359.
35. Lai, C-J., Dhar, R., and Khoury, G. (1978).
    Cell 14, 971.
36. Bachenheimer, S., and Darnell, J. E. (1975).
    PNAS 72, 4445.
37. Derman, E., Goldberg, S., and Darnell, J. E.
    (1976). Cell 9, 465.
38. Fraser, N., Sehgal, P. B., and Darnell, J. E.
    Nature 274, 590.
39. Evans, R., Weber, J., Ziff, E., and Darnell,
    J. E. (1979). Nature, submitted.
40. Nevins, J. R., and Darnell, J. E. (1978).
    Cell 15, 1477.
41. Tamm, I. (1977). PNAS 74, 5011.
42. Sehgal, P. B., Fraser, N. W., and Darnell,
    J. E. (1979). Virology, in press.
43. Fraser, N., Nevins, J. R., Ziff, E., and
    Darnell, J. E. (1978). J. Mol. Biol., in
    press.
44. Knapp, G., Beckmann, J. S., Johnson, P. F.,
    Fuhrman, S. A., and Abelson, J. (1978). Cell
    14, 221.
45. Puckett, L., and Darnell, J. E. (1977).
    J. Cell. Phys. 90, 521.
46. Chen-Kiang, S., Nevins, J. R., and Darnell,
    J. E. (1979). Cell, submitted.
47. Nevins, J. R. (1979). J. Mol. Biol., submitted.
48. Berk, A. W., and Sharp, P. A. (1977). Cell
    12, 45.

49.   Wilson, M., Fraser, N. W., and Darnell, J. R.
      (1979). Virology, in press.
50.   Sommers, S., Salditt-Georgieff, M.,
      Bachenheimer, S., Darnell, J. E., Furuichi,
      H., Morgan, M., and Shatkin, A. J. (1976).
      Nuc. Acids Res. 3, 749.
51.   Soeiro, R., Vaughan, M., Warner, J. R., and
      Darnell, J. E. (1968). J. Cell Biol. 39, 112.
52.   Lengyel, J., and Penman, S. (1975). Cell 5,
      281.
53.   Chikaraishi, D. M., Deeb, S. S., and Sueoka,
      N. (1978). Cell 13, 111.
54.   Bantle, J. A., and Hahn, W. E. (1976). Cell
      8, 139.
55.   Scherrer, K., and Marcaud, L. (1968). J. Cell
      Physiol. 72 (supplement), 181, 212.
56.   Georgiev, G. P. (1969). J. Theor. Biol. 25,
      473.
57.   Lewin, B. (1975). Cell 4, 11.

DISCUSSION

C. WEISSMANN: I would like to direct this question both to you and to Bob Perry. In talking about the normal eukaryotic cells, would you say that the class of heterogeneous nuclear RNA can now be explained - as being all messenger precursors from which introns are removed?

J.E. DARNELL: My guess would be, potentially but not actually. Let me tell you what I mean. Michael Harold in our lab has studied nine cloned Chinese hamster cDNAs. The concentration of sequences complementary to those messenger RNAs in the cytoplasm is 20 to 50 times less than the concentration of those sequences in the nucleus. The primary transcripts are only about 5-fold different in size. Therefore, we believe that there are a larger number of "diluting" sequences in the nucleus which might be potential messenger RNAs but are not polyadenylated and not shifted to the cytoplasm at this time.

R.P. PERRY: I do not really agree to that. I think the efficiency is much greater for the bulk of the Hn RNA. Transcripts for messenger RNA which existed in very few copies in the cytoplasm might be under a more inefficient processing mechanism and, if you had chosen one of those to study, you might have found there would be some extra transcription which did not subsequently appear in the cytoplasm. From the study of CAP conservation it is my feeling that the efficiency is pretty much explained by excision of intron sequences from the bulk of the population and only in so far as there is a small class of nuclear transcripts which would be those coding for sparse cytoplasmic messages, those might in fact be the ones where there would be an excess in the nucleus.

T. HUNT: It seems to me that what we have heard this morning goes a long way to explain what Hn RNA is, but there is one class of nuclear RNA which you are tending to brush under the carpet, and that is the small nuclear RNA. What fraction of those are tRNA and 5S RNA precursors and what fraction has some more interesting role?

J.E. DARNELL: The fraction of the nuclear RNA that Bob and I normally work with does not contain pre-tRNA or 5s RNA. The small RNA that I was talking about is on average slightly larger, (about 300-900 nucleotides in size), and appears to be mainly prematurely terminated transcripts. The specific small nuclear RNAs of 100-200 bases that are species specific have no definitely assigned role to my knowledge at the present time. They were thought, at one point, perhaps to be primers for RNA synthesis but I think that is wrong. The caps are exclusively $m^{2,2,7}G$ and not m^7G, for example, and so I don't think that they are primers which are extended in RNA synthesis. One popular present notion, without so far as I know any real substantial evidence, is that they have a role in splicing. Perhaps Dr. Zain would comment on a possible role for the A α-2 VA RNA.

S. ZAIN: The VA RNA sequencing was not my work. I do have the adenovirus-2 fiber mRNA sequence and the leader sequence. Originally we thought VA RNA plays a role in mRNA processing and were fooled for a long time before splicing was discovered. The Ad-2 late leader sequence does not show very striking resemblance to VA RNA sequence. At the moment I really don't know if there is any conclusive evidence to suggest VA RNA involvement in mRNA processing.

5'-CAPPING AND EUKARYOTIC mRNA FUNCTION

Aaron J. Shatkin
Nahum Sonenberg
Yasuhiro Furuichi

Roche Institute of Molecular Biology
Nutley, New Jersey

I. INTRODUCTION

Recent studies of mRNA structure and synthesis have revealed unanticipated complexities in the organization and expression of eukaryotic genes (1). In contrast to their prokaryotic counterparts, most eukaryotic mRNAs share several distinctive structural features. They include 3'-terminal polyadenylic acid, methylated nucleotides at internal positions, and a blocked and methylated 5'-end, $m^7G5'ppp5'N$, called a "cap" (2,3). Because mRNAs are synthesized and translated in a 5' to 3' direction, the presence of a 5'-terminal cap is of special interest. We have concentrated on understanding the mechanisms of synthesis of the cap and its effects on mRNA stability and function.

In cellular mRNAs and mRNAs of viruses that replicate in the nucleus of infected cells, caps are present on nuclear messenger precursors but not on ribosomal or tRNA transcripts (4). In some mRNAs, caps may be synthesized as part of the initiation of transcription and involved in its regulation. An example of tight coupling between cap synthesis and initiation of transcription has already been reported in the case of cytoplasmic polyhedrosis virus mRNA synthesis (5). For cellular mRNAs, caps apparently are conserved in the cytoplasmic messengers (6) that are formed from nuclear transcripts by RNA·RNA splicing. Retention of capped 5'-terminal sequences, in contradistinction to splicing out of intervening sequences during mRNA processing, is consistent with the stabilizing effect of caps observed in vitro (7,8). Capped mRNAs are degraded more slowly than the corresponding uncapped mRNAs in HeLa cell nuclear and cytoplasmic fractions, in cell-free protein synthesizing systems from wheat germ and mouse L cells, and in microinjected Xenopus oocytes. In addition, the presence of a cap promotes binding of mRNA to ribosomes, and capped viral and cellular mRNAs are translated less effectively following decapping (2). An effect of 5'-terminal modification on mRNA function might have been anticipated on the basis that efficient utilization of the information content of a monocistronic mRNA would require initiation of translation near the 5'-end. In fact, the 5'-terminal nucleotide sequences of many eukaryotic mRNAs have been established recently, and in most cases the 5'-proximal AUG triplet functions as the initiator codon for protein synthesis (9). Thus, addition of the cap at an early stage, perhaps the first step in mRNA formation, would serve to protect the 5'-terminal regions that are involved subsequently in the initiation of polypeptide synthesis. In this talk, I summarize aspects of cap structure, distribution among eukaryotic mRNAs, mechanisms of synthesis and functional effects on mRNA stability and initiation of translation (2). Because of space limitations, the manuscript includes only more recent experimental results concerning the influence of cap on the initiation of eukaryotic transcription and translation.

II.   RESULTS AND DISCUSSION

5'-Capping during Initiation of Transcription and Stimulation of Cytoplasmic Polyhedrosis Virus mRNA Synthesis by S-Adenosylmethionine

Elucidation of the mechanisms of eukaryotic mRNA synthesis and function has been facilitated by the availability of purified animal viruses that contain RNA polymerase and the several other enzymes involved in capped mRNA formation. Thus human reovirus (10), vaccinia virus (11), and vesicular stomatitis virus (12) have been used to analyze in detail the enzymatic steps involved in cap synthesis by chymotrypsin or detergent-treated virions. Purified cytoplasmic polyhedrosis virus (CPV) also catalyzes viral mRNA synthesis in vitro but in the absence of particle-disrupting agents (13). In this viral system the presence of S-adenosylmethionine (SAM) or one of its analogs such as S-adenosylhomocysteine (SAH) was required for activation of mRNA synthesis (5,13). Analysis of the 5'-terminal structures of the nascent CPV mRNA products indicated that the formation of cap occurs during ini-

tiation by a series of reactions similar to those described
previously for reovirus and vaccinia virus.

   To determine if the stimulatory effect of SAM and SAH was
exclusively on the initiation of CPV transcription, a two-
step reaction was carried out. In the first step, purified
CPV was incubated briefly in a reaction mixture containing
RNA precursors and SAM or SAH. The synthesis of mRNA was
stopped after 2 min by chilling and dilution. The virions
containing nascent RNAs were collected by centrifugation and
reincubated in a second reaction mixture containing no SAM or
SAH. As shown in Figure 1, CPV containing pre-initiated RNA
molecules elongated the nascent chains, but reinitiation did
not occur and synthesis stopped after about 10 min. By contrast, CPV that was incubated in the first reaction without

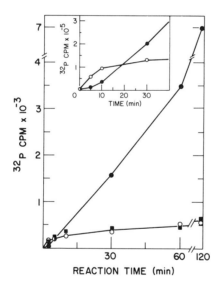

FIGURE 1. Requirement of SAM for the initiation of CPV
mRNA synthesis. CPV (230 µg) was incubated in the first
reaction mixture of 0.1 ml containing 50 mM Tris-HCl (pH 8),
4 mM $MgCl_2$, 0.8 mM ATP, 0.4 mM each of GTP, CTP and UTP with
either 0.1 mM SAM or SAH or neither compound at 31°C. After 2
min, mixtures were chilled, diluted, centrifuged, and the pelleted particles were resuspended in 10 mM Tris-HCl buffer
(pH 8). Aliquots (100 µg) were incubated in the second RNA
synthesis mixture with α-^{32}P-UTP ± SAM and RNA synthesis was
determined by acid precipitation. ●, none (1st), SAM (2nd);
O, SAM (1st), none (2nd); ■, SAH (1st), none (2nd); inset,
RNA synthesis in the 2nd reaction with 1/20 concentration of rNTPs.

SAM or SAH and then transferred to an incubation mixture containing SAM synthesized mRNA for more than 2 hr. The initial rate of mRNA synthesis with CPV containing pre-initiated RNA molecules was several-fold greater than in particles without nascent RNAs (inset, Fig. 1). After 10 min, the rate of RNA synthesis in the presence of SAM became equal to the initial rate of mRNA synthesis by pre-incubated CPV. These results indicate that SAM (or its analog) is required for initiation of CPV mRNA synthesis and that this effect on initiation is reversible.

Further evidence that a process involved in capping is related to the initiation of CPV transcription was obtained in other experiments with imido derivatives of ribonucleoside triphosphates (5). Replacement of the initiating nucleotide, ATP by the nonhydrolyzable analog, pNHppA prevented initiation but not elongation of RNA chains. In addition, all the nascent CPV RNA molecules initiated in the presence of SAM contained a 5' cap. These included very short chains consisting of only a few nucleotides. Similarly, in cellular transcripts caps were also detected in various-sized nuclear polyadenylated RNA molecules, suggesting that auxiliary capping activities associated with RNA polymerase II but not with polymerases I and III modify the 5' ends of nascent mRNA molecules (14,15) and low mol. wt. nuclear RNAs U1 and U2 (16).

Table 1

Reduction of the apparent Km for ATP required for CPV mRNA synthesis
by the presence of SAM

Concentrations of SAM (μ M)	Km for ATP (mM)
5	10
50	1
100	0.6
200	0.4
500	0.25

Allosteric Activation by SAM of the
Initiation of CPV Transcription

Because the relationship observed between capping and initiation of mRNA synthesis in the relatively simple CPV transcription system may also apply to more complex systems, we have studied the mechanism of SAM activation in detail. Enzymological constants were measured for SAM, ATP and other ribonucleoside triphosphates required for the synthesis of mRNA containing the 5'-terminal cap sequence, m^7GpppA^m. Efficient RNA synthesis was dependent on the relative concentrations of SAM and the initiating nucleotide, ATP. For example, at a low level of ATP (0.2 mM), a high concentration of SAM (1 mM) was required for maximum stimulation of CPV mRNA synthesis while at high ATP (2 mM), a low level of SAM (50 μM) was sufficient for optimum transcription. This inverse relationship between SAM and ATP concentration prompted us to measure the Km for each compound during RNA synthesis. As shown in Table 1, there was a striking decrease in the apparent Km for ATP as the SAM concentration was increased. This effect was observed only for ATP; the other ribonucleoside triphosphates had Km values of ∿50 μM.

The double-reciprocal plots used for estimating the Km of ATP deviated from linearity at low ATP concentrations (50-100 μM) and at SAM levels of 5 μM to as high as 0.5 mM, forming concave upwards curves (Fig. 2). At 2 mM SAM or SAH, the

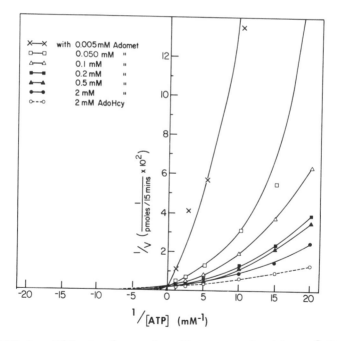

FIGURE 2. Effect of varying the concentration of SAM and ATP on the initiation of CPV transcription. Reaction mixtures similar to those in Fig. 1 were incubated at 31°C for 15 min and the newly synthesized acid-precipitable RNA was collected on Millipore filters and counted. (Adomet=SAM; Adohcy=SAH)

plots were close to linear. Anomalous kinetic behavior of
this type is indicative of a positive cooperative interaction
between enzyme, substrate and effector molecules, i.e. an allosteric effect (17). SAM apparently interacts with a methyl
transferase activity in CPV, the resulting allosteric conformational change within the transcription complex lowering the
apparent Km for the initiating nucleotide and allowing RNA
chain polymerization to proceed. This hypothesis was further
supported by the finding that SAM did not affect the Km for
ATP during chain elongation. RNA synthesis by particles containing pre-initiated nascent RNA chains gave an apparent Km
for ATP during elongation of 50 μM in the presence or absence
of SAM. This value is lower than that for initiation but
similar to the values for the other three ribonucleoside triphosphates.

These results with CPV are consistent with eukaryotic
transcription involving distinct initiation and elongation
steps, the former requiring more energy. The Km lowering
effect of SAM for the CPV initiating nucleotide, facilitating
the passage of the viral polymerase through the initiation
barrier, may be a model for other eukaryotic systems in which
transcription of certain cistrons is induced by hormones,
other physiologically active compounds, or by treatments such
as heat shock.

Positive Effect of the Cap on
Initiation of mRNA Translation In Vitro

Several different kinds of experiments have provided support for a facilitating effect of the cap on mRNA translation
in cell-free systems (2). Messages with 5'-terminal m^7GpppN
were translated more effectively than the comparable unmethy-
lated mRNAs containing 5'-GpppN. Removal of the m^7G from
capped mRNAs by chemical (18) or enzymatic treatment (19) decreased translation in each instance, including viral and cellular mRNAs incubated in wheat germ extracts and rabbit reticulocyte lysates and Xenopus oocytes injected with globin mRNA
(20). The cap apparently influences translation at the initiation stage of protein synthesis by promoting the formation
of stable initiation complexes. For example, capped reovirus
mRNAs bound extensively to wheat germ ribosomes while the un-
methylated or decapped mRNAs formed few or no stable complexes
with ribosomes.

Ribosome binding experiments carried out with mRNA fragments also indicate an important effect of the cap (9). In
the case of the multiple species of reovirus mRNA, the cap
comprised part of the wheat germ 40S ribosome binding site as

defined by protection against RNase digestion. The 80S ribosome-protected fragments were a subset of the 40S-derived sequences. They retained the single initiator codon but not the
cap from the 40S-protected fragments. The ability of the 40S-
protected (capped) and 80S-derived fragments to rebind to
either wheat germ or reticulocyte ribosomes differed markedly.
For each pair of mRNA fragments, the capped fragment rebound
extensively (>80%) while most of the uncapped material failed
to form stable complexes with ribosomes (21).

Another approach to assess the role of 5'-terminal structure in ribosome binding involves the use of cap analogs such
as m^7G^5 pp which effectively inhibit mRNA binding to 40S ribosomal subunits (22). Studies with a variety of other cap-
related, modified guanosine nucleotides showed that 7-substi-
tution, but not a specific substituent, was required for this
effect (23). Binding of capped reovirus mRNA to wheat germ
ribosomes was decreased to the same extent (70-80%) by 0.1 mM
7-methyl-, 7-ethyl- or 7-benzyl-$G^5{}'$pp. Each of these compounds carries an extra plus charge on the imidazole portion
of the G as a consequence of the N^7 alkylation. Elimination
of the extra charge by treatment of the compounds with alkali
or by reduction with borohydride was accompanied by a loss of
inhibitory activity. From these and other findings, it was
suggested that the positively-charged imidazole ring of the 7-
substituted G interacts with the negatively-charged phosphate
groups to form a preferred conformation for ribosome binding.
Consistent with these indirect inhibition studies with cap
analogs, reovirus mRNA with an ethylated cap was translated in
vitro as well as methylated mRNA, was strongly inhibited by
0.1 mM $m^7G^5{}'$p, and yielded essentially the same products as
methylated mRNA in wheat germ extracts and reticulocyte lysates (data not shown).

It has been proposed that during initiation of eukaryotic
protein synthesis 40S ribosomal subunits bind initially to the
5'-end of mRNA and only subsequently move to the 5'-proximal
AUG which in most mRNAs functions as the initiator codon (9,
24). In capped mRNAs, the rigid conformation and base stacking of the cap (22) may enhance binding of the ribosomal subunits to the 5' end of mRNA, like a needle facilitating the
threading of beads on a string.

A Polypeptide in Eukaryotic Initiation Factor Preparations
that Cross-Links Specifically to 5'-Terminal Cap in mRNA

Since the presence of the cap facilitates initiation complex formation by capped mRNA, we considered the possibility
that one or more of the protein synthesis initiation factors

may be a protein that recognizes the cap. To test for factors that bind near or at the 5' end of mRNA, we used ^{3}H-methyl-labeled reovirus mRNA that had been oxidized to convert the 5'-terminal m^7G to a reactive dialdehyde. Complexes between factor and oxidized mRNA were reduced with sodium cyanoborohydride to stabilize any putative Schiff bases involving the m^7G dialdehyde and free amino groups in bound proteins. The complexes were digested with RNase to degrade all the polynucleotide chain but the cap, and proteins that were ^{3}H-labeled as a consequence of cross-linking of caps were resolved by SDS-polyacrylamide gel electrophoresis and located by fluorography (25). In a mixture of factors analyzed by this procedure only one cap-specific polypeptide (M_r = 24K) was detected (Fig. 3, left). Cross-linking of cap to the additional polypeptides (M_r = 50-55K) observed at 30° was not cap-specific since it was inhibited by GDP but not by m^7GDP. The 50-55K material was shown to correspond to EF-1 (data not shown). Among the several purified initiation factors surveyed, the 24K putative cap-binding polypeptide was present in factors eIF-3 (Fig. 3, center) and -4B (Fig. 3, right), but in less than stoichiometric amounts. Because the 24K polypeptide may be a new

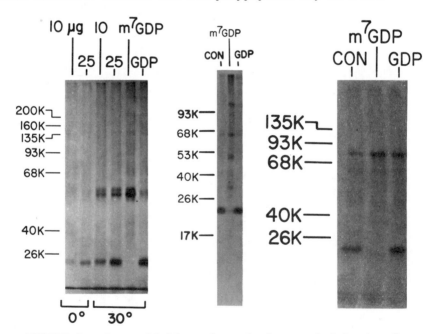

FIGURE 3. Cross-linking of reticulocyte initiation factors to cap in reovirus mRNA. Left: DEAE-cellulose purified factors. 15 µg protein in samples with 1 mM m^7GDP. Center: 15 µg eIF-3, 0.4 mM m^7GDP or GDP. Right: 2 µg eIF-4B, 0.4 mM m^7GDP or GDP. CON = control without G nucleotide (25).

protein synthesis initiation factor with a higher affinity for capped than uncapped mRNA, we have purified it for additional functional studies.

Affinity chromatography procedures were carried out in collaboration with Dr. Sidney Hecht. m^7GDP-Sepharose and GDP-Sepharose resins were prepared by carbodiimide mediated coupling of levulinic acid-(7-methyl guanosine-$O5'$-diphosphate-$O2',3'$-acetal) and levulinic acid-(guanosine-$O5'$-diphosphate-$O2',3'$-acetal), respectively, to AH-Sepharose-4B (26). Reticulocyte initiation factors prepared from high salt wash

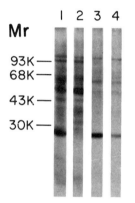

FIGURE 4.   Cross-linking pattern of proteins separated by affinity chromatography. An ammonium sulfate fraction (0-40%) of reticulocyte initiation factors was centrifuged in a 10-30% sucrose gradient to separate eIF-3 from lower molecular weight proteins (33). An aliquot of the pooled top fractions (1 ml, 19 mg) was loaded onto a 1 x 0.7 cm column of m^7GDP-Sepharose or GDP-Sepharose equilibrated with 20 mM Hepes buffer (pH 7.6), 0.1 M KCl, 1 mM DTT, 0.2 mM EDTA and 10% glycerol. The columns were washed with 50 ml of the same buffer and eluted in buffer containing 1 M KCl. Analysis of cross-linked proteins by gel electrophoresis and fluorography was as described (25) with the following amounts of protein. (1) 62 μg of material not bound to GDP-Sepharose;   (2) 96 μg of material not bound to m^7GDP-Sepharose;   (3) 6 μg of material eluted from m^7GDP- Sepharose in 1 M KCl;   (4) 6 μg of material eluted from GDP-Sepharose in 1 M KCl.

of ribosomes were applied to the resins in 0.1 M KCl and
eluted in 1 M KCl. The resulting protein fractions were
assayed for the ability to cross-link to the 5'-end of reovirus mRNA. As seen in the fluorograms of polyacrylamide
gels, almost all of the 24K polypeptide bound to the m^{7}GDP-
Sepharose with little of this polypeptide in the flow through
fraction (Fig. 4). By contrast, GDP-Sepharose retained only
a fraction of the 24K polypeptide, most of it eluting during
column loading.
    The specificity of the affinity purification procedure was
improved by changing the elution conditions. Samples were
again loaded in 0.1 M KCl but eluted with m^{7}GDP before addition of 1 M KCl. Stained gel analysis showed that most of the
proteins in the sample did not bind to the resin, and the gel
profile of the unbound material was similar to that of the
initial sample (Fig. 5). The fractions that eluted with m^{7}GDP

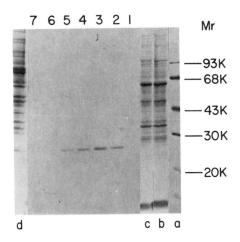

FIGURE 5. Stained gel profile of m^{7}GDP-Sepharose purified
24K polypeptide. Protein in the top fractions from the sucrose gradient (38 mg) was loaded onto m^{7}GDP-Sepharose, and the
column was washed as described in the legend to Fig. 4. Elution was in the same buffer but also containing 70 μM m^{7}GDP.
Fractions of 0.3 ml were collected, and 30 μl aliquots added
to wells 1-7 for polyacrylamide-SDS gel analysis. Lane a contains marker proteins. The other fractions and amounts
applied were: b, load, 19 μg from 38 mg total; c, flow
through, 17 μg/30 mg; d, 1 M KCl eluate, 10 μg/4 mg. From
the initial 38 mg protein, 40 μg of 24K polypeptide was
obtained in pooled fractions 2-7 after dialysis.

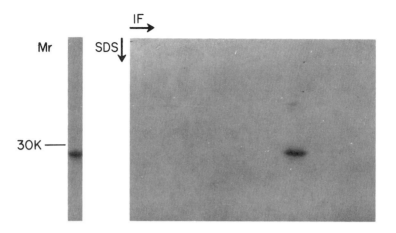

FIGURE 6. Autoradiogram of ^{125}I-labeled 24K polypeptide analyzed in one- and two-dimensional gel electrophoresis systems. m^{7}GDP-Sepharose purified 24K protein was iodinated by the Bolton-Hunter technique to a specific activity of 3 x 10^6 cpm/µg. Radioactive protein (20,000 cpm) was analyzed by electrophoresis in a 10-18% polyacrylamide-SDS gel and in a 2-D system (34).

contained almost exclusively a 24K polypeptide. High salt eluted the remaining spectrum of polypeptides. The polypeptide that eluted in m^{7}GDP cross-linked specifically to the 5'-cap of oxidized reovirus mRNA (data not shown). It thus corresponds to the polypeptide of the same M_r that contaminates factors eIF-3 and eIF-4B. As another test of purity, the material that bound to m^{7}GDP-Sepharose and eluted in m^{7}GDP was analyzed in one- and two-dimensional gels after iodination with I^{125}. The patterns consisted of a single polypeptide of molecular weight 24,000 (Fig. 6).

Some eukaryotic mRNAs including polio and EMC virus RNAs do not contain a cap and are translated by cap-independent mechanisms (27-29). In cells infected with these viruses capped mRNA translation is decreased in favor of the viral RNAs. This conversion has been ascribed to an inactivation of eIF-4B activity in poliovirus-infected cells (30). Furthermore, in a mouse cell-free translating system, EMC RNA has a competitive advantage over globin mRNA, apparently because it

has a higher affinity for eIF-4B and less of this activity is
needed to translate viral RNA that lacks a cap (31). eIF-4B
has also been reported to have cap-recognizing activity (32).
However, the present studies indicate that the cap-related activities found in purified eIF-4B may be due to the presence
of contaminating 24K polypeptide which we have purified to
homogeneity by affinity chromatography. It will be of interest to test its effects on the initiation of translation of
capped and uncapped mRNAs in protein synthesizing extracts
prepared from normal and virus-infected cells.

REFERENCES

1. Darnell, Jr., J.E., Science 202, 1257 (1978).
2. Shatkin, A.J., Cell 9, 645 (1976).
3. Rottman, F.M., "MTP International Review of Science-
    Biochemistry", ser. 3, Nucleic Acids, Butterworths,
    London (1977).
4. Perry, R.P., Ann. Rev. Biochem. 45, 605 (1976).
5. Furuichi, Y., Proc. Nat. Acad. Sci. USA 75, 1086 (1978).
6. Perry, R.P. and Kelley, D.E., Cell 8, 433 (1976).
7. Furuichi, Y., La Fiandra, A. and Shatkin, A.J., Nature
    266, 235 (1977).
8. Shimotohno, K., Kodama, Y., Hashimoto, J. and Miura,
    K-I., Proc. Nat. Acad. Sci. USA 74, 2734 (1977).
9. Kozak, M., Cell 15, 1109 (1978).
10. Furuichi, Y., Muthukrishnan, S., Tomasz, J. and Shatkin,
    A.J., J. Biol. Chem. 251, 5043 (1976).
11. Ensinger, M.J., Martin, S.A., Paoletti, E. and Moss, B.,
    Proc. Nat. Acad. Sci. USA 72, 2525 (1975).
12. Abraham, G., Rhodes, D.P. and Banerjee, A.K., Cell 5, 51
    (1975).
13. Furuichi, Y., Nuc. Acids Res. 1, 809 (1974).
14. Salditt-Georgieff, M., Jelinek, W., Darnell, J.E., Furuichi,
    Y., Morgan, M. and Shatkin, A., Cell 7, 227 (1976).
15. Groner, Y., Gilboa, E. and Aviv, H., Biochemistry 17, 977
    (1978).
16. Ro-Choi, T.S., Raj, N.B.K., Pike, L.M. and Busch, H.,
    Biochemistry 15, 3823 (1976).
17. Mahler, H.R. and Cordes, E.H., in "Biological Chemistry"
    p. 267. Harper and Row, New York (1971).
18. Muthukrishnan, S., Both, G.W., Furuichi, Y. and Shatkin,
    A.J., Nature 255, 33 (1975).
19. Zan-Kowalczewska, M., Bretner, M., Sierakowska, H.,
    Szczesna, E., Filipowicz, W. and Shatkin, A.J., Nuc.
    Acids Res. 4, 3065 (1977).

20. Lockard, R.E. and Lane, C., Nuc. Acids Res. 5, 3237 (1978).
21. Kozak, M. and Shatkin, A.J., Cell 13, 201 (1978).
22. Hickey, E.D., Weber, L.A., Baglioni, C., Kim, C.H. and Sarma, R.H., J. Mol. Biol. 109, 173 (1977).
23. Adams, B.L., Morgan, M., Muthukrishnan, S., Hecht, S.M. and Shatkin, A.J., J. Biol. Chem. 253, 2589 (1978).
24. Kozak, M. and Shatkin, A.J., J. Biol. Chem. 253, 6568 (1978).
25. Sonenberg, N., Morgan, M.A., Merrick, W.C. and Shatkin, A.J., Proc. Nat. Acad. Sci. USA 75, 4843 (1978).
26. Seela, F. and Waldek, S., Nuc. Acids Res. 2, 2343 (1975).
27. Nomoto, A., Lee, Y.F. and Wimmer, E., Proc. Nat. Acad. Sci. USA 73, 375 (1976).
28. Hewlett, M.J., Rose, J.K. and Baltimore, D., Proc. Nat. Acad. Sci. USA 73, 327 (1976).
29. Frisby, D., Eaton, M. and Fellner, P., Nuc. Acids Res. 3, 2771 (1976).
30. Rose, J.K., Trachsel, H., Leong, K. and Baltimore, D., Proc. Nat. Acad. Sci. USA 75, 2732 (1978).
31. Golini, F., Thach, S.S., Birge, C.H., Safer, B., Merrick, W.C. and Thach, R.E., Proc. Nat. Acad. Sci. USA 73, 3040 (1976).
32. Shafritz, D.A., Weinstein, J.A., Safer, B., Merrick, W.C., Weber, L.A., Hickey, E.D. and Baglioni, C., Nature 261, 291 (1976).
33. Benne, R. and Hershey, J.W.B., Proc. Nat. Acad. Sci. USA 73, 3005 (1976).
34. O'Farrell, P.H., J. Biol. Chem. 250, 4007 (1975).

DISCUSSION

R. PERRY: I was impressed by the sequence data indicating that the reovirus messages have a common tetranucleotide at the 5' end. The sequence contiguous to the cap is the same, and yet these messenger RNAs are presumably all transcribed independently. Do you have any speculation as to the significance of that tetranucleotide? Why is it so strictly conserved?

A. SHATKIN: What one could think with these multipartite genome viruses is that originally the viral genome consisted of a single piece of RNA which then evolved into more fragments. Some regions were conserved as the virus became more and more cell-independent. As new cistrons developed from the previously existing ones, perhaps those portions of the previous ones that allowed efficient replication were conserved. We are more impressed, however, by the divergence of the sequences after the first four nucleotides. We could not find anything that was strikingly complementary to the 3'-terminal sequence of the 18S ribosomal RNA that would provide a Shine–Dalgarno base pairing, for example.

K. GUPTA: Dr. Banerjee at your Institute with VSV and we, working with spring viremia of Casp virus, another rhabdo virus, observed that methylation of penultimate nucleoside preceded guanosine. Do you think this is specific only to rhabdoviruses or any other examples known?

A. SHATKIN: You are asking whether methylation of the 2'-position of the penultimate residue occurs before N^7 methylation of the terminal G in the cap? I know of no case except the in vitro situation with VSV and very low S-adenosylmethionine concentrations. This observation would be consistent with VSV capping occurring by a mechanism different from that of reovirus and CPV. If VSV messenger is made from a larger precursor, we could think of the possibility that the 2'-O-methylated residue is derived from a cleavage site that is recognized as a special nucleotide. This is pure speculation.

K. GUPTA: We had a situation where a large number of penultimate nucleosides were methylated as compared to guanosine even in the presence of large amounts of s-adenosyl-L-methionine (SAM).

A. SHATKIN: I assume that is a comment and not a question so perhaps we should go on to another question.

J. DARNELL: Do you know anything about the distribution of the 24-K protein in the cell yet? Is there any in the nucleus? Is there any in post-polysomal particles?

A. SHATKIN: N. Sonenberg looked at S100 fractions and found very little. So, at least for S100 vs. the total cytoplasmic extract, most of the 24K protein is ribosome-associated. It may also be in the nucleus and have something to do with mRNA transport from the nucleus to the cytoplasm. We do not know at this stage.

T. HUNT: Aaron, I would like to press you on the function of the 24K protein. Would you be prepared to say more about that?

A. SHATKIN: In collaboration with John Bergmann, Hans Trachsel and Harvey Lodish at M.I.T., we tested for the presence of 24K protein in initiation factor fractions which they found contained eIF-3 and -4B activity and stimulated the translation of chemically decapped VSV mRNA. By sucrose gradient centrifugation, the eIF-3 and -4B activities could be separated from each other and from the ability to stimulate decapped mRNA translation. The latter activity co-purified with the 24K polypeptide. One of our current ideas is that the diminished translation of decapped mRNA can be compensated for by addition of an excess of initiation factors including the 24K polypeptide. Thus, decapped RNA would have a lower affinity for the cap-recognizing protein(s). In the presence of an excess of such protein(s), there is partial restoration of the translation of decapped RNA. The restoring activity collected from the gradient included in addition to the 24K protein two other polypeptides that also differed from eIF-3 and -4B. When we put the purified material consisting of the three polypeptides through a column of m⁷GDP-Sepharose, the 24K was specifically bound. Upon elution, it was inactive, and we could not reconstitute the stimulatory activity by mixing the fractions. We need to try other preparations and different conditions.

P. CHAMBON: Is there any genetic evidence that a mutant deficient in methyl transferase could be impaired in transcription?

A. SHATKIN: My impression is that those mutants have many phenotypic defects, making it difficult to sort out the primary effect. Perhaps an in vitro nuclear system will be more useful to test possible effects of S-adenosylmethionine or capped structures on transcription.

SPLICE PATTERNS OF ADENOVIRUS
AND ADENOVIRUS-SV40 MOSAIC RNA'S

Heiner Westphal
Geoffrey R. Kitchingman
Sing-Ping Lai

Laboratory of Molecular Genetics
National Institute of Child Health and Human Development
National Institutes of Health
Bethesda, Maryland

Charles Lawrence
Tony Hunter
Gernot Walter

Tumor Virology Laboratory
The Salk Institute
San Diego, California

I.   INTRODUCTION

The phenomenon of RNA splicing has profoundly influenced
current concepts of transcriptional controls in eukaryotic
systems, and is extensively discussed in several contributions presented at this Conference. Following the discovery
of intervening sequences that interrupt early adenovirus
genes (Westphal and Lai, 1977 a&b), our laboratory has continued to examine, by electron microscopy, nuclear and cytoplasmic viral RNAs in an effort to learn more details about
the process of eukaryotic messenger RNA maturation. In the
first part of this paper (G.R. Kitchingman and H. Westphal),
we present experiments which support the notion that intervening sequences are incorporated in nuclear precursor RNA
during primary transcription of early adenovirus gene blocks,
and that they are removed during formation of cytoplasmic
messenger RNA. The second study (H. Westphal, S.-P. Lai,
C. Lawrence, T. Hunter and G. Walter) deals with Adenovirus-
SV40 mosaic messenger RNAs specified by the nondefective
hybrid virus Ad2[+]ND$_4$. Intricate pathways of RNA maturation
are signaled by the complex structures of these messages.

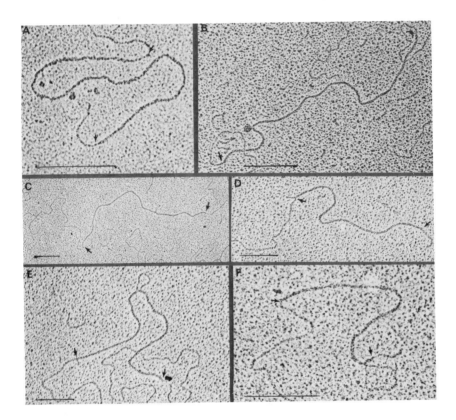

Figure 1. *Electron micrographs of hybrids formed between the separated strands of Ad2 DNA and poly(A)-containing early RNA. Cells were harvested 6 hours after infection (10⁴ particles/cell), and nuclear and cytoplasmic RNA was prepared by the procedure of Penman et al. (1969). Details of the hybridization and visualization procedures have been published (Westphal and Lai, 1977a). Depicted are regions of individual DNA molecules that are hybridized to RNA (between arrows). Panels A-D, continuous RNA/DNA hybrids of early regions 1, 3, 2, and 4, respectively. Panels E and F, RNA/ DNA hybrids interrupted by loops of single-stranded, intervening DNA. The hybrid of panel E was generated by an RNA of the type 2b (Figure 2), the hybrid of panel F by an RNA of the type 4b. The bar in these electron micrographs as well as those shown in Figure 4 correspond to 0.25 μm.*

II. RESULTS

A. The Structures of Early Adenovirus RNA

In cells lytically infected with adenovirus, four
separate early gene blocks are transcribed before the onset
of viral DNA synthesis (for review, see Flint, 1977). During
the course of mapping the transcripts by electron microscopy,
we discovered that the early viral mRNAs are composed of
nucleotide sequences which are not contiguous on the template
DNA. The early mRNAs give rise to RNA/DNA duplex structures
which are interrupted at specific locations by loops of
intervening DNA (Westphal and Lai, 1977 a&b; Kitchingman et.
al., 1977). At the same time, other laboratories reported
that hybrids of late adenovirus mRNAs and complementary DNA
revealed even more complex structures of noncontiguous transcripts (Berget et al., 1977; Chow et al., 1977), and the
theory was advanced that the viral mRNAs may be derived from
continuous nuclear precursor RNA by a cut-and-splice mechanism (see Klessig, 1977, for details of this model).

Adenovirus Early Nuclear RNAs Contain Intervening
Sequences which are Missing from Cytoplasmic RNA   The cut-
and-splice model predicts that the nuclear precursor RNAs
contain internal sequences which are absent from cytoplasmic
mRNAs. In the examples of Figure 1, we show poly(A)-contain-
ing early adenovirus type 2 (Ad2) nuclear RNA hybridized to
template DNA (between arrows). Panels A-D show RNAs that
formed continuous duplexes with the DNA in each of the four
early regions. In panel E, we see a nuclear RNA molecule of
region 2 which bridges a loop of intervening DNA sequences.
One example of a cytoplasmic RNA hybridizing to region 4 DNA
is shown in panel F (for electron micrographs of the other
cytoplasmic RNAs, see Kitchingman et al., 1977). Figure 2
lists the early RNA molecules most freqently observed in the
nucleus (solid boxes) or cytoplasm (hatched boxes). Wher-
ever the ends of molecules or the positions of intervening
sequences (white boxes) fall within one standard deviation of
each other, they have been aligned in this display. Table I
gives the actual measurements, including error limits.

We notice continuous nuclear transcripts in each of the
four early gene blocks, as well as shorter cytoplasmic RNAs
which lack internal sequences. Within the limits of accuracy
of electron microscopy, the map positions of corresponding

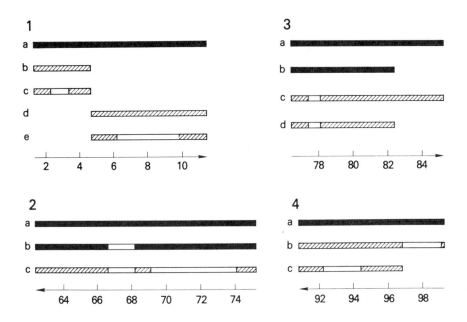

*FIGURE 2. Positions of nuclear and cytoplasmic early Ad2
RNAs on the Ad2 map. Arrows indicate direction of transcription. Nuclear RNAs are indicated by solid boxes,
cytoplasmic RNAs by hatched boxes. White boxes mark the
positions of intervening DNA sequences which are not represented in the RNA.*

nuclear and cytoplasmic RNAs appear very similar, if not
identical. This is exactly what we expect if the cytoplasmic
RNAs are derived from nuclear precursors by a cut-and-splice
mechanism. Moreover, the data of Figure 2 allow us to develop some ideas concerning the temporal sequence of RNA processing. We may postulate, for instance, that in region 2
a precursor molecule 2a is processed in the nucleus to give
rise to molecule 2b, and that a subsequent step generates
the mature mRNA 2c. Likewise, in region 1, cleavage of
molecule 1a near map position 4.5 may well generate molecules
1b and 1d which, in turn, may be considered as precursors of
the mature mRNAs 1c and 1e. Similar pathways can be constructed for the molecules in regions 3 and 4. It should be
emphasized that no direct evidence is available for the proposed processing pathways of early Ad2 RNAs, but several
experiments reported in the recent literature lend strong
support to these models. In the case of mouse β-globin mRNA,

TABLE I. Map Positions of Ad2 Nuclear and Cytoplasmic RNAs[a]

	RNA		Intervening sequences			RNA		
RNA species	5' end	Length (bases)	Start	Loop length (bases)	End	Length (bases)	3' end	N
1a	1.5 ± 1.0	3340 ± 440					10.8 ± 1.0	15
1b	1.4 ± 0.7	1185 ± 245					4.8 ± 0.7	36
1c	1.1 ± 0.3	400 ± 145	2.3 ± 0.5	375 ± 140	3.4 ± 0.4	500 ± 180	4.9 ± 0.6	15
1d	4.5 ± 0.8	2290 ± 350					11.0 ± 0.8	33
1e	4.6 ± 0.4	590 ± 135	6.3 ± 0.6	1320 ± 190	10.1 ± 0.6	525 ± 110	11.6 ± 0.6	24
2a	74.7 ± 1.3	4440 ± 410					62.1 ± 1.4	22
2b	74.6 ± 1.6	2230 ± 315	68.1 ± 1.1	665 ± 80	66.4 ± 1.1	1730 ± 300	61.4 ± 1.5	15
2c	74.9 ± 0.9	<100	74.6 ± 0.9	2240 ± 220	68.2 ± 1.1			
		<100	67.9 ± 1.1	660 ± 60	66.0 ± 1.1	1600 ± 245	61.5 ± 1.0	25
3a	76.9 ± 1.1	3120 ± 560					85.9 ± 1.4	17
3b	76.1 ± 1.3	2175 ± 440					82.2 ± 0.9	46
3c	76.7 ± 0.9	370 ± 45	77.7 ± 0.8	420 ± 50	78.9 ± 0.8	2130 ± 340	85.0 ± 1.5	10
3d	76.0 ± 1.4	385 ± 120	77.0 ± 1.3	450 ± 70	78.2 ± 1.4	1430 ± 215	82.5 ± 1.5	118
4a	99.0 ± 0.6	2720 ± 415					91.1 ± 0.7	10
4b	99.3 ± 0.4	110 ± 80	99.0 ± 0.4	960 ± 190	96.2 ± 0.7	1830 ± 330	91.2 ± 1.3	21
4c	96.8 ± 0.8	800 ± 300	94.5 ± 0.5	760 ± 120	92.2 ± 1.0	410 ± 210	91.0 ± 1.1	17

[a] RNA species correspond to the RNAs depicted in Fig. 2. The 5' (column 2) and 3' (column 8) ends refer to the map positions of the start and stop of the RNA on the standard Ad2 map. The ± designation is one standard deviation. RNA lengths, in nucleotides, refer to the distance between two distinct structural features. For uninterrupted RNA/DNA duplexes, this figure (column 3) represents the total length of the RNA. For RNAs which generate loops of intervening DNA sequences, the figures refer to the number of nucleotides between the 5' end of the RNA and the start of the intervening sequence (column 3) on the one hand, and between the end of the intervening sequence and the 3' end of the RNA (column 7) on the other. Columns 4 and 6 contain the map positions for the start and end of the intervening sequence of DNA. The number of nucleotides of DNA which separate the two segments of DNA to which the RNA is complementary is found in column 5. N refers to the number of molecules that were used in the statistical analysis for each RNA species.

pulse-chase experiments have demonstrated that a 15S nuclear
RNA is converted to 9S cytoplasmic RNA (Curtis and Weissman,
1976; Ross, 1976; Bastos and Aviv, 1977). Correspondingly,
the nuclear globin RNA contains the intervening sequences
which are missing from the mature mRNA (Tilghman et al.,
1978). Further, Blanchard et al. (1978) have shown that 28S
nuclear RNA from early region 2 of Ad2 (corresponding to
molecule 2a) can be converted in isolated nuclei to two smaller RNAs, one of which has the characteristics of molecule 2c.

In conclusion, processing of early Ad2 RNA by a cut-and-
splice mechanism is very likely. However, the picture is
certainly more complicated than indicated by Figure 2 because
several additional types of early RNA molecules which may
play a role as intermediates in early Ad2 RNA processing have
been observed (Berk and Sharp, 1978a; our unpublished results).

B. Mosaic $Ad2^+ND_4$ RNA

In the following, we will summarize our characterization of mosaic Ad2-SV40 messenger RNAs extracted from the
cytoplasm of human cells infected with the nondefective hybrid virus $Ad2^+ND_4$. A more extensive account of our experiments will be published elsewhere.

In $Ad2^+ND_4$ (for review, see Lewis, 1977), a deletion of
Ad2 sequences which extends from position 81 to 86 has been
replaced by a segment of SV40 DNA representing almost the
entire early region of that virus. The SV40 sequences of
$Ad2^+ND_4$ are transcribed in rightward polarity, that is, in
the same orientation as late Ad2 RNA. The expression of at
least part of the integrated early SV40 information, including the C-terminal region of the SV40 T antigen (Fey et al.,
in press) is required for the growth of $Ad2^+ND_4$ in monkey
cells. But although the hybrid virus grows well on monolayers of monkey kidney cells, it is more conveniently
propagated in spinner cultures of human tumor cells. It is
for this reason that gene expression of $Ad2^+ND_4$ has been
mostly studied in human cells. In HeLa cells, $Ad2^+ND_4$ specifies a series of polypeptides that are related to the SV40
T antigen and form an overlapping set with a common C-terminal region. A similar set of polypeptides is obtained when
SV40-specific $Ad2^+ND_4$ mRNA is translated in vitro (Mann et.
al., 1977). Earlier reports suggested that RNA giving rise
to SV40-specific polypeptides in cells infected with non-

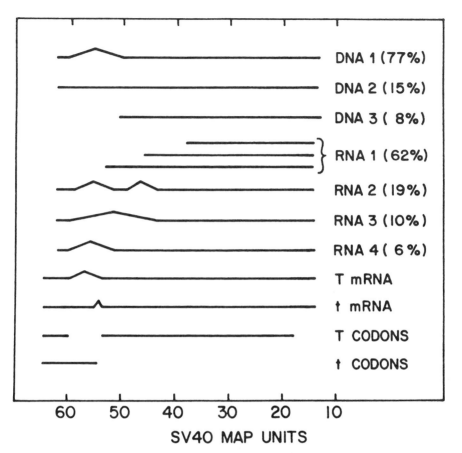

FIGURE 3. Map of SV40 sequences in Ad2+ND4 DNA and RNA.
The map positions of individual populations of DNA and RNA
(see text) are indicated by horizontal lines. Interruptions
in these lines indicate the position of internal sequences
that are missing from the molecules. Exact map coordinates
and error limits are listed in Table II. This Figure gives
mean values. Since RNA 1 is a heterologous population, three
representative RNA molecules have been drawn as examples.
The relative abundance of DNA and RNA populations (percent
values) are based on the observation of 600 DNA and 400 RNA
molecules. The structures of the mRNAs of the SV40 T and
t antigens are copied from Berk and Sharp (1978b). The positions of the codons for T and t antigens have been derived
from Fiers et al. (1978) and Reddy et al. (1978). The
direction of transcription is from left to right.

defective hybrid viruses consists of covalently linked adenovirus and SV40 sequences (Oxman et al., 1974; Dunn and Hassell, 1977). We wished to determine the structure of Ad2$^+$ND$_4$
mRNAs in an effort to learn about the controls which guide
their synthesis and expression.

1. The SV40 sequences of Ad2$^+$ND$_4$ undergo modifications
in human cells. We began our study with an analysis of the
SV40 DNA sequences contained in Ad2$^+$ND$_4$ stocks which had been
propagated in human cells. Heteroduplexes of complementary
Ad2$^+$ND$_4$ and SV40 DNA strands were formed and observed in the
electron microscope. Figure 3 indicates the structures which
we determined. DNA 1, the most abundant form, corresponds to
Ad2$^+$ND$_{4del}$ previously described by Morrow et al. (1973).
SV40 sequences bounded by map positions 60 and 50 are deleted
from this DNA. Less abundant is DNA 2, containing the SV40
segment originally described for Ad2$^+$ND$_4$ DNA (Kelly and Lewis,
1973). Molecules of DNA 3 are similar to those of DNA 1,
except no SV40 DNA was found hybridized to them beyond map
position 50 (see Table II for coordinates).

Remarkably, the SV40 sequences deleted from Ad2$^+$ND$_4$
DNAs 1 and 3 overlap with intervening sequences that are
found removed from the early genes of SV40. Part of these
sequences, near position 54, contain stop signals for protein
synthesis in all reading frames (Fiers et al., 1978; Reddy
et al., 1978), and are not present in any of the mature early
SV40 mRNAs (Berk and Sharp, 1978b). The major part of the
intervening sequences is preserved in the mRNA coding for
SV40 t antigen, but is absent from the T antigen message.
Figure 3 illustrates the position of the t and T codons
(Fiers et al., 1978; Reddy et al., 1978), and of the corresponding SV40 mRNAs (Berk and Sharp, 1978b).

Since Ad2$^+$ND$_{4del}$ is absent when cloned Ad2$^+$ND$_4$ is propagated in monkey cells (A.M. Lewis, Jr., personal communication), but appears with high frequency in human cells, it
seems possible that this deletion offers a selective advantage for the growth of the hybrid virus in the human host
cell. We may speculate, for instance, that intervening SV40
sequences, located within the integrated SV40 segment, are
detrimental to the growth of Ad2 in human cells, and are
therefore removed by recombination.

2. SV40 sequences in Ad2$^+$ND$_4$ mRNA are arranged in a
variety of structures. The heterogeneity of SV40-specific
macromolecules found in Ad2$^+$ND$_4$ infected human cells is not
restricted to virus DNA and proteins. When we annealed

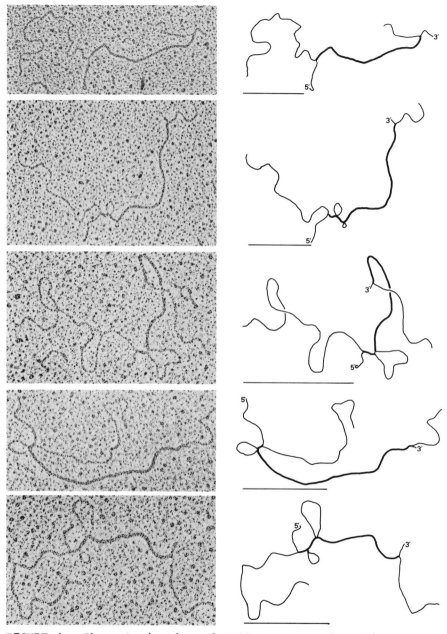

FIGURE 4. Characterization of SV40 sequences in Ad2⁺ND₄ RNA.
SV40-specific, poly(A)-containing cytoplasmic RNA (Mann et
al., 1977) was hybridized to the separated (Hayward, 1972)
strands of SV40 DNA cut by EcoRI endonuclease. Heavy
contours indicate the position of the DNA/RNA helix, light
contours the ends of unhybridized DNA and RNA sequences.

SV40-specific RNA, derived from cells harvested late in infection, to the complementary regions of SV40 DNA, we observed
a variety of structures in the electron microscope (Figure 4).
The most abundant forms of RNA (classes 1-4) are schematically displayed in Figure 3, and their coordinates are listed in
Table II. Molecules of class 1 form continuous hybrids with
the DNA. This RNA population is heterogenous in size, mainly
with respect to the location of the 5' terminal SV40 sequences. In contrast, classes 2-4 are seemingly homogeneous
populations. Class 4, like DNA 1 or like the mRNA for SV40
T antigen, excludes the main intervening sequence interrupting early SV40 genes. At present, we do not know which of
the various forms of Ad2+ND4 DNA served as the template for
the individual classes of RNA which we observe. For instance, RNA 4 may be derived either from DNA 1 or from DNA 2.
However, whatever the template, we favor the idea that precise splicing on the RNA level, characteristic for many eukaryotic messenger RNAs, has generated the reading frames for
individual polypeptides which are expressed by the various
composite RNAs. We know, for example, that the largest of
the SV40-specific AD2+ND4 proteins, a polypeptide of 95,000
daltons, is coded in part to the left of position 60, that
is, upstream of the intervening sequence which interrupts the
gene continuity of T antigen, and that it is colinear in
amino acid sequence with the T antigen. Likewise, a number
of smaller proteins are initiated, most probably on individual classes of mRNA, at different sites along the T antigen coding sequence. All of these SV40-specific Ad2+ND4
polypeptides, like the SV40 T antigen, terminate in a common
carboxy-terminal sequence near map position 14 (C. Lawrence,
T. Hunter, and G. Walter, unpublished).

Some contact points of noncontiguous SV40 sequences in
Ad2+ND4 RNAs, one at position 49 (RNA 2), the other at
position 44 (RNAs 2 and 3), plus rare ones such as in Figure
4, panel 5, clearly distinguish these transcripts from SV40
RNA. It is possible that the contact points unique to
Ad2+ND4 RNA result from processing of SV40 RNA sequences in
the altered genetic context of an Ad2-SV40 mosaic RNA precursor.

---

*The polarity of RNA is indicated in the explanatory line
drawings. Panels 1-4 show examples of RNA populations 1-4
(see Fig. 3). A rare RNA molecule is shown in panel 5.
Single-stranded loops interrupting the RNA/DNA contour in
panels 2-5 represent intervening sequences.*

Table II. Map Coordinates of SV40 Sequences
in Ad2$^+$ND$_4$ DNA and RNA[a]

	N	Left Terminus	Position of Displacement Loop(s)	Right Terminus
DNA 1	70	62.4 $\pm$ 2.9	60.7 $\pm$ 2.8/49.9 $\pm$ 2.6	13.0 $\pm$ 2.0
DNA 2	32	62.0 $\pm$ 3.9	--	13.4 $\pm$ 2.1
DNA 3	17	50.8 $\pm$ 1.5	--	12.1 $\pm$ 1.6
RNA 1	100	46.1 $\pm$ 7.8	--	16.2 $\pm$ 4.4
RNA 2	42	62.5 $\pm$ 2.1 {	59.3 $\pm$ 2.0/52.0 $\pm$ 2.2 49.4 $\pm$ 2.3/44.3 $\pm$ 2.5	14.1 $\pm$ 2.4
RNA 3	18	62.3 $\pm$ 1.8	60.4 $\pm$ 2.0/44.5 $\pm$ 2.1	14.2 $\pm$ 1.5
RNA 4	35	62.1 $\pm$ 2.1	60.2 $\pm$ 1.9/52.0 $\pm$ 2.8	14.4 $\pm$ 2.0

[a]*N = number of molecules measured. Coordinates are expressed
as percent of contour length of SV40 DNA cut by EcoRI restriction endonuclease. Error limits indicate one standard
deviation.*

3. Composite Ad2 leader sequences are covalently
attached to the 5' ends of SV40 sequences in Ad2$^+$ND$_4$ mRNA.
In their original study of Ad2-SV40 mosaic RNAs, Oxman et al.
(1974) showed that the synthesis of SV40 T antigen in
Ad2$^+$ND$_4$ infected cells is under adenovirus gene control.
They suggested that the synthesis of RNA giving rise to this
protein was initiated in Ad2 sequences. Consistent with this
idea were results of Dunn and Hassell (1977) who demonstrated
that RNA specified by Ad2$^+$ND$_4$ hybridized both to SV40 DNA
and to Ad2 DNA restriction fragments located upstream of the
SV40 insertion in Ad2$^+$ND$_4$ DNA. In the examples of Figure 4,
these adenovirus sequences may be discerned as small tails,
protruding from the 5' end of the Ad2$^+$ND$_4$ mRNAs. We hybridized these mRNAs to Ad2 or to Ad2$^+$ND$_4$ DNA in order to establish the structure of the leader sequences. The result is
shown in Table III. Up to five different Ad2 sequences,

TABLE III. Ad2 Sequences Attached to SV40 RNA
in Ad2+ND4-Infected HeLa Cells

	Ad2 Map Position	Number of Observations
A	16.8 ± 0.7	93
B	19.9 ± 0.7	93
C	26.9 ± 0.8	93
D	75.0 ± 0.5 to 75.6 ± 0.5	18
E	76.8 ± 0.5 to 77.5 ± 0.4	57
F	78.5 ± 0.6 to 79.2 ± 0.6	42

Observed Composition of Ad2-SV40 Mosaic RNAs

Composition	Number of Observations
A + B + C + SV40	9
A + B + C + E + SV40	24
A + B + C + D + E + SV40	18
A + B + C + F + SV40	27
A + B + C + E + F + SV40	15

located in six nonadjacent regions of the genome, are found
attached, in various combinations, to the 5' end of SV40
sequences in individual Ad2+ND4 RNAs. Sequences A-C form the
familiar tripartite leader of late Ad2 cytoplasmic RNAs
(Berget et al., 1977; Chow et al., 1977). Between this
tripartite leader and SV40 RNA we find one or the other of
sequences D, E, or F, located in the immediate vicinity of
the 5' end of two early Ad2 gene blocks transcribed in opposite directions (Kitchingman et al., 1977). In Ad2+ND4 mRNA
class 1, the Ad2 leader sequences are attached to various
points within the SV40 sequence, whereas in RNAs 2-4, attach-

ment is close to the left Ad2/SV40 junction at Ad2 map position 81, either at the 5' end of the SV40 sequences or in the Ad2 sequences immediately upstream of the junction.

The presence of the tripartite leader at the 5' ends of the SV40-specific Ad2$^+$ND$_4$ mRNAs suggests that the SV40 sequences are included in the late primary transcript which initiates near position 16 on the Ad2 map and terminates to the right of the fiber gene (Goldberg et al., 1977; Ziff and Evans, 1978). Less well understood is the fact that sequences bounded by Ad2 map positions 75 and 79 (D, E, and F of Table III) are inserted between the tripartite leader and SV40 RNA sequences in a high proportion of the mosaic RNAs. Remarkably, sequences corresponding to positions E and F have been detected in Ad2 fiber mRNA by Chow and Broker (1978). Their abundance in that mRNA increases late in infection. The similarities between the 5' ends of the fiber mRNA and of the Ad2$^+$ND$_4$ mRNAs suggest that the Ad2 RNA coded to the left of Ad2 map position 81 is processed independently of the sequences present on the 3' end of the RNA molecules, whether they be Ad2 or SV40 coded.

ACKNOWLEDGMENTS

We thank Terri Broderick for expert editorial assistance. G.R.K. was supported by a fellowship from the Damon Runyon-Walter Winchell Cancer Fund. C.L. is a Special Fellow of the Leukemia Society of America. C.L., T.H. and G.W. were supported by grants nos. CA 17096 and CA 15365 from the National Cancer Institute.

REFERENCES

Bastos, R.N., and Aviv, H. (1977). Cell 11, 641.
Berget, S.M., Moore, C., and Sharp, P.A. (1977). Proc. Natl.
    Acad. Sci. USA 74, 3171.
Berk, A.J., and Sharp, P.A. (1978a). Cell 14, 695.
Berk, A.J., and Sharp, P.A. (1978b). Proc. Natl. Acad. Sci.
    USA 75, 1274.
Blanchard, J.-M., Weber, J., Jelinek, W., and Darnell, J.E.
    (1978). Proc. Natl. Acad. Sci. USA 75, 5344.
Chow, L.T., and Broker, T.R. (1978). Cell 15, 497.
Chow, L.T., Gelinas, R.E., Broker, T.R., and Roberts, R.J.
    (1977). Cell 12, 1.
Curtis, P.J., and Weissmann, C. (1976). J. Mol. Biol. 106,
    1061.
Dunn, A.R., and Hassell, J.A. (1977). Cell 12, 23.
Fey, G., Lewis, J.B., Grodzicker, T., and Bothwell, A. (1979).
    J. Virol., in press.
Fiers, W., Contreras, R., Haegeman, G., Rogiers, R., van de
    Voorde, A., van Heuverswyn, H., van Herreweghe, J.,
    Volckaert, G., and Ysebaert, M. (1978). Nature 237, 113.
Flint, J. (1977). Cell 10, 153.
Goldberg, S., Weber, J., and Darnell, J.E., Jr. (1977).
    Cell 10, 617.
Hayward, G.S. (1972). Virology 49, 342.
Kelly, T.J., Jr., and Lewis, A.M., Jr. (1973). J. Virol.
    12, 643.
Kitchingman, G.R., Lai, S.-P., and Westphal, H. (1977). Proc.
    Natl. Acad. Sci. USA 74, 4392.
Klessig, D.F. (1977). Cell 12, 9.
Lewis, A.M., Jr. (1977). Prog. Med. Virol. 23, 96.
Mann, K., Hunter, T., Walter, G., and Linke, H. (1977).
    J. Virol. 24, 151.
Morrow, J.F., Berg, P., Kelly, Th.J., Jr., and Lewis, A.M.,
    Jr. (1973). J. Virol. 12, 653.
Oxman, M.N., Levin, M.J., and Lewis, A.M., Jr. (1974).
    J. Virol. 13, 322.
Penman, S., Greenberg, H., and Williams, M. (1969). In
    "Fundamental Techniques in Virology" (K. Habel and .
    N.P. Salzman, eds.), p. 49. Academic Press, New York.
Reddy, V.B., Thimmappaya, B., Dhar, R., Subramanian, K.N.,
    Zain, B.S., Pan, J., Gosh, P.K., Celma, M.L., and
    Weissman, S.M. (1978). Science 200, 494.

Ross, J. (1976). J. Mol. Biol. 106, 403.
Tilghman, S.M., Curtis, P.J., Tiemeier, D.C., Leder, P., and Weissmann, C. (1978). Proc. Natl. Acad. Sci. USA 75, 1309.
Westphal, H., and Lai, S.-P. (1977a). J. Mol. Biol. 116, 525.
Westphal, H., and Lai, S.-P. (1977b). Cold Spring Harbor Symp. Quant. Biol. 42, 555.
Ziff, E.B., and Evans, R.M. (1978). Cell 15, 1463.

PROCESSING OF BACTERIOPHAGE T7 RNAs BY RNase III[1]

F. William Studier
John J. Dunn
Elizabeth Buzash-Pollert[2]

Department of Biology
Brookhaven National Laboratory
Upton, New York

I. INTRODUCTION

RNase III of Escherichia coli was originally described as an enzyme that nonspecifically degrades double-stranded RNA (1), but its physiological role appears to be as an RNA processing enzyme (2). In uninfected E. coli, RNase III specifically cuts primary transcripts for ribosomal RNAs, some tRNAs and perhaps other RNAs (3-5). During infection by bacteriophage T7, RNase III cuts the T7 RNAs at specific sites to produce both early and late T7 messenger RNAs (2). The ribosomal and tRNAs are further processed by other enzymes (6), but RNase III seems to be the only enzyme involved in processing the T7 messenger RNAs; the ends produced in vitro by purified RNase III are identical to those found in messenger RNAs isolated from infected cells (7).

Cutting by RNase III in vivo is very rapid and specific. In cells that have normal levels of RNase III, no uncut precursor RNAs are observed, even with short pulses of label, indicating that the cuts are made as soon as cleavage sites are exposed during transcription. Only five sites are

---

[1]Research carried out under the auspices of the United States Department of Energy under Contract No. EY-76-C-02-0016.

[2]Supported by NIH grant 5-T32-CA09121-03.

cut in the 7000-base long T7 early messenger RNA, and only a
few sites are cut in the primary transcript for E. coli
ribosomal RNAs. These are also the preferred cleavages in
vitro by purified RNase III, and are referred to as primary
cuts (8). However, at low ionic strength the purified
enzyme will also cut at secondary sites (8). These are also
specific, but are cut much less efficiently than primary
sites. Purified RNase III acts directly on purified RNA,
requiring neither DNA nor auxiliary proteins.

Most RNAs that are not natural substrates for RNase III
seem to lack primary cleavage sites, but many different RNAs
can be cut at specific secondary sites by using an
appropriate ionic strength and enzyme concentration. This
has made RNase III a useful tool for dissecting RNA
molecules. Specific regions of an RNA can be located and
isolated for study and sequence analysis, in much the same
way as specific DNA fragments can be isolated by using
restriction endonucleases (9).

RNase III does not appear to release small RNA fragments
from either primary or secondary cleavage sites, indicating
that the enzyme makes a single cut rather than a cluster of
cuts (10). A 29-base fragment is released from one of the
cleavage sites in T7 early RNA, but this arises from the
combination of a primary cut and a relatively strong
secondary cut located 29 bases apart (8,11). The two cuts
appear to be made sequentially in vitro, and the secondary
cut is relatively inefficient in vivo. Two primary cuts are
made in a cleavage site in E. coli ribosomal RNA, but it is
not known whether these are sequential or concerted (6).
All cleavages by RNase III release RNAs with 5' phosphate
and 3' OH ends (12).

II. NUCLEOTIDE SEQUENCES OF RNase III CLEAVAGE SITES

The availability of deletion mutants of T7 has made it
possible to locate the five sites at which T7 early RNAs are
cut by RNase III and to determine the nucleotide sequence
around each (11,13,14, and unpublished). The published
sequences for two of the sites are representative (Fig. 1).
All five sites can be drawn in the form of a base-paired
hairpin loop having a characteristic upper stem of 9-11
uninterrupted base pairs, an unpaired bubble, and a lower
stem of 8-9 base pairs. In four of the five sites the lower
stem is interrupted in one or two places by unpaired bases.
In all five sites RNase III cuts at a base within the
unpaired bubble, to the 3' side of the upper stem. A few

```
              A                                    C A
        U         G                          G         A
          C     C                              C=G
           C=G                                 U-A
           U-A                                 U•G
           C=G                                 A-U
           G•U                                 C=G
           C=G                                 U•G
           U•G                                 G=C
           G=C                                 G=C
           G=C                                 A-U
           A-U                                 A-U
           A-U                              C         U
        C         U                      U              A
      U             A                                    U
                 U                     C                  G  ◄—
      C             G  ◄—              A         A
        A         A                        A-U
           G•U                             A-U
           A-U                             C=G
           U-A                          A         A
           A-U                             G=C
           G=C                             G=C
           U-A                             G•U
           G=C                             A-U
           A-U                             G=C
           G•U                             A-U
        U         U                     U         U
 ......GC        AC.......        ......UA        CC.......
 0.3 RNA          0.7 RNA         1 RNA            1.1 RNA
```

FIGURE 1. Two sites of RNase III cleavage in T7 early
RNA (13,14). The arrows indicate where the cuts are made.

sequence elements are conserved in all five sites, but as
indicated below, these do not seem to be sufficient to
define a cleavage site.

We have analyzed two different mutations of T7 that
interfere with RNase III cleavage. The first is actually a
double base change in the upper stem of the first RNase III
cleavage site in the early RNA, changing two GC base pairs
to GU. This change would be expected to diminish the
stability of the upper stem, and the effect of this double
base change is to greatly reduce the efficiency of cleavage
at this site. Thus, it seems likely that pairing of the
upper stem is important for cleavage. The second mutation
is a deletion that resulted from a crossover in a 10-base

region of homology between the first and second cleavage
sites, a region that includes the point of cleavage. This
generated a hybrid site in which the sequence to the 5' side
of the normal point of cleavage comes from the first
cleavage site and the 3' portion from the second cleavage
site. The hybrid site would therefore include the entire
upper stem from the first cleavage site and a hybrid lower
stem. The hybrid lower stem could form five uninterrupted
base pairs, and perhaps two additional pairs beyond a
three-base bubble. Even though the hybrid site contains all
the sequences common to the five natural cleavage sites and
can be drawn in a similar structure, it is not cut by RNase
III either in vivo or in vitro. The critical difference
between the natural sites and the hybrid site would seem to
be in the pairing of the lower stem, suggesting that this
too is an important recognition element for cleavage.

The T7 cleavage sites, although similar to each other,
are distinctly different from a recently sequenced RNase III
cleavage site in the 30S precursor to E. coli ribosomal RNA
(6). In this RNA, two sequences from opposite ends of the
16S ribosomal RNA can come together to form 26 base pairs in
a row, and RNase III makes a cut in each RNA strand of this
paired structure. Thus, this ribosomal cleavage site more
closely resembles completely double-stranded RNA, which
RNase III cuts into fragments about 10-18 bases long (12).

It seems clear that the specific structure of the RNA is
important for cleavage by RNase III, but it is still not
certain just what is needed to make a primary cleavage site
in single-stranded RNA. Perhaps analysis of other mutant
RNAs and of modified RNA fragments from primary cleavage
sites may be able to define this further. Knowing the
sequences around secondary cleavage sites may also be
helpful.

III. EFFECT OF RNase III CLEAVAGE ON TRANSLATION OF T7 mRNAs

In a normal infection, all of the early and many of the
late T7 messenger RNAs are products of RNase III cleavages.
However, T7 is also able to grow in RNase III-deficient
hosts, where the usual RNase III cleavages are not made
(15). In such hosts, synthesis of T7 proteins is slightly
delayed, but most are synthesized at close to normal rates.
One striking exception is the gene 0.3 protein of T7, which
is synthesized much less efficiently in RNase III-deficient

hosts. A similar effect of RNase III cleavage on synthesis of 0.3 protein is observed using purified T7 early RNA in cell-free protein synthesizing systems (15).

The early region of T7 DNA is transcribed from a cluster of three strong promoters near the genetic left end of the DNA, and the 0.3 RNA is the first messenger RNA to be transcribed (2,16). The 0.3 RNA is released from the primary transcripts by two RNase III cleavages, which separate it from the initiator RNAs to the 5' side and the 0.7 RNA to the 3' side. The resulting RNA is about 600 bases long and specifies two different proteins, the gene 0.3 and 0.4 proteins (17,18). The 0.3 protein comes from the 5' portion of the RNA; it is 120 amino acids long, is made in large amounts, and is responsible for overcoming host restriction. The 0.4 protein comes from the 3' portion of the 0.3 RNA; it is about 60 amino acids long, is made in relatively small amounts, and its function is unknown. Each protein is synthesized from its own ribosome binding and initiation site, and the ribosome-protected regions have been identified and sequenced (17). The initiation codon for the 0.3 protein is 35 bases from the RNase III cut at the 5' end of the 0.3 RNA; the initiation codon for the 0.4 protein is about two-thirds of the way down the RNA (17,18). Electrophoretic analysis of proteins made after infection of normal and RNase III-deficient hosts shows that, although synthesis of 0.3 protein is perhaps five fold lower in the RNase III-deficient host, synthesis of 0.4 protein is about the same in both. Thus, only one of the two proteins specified by the 0.3 RNA is affected by RNase III cleavage.

Because the AUG initiation codon for the 0.3 protein lies near the 5' end of the 0.3 RNA, it seemed possible that an RNA remaining attached to this end of the 0.3 RNA might interfere with initiation of synthesis of the 0.3 protein, perhaps at the ribosome binding step. To investigate this, we prepared, by partial digestion with RNase III followed by gel electrophoresis, 0.3 RNA that had been cut by RNase III at its 3' end but which remained attached at its 5' end to I3 RNA, the smallest of the three initiator RNAs from the promoter region. The I3 portion of the I3-0.3 RNA contains no efficient ribosome binding sites, and the initiation sites for the 0.3 and 0.4 proteins are readily distinguished by characteristic oligonucleotides in the ribosome-protected fragments (17). We analyzed the protected fragments after binding to ribosomes prepared from normal and RNase III-deficient cells. Ribosomes from normal cells contain RNase III, which rapidly cuts the added RNA, but ribosomes

TABLE I. Ribosome binding and protection of I3-0.3 RNA

| I3-0.3 RNA | Ribosomes | Oligonucleotides protected in initiation region for: | |
		0.3 protein	0.4 protein
RNase III cut	RNase III$^+$	+	+
RNase III cut	RNase III$^-$	+	+
Uncut	RNase III$^+$	+	+
Uncut	RNase III$^-$	0	+

from RNase III-deficient cells contain no detectable RNase III. The I3-0.3 RNA was used directly or after cutting to completion with RNase III.

The results of the ribosome protection analyses are summarized in Table I. The initiation region for the 0.4 protein was protected whether or not the I3-0.3 RNA was cut by RNase III, but the initiation region for the 0.3 protein was protected only when the RNA was cut. Clearly, ribosomes do not bind efficiently to the initiation site for the 0.3 protein unless the 0.3 RNA has been separated from the initiator RNA at its 5' end.

Why does the attached I3 RNA interfere with ribosome binding? It seems likely that it does so by base pairing to a region in the 0.3 RNA that needs to be free in order to bind ribosomes. The entire nucleotide sequence of the I3 RNA (141 bases) has been determined, as has the sequence of the first 176 nucleotides from the 5' end of the 0.3 RNA (18, unpublished). In analyzing these sequences, we find a potential pairing that involves 10 of 12 bases surrounding the AUG codon used to initiate synthesis of the 0.3 protein (Fig. 2). It seems quite possible that such a pairing might interfere with ribosome binding at this site as long as the two RNAs are attached, but that it would not be stable enough to hold the two RNAs together after RNase III cleavage. Thus, by separating the I3 and 0.3 RNAs, RNase III cleavage would remove the inhibition of 0.3 protein synthesis.

```
                    fMet    Ala    Met      -          0.3 protein

        30                        40
 ,- - - A C A A G A U G G C U A U G - - - 3'   0.3 RNA
 |           | | | | |       | | | | |
 `- - -   G U U C U ——— C G A U A            A - - - 5'   I3 RNA
        A         90              85
```

FIGURE 2. Possible base pairing between the attached I3
and 0.3 RNAs, near the initiation codon for the gene 0.3
protein. Bases 30-43 of the 0.3 RNA and 83-94 of the I3 RNA
are shown, as are the first amino acids of the 0.3 protein.
The UG of the initiation codon would be expected to loop out
in this paired structure.

IV.   CONCLUSIONS

     T7 messenger RNAs are produced by RNase III cleavages of
the   primary   transcripts.   Analysis   of   the   nucleotide
sequences around the five points of cleavage in T7 early
RNA, together with the effects of mutations on cleavage,
suggests that efficient cleavage by RNase III may require a
particular structure in the RNA. The proteins specified by
most T7 messenger RNAs seem to be made about equally well
whether or not the RNAs have been cut by RNase III.   An
exception is the gene 0.3 protein, which is made in much
greater amounts if the messenger RNA has been cut. If the
attached RNA to the 5' side is not removed by RNase III
cleavage, it interferes with ribosome binding at the
initiation site for the 0.3 protein, probably by base
pairing to a sequence in the 0.3 RNA that needs to be free
to interact with the ribosome. Synthesis of the gene 0.4
protein, also specified by the 0.3 RNA, is unaffected by
cleavage. The effect of RNase III cleavage on synthesis of
0.3 protein provides an example of specific control of
translation by modification of the messenger RNA, although
T7 does not appear to use this mechanism to control
synthesis of 0.3 protein during a normal infection.

REFERENCES

1. Robertson, H. D., Webster, R. E., and Zinder, N. D., J. Biol. Chem. 243, 82 (1968).
2. Dunn, J. J., and Studier, F. W., Proc. Nat. Acad. Sci. USA 70, 1559 (1973).
3. Dunn, J. J., and Studier, F. W., Proc. Nat. Acad. Sci. USA 70, 3296 (1973).
4. Nikolaev, N., Silengo, L., and Schlessinger, D., Proc. Nat. Acad. Sci. USA 70, 3361 (1973).
5. Lund, E., Dahlberg, J. E., Lindahl, L., Jaskunas, S. R., Dennis, P. P., and Nomura, M., Cell 7, 165 (1976).
6. Young, R. A., and Steitz, J. A., Proc. Nat. Acad. Sci. USA 75, 3593 (1978).
7. Rosenberg, M., Kramer, R. A., and Steitz, J. A., J. Mol. Biol. 89, 777 (1974).
8. Dunn J. J., J. Biol. Chem. 251, 3807 (1976).
9. Harris, T. J. R., Dunn, J. J., and Wimmer, E., Nuc. Acids Res. 5, 4039 (1978).
10. Dunn, J. J., and Studier, F. W., Brookhaven Symp. Biol. 26, 267 (1975).
11. Robertson, H. D., Dickson, E., and Dunn, J. J., Proc. Nat. Acad. Sci. USA 74, 822 (1977).
12. Robertson, H. D., and Dunn, J. J., J. Biol. Chem. 250, 3050 (1975).
13. Rosenberg, M., and Kramer, R. A., Proc. Nat. Acad. Sci. USA 74, 984 (1977).
14. Oakley, J. L., and Coleman, J. E., Proc. Nat. Acad. Sci. USA 74, 4266 (1977).
15. Dunn, J. J., and Studier, F. W., J. Mol. Biol. 99, 487 (1975).
16. Minkley, E. G., and Pribnow, D., J. Mol. Biol. 77, 255 (1973).
17. Steitz, J. A., and Bryan, R. A., J. Mol. Biol. 114, 527 (1977).
18. Dunn, J. J., Buzash-Pollert, E., and Studier, F. W., Proc. Nat. Acad. Sci. USA 75, 2741 (1978).

DISCUSSION

M. PHILIPP: I would like to note that the kind of structure
you drew for this prokaryotic system may also exist in the
early gene transcripts of the SV-40 system.   A similar
structure can be drawn in the area surrounding the points
where the introns are deleted.

C. WEISSMANN: What is the relative cleavage rate of double-
stranded RNA vs. the 0.3 RNA of T7?   There you have a
completely double-stranded molecule without the bulges which
apparently are part of the cleavage site.

F. STUDIER: I do not know whether I can tell you the
relative rate, but I can say that double-stranded RNA very
efficiently competes against these cleavages, so if you have
a small amount of double-stranded RNA you can compete away
these cleavages in the single-stranded RNA.

REGULATION OF PROMOTER SELECTION BY RNA POLYMERASES

A.A. TRAVERS, R. BUCKLAND and P.G. DEBENHAM
MRC Laboratory of Molecular Biology,
Hills Road, Cambridge CB2 2QH,
England.

Abstract: We show that E. coli RNA polymerase can exist in a
number of distinct structural forms each of which has a
different promoter preference. Promoter selection is
regulated by varying the number and proportions of forms
present. We suggest that the σ initiation factor and
analogous polypeptides determine the range of forms which
the enzyme can assume. Other polymerase effectors alter
the distribution of forms within a given range. Thus
ppGpp and ppApp act as functionally opposing effectors
respectively inhibiting and stimulating the synthesis of
rRNA. Certain macromolecules, which also function in
protein synthesis, act in a similar manner to the low
molecular weight effectors. We suggest that these interactions allow RNA polymerase to monitor the state of translation initiation and elongation and thus to respond
appropriately to any imbalance in the necessary components
for continued efficient protein synthesis. We also show
that RNA polymerase IIa from wheat germ undergoes similar
structural transitions to the prokaryotic enzyme in
response to the purine nucleoside tetraphosphates and that
these transitions are reflected in analogous changes in
the transcription of ad-5 DNA. We suggest that this
eukaryotic enzyme has the potential to be regulated in the
same manner as the bacterial RNA polymerase.

INTRODUCTION

In Escherichia coli regulation of promoter selection by
RNA polymerase is believed to complement the function of
operon specific regulatory proteins in the control of transcription (1). Such a role for RNA polymerase implies that
the enzyme must both recognise and discriminate between a
wide variety of promoters. Thus any factor altering the
promoter preference of RNA polymerase may influence the
initiation of RNA synthesis from different promoters in
disparate ways. In vivo this type of phenomenon occurs when
bacterial cells accumulate the nucleotide guanosine 3'
diphosphate 5' diphosphate, ppGpp (2). Cells containing
ppGpp transcribe rRNA cistrons $\sim$ 10-20 fold less efficiently
than cells lacking ppGpp (3). Conversely the level of lac
RNA transcription is up to 20 fold greater in cells containing
ppGpp (4,5). In vitro ppGpp interacts directly with RNA polymerase (6-8), altering the pattern of transcription in a
manner qualitatively similar to that observed during the
accumulation of the nucleotide in vivo (7,9,10). In this
paper we describe the molecular basis for this switch in
transcriptional specificity.

RNA polymerase is functionally heterogeneous.

By what mechanism does RNA polymerase discriminate
between different promoters? One possibility is that the
enzyme can exist in distinct states each of which possesses a
different promoter preference (1). Indeed crude or partially
purified preparations are demonstrably heterogeneous with
respect to template specificity after zone sedimentation (11-
13). To test whether this heterogenity is a property of RNA

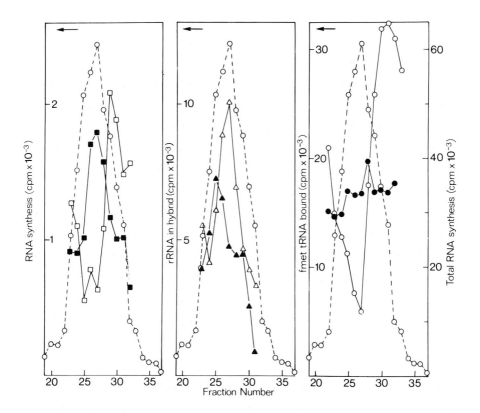

Fig. 1. Functional heterogeneity of E. coli RNA polymerase. RNA polymerase holoenzyme was sedimented at 39000
rpm in an SW40 rotor for 30 hrs at 2°C through a 15-30%
glycerol gradient containing 0.01M Tris HCl pH 7.9, 0.01M
MgCl$_2$, 0.2M KCl 0.0001M dithiothreitol. 75 fractions were
collected. L.h. panel; RNA synthesis from Cla DNA (■-■)
and lac 205 DNA (□-□). Centre panel: rRNA synthesis primed by
ApC (△-△) and ApU (▲-▲). R.h. panel fMet tRNA bound after 1'
(o-o) and 10' (●-●) incubation at 30°. Arrow indicates
direction of sedimentation. o--o indicates polymerase
activity profile.

polymerase itself rather than simply being a consequence of
different polypeptides associating with the enzyme a highly
purified preparation of polymerase was sedimented through a
15 - 30% glycerol gradient. The promoter selectivity of each
fraction from the resulting sedimentation profile was then
assayed. DNA restriction fragments each containing a single
promoter were used as templates for equal weights of enzyme
from each fraction. The fragments used were Cla (14) and lac
205 (15) containing respectively the su_{III}^+ tRNA and lac UV5
promoters and no efficient terminators distal to the promoter
sites. Both these fragments support the transcription of a
single major RNA product respectively ∿ 150 and ∿ 65 nucleotides long. The Cla fragment was efficiently transcribed by
fractions from the trailing shoulder of the polymerase
sedimentation profile while other fractions exhibited only
low activity (Figure 1). A similar profile was apparent when
the lac 205 fragment was used as template. In this case
however maximal transcription was observed with fractions
from the trailing edge together with a minor peak of activity
at the leading edge. These maxima did not correspond to that
observed with Cla DNA.

This functional heterogeneity with respect to promoter
preference is also true of the promoters for rRNA synthesis.
At least 3 of the rRNA cistrons found in E. coli have tandem
promoters separated by ∿ 100 base pairs (16,17). The proximal
and distal promoters of the rrnX cistron carried by the trans-
ducing phage λ d_5 ilv, can be selectively primed by ApU and
ApC respectively (17). When the fractions from a polymerase
sedimentation profile were assayed in this way for their
capacity to initiate at the individual rRNA promoters the
proximal promoter was utilised most efficiently by fractions

from the leading shoulder while the distal promoter was
favoured by fractions from the trailing shoulder (Figure 1).
The data show that the profile for the distal rrnX promoter
probably parallels that of the su$_{III}^{+}$ tRNA promoter. We
conclude that promoter utilisation by individual polymerase
fractions is highly selective and differs from that of most
other fractions. RNA polymerase can thus exhibit functional
heterogeneity with respect to promoter preference, this
heterogeneity being apparent as a correlation between the
sedimentation position of the enzyme and its template
preference. Since the sedimentation coefficient must reflect
the structure of the enzyme we infer that RNA polymerase can
exist in different structural states which differ also in
promoter preference. In particular those forms which
recognise promoters whose expression is inhibited in the
presence of ppGpp in vivo, i.e. rRNA and su$_{III}^{+}$ tRNA promoters,
are distinguishable from the form(s) which recognise promoters
which do not respond in that manner, i.e. lac UV5 promoter.

Regulation of promoter selectivity.

    The nucleotides ppGpp and ppApp alter the promoter
preference of RNA polymerase in opposite ways, respectively
inhibiting and stimulating the synthesis of rRNA at high ionic
strength (18). To test whether the previously observed
correlation between sedimentation position and template
specificity was maintained in the presence of these nucleotides RNA polymerase was sedimented through a glycerol
gradient containing 10 μM ppGpp. Without the nucleotide the
enzyme sedimented as a broad peak of activity at ∿ 14S. With
ppGpp the peak of polymerase activity and protein was shifted
significantly to 13 - 13.5 S (Figure 2). By contrast

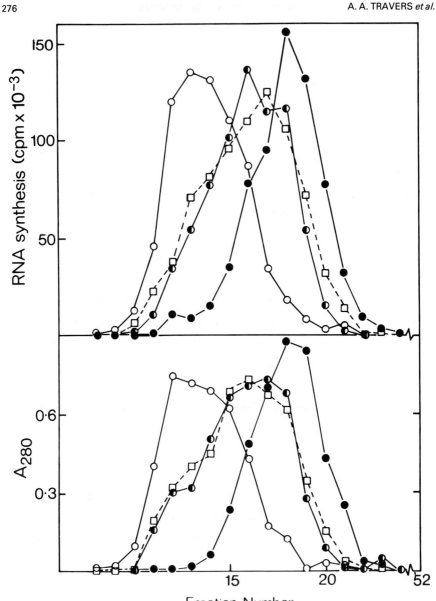

Fig. 2. Effect of ppApp and ppGpp on sedimentation of RNA polymerase activity and protein. ▢--▢ , holoenzyme alone, ●-●, holoenzyme + 10 μM ppGpp; o-o, holoenzyme + 10 μM ppApp + 10 μM ppGpp.

sedimentation through a parallel gradient containing 10 μM
ppApp resulted in a shift in apparent S value of about the
same magnitude in the opposite direction, i.e. to 14.5 - 15 S.
With both nucleotides present at 10 μM the peak of enzyme
activity was coincident with that of the control. In all
cases the nucleotides did not significantly alter the recovery
of enzyme activity. Thus the alterations in the sedimentation
coefficient of RNA polymerase induced by ppGpp and ppApp are
base specific and opposite. Furthermore the direction of the
shift in each case correlates with the change in promoter
preference elicited by the nucleotide. Thus in the presence
of ppGpp the apparent S value of the enzyme corresponds to
that of forms of untreated enzyme which preferentially
initiate at the lac UV5 promoter. Conversely in the presence
of ppApp the apparent S value of the enzyme corresponds to
that of forms of the enzyme which selectively utilise the
proximal rrnX promoter. We suggest therefore that promoter
selection by E. coli RNA polymerase can be regulated by varying the number of proportions of interconvertible but
functionally distinct states of the enzyme.

This correlation between the pattern of promoter
preference of RNA polymerase and the effects of ppApp and
ppGpp on enzyme structure and function would imply that the
effect of the nucleotides on the capacity for rRNA synthesis
should depend on sedimentation position. This is indeed the
case (Figure 3), ppGpp inhibiting rRNA synthesis by molecules
sedimenting on the leading edge to a significantly greater
extent than that by molecules sedimenting on the trailing
edge.

Another in vitro regulator of polymerase selectivity,

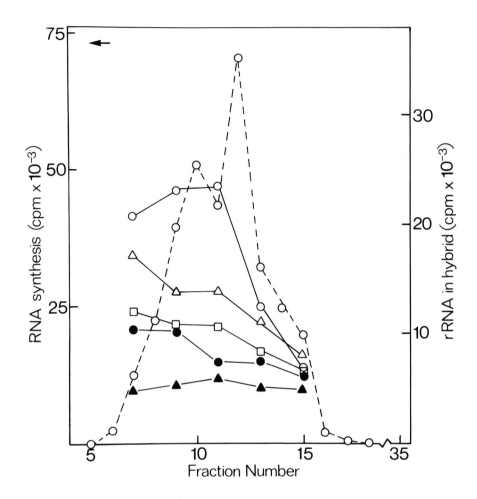

Fig. 3. Effect of ppGpp on rRNA synthesis by equal
weights of enzyme from different fractions of a sedimentation
profile. rRNA synthesis from λd₅ ilv DNA in the presence of
0 (o-o), 10 (Δ-Δ), 50 (□-□), 200 (●-●) and 1000 μM (▲-▲-▲)
ppGpp. o--o, polymerase activity profile.

fMet tRNA$_f^{Met}$, has a functional effect very similar to that elicited by ppGpp, i.e. the charged tRNA preferentially inhibits the initiation of stable RNA synthesis and stimulates lac RNA synthesis (19,20). Analysis of the capacity of different polymerase fractions from a normal sedimentation profile to bind this tRNA again revealed a bimodal pattern of binding. The peaks of binding activity were at the leading shoulder and the trailing edge and corresponded to fractions which did not initiate efficiently from either of the rrnX promoters (Figure 1). Thus those polymerase molecules which have the highest affinity for fMet tRNA also exhibit this aspect of the promoter preference normally elicited by the tRNA. We conclude that in addition to the observed correlation between sedimentation position and promoter preference there is an additional correlation between the former and the capacity to respond to regulators of polymerase selectivity.

Functional equilibration and autoregulation of RNA polymerase.

The variation in the binding of fMet tRNA by different fractions from a normal polymerase sedimentation profile is observed only when the charged tRNA and the enzyme interact for a short time, i.e. $\sim$ 1' at 30°. After 10' incubation in the presence of limiting amounts of fMet tRNA there is no significant difference between polymerase fractions in their ability to bind the charged tRNA (Figure 1). This change involves both an increase in binding by previously low binding fractions and a decrease in binding by previously high binding fractions. This result suggests that under the appropriate conditions the different RNA polymerase fractions can equilibrate to functional equivalence.

The preferential binding of fMet tRNA by particular
structural forms of the enzyme provides a convenient assay
for measuring the effect of various parameters on the position
of the equilibrium between these forms. One parameter known
to affect the promoter selectivity of RNA polymerase is polymerase concentration (21). To test whether this parameter
also influences polymerase structure, as determined by its
capacity to bind fMet tRNA, the enzyme was preincubated at
different concentrations for 60'. Equal weights of enzyme
from each concentration were then assayed by incubation with
fMet tRNA for 1' in the standard binding assay. Enzyme
preincubated at low concentration had a high capacity for
fMet tRNA. This capacity declined to a minimum and then rose
again to an intermediate value as the preincubation concentration of enzyme increased (Figure 4). This variation in
the pattern of fMet tRNA binding suggests that polymerase
concentration directly influences the structure of the enzyme.
To test the effect of varying polymerase concentration on
transcription we determined the salt dependence of rRNA
synthesis by enzyme preincubated at different concentrations.
The data show (Figure 5) that the salt optimum of rRNA
synthesis increased as the protein concentration at which the
enzyme was preincubated increased. That is, enzyme preincubated at low concentration exhibits the pattern of transcription elicited by ppGpp while the pattern for enzyme
preincubated at high concentration approaches that elicited
by ppApp (18). This, together with the corresponding
variations in fMet tRNA binding, suggests that at low
concentrations the structure of the enzyme is that normally
assumed in the presence of ppGpp, while at high concentrations
the enzyme is shifted towards the 'ppApp' form.

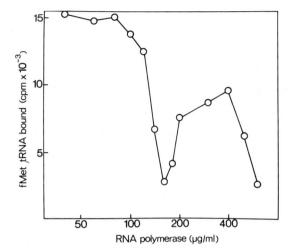

Fig. 4. Effect of polymerase concentration on the capacity of the enzyme to bind fMet tRNA. Data plotted is amount of fMet tRNA bound after 1' incubation by equal weights of enzyme preincubated at the indicated concentrations.

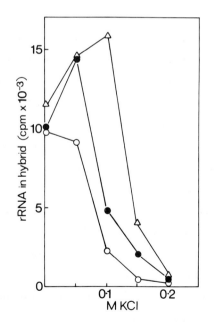

Fig. 5. rRNA synthesis from λd5 ilv DNA by equal weights of RNA polymerase preincubated at 100 (o-o), 300 (●-●) and 1000 μg/ml (Δ-Δ). RNA synthesis was for 15' at 30°.

We show elsewhere (Travers, Buckland, Butler and Debenham, manuscript in preparation) that in addition to enzyme concentration both temperature and ionic strength influence the amount of fMet tRNA bound to RNA polymerase. Variation of both these latter parameters reveals a bimodal pattern of fMet tRNA binding similar to that observed for variation in enzyme concentration. In particular RNA polymerase from a low temperature or high ionic strength environment has the highest affinity from fMet tRNA. Enzyme preincubated under these same conditions has a low salt optimum for rRNA synthesis, similar to that normally observed in the presence of ppGpp. Conversely enzyme preincubated at high temperature or low ionic strength has high salt optimum for rRNA synthesis resembling that normally elicited by ppApp.

We conclude that the parameters of temperature, ionic strength and enzyme concentration all influence both the transcriptional properties of the enzyme and its capacity to bind fMet tRNA. In general, low polymerase concentration, low temperature and high ionic strength mimic the effect of ppGpp and the converse conditions those of ppApp. This in turn implies that these parameters alter the structure of the enzyme such that, for example, at low temperature RNA polymerase corresponds to the form of the enzyme observed at the trailing edge of sedimentation profile while at high temperature the enzyme assumes the form normally sedimenting at the leading edge. We suggest therefore that RNA polymerase is a multistate system in which the equilibrium distribution of enzyme forms is variable.

The effect of polymerase concentration suggests that the

enzyme regulates its own activity, such that at low concentrations it assumes the state normally induced by ppGpp and
at high concentrations the state normally induced by ppApp.
This argues that the structural transition from a 'ppGpp'
form to a 'ppApp' form of the enzyme is mediated through
polymerase-polymerase interactions. We propose that two
holoenzyme monomers in the 'ppGpp' form dimerise, inducing
a conformational change in one or both of the monomers. The
altered monomer(s) might be in either the 'ppApp' form on an
intermediate configuration. On dissociation of the dimer the
altered monomers would be metastable and would slowly revert
to the 'ppGpp' form. It follows that conditions which should
promote polymerase-polymerase interactions, i.e. low salt
concentrations (22) or higher temperatures should also shift
the equilibrium in favour of the 'ppApp' forms as is indeed
observed. Such a cooperative mechanism would account for the
substantial changes in polymerase function elicited by only
small changes in polymerase concentration. We note that
in vitro the effective concentration of RNA polymerase
molecules is likely to be that in the immediate vicinity of
the template DNA rather than the notional concentration in
the reaction mixture as a whole. Hence the properties of the
enzyme will depend not only on the total amount of polymerase
present but also on the ratio of enzyme to template.

Variations in polymerase selectivity with ionic strength
or temperature have been previously interpreted solely in
terms of a direct effect of these parameters on the melting
of the DNA double helix at a promoter site (23). The
variation of polymerase structure and hence promoter
preference with these same parameters suggests that this
interpretation is an oversimplification and that modulations

in polymerase structure, as opposed to DNA structure, may be
a major determinant of the salt and temperature dependencies
of initiation from a particular promoter. We note that the
observed promoter specific decrease in initiation frequency at
temperatures > 25° (24,25) is simply explicable on the basis
of a preferential reduction of particular polymerase forms at
high temperatures. By contrast models invoking a direct
effect of elevated temperature on DNA structure fail to account
for the observed phenomena. A further corollary of this
argument is that the observed variation of promoter preference
with increasing ionic strength is also consequent, at least in
part, on modulations of polymerase structure. However, this
modulation does not account for the observed difference in the
salt dependence of transcription of promoters under the same
control in vitro, for example the su$^{+}_{III}$ tRNA and distal rrnX
promoters. The latter effect can be explained by a direct
effect of increasing salt on the electrostatic interaction
between RNA polymerase and promoter DNA, an effect which would
influence initiation frequency at strong promoters to a
greater extent than weak promoters. It follows that a
promoter must encode at least two types of information, that
specifying the regulation of promoter recognition and that
specifying the initiation frequency. Inspection of promoter
sequences suggests that the former is determined in part by
the sequence between the Pribnow box and the start point for
transcription while the latter is determined primarily by the
Pribnow box itself.

A model for promoter selection.

We have shown that sedimentation of RNA polymerase holoenzyme on a glycerol gradient reveals a separation of

polymerase molecules into functionally distinct populations.
These populations differ in promoter preference, the response
to the regulatory nucleotide ppGpp and the capacity to bind
fMet tRNA. These observations confirm directly that the
enzyme can exhibit functional heterogeneity, a conclusion
which had previously been inferred from template competition
and rifampicin binding experiments (26,27).

What is the structural basis for this heterogeneity?
The correlation of polymerase function with sedimentation
position could reflect differences in polypeptide composition,
the extent of dimerisation or protein conformation. Polypeptide analysis by polyacrylamide gel electrophoresis of
fractions across a sedimentation profile revealed no variation
in the proportion of the core polymerase subunits α, β, β'
and ω, nor was there any indication of additional polypeptides
that might be generated by proteolytic degradation. In some
experiments there was 20 - 30% less σ subunit at the trailing
edge compared to the leading edge but there was, in general,
no correlation between σ content and polymerase function in
these experiments. The ability of distinct gradient
fractions to attain structural and functional equivalence
also argues strongly that the observed differences in
sedimentation position are not a consequence of differences
in polypeptide composition. The functional heterogeneity of
RNA polymerase is observed when the enzyme sediments
predominately in the protomeric state rather than the dimeric
aggregate, the $S_{20,w}$ of these forms being $\sim$ 15 S and $\sim$ 23 S
respectively (28). It is difficult to envisage a purely
aggregational mechanism for generating the diversity of
function characteristic of enzyme sedimenting between $\sim$ 13.5
S and 15 S. Consequently we propose that the differences in

sedimentation position reflect, at least in part, conformational differences.

The pattern of functional heterogeneity observed within
a polymerase sedimentation profile depends on both the
temperature and concentration at which the enzyme is sedimented. An increase in either of these parameters results in
a duplication of the pattern of promoter preference with the
profile (Figure 6) such that each tested promoter is utilised
by polymerase molecules sedimenting in two different
positions. This bimodality is also characteristic of the
pattern of fMet tRNA binding. We suggest that this property
may reflect an additional aspect of polymerase function. For
example, it could allow the efficient recognition of different
DNA conformations at a single promoter site.

The position of the equilibrium between the different
structural states of the polymerase can be varied either by a
mutation affecting a component polypeptide (18) or by
effectors interacting directly with the enzyme. Our working
hypothesis is that the range of states which the enzyme can
assume is determined by the polymerase itself. This range
can be changed by alterations in the primary structure of the
enzyme. In particular, the alt-1 mutation affecting the σ
subunit appears to restrict the normal range to forms which
are normally favoured in the presence of ppGpp (18). Within
a given range the distribution of forms can be altered either
by low molecular effectors or by macromolecular effectors
such as fMet tRNA or EF-TuTs (29).

To what extent does this type of regulation determine
in vivo transcription patterns? The polymerase mutation

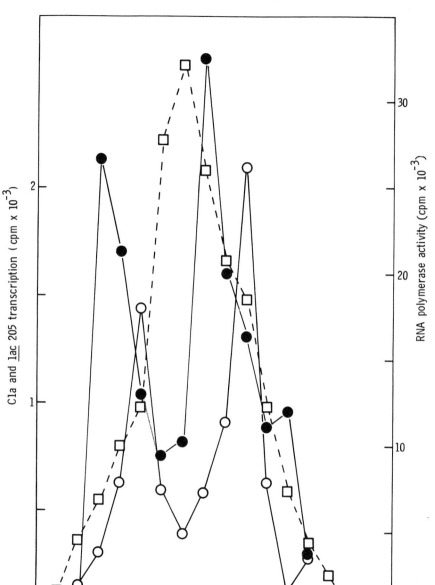

Fig. 6. Bimodal pattern of promoter preference. RNA
polymerase was sedimented for 24 hr at 7° at 39000 rpm in an
SW40 rotor. •-• and o-o indicate activity with Cla and lac
205 DNA respectively. ▢--▢ indicates polymerase activity.

alt-1 alters both the in vitro promoter preference and
sedimentation profile of the enzyme in a manner paralleling
that of 10 μM ppGpp (18). Further the in vitro pattern of
RNA synthesis by the mutant enzyme correlates with the in
vivo phenotype (30). Thus in this case the functional
properties of RNA polymerase apparently reflect the in vivo
situation. Similarly gradient analysis of crude preparations
of RNA polymerase from E. coli in stationary phase (13) or
infected with phage T4 (12), i.e. cells in which the
initiation of rRNA synthesis has been inhibited, reveals that
polymerase molecules normally sedimenting at the leading edge
of the activity profile are no longer detectable. In the
latter case this loss is prevented by a mutation in the β
subunit of RNA polymerase, a change which also delays the
shut off of host RNA synthesis by phage T4. These results
thus parallel both the gross changes in the in vivo pattern of
transcription and the effect of ppGpp on highly purified polymerase suggesting that the structural transitions observed
in vitro may also occur in vivo. The in vivo role of the
macromolecular effectors is less well documented. However
the addition of kirromycin, a specific inhibitor of EF-Tu, to
growing E. coli preferentially increases the rate of rRNA
transcription (31). One consequence of kirromycin addition
would be a rise in the intracellular concentration of EF-Ts,
a protein which in vitro affects polymerase structure and
function in a similar manner to ppApp, i.e. rRNA synthesis is
stimulated. Thus in this case the in vitro response is again
fully consistent with the in vivo observations. We suggest
that by interacting with macromolecular components of the
protein synthesising machinery of the cell RNA polymerase can
monitor the state of translation initiation and elongation
and respond appropriately and rapidly to any imbalance in

these processes.

Control of eukaryotic RNA polymerases.

To what extent does the direct regulation of promoter
selection by RNA polymerases determine transcriptional
patterns in eukaryotes? The transcriptional machinery of
eukaryotes differs in at least one major respect from that of
prokaryotes. In place of the single prokaryotic RNA polymerase the nucleus of eukaryotes contains three structurally
distinct polymerases, I, II and III, responsible respectively
for the synthesis of rRNA, some mRNA precursors and tRNA
together with 5 S RNA (32). These enzymes are all hetero-
multimers whose subunit structures, although more complex,
bear strong resemblances to those of the bacterial polymerase.

In highly purified in vitro systems the eukaryotic polymerases fail to initiate efficiently at promoter sites believed
to be utilised in vivo. However, despite this lack of transcriptional fidelity we can test the structural response of
the eukaryotic enzymes to nucleoside polyphosphates. Figure
7 shows that the sedimentation coefficient of RNA polymerase
II responds to ppApp and ppGpp in a similar manner to that of
E. coli polymerase, i.e. ppApp increases and ppGpp decreases
the apparent S value. These alterations in the sedimentation
characteristics of the enzyme are mirrored in its ability to
transcribe native adenovirus-5 DNA (Figure 8), a template
similar to adenovirus-2 DNA to which the enzyme is known to
bind specifically (33). With this template ppApp increases
and ppGpp decreases the optimum salt concentration for
transcription. These responses suggest that polymerase II has
the capacity to be regulated in the same way as the bacterial

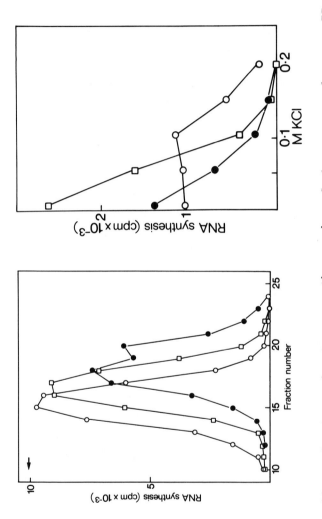

Fig. 7. Effect of ppApp and ppGpp on sedimentation of wheat germ polymerase IIa. In the experiment shown 56 fractions were collected from each gradient. □–□, polymerase alone; o–o polymerase + 10 μM ppApp; ●–●, polymerase + 10 μM ppGpp.

Fig. 8. Effect of 20 μM ppApp and 20 μM ppGpp on transcription of ad-5 DNA by polymerase IIa. Symbols as for Fig. 7.

enzyme.

What is the _in vivo_ relevance of this capacity for
structural modulation? One possibility is that as in
prokaryotes the structure may vary as a function of growth
rate. Polymerase II is normally isolated as a mixture of two
major forms, II_A, found principally in quiescent tissues, and
II_B, found in rapidly growing tissues (34). These two forms
differ only in the molecular weight of the largest subunit
which in II_A has a molecular weight of $\sim$ 205,000 daltons and
in II_B of 170,000 daltons (34). Yet despite being of
apparently lower molecular weight II_B sediments more rapidly
than II_A (35). We suggest therefore that the II_A and II_B
forms may be conformationally analogous to the 'ppGpp' and
'ppApp' forms of the prokaryotic enzyme.

Finally we point out that two properties of the response
of RNA polymerase to regulatory effectors correspond to
requirements often invoked for the control of differentiation
in higher eukaryotes. First, different promoters are selected
in response to varying concentrations of effectors such as
ppGpp, that is, the enzyme can detect and respond appropriately to a spatial or temporal gradient of effector
molecules. Second, an effector can elicit different
alterations in the pattern of promoter selection depending on
which particular polymerase form is its target. We suggest
therefore that RNA polymerase II may play an active role in
the control of transcription during differentiation.

REFERENCES

(1) A. Travers, Nature, 263 (1976) 641.

(2) M. Cashel and J. Gallant, Nature, 221 (1969) 838.

(3) T.D. Stomato and D.E. Pettijohn, Nature New Biol., 234, (1971) 99.

(4) D.W. Morris and N.O. Kjeldgaard, J. Mol. Biol., 31 (1968) 145.

(5) S. Pedersen and N.O. Kjeldgaard, in: Control of Ribosome Synthesis, eds. N.O. Kjeldgaard and O. Maaløe (Munksgaard, Copenhagen, 1976) p. 164.

(6) M. Cashel, Cold Spr. Harb. Symp. Quant. Biol., 35 (1970) 415.

(7) A. Travers, Mol. Gen. Genet., 147 (1976) 225.

(8) M. Cashel, E. Hamel, P. Shapshak and M. Bouquet, in: Control of Ribosome Synthesis, eds. N.O. Kjeldgaard and O. Maaløe (Munksgaard, Copenhagen, 1976) p.279.

(9) A.J.J. van Ooyen, M. Gruber and P. Jorgensen, Cell, 8 (1976) 123.

(10) H.L. Yang, G. Zubay, E. Urm, G. Reiness and M. Cashel, Proc. Nat. Acad. Sci. USA, 71 (1974) 63.

(11) J.J. Pene, Nature, 223 (1969) 705.

(12) L. Snyder, Nature New Biol., 243 (1973) 131.

(13) A. Travers and R. Buckland, Nature New Biol., 243 (1973) 257.

(14) A. Landy, C. Foeller and W. Ross, Nature, 249 (1974) 738.

(15) K. Backman, M. Ptashne and W. Gilbert, Proc. Nat. Acad. Sci. USA, 73 (1976) 4174.

(16) G. Glaser, L. Enquist and M. Cashel, Gene, 2 (1977) 159.

(17) R. Young and J.A. Steitz, submitted for publication.

(18) A.A. Travers, FEBS Lett., 94 (1978) 345.

(19) O. Pongs and N. Ulbrich, Proc. Nat. Acad. Sci. USA, 73 (1976) 3064.

(20) P. Debenham, O. Pongs and A.A. Travers, submitted for publication.

(21) P. Jorgensen and N.P. Fiil, in: Control of Ribosome Synthesis, eds. N.O. Kjeldgaard and O. Maaløe Munksgaard, Copenhagen, 1976) p.370.

(22) J.P. Richardson, Proc. Nat. Acad. Sci. USA, 55 (1966) 1616.

(23) M. Chamberlin, Ann. Rev. Biochem., 43 (1974) 775.

(24) P.G. Debenham and A. Travers, Eur. J. Biochem., 72

(1977) 515.

(25) J.S. Miller and R.R. Burgess, Biochemistry, 17 (1978) 2064.

(26) A. Travers, FEBS Lett., 69 (1976) 195.

(27) W. Bahr, W. Stender, K.H. Scheit and T.M. Jovin, in: RNA polymerase, eds. R. Losick and M. Chamberlin (Cold Spring Harbor, New York, 1976) p.329.

(28) D. Berg and M. Chamberlin, Biochemistry, 9 (1970) 5055.

(29) A. Travers, R. Kamen and M. Cashel, Cold Spr. Harb. Symp. Quant. Biol., 35 (1970) 415.

(30) A.E. Silverstone, M. Goman and J.G. Scaife, Mol. Gen. Genet., 118 (1972) 223.

(31) F.S. Young and F.C. Neidhardt, J. Bact., 135 (1978) 675.

(32) R.G. Roeder, in: RNA polymerase, eds. R. Losick and M. Chamberlin (Cold Spring Harbor, New York, 1976) p.285.

(33) S. Seidman, S. Surzycki, W. DeLorbe and G.N. Gussin, submitted for publication.

(34) T.J. Guilfoyle and J.J. Jendrisak, Biochemistry, 17 (1978) 1860.

(35) G. Krebs and P. Chambon, Eur. J. Biochem., 61 (1976) 15.

DISCUSSION

T. HUNT: Have you tested the effect of ppp5'A2'p5'A2'p5'A on the eukaryotic polymerase?

A.A. TRAVERS: No.

M.F. MILES: Have you bothered comparing the ppApp or ppGpp forms of the polymerase with a form of the polymerase that has been phosphorylated by a cyclic AMP-dependent protein kinase?

A.A. TRAVERS: No we have not. One experiment we have done is to look at E. coli RNA polymerase in which the alpha subunits have been ADP-ribosylated. In this case the addition of the ADPribosyl-residues appears to mimick the effect of low concentrations of ppGpp.

K. NATH: I have three questions. One, have you followed the general synthesis by initiation studies alone, i.e. by γ^{32}P-ATP or GTP when you are talking of promoter selection. Have you looked at the initiation rather than the general synthesis? Two, have you looked at the subunit composition at different fractions of your gradient? Three, when you talk about different kinds of RNA polymerases, are you talking about two specific kinds?

A.A. TRAVERS: We have looked at the ability of the different fractions to bind single restriction fragments in the presence of two triphosphates and we find that this ability corresponds to the pattern of transcription that we observe. Second, we have looked at the polypeptide composition – there is no difference in the portions of the α, β, β prime and the ω subunits across the gradients. There is some variation in the amount of σ that is present, but that does not correlate with any change or difference in functions that we might observe. In reply to the third question, the data suggest that there are a minimum of three forms.

J. LEBOWITZ: If you shift the enzyme to high salt before you add your nucleotide, ppGpp or ppApp, do you get stabilization of one over the other?

A.A. TRAVERS: Yes, it becomes virtually insensitive to ppGpp. High salt appears to mimick the effect of ppGpp.

SYNTHESIS AND MATURATION OF TRANSMEMBRANE
VIRAL GLYCOPROTEINS[1]

Harvey F. Lodish
Dyann Wirth[2]
Mary Porter

Department of Biology
Massachusetts Institute of Technology
Cambridge, Massachusetts

INTRODUCTION

The biosynthesis of integral membrane glycoproteins has
been studied using RNA-containing enveloped viruses such as
VSV and Sindbis virus as model systems (1 - 10). The advantage of using such systems is that the viruses contain few
structural proteins, including usually one (VSV: G protein)
or two (Sindbis: E_1 and E_2) integral membrane glycoproteins
and in some cases, an internal peripheral membrane protein
(VSV: M protein). In contrast, an average mammalian cell contains several hundred membrane proteins. Further, these viruses do not contain enough genetic information to encode all
of the enzymes necessary for their biogenesis and therefore
must rely on the host cell for many functions. In general,
host cell protein synthesis is inhibited during viral infection, making the study of specific proteins much easier.

We have utilized both viruses to probe two fundamental
problems in membrane biogenesis: (1) How are glycoproteins
synthesized on, glycosylated by, and inserted into, the rough
endoplasmic reticulum and, in the case of VSV G and Sindbis
pE_2 (the precursor to E_2), achieve a transmembrane disposition
in the membrane? (2) How do these proteins move from their
site of synthesis in the rough endoplasmic reticulum to the
cell surface and how is the budding virus particle formed?

[1]Supported by NIH grant AI08814
[2]Present address: Biological Laboratories, Harvard University, Cambridge, Massachusetts.

It is assumed that the mode of biogenesis of these viral proteins will be similar to the manner in which normal cells produced those surface glycoproteins which have a transmembrane
disposition similar to that of the viral glycoproteins, such
as the red cell protein glycophorin and the major histocompat-
ability antigen (11, 12).

    In this paper we shall summarize our progress in these
areas; emphasis will be placed on the use of viral temperature
sensitive mutants to elucidate both processes.

TRANSMEMBRANE BIOSYNTHESIS OF THE VSV G PROTEIN

    VSV encodes five polypeptides, all of which are structural
components of the virion, and directs the synthesis of five
corresponding mRNA species. Messenger RNAs for four of them
(N, NS, M and L) are translated on free polyribosomes, and the
newly made polypeptides are soluble in the cell cytoplasm. G
mRNA, by contrast, is exclusively bound to the endoplasmic reticulum (er) and at all stages of its maturation G itself is
bound to membranes (2). Immediately after its synthesis G
spans the er, and is thus transmembrane. About 30 amino acids
at the very COOH-terminus remain exposed to the cytoplasm (3,
13). The balance of the polypeptide, including the NH_2-ter-
minus and the two asparagine-linked carbohydrate chains, face
the lumen of the er, and are protected from extravesicular
protease digestion by the permeability barrier of the endoplasmic reticulum membrane (14, 15).

    The G mRNA appears to be bound to the membrane <u>via</u> the
nascent G chain, since treatment with puromycin, which causes
premature termination of the growing polypeptide, causes release of the G mRNA from the er. Unlike the case of mRNAs encoding secretory proteins, there apparently is not an ionic
linkage between the ribosomes and membranes, since dislodging
of the mRNA does not require solutions of high salt (4).

    Several recent experiments established that the growing G
chain is extruded across the er membrane into the lumen, similar to the way in which a nascent secretory protein is processed.

    1) G protein synthesized <u>in vitro</u> in the absence of mem-
branes (G_0) contains 16 amino acids at the NH_2-terminus that
are absent from the form of G made either in the presence of
er membranes, or made by the cell (G_1) (Fig. 1)(15). Most of
these residues are highly hydrophobic, and resemble in structure "signal" peptides found at the NH_2-terminus of presecretory proteins (16).

    2) During synchronized <u>in vitro</u> protein synthesis in
wheat germ extracts er membranes must be added to the protein
synthesis reaction before the nascent chain is about 80 amino

acids in length in order for the G molecule to be subsequently
inserted into the er bilayer and to be glycosylated. Since
about 30-40 of these 80 NH_2-terminal residues would, at this
key time, still be imbedded in the large ribosome subunit,
this result establishes that it is the 40-50 most NH_2-termin-
al residues, which contain the 16 cleaved amino acids, which
are crucial in directing proper interaction of the nascent
chain with the membrane (5). Presumably, if the nascent chain
is of larger length when the membranes are added, the NH_2-ter-
minus is folded in such a fashion that it cannot interact with
the postulated er receptors.

BIOGENESIS OF TWO SINDBIS-VIRUS GLYCOPROTEINS

Sindbis provides a rather different and very important
system with which to study biogenesis of transmembrane glycoproteins. Sindbis, like VSV, is a lipid-enveloped virus, but
contains only three structural proteins. Two are envelope
glycoproteins which are integral membrane proteins (E_1 and
E_2); one (E_2), like VSV G, spans the lipid membrane with a few
amino acids, at the COOH-terminus, exposed to the cytoplasmic
surface. E_1, is also imbedded in the lipid membrane near the
COOH-terminus, but may not be transmembrane (6, 17). The
third protein, core (C) is internal to the membrane and like
VSV N, is complexed to the viral RNA genome. In marked contrast to the case of VSV infected cells, all three proteins
are synthesized from one polyadenylated mRNA, 26S, which contains the nucleotide sequences found at the 3' end of the
virion 42S RNA (18, 19). Both in cell-free systems and in infected cells, a single initiation site is used for the synthesis of all three proteins encoded by 26S RNA (20, 21). The
core protein is synthesized first, followed by the two envelope glycoproteins (E_1 and PE_2, a precursor to E_2). C, E_1 and
PE_2 are derived by proteolytic cleavage of the nascent chain.
The gene order in Semilki Forest virus, a close relative of
Sindbis, is core-PE_2-E_1 (22). Thus one RNA encodes two very
different types of proteins: a soluble cytoplasmic protein
(C) and two integral membrane proteins.
During infection the 26S RNA is found mainly in membrane-
bound polysomes which synthesize all three virion proteins
(6). Attachment of these polysomes to membranes, like those
synthesizing VSV G, is mediated by forces similar to those
which bind polysomes synthesizing secretory proteins to membranes, since a combination of puromycin treatment and high
salt concentration, but not either alone, will dislodge the
polysomes from the membrane vesicles. Vesicles containing
Sindbis 26S RNA in polysomes will direct cell-free synthesis
of all three Sindbis structural proteins, C_1, PE_2 and E_1.

All of the newly made C protein is on the outside (cytoplasmic side) of the vesicles, while E_1 and E_2 are sequestered in the vesicles. About 30 amino acids of E_2 remain exposed to the cytoplasm, as is the case with synthesis of VSV G (6, 10).

A model (6) explaining these and other results on the translation of 26S RNA is shown in Fig. 2. A ribosome begins translating the core protein at the 5' end of a free 26S mRNA. As soon as the core protein is finished, it is removed by a protease, thus exposing the amino terminus of the nascent envelope protein, PE_2. The amino terminus of the PE_2 protein initiates an interaction with the membrane, and is subsequently transferred into and across the membrane presumably through a protein channel. As in the case of VSV G protein, this interaction leads to the binding of the polysome to the endoplasmic reticulum. As the ribosome continues to transverse the mRNA, completed PE_2 is cleaved and remains imbedded in or sequestered by the membrane. E_1 is then translated and inserted into the membrane, again attaching the polysome to the membrane (see figure 2).

Since the same ribosome synthesizes both soluble (C) and membrane (E_1 and PE_2) proteins, it is clear that the specificity of binding the 26S RNA to membranes in such a way as to transfer only the glycoprotein into the membranes cannot reside in the ribosomes alone. The 60S ribosome subunit may interact directly with membrane proteins, but this is neither the primary interaction nor is this sufficient to result in specific insertion of the membrane proteins. Similarly, binding of the mRNA directly to the membrane cannot alone account for the specificity of this interaction. We conclude that the nascent chain of the two glycoproteins - the only other possibility - determines the specific insertion of the membrane proteins and has a major role in the binding of polysomal 26S RNA to membranes.

Strong support for this model has come recently from two sources. First, Garoff et al. (10) have done in vitro translation and er membrane addition experiments on the 26S RNA similar to those done with VSV G mRNA (5). In synchronized in vitro translation of 26S mRNA, er membranes can be added as late as the time when the soluble C protein has just been completed, and still allow normal membrane insertion of the E_2 and E_1 proteins. If membrane addition is delayed, however, beyond this point, the E_1 and E_2 protein subsequently made are not inserted into the er phospholipid bilayer, nor is E_1 cleaved from PE_2. This is consistent with the notion that it is the NH_2-terminus of nascent PE_2 which is crucial in directing interaction of the complex of ribosomes, mRNA, and growing polypeptide to er membranes.

Secondly, we have investigated the properties of a temperature sensitive mutant of Sindbis virus, ts2 (23). This mutant fails to cleave the structural proteins at the non-permissive temperature, resulting in the production of a polyprotein of 130,000 molecular weight (referred to as the ts2 protein). The order of the proteins in this polypeptide is presumed to reflect the gene order which is NH_2-core-PE_2-E_1-COOH (21, 22, 24). Therefore, in the ts2 protein, the amino terminal sequence of each glycoprotein, E_1 and PE_2 is internal.

Figure 3 shows that the majority of viral specific RNA in cells infected at 41°C, the nonpermissive (Fig. 3a) or 30°C, the permissive (Fig 3b) temperature with ts2 was membrane bound. Here cells were labeled with [^{3}H] uridine, lysed and fractionated on a linear sucrose gradient, with a dense sucrose cushion. In this system, the membranes migrate to the 55%/40% sucrose interface and the soluble material, including the polysomes, is distributed in the linear sucrose gradient. Further studies established that indeed the 26S RNA, representing the majority of the virus-specific RNA, also is bound to membranes (23). Moreover, as shown in Figure 4, about 85% of the [^{35}S] methionine radioactivity in the ts2 protein is also associated with the membrane fraction. In this study, infected cells were labeled with [^{35}S] methionine and fractionated as before; the labeled polypeptides from membrane and soluble fractions were analyzed by SDS gel electrophoresis (inset).

These results were surprising to us, as the NH_2-terminus of PE_2 is buried within the ts2 protein, and thus the 26S RNA, might not be expected to bind to membranes.

Thus, the crucial question was to determine the topological relationship between the ts2 protein and the isolated membrane vesicles. The ts2 protein could be localized on the cytoplasmic side of the endoplasmic reticulum bilayer as is the core protein, it could be entirely inserted into the lipid bilayer as are the envelope proteins or it could be arranged as are the normal wild-type proteins - the core segment being on the cytoplasmic side of the er and the envelope proteins inserted into it. To test these possibilities, membrane vesicles were isolated as described above from cells infected with ts2 at 41°C and pulse-labeled with [^{35}S] methionine for 10 min prior to cell lysis. We have previously shown that in vesicles isolated from the endoplasmic reticulum, the polysomes are found on the outside of the vesicles while the lumen is inside (6). These isolated vesicles were digested with low levels of chymotrypsin. If the ts2 protein is entirely on the cytoplasmic side of the endoplasmic reticulum, it should be removed by proteolytic digestion. If the entire protein is inserted through the bilayer into the lumen, then it should be resistant to proteolysis, unless the membrane vesicles are destroyed by pretreatment with detergent. If the protein is partially inserted, then it should be cleaved to a smaller size by proteolysis.

Figure 5 shows that the ts2 protein is absent after proteolytic digestion of the membrane vesicles from pulse-labeled infected cells. No smaller fragments are found after proteolysis which could account for the radioactivity associated with the ts2 protein before proteolysis.

As an internal control for the proteolysis experiment, cells were coinfected with wild-type Sindbis and the ts2 mutant at the nonpermissive temperature. As can be seen in Figure 5, in coinfected cells, the ts2 protein, PE_2, E_1 and core protein are synthesized. If membrane vesicles from [35S] methionine pulse-labeled cells are digested with chymotrypsin, some 90% of the radioactivity associated with the ts2 and core proteins is destroyed. By contrast, over 80% of the radioactivity associated with PE_2 and E_1 is protected from proteolysis. PE_2 has an increased electrophoretic mobility, as described previously (6) due to a loss of the COOH-terminal segment (17). If the vesicles are permeabilized with detergent before proteolysis, all of the proteins are digested. Based on these results, we conclude that the ts2 protein is located on the cytoplasmic side of the endoplasmic reticulum membrane. In this respect, the ts2 protein resembles the core protein and not PE_2 or E_1 from cells infected with wild-type virus. Although the envelope protein sequences are present, they are not inserted into the membrane and are not protected from proteolysis. This result is consistent with the notion that the proteolytic cleavage between core and PE_2 exposes a sequence, presumably at the amino terminus of PE_2, which is essential for proper insertion of PE_2 into the endoplasmic reticulum membrane. At least one other proteolytic cleavage does not occur in ts2 infected cells: that between PE_2 and E_1. Presumably the principal defect of the ts2 mutation is the inhibition of the C-PE_2 cleavage; the cleavage between E_1 and PE_2 probably does not occur since PE_2 is not properly inserted into the membrane. Likewise, it may be the prevention of this second cleavage which prevents the proper insertion of the E_1 segment in the rough endoplasmic reticulum.

The majority of the 26S mRNA is bound to membranes in cells infected with ts2 at the nonpermissive temperature. Presumably this is due to an indirect binding of the mRNA via the nascent ts2 protein which, itself, binds to membranes (due to its content of hydrophobic sequences). Thus, the binding of mRNA to membranes may, in fact, be only indirectly related to function. In the case of the ts2 26S mRNA, there is no detectable insertion of the protein into membranes, yet the mRNA exhibits characteristics similar to the wild-type 26S RNA

where insertion of the envelope protein does occur. Thus, the
presence of mRNA in membrane polysomes may accompany the insertion of a nascent polypeptide into the endoplasmic reticulum, but binding alone is not sufficient to direct this insertion.

MOVEMENT OF GLYCOPROTEINS FROM THE ROUGH
ENDOPLASMIC RETICULUM TO THE PLASMA MEMBRANE

This remains one of the most mysterious, least-understood
problems in membrane biogenesis. These proteins do not appear
on the cell surface until 30-45 min after their synthesis
(25). The polypeptides move first to an intracellular smooth
membrane, presumably the Golgi, where glycosylation is completed (2). This "terminal" glycosylation is, in fact, an extremely complex process, involving first the removal of certain sugars - mannose and glucose - from the oligosaccharide,
followed by stepwise addition of the peripheral sugars N-
acetylglucosomine, galactose, and sialic acid (9, 26-28). In
what manner, and by what force, these integral membrane proteins are channeled, first to the Golgi, and then to the surface, is obscure.

One approach to elucidating this process is by studying ts
viral mutants in the complementation group corresponding to
the viral glycoprotein. A number of ts VSV mutants in complementation group V have been isolated. In the ones that have
been studied in detail, it is clear that the viral G protein
is synthesized in normal amounts at the nonpermissive temperature. The protein receives the normal "core" sugars (N-
acetylglucosomine, and mannose) and is metabolically stable.
However, it does not move to the cell surface and is not
found in budding virus particles (29-32). Cell fractionation
studies on cells infected by one such mutant, ts045 - showed
that the G polypeptide remained localized in the rough endoplasmic reticulum (30). Hopefully, sequence analysis of the
wild-type and mutant G proteins, and of the G protein from
temperature-resistant revertants, will shed light on the cause
of the inability of the mutated G to move from the rough er.

BUDDING OF VIRUS PARTICLES

Late in infection VSV G becomes the major cell-surface
proteins, and VSV particles are formed by budding from the
plasma membrane. This complex process is not well understood
but is thought to involve the concerted interaction of the
transmembrane viral glycoprotein imbedded in the plasma
membrane; the viral matrix protein (M); and the viral nucleocapsid, consisting of one molecule of viral RNA complexed with

the other three virus-encoded proteins, N, NS, and L.   The M
protein is localized to the inner surface of the virus membrane and may serve as a "bridge" between the nucleocapsid
and G proteins.

The VSV G protein is, by far, the predominant protein exposed on the surface of a virus particle.   However, the VSV
budding process is not totally specific for the VSV G protein,
since budding VSV cores can incorporate into virions glycoproteins of other, unrelated viruses, such as retroviruses, and
possibly also specific cell surface antigens (32-36).   We have
recently accumulated evidence establishing that budding VSV
will incorporate a discrete subset of host cell surface proteins into virions.   In these studies, the proteins exposed
on the cell surface were labeled by lactoperoxidase-catalyzed
iodination, and the cells were subsequently infected by wild-
type VSV.   About 1% of the cell surface [^{125}I] radioactivity
was incorporated into particles which had both the same sedimentation velocity (Fig. 6) and equilibrium density (Fig. 7)
as normal VSV.   Gel analysis of the purified, [^{125}I] labeled
virus (Fig. 8) showed that the VSV contained only some of the
normal cell surface proteins.   Because of the difficulty of
estimating the specific activity of [^{125}I] labeled plasma membrane, it is not possible to determine how many host proteins
are incorporated into each budding virion.

Additional information concerning the interaction involved
in virus budding have come from studies of VSV temperature-
sensitive mutants.   Mutants in complementation group V are, as
noted above, defective in maturation of the virus G protein;
the mutant G does not mature to the plasma membrane.   Budding
from these mutant-infected cells are VSV particles which contain no VSV glycoprotein but normal amounts of the other viral
structural proteins N, M, NS, and L and of RNA (31, 32) (Fig.
9b, compare to c).   Thus there is no absolute requirement for
G in the formation of a budding particle.   Since the amount of
these particles produced is between 0.5 and 3.0% that of normal, wild-type VSV, depending on the ts mutant studied, it
does appear that G can facilitate, or accelerate, the budding
process.

We also showed recently that several temperature-sensitive
mutants of VSV in complementation group III, the structural
gene for the virus matrix protein, produce, at the non-permissive temperature, non-infectious particles which contain the
viral M (matrix) and G proteins but which contain less than
10% of the normal proportion of N protein or RNA (37).   These
findings demonstrate that an interaction between M and the
nucleocapsid is of importance in virus budding and, taken together with the above results on group V mutants, they suggest
that M is the key protein in formation of a bud.

In these experiments cells were labeled with [^{35}S] methionine after VSV infection. The clarified cell supernatant
was subsequently layered atop a cushion of 20% sucrose, centrifuged, and the pellet examined by SDS-polyacrylamide gel
electrophoresis. Figure 9a shows that cells infected at
39.5°C by tsM301 (III) produce particles which contain M and
G protein in the normal proportion but only 1/10 the relative
amount of N characteristic of normal virions (Fig. 9c) The a-
mount of radioactive virus proteins in the particulate fraction from tsM301-infected cells was about 0.3% that produced
by cells infected with wild-type VSV. This material is essentially noninfectious, since the yield of infectious particles
(pfu) from these cells is only 0.01% that from cells infected
with wild-type VSV. At the permissive temperature (32°C),
cells infected by all group III mutants produced normal a-
mounts of infectious particles having the normal composition
of virus polypeptides (data not shown).

Figure 10 shows that the particles produced by cells infected at 39.5°C by tsM301 (III) have a sedimentation velocity
of about 61% that of wild-type VSV. These particles also band
in an equilibrium sucrose gradient at a much lower density
than do wild-type VSV: 1.149 g/cc versus 1.176 (Fig. 10).
Thus these particles have a slightly greater density than do
plasma membranes of VSV-infected cells (2).   SDS gel analysis
of the peak fractions from these gradients indicates a polypeptide composition identical to that of the crude particulate
fraction layered on the gradient (Fig. 1a). We also showed
that these particles contain less than 10% of the amount of
viral RNA, relative to total VSV protein, as do normal VSV
virions.

Thus the physical and analytic properties of the particles
produced by tsM301 are those of a phospholipid membrane containing the VSV G and M proteins, but lacking the internal
nucleocapsid of N protein and RNA.

It is known that, in cells infected by ts mutants in complementation group III, G protein matures normally to the cell
surface, and viral nucleocapsids are produced in normal a-
mounts. Although the tsM301 (III) defective particle contain
essentially only the two virus-encoded membrane proteins G
and M, a small amount of N (or NS) is invariably present.
This result might suggest that a small amount of N protein -
presumably a piece of viral nucleocapsid - is essential for
the budding process. Alternatively, 10% of the defective particles could have a normal complement of N, and the remainder
could have none. This would imply that a virus bud could form
in the total absence of N. What is clear is that some interaction between M and N is essential for normal budding to take
place, and that the lesions in the M polypeptide in tsM301

(III), and other ts (III) mutants prevent, to a large extent, this interaction. These results suggest that M and G are sufficient for some virus budding to occur.

Taken together with earlier studies of ts mutants in group V which demonstrated that particles can be formed which lack G but which contain M, N, NS, and L in normal proportions (31, 32), and that host glycoproteins are incorporated into VSV particles, it appears that M itself is the key polypeptide in forming a virion. Possibly it can self-aggregate under the plasma membrane in a two-dimensional polypeptide network. This alone might be sufficient to induce a bud in the over-lying plasma membrane, or a small amount of nucleocapsid might also be essential.

Ribosome binding site, mRNA

```
met lys cys leu leu tyr leu
 1   2   3   4   5   6   7
```

G_0:

```
met lys cys leu leu tyr leu ala phe leu phe ileu his val asn cys lys phe ___ ileu val phe pro
 1   2   3   4   5   6   7   8   9  10  11  12  13  14  15  16  17  18  19  20  21  22  23
```

```
                                            G₁:  lys phe ___ ileu val phe pro
                                                  1   2   3   4   5   6   7
```

```
                                            G₁': lys phe ___ ___ ___ phe ___
                                                  1   2   3   4   5   6   7
```

```
                                            G₂:  lys phe trp ileu val phe pro
                                                  1   2   3   4   5   6   7
```

FIGURE 1. Amino-terminal sequences of different forms of the VSV glycoprotein (from ref. 15).

VSV G_0 is the form synthesized in a wheat germ cell free system in the absence of membranes; its amino-terminal sequence is identical to that determined from the sequence of the ribosome binding site of G mRNA (38).

G_1 is the form synthesized in the wheat germ cell free system in the presence of endoplasmic reticulum vesicles; G_1' is the product of digestion of G_1, contained in er vesicles, with extravesicular protease (3, 13).

G_2 is the virion form of G with the finished carbohydrate chains.

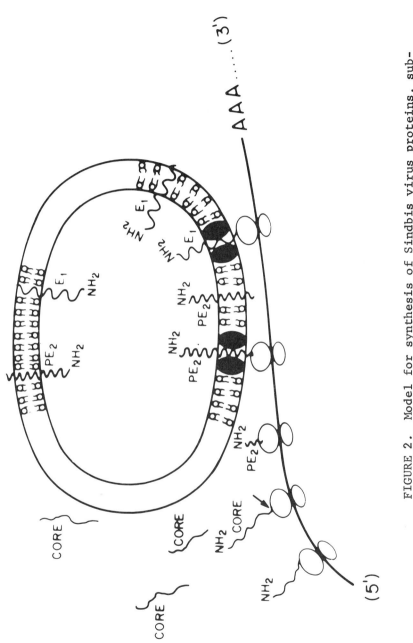

FIGURE 2. Model for synthesis of Sindbis virus proteins, subsequent nascent cleavage, and sequestration of the envelope proteins PE$_2$ and E$_1$ by the endoplasmic reticulum (from ref. 6).

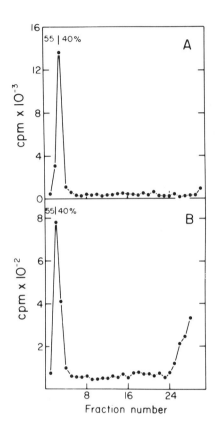

FIGURE 3. Subcellular fractionation of Sindbis specific RNA.
     Two roller bottles of chick embryonic fibroblasts (CEF)
(8 X 10⁷ cells/bottle) were infected with Sindbis ts2 virus
at a m.o.i. of 10-50 in medium containing 5% calf serum and
1.5 μg/ml of actinomycin D. After absorption (30 min) and an
incubation at 30°C, one of the infected cultures (panel a) was
shifted to 41°C; the other remained at 30°C (panel b). ³H-
uridine (5 μCi/ml) was added to each culture. The cells were
harvested 4.5 h post-infection, lysed by Dounce homogenization
and the post-nuclear supernatant was layered onto a 15-40%
linear sucrose gradient with a 55% sucrose cushion in SS buffer (6) (15-40%/55%). After a 4.5 h centrifugation at 26,000
rpm in an SW27 rotor, the gradients were fractionated. 1.5 ml
fractions were collected and the TCA insoluble radioactivity
was determined for 100 μl aliquots of each fraction. Sedi-
mentaion is from right to left.

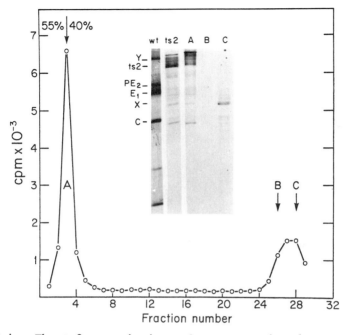

FIGURE 4. The ts2 protein is membrane associated
    CEF (5 X 10^6 cells/100 mm plate) were infected with ts2
virus at a m.o.i. of 10, in MEM containing 5% calf serum and
1 μg/ml actinomycin D. After absorption (30 min) and 1 h incubation at 30°C, the infected monolayers were transferred to
41°C. At 4.5 h after infection, the culture medium was removed and the monolayer washed twice with MEM containing 2%
dialyzed calf serum but without methionine. The cultures were
incubated for 10 min at 41°C in the above medium containing 20
μCi/ml of ^35S-methionine. The cells were harvested, lysed by
Dounce homogenization and the post-nuclear supernatant was
layered onto a 15-40%/55% sucrose gradient in SS buffer.
After centrifugation for 4.5 h at 26,500 rpm in a SW27 rotor
at 4°C, the gradient was fractionated. 1.5 ml fractions were
collected and the TCA insoluble radioactivity from 100 μl
aliquots was determined. The direction of sedimentation is
from right to left.
Inset: The proteins in the fractions indicated were concentrated by acetone percipitation and resolved by 10% PAGE.
Shown is an autoradiogram of the dried gel. Exposure time:
7 days. X and Y indicate the positions of host cell proteins.
    wt) Proteins made in wild-type infected cells after a 10
min pulse of ^35S-methionine
  ts 2) cytoplasmic extract of cells infected with ts2 and
labeled with ^35S-methionine for 10 min at 41°C
    A-C) Proteins contained in the membrane and soluble frac-
tions of the cytoplasmic extract of cells infected with ts2 as
described above.

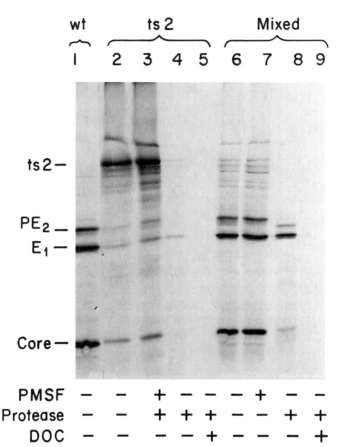

FIGURE 5. Topology of ts2 protein in infected cells
    One culture of CEF (5 X 10^6 cells/100 mm dish) was infec-
ted with ts2, as described in Figure 4. A second culture was
infected with both ts2 and wild-type at a m.o.i. of 10 for
each virus. After absorption (30 min) and a 1 h incubation at
30°C, the culture was shifted to 41°C. At 4.5 h post infection the culture was pulse-labeled with ^{35}S-methionine (10',
50 μCi/ml) and fractionated, as described in the legend to
Figure 4. The membrane fraction, sedimenting at the 40/55%
sucrose interface was isolated, diluted with PBS buffer and
pelleted for 30 min at 4°C (26,500 rpm in an SW27 rotor). The
membrane fraction were resuspended in PBS, and aliquots were
incubated as described below. All incubation were for 30 min
at 30°C. Chymotrypsin (Worthington, final concentration 10
μg/ml), PMSF (Sigma, final concentration 20 μg/ml) and DOC
(Sigma, final concentration 1% w/v) were added to the reaction
as indicated by "+". At the end of the incubation period, the
reactions were heated for 5 min at 100°C. The proteins were
concentrated by acetone precipitation and resolved by 10%

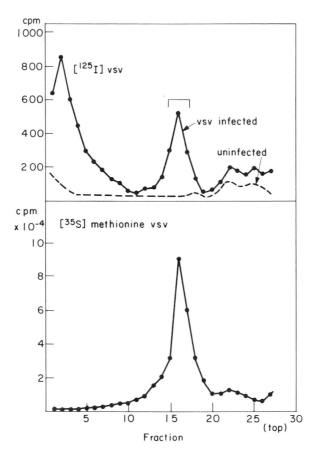

FIGURE 6.   Purification by velocity gradient centrifugation of
VSV containing labeled host surface proteins.
       Monolayers (7 cm²) of Vero cells were iodinated with
[^{125}I], catalyzed by lactoperoxidase, glucose, and glucose oxidase (2). They were then infected with 10 plaque forming
units of VSV, and incubated for 8 hrs at 37°C. Control cultures, (top panel) were not infected. In parallel, cultures
were infected with VSV and labeled with [^{35}]-methionine from
4 to 8 hrs. The cell medium was filtered through a 0.22 μ
Millipore filter, then virus was pelleted through 20% sucrose
by centrifugation for 60 min at 135,000 g and 4°C using a
SW50.1 rotor in a Beckman L5-65 ultracentrifuge. The viral
pellets were then resuspended in PBS, and layered on a linear
15-40% sucrose gradient, made up in PBS. Centrifugation,
right to left, was for 45 min at 40,000 rpm and 4°C in the
SW40 Beckman rotor. Fraction were collected from the bottom
and aliquots counted directly in a scintillation counter.

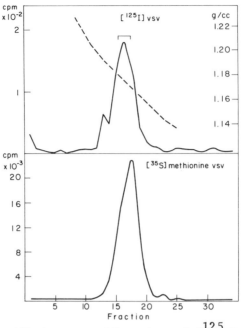

FIGURE 7. Equilbrium centrifugation of [^{125}I] labeled VSV
particles.

Peak virus fractions (Fig. 6) were pooled, and layered
atop 20-50% sucrose gradients, and centrifuged to equilibrium
(18 hr at 35,000 rpm in the SW40 rotor).

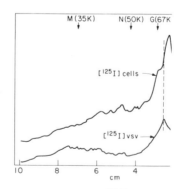

FIGURE 8. SDS gel analysis of [^{125}I] labeled proteins (top)
isolated from labeled, iodinated, uninfected cells or (bottom)
isolated from gradient purified VSV grown on [^{125}] labeled
cells (Fig. 7). Shown are scans of the radioautograms of the
fixed, dried, gel. We utilized a Joyce-Loebl microdensitome-
ter with a full scale pen deflection of 1.0 OD unit, within
the linear range of the X-ray film. Markers indicated are
the [^{35}S] methionine-labeled VSV proteins, also from purified
virions.

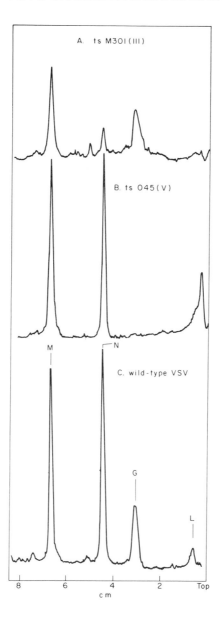

FIGURE 9. Comparison of viral proteins in the particulate fraction of cells infected at 30.5°C with tsM301 (III) (panel a), ts045 (V) (b), or wild-type VSV (c). Vero cells were in-fected at 39°C with 10 plaque-forming units per cell of virus. [^{35}S] methionine was added at 2 hrs, and after a total of 8 hrs of incubation the particulate fraction from the medium was obtained and analyzed by SDS gel electrophoresis as in Figs. 6 and 8.

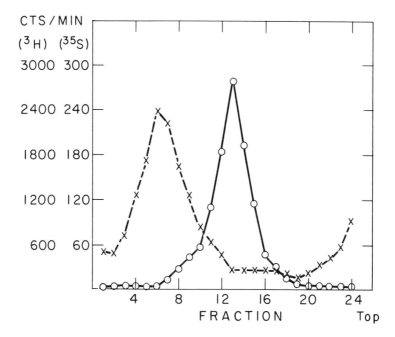

FIGURE 10. Comparison of sedimentation rate of gradient purified particles produced at 39.5° by VSV ts mutant M301 (III)
with VSV wild-type particles. Virus particles were grown and
purified by equilibrium density centrifugation as described in
the legends to Figs. 9 and 11. Fractions containing labeled
particles were combined, diluted with PBS, and virus pelleted
by centrifugation at 135,000 for 45 min in a SW50.1 rotor.
The virus pellet was resuspended in 100 μl PBS, aliquots then
layered over a preformed 10-30% sucrose gradient and centrifuged (right to left) at 135,000 g for 80 min in a SW50.1
rotor. Fractions were collected and radioactivity determined
as described above.
    X [^{3}H] uridine-labeled wild-type VSV
    0 [^{35}S] methionine-labeled tsM301 (III) particles.

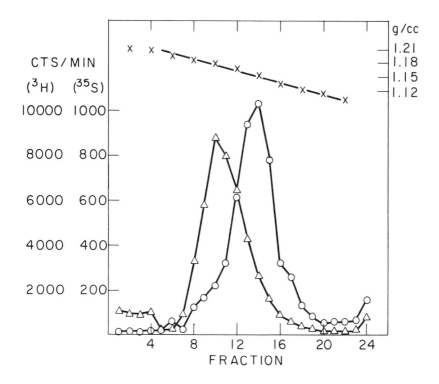

FIGURE 11.   Comparison of buoyant density of particles produced at 39.5° by VSV ts mutant M301 (III) with VSV wild-type
particles. Virus was inoculated onto a confluent monolayer of
Vero cells at an m.o.i. of 3-5 and allowed to adsorb for 1 h
at room temperature. The inoculum was removed, the cells
washed once with medium and then fed with either medium containing ^{3}H-uridine (WT VSV infected) or ^{35}S methionine (tsM301
(III)-infected) and incubated at 39.5° for 16 h. Supernatant
fluids were removed, clarified by centrifugation at 2,000 rpm
for 10 min, layered over 20% sucrose and virus pelleted by
centrifugation in a Beckman SW50.1 rotor at 135,000 for 45
min. The virus pellet was resuspended in 100 μl PBS and run
on a 20-50% sucrose gradient in the same rotor overnight at
40,000 g. Fractions were collected by pumping from the bottom of the tube and aliquots taken for determination of re-
fractive index. The remainder of the samples were then precipitated and counted.
     Δ [^{3}H] uridine-labeled wild-type VSV
     0 [^{35}S] methionine labeled tsM301 (III) particles
     X density.

HARVEY

HARVEY HARVEYF

HARVEY

REFERENCES

1. Katz, F. N., Rothman, J. E., Knipe, D. M., and Lodish, H. F. (1977). J. Supramol. Structure, 7, 353.
2. Knipe, D. M., Baltimore, D,, and Lodish, H. F. (1977). J. Virol., 21, 1128.
3. Katz, F. N., Rothman, J. E., Lingappa, V., Blobel, G., and Lodish, H. F. (1977). Proc. Nat. Acad. Sci. USA, 74, 3278.
4. Lodish, H. F., and Froshauer, S. (1977). J. Cell Biol. 74, 358.
5. Rothman, J. E., and Lodish, H. F. (1977). Nature, 269, 775.
6. Wirth, D. F., Katz, F., Small, B., and Lodish, H. F. (1977). Cell, 10, 253.
7. Sefton, B. M. (1977). Cell, 10, 659.
8. Leavitt, R., Schlesinger, S., and Kornfeld, S. (1977). J. Biol. Chem. 252, 9018.
9. Robbins, P. W., Hubbard, S. C., Turco, S. J., and Wirth, D. F. (1977). Cell, 12, 893.
10. Garoff, H., Simons, K., and Dobberstein, B. (1978). J. Mol. Biol., 124, 587.
11. Tomito, M., and Marchesi, U. (1975). Proc. Nat. Acad. Sci., 71, 2964.
12. Springer, T. A., and Strominger, J. L. (1976). Proc. Nat. Acad. Sci., 73, 2481.
13. Katz, F. N., and Lodish, H. F. (1979) J. Cell Biol. (in the press).
14. Rothman, J. E., Katz, F. N., and Lodish, H. F. (1978). Cell, (in the press).
15. Lingappa, V., Katz, F. N., Lodish, H. F., and Blobel, G. (1978) J. Biol. Chem. (in the press).
16. Devillers-Thiery, A., Kindt, T., Scheele, G., and Blobel, G. (1975). Proc. Nat. Acad. Sci., 72, 5016.
17. Garoff, H. and Soderland, H. (1978). J. Mol. Biol., 124, 535.
18. Mowshowitz, D. (1973). J. Viology, 11, 535.
19. Rosemond, H. and Sreevalson, T. (1973). J. Virol., 11, 399.
20. Cancedda, R., Villa-Komaroff, L., Lodish, H. F., and Schlesinger, M. J. (1975). Cell, 6, 215.
21. Clegg, J. C. S. and Kennedy, S.T. (1975). J. Mol. Biol., 97, 401.
22. Lachmi, B. E., and Kaariainen, L. (1976). Proc. Nat. Acad. Sci. USA, 73, 1936.
23. Wirth, D. F., Lodish, H. F., and Robbins, P. W. (1979). J. Cell Biol. (in the press).
24. Simmons, D. T., and Strauss, J. H. (1974). J. Mol. Biol.,

86, 397.
25. Knipe, D., Lodish, H. F., and Baltimore, D. (1977). J.
     Virol. 21, 1121.
26. Hunt, L. A., Etchinson, J. R., and Summers, D. F. (1978).
     Proc. Nat. Acad. Sci. USA, 75, 754.
27. Li, E., Tabas, I., and Kornfeld, S. (1978). J. Biol.
     Chem., 253, 7762.
28. Kornfeld, S., Li, E., and Tabas, I. (1978). J. Biol.
     Chem., 253, 7771.
29. Lafay, F. (1974). J. Virol., 14, 1220.
30. Knipe, D. M., Baltimore, D., and Lodish, H. F., (1977).
     J. Virol. 21, 1149.
31. Lodish, H. F., and Weiss, R. A. (1979). J. Virol. (in the
     press).
32. Schnitzer, T., Dickson, C., and Weiss, R. (1979). J.
     Virol. (in the press).
33. Zavada, J. and Zavodska, E. (1973/4). Intervirology,
     2, 25.
34. Weiss, R. A., Boettinger, D., and Murphy, H. M. (1977).
     Virology, 76, 808.
35. Weiss, R., and Bennett, P. (1979). J. Virol. (in the
     press). ·
36. Hecht, T., and Summers, D. F. (1976). J. Virol. 19, 833.
37. Schnitzer, T., and Lodish, H. F. (1979). J. Virol. (in
     the press).
38. Rose, J. K. (1977). Proc. Nat. Acad. Sci. USA, 74, 3672.

DISCUSSION

A. SCHULTZ: I would like to return to those experiments you
described which demonstrated a time constraint for the
addition of membranes to a cell-free translation system in
order to obtain proper glycosylation. Could you comment on
why this is so, that the signal sequence could not at its
leisure attach to the glycosylating membranes?

H. LODISH: If the signal sequence is too long, it cannot
interact productively with the endoplasmic reticulum membrane
and direct the nascent peptide into the lumen; unless the
polypeptide extends into the lumen glycosylation probably
cannot occur.

A. SCHULTZ: Is it possible that the signal sequence is
cleaved off if too much time elapses before attachment?

H. LODISH: This raises the question as to on which side of
the endoplasmic reticulum membrane the signal sequence is
cleaved off. I think what you suggest is certainly possible.
I think the general feeling is that the presumed signal
peptidase is cryptic; that it probably resides on the lumenal
side of the endoplasmic reticulum; and thus that the signal
peptide has to get across the membrane normally. We also
know that if the protein is synthesized and completed in the
total absence of membranes, the signal peptide is still on.
In fact, that is how we detected it.

A. SHATKIN: This raises the issue of whether the ts2 protein
has some autocatalytic digestion activity. Can you comment
on that; for example, can you do complementation with wild
type extracts?

H. LODISH: The question is whether or not the ts2 protein
itself acts as its own protease. In this particular case,
there is no good evidence one way or the other. With respect
to polio virus, where there is a multitude of cleavages of
polypeptide chains, I think it is believed that the nascent
chain cleavages, which form the three precursors to the major
polio structural proteins, take place on nascent chains.
Since there occur in the protein synthesizing extract, I
suspect this is due to cellular protease. It is true in the
case of the oncornavirus that at least one of the major
proteolytic cleavages takes place within the capsid and
probably is caused by a viral protease. The only point with
respect to the primary defect of the ts2 mutants is that it
does seem to be in cleavage. There is a report from Simmons
and Strauss (1974) that upon translating the ts2 message in
vitro at the high temperatures they did not get cleavage and
they accumulated the ts2 polyprotein. Translating at permissive temperatures did result in cleavage. So we tend to
argue that the primary defect is in cleavage per se but that
still does not answer the specifics of your question.

T. HUNT: There is one point and that is that when the
primary cleavage does not go, that other cleavage never
goes. There is no obvious reason why that takes place. It
is cellular protein on the outside of the membrane.

H. LODISH: That is a very good point. In fact, in the
absence of the first cleavage between capsid and pE_2, the
second cleavage does not go. I suspect what is happening

is that because the pE_2 is not inserted normally into the
membrane, the key proteolytic site is not exposed to the
appropriate protease, or it is not in the right conformation.
It is difficult to be more specific because, in the normal
case, with this cleavage between pE_2 and E_1 occurring, we
do not know on which side of the membrane the cleavage takes
place. One could envisage that the cleavage takes place on
the lumenal side and because the pE_2 never makes it to the
lumen, obviously you must expect the cleavage not to take
place.

T. HUNT: But surely the carboxy terminus is sticking out the
membrane. What you just said would imply that that would
have to first go in, get cleaved, and then stick out of the
membrane again, which is very horrible.

H. LODISH: No, it certainly seems improbable. I do not like
it but it is a probable possibility. But even if the
cleavage between E_1 and pE_2 occurs on the cytoplasmic
surface, it is still true that in this polyprotein, the
junction between pE_2 and E_1 is in no way in the normal
configuration; it may not even be bound to the membrane, and
therefore the protease(s) may simply be unable to recognize
the site.

K. MCCARTY: Does the protease retain its activity in the
absence of the glycolysated residues?

H. LODISH: This is really Stu Kornfeld's story and I think
he is going to talk about it. I will answer your question as
briefly as possible. If you inhibit synthesis of the
carbohydrate chain, depending on the temperature and on the
strain of VSV, the protein is or is not inhibited in its
movement to the surface. As I said very clearly, in all
cases, you can get virus in different amounts with
non-glycosylated viral glycoprotein and this virus is
completely infectious. The implication is that the
carbohydrate is involved in some way in the movement to the
protein from the rough membranes on the surface.

K. MCCARTY: Is the peptide cleaved in the absence of the
glycolysated residues?

H. LODISH: You are asking whether the signal sequence is
cleaved?

K. MCCARTY: Yes.

H. LODISH: It is believed to be, but I do not know of any
direct evidence.   We have not actually sequenced it, but from
its mobility and things like that, it appears to have lost
the signal sequence.

THE CONTROL OF PROTEIN SYNTHESIS
IN RABBIT RETICULOCYTE LYSATES

Tim Hunt[1]

University of Cambridge
Department of Biochemistry
Tennis Court Road
Cambridge
England

I. INTRODUCTION

Over 20 years ago it was discovered that reticulocytes
need iron in order to make globin (1). Ten years later it was
shown that the supply of hemin regulated protein synthesis in
these cells (2, 3) and shortly afterwards, cell-free systems
which responded to added hemin were developed (4, 5) thus
allowing the molecular physiology of this regulatory system
to be studied. It was immediately recognized that hemin cont-
roleed the initiation of protein synthesis, but since very
little was known about the pathways of mammalian initiation
at the time (the fact that all eukaryotic proteins are initiated with methionine supplied by Met-tRNA$_f$ was yet to be
discovered), studies of 'the hemin effect' did not make much
headway. However, a simple but incisive experiment of Maxwell
and Rabinovitz (6) showed that the inhibition of protein synthesis was due to the activation of an inhibitor, rather than
that some initiation factor required hemin as an essential
cofactor in these particular cells. Subsequent work by Rabinovitz and his group showed that the inhibitor was a protein
which acted catalytically; one molecule could inhibit over
100 ribosomes (34). They also defined a pathway of activation

[1]Alan Johnston, Lawrence and Moseley Research Fellow of the
Royal Society. This work was supported by the MRC and CRC.

of the inhibitor which they called HCR (hemin controlled repressor) showing that it normally existed as an inactive entity
which could be activated either by incubation at physiological
temperature in the absence of hemin, or by brief incubation
with N-ethylmaleimide or other -SH reactive reagents. At early
stages of activation it can be deactivated by the addition of
hemin, but at later stages hemin has no effect on it. As
presently understood there are four forms of HCR connected as
follows:

Brief warming	Prolonged warming	NEM
ProHCR $\rightleftharpoons$ reversible HCR $\rightleftharpoons$ intermediate HCR $\longrightarrow$ HCR		
hemin, GTP	GTP	

Not until 1975 was the basis of the action of HCR to become
clear; and the molecular basis of the activation of HCR is
still very obscure.

Meanwhile, another two conditions which inhibited protein
synthesis in reticulocyte lysates were discovered: the additions of oxidized glutathione (GSSG) and very low levels of
double-stranded RNA (dsRNA) (7, 8). What immediately struck us
was that these agents inhibited protein synthesis in a manner
strongly resembling the absence of hemin or the addition of
HCR, though hemin was ineffective at preventing these inhibitions. As time went by, the list of resemblances grew so
long as to force the conclusion that this disparate set of
agents acted at the same proximal step in initiation, though
presumably arriving by different routes. The characteristics
of these inhibitions are as follows:

1. Protein synthesis proceeds at normal rates for a few
   minutes, whereupon there is an abrupt transition to the
   low inhibited rate (4,5,7,8)

2. Just before this transition, $40S/Met\text{-}tRNA_f/GTP$ initiation
   complexes disappear, although 40S ribosomes, $Met\text{-}tRNA_f$ and
   GTP are still present at absolutely normal levels (10, 11)

3. Accompanying the transition, polysomes break down to inactive 80S ribosomes. This process is prevented by elongation inhibitors of protein synthesis (4, 5).

4. The synthesis of all proteins, not just globin, and including synthesis programmed by added mRNA is inhibited (12,
   13, 14).

5. The inhibitions can be prevented or reversed by the addition of the initiation factor eIF-2 and certain other less
well-characterized initiation factors (15, 16, 17).

6. High levels (5 - 10mM) of cAMP, 2-aminopurine, caffeine,
theophylline, adenine and certain other purines can prevent
and sometimes reverse the inhibitions (18, 19).

7. High levels (1 - 5mM) of ATP tend to potentiate, whereas
high levles of GTP tend to alleviate the inhibitions (18,
19).

8. When inhibited lysates are mixed with fresh lysates the
resulting mixture is immediately though often only transiently inhibited (10, 11).

These characteristics are interpretable in terms of a model in
which protein synthesis is inhibited by activation of dominant
inhibitory elements which act by interfering with the function
of the initiation factor eIF-2. The important breakthrough
came with the discovery that ATP hydrolysis was required for
the action (and in the case of the dsRNA-activated inhibitor
(DAI) activation) of the inhibitors (21). This was only
apparent when the partial reaction of Met-tRNA$_f$ binding to
crude ribosomes was studied, since it only requires GTP as
cofactor. Since protein synthesis has an absolute requirement
for ATP, and it was previously only possible to assay HCR and
DAI by their property of inhibiting protein synthesis, it is
not surprising that the role of ATP took so long to discover.

Once this ATP requirement was appreciated, it was a simple
matter to show that both HCR and DAI contained protein kinase
activity with eIF-2 as substrate. Neither appeared to be
cAMP-dependent protein kinases, and apart from an interesting
ability to phosphorylate themselves (of which more anon), the
only substrate we could find in ribosomes whose binding of
Met-tRNA$_f$ could be inhibited was the smallest component (α
subunit) of eIF-2, which was strongly labelled by γ^{32}P-ATP
but not by ring-labelled ATP (21).

Evidence that Phosphorylation of eIF-2 is the Basis of Lesion

The following observations support the argument that HCR and
DAI inhibit protein synthesis by virtue of their protein
kinase activity directed towards the α subunit of eIF-2. This
is, of course, the crux of the matter which somewhat to our
embarrassment cannot be considered as completely settled.

1. The kinase activities copurify with the inhibitory activity

2. The kinase activity and the inhibitory activity of HCR show
   identical heat inactivation kinetics.

3. During the activation of HCR and DAI their kinase activities
   and inhibitory activities appear in parallel.

4. Agents like the purines (see 6 above) which prevent the
   inhibitors from working inhibit the kinase activities.

5. The inhibitors require ATP as cofactor.

6. Inhibition is overcome by the addition of purified eIF-2.

7. Antibodies against HCR inhibit both kinase and inhibitory
   activities.

      It will be appreciated that these arguments provide only
circumstantial evidence that the phosphorylation of eIF-2 is
the basis of the inhibition of initiation caused by HCR and
DAI. It has not been shown directly that purified phosphorylated eIF-2 is inactive in any of its known, assayable functions. Nor has it yet proved possible to obtain a phosphatase
which will specifically dephosphorylate eIF-2 and restore the
function of inhibited ribosomes, although the lysate clearly
posesses such an enzyme (our unpublished observations). There
have been several attempts to demonstrate differences between
phosphorylated and non-phosphorylated eIF-2, but the results
of the following tests have been equivocal or negative:

1. Formation of the ternary complex eIF-2/Met-tRNA$_f$/GTP (22,23)

2. Formation of the binary complexes with GTP or GDP.

3. Binding of phosphorylated eIF-2 to ribosomes (24).

4. Restoration of initial rates of protein synthesis in an
   inhibited lysate by added phosphorylated eIF-2 (25, and
   our unpublished experiments).

      In the case of the first test, results have been variable.
While some groups have been unable to find differences between
phospho-eIF-2 and eIF-2 (22, and our experience), others have
reported convincing inhibitions of function when HCR and ATP
are present in their assays (23, 26). In general it seems that
sensitivity to phosphorylation is only found when additional
factors obtained from ribosomes are also present, which may

correspond with our experience using crude ribosomes (21), in which we could obtain almost complete inhibition of Met-tRNA$_f$ binding by the addition of low doses of HCR and ATP. It is thus possible to rationalize these conflicts by saying that purified eIF-2 is unaffected by phosphorylation, whereas impure eIF-2 is inhibited. This is tantamount to admitting rather serious ignorance about the physiology of eIF-2, and the tacit admission that its several in vitro assays do not completely or accurately reflect its function in protein synthesis.

It is worth mentioning here an experiment which we did in collaboration with Hans Trachsel. We compared the sensitivity of the highly fractionated reticulocyte system with that of the unfractionated lysate to the addition of HCR. We found that the fractionated system was unaffected. However, the fractionated system had only 5% of the activity per ribosome as the lysate, and when we deliberately reduced the activity of the lysate to that of the purified system by adding cycloheximide, it was equally insensitive to HCR. This suggests that a possible explanation for the results mentioned in the previous paragraph is that phosphorylation of eIF-2 does not completely abolish its activity, so that if some other step in protein synthesis is rate limiting it is impossible to detect any deficiency in eIF-2 function.

Even so, it is difficult to understand how it is possible for added phospho-eIF-2 to rescue protein synthesis in a crude system inhibited by the absence of hemin or the presence of dsRNA (25). The only suggestive pointer to emerge from these experiments is that the number of globin chains synthesised after the addition of the eIF-2 corresponds closely to the number of molecules of eIF-2 added, whereas eIF-2 acts catalytically in an uninhibited situation. This result has suggested that some kind of recycling step occurs during the normal function of eIF-2, and that phosphorylation may slow down that step. It is known that eIF-2 is released from the ribosome after the binding of mRNA, accompanying the joining together of the two subunits and the hydrolysis of GTP to GDP. It is also known that eIF-2 has a 100-fold higher affinity for GDP than GTP, in which case one might suppose that removal of that GDP or its rephosphorylation to GTP could be a recycling step analogous to the Tu.Ts cycle in prokaryotic elongation. It is possible that one of the accessory proteins which has been claimed to enhance eIF-2 function (ESP of de Haro (26), Co-EIF-1 or RF of Gupta's group (27)) might conceivably play such a part, and might fail to function with phospho-eIF-2. The fact that the α subunit of eIF-2 is claimed to be the GTP-binding subunit makes this idea all the more attractive. On the other hand, simple experiments

designed to explore this hypothesis, like measuring the affinity
of phospho-eIF-2 for GDP, or preincubating eIF-2 with GDP before
testing its activity have yielded negative results.

The results obtained by Gupta's and Ochoa's groups are
encouraging that it will be possible to show impairment of
eIF-2 function in a highly defined system, but one would like
to know what the role of the accessory proteins is in natural
protein synthesis. It would also be easier to interpret their
data if reactions contained pre-phosphorylated eIF-2 rather
than relying on the addition of HCR and ATP to the reaction
mixes.

II. RESULTS

I now turn to the properties of eIF-2 kinase. For the
purposes of this article I am going to ignore the dsRNA activated kinase, and concentrate on HCR. Studies of the dsRNA
activated kinase are now being made by workers in the interferon
field, since it seems that this kinase is one of the antiviral
proteins induced by interferon treatment of cells (28, 29, 30).
As far as I can see the properties of the interferon-induced
kinase and the reticulocyte DAI are very similar, if not identical, which raises interesting questions as to its role in reticulocytes as well as its role in the interferon-induced antiviral state. Suffice it to point out that there are good reasons
for believing that DAI is a completely different entity from
HCR:

1. HCR and DAI elute in different positions on gel-filtration
   columns.

2. Their substrate specificity is distinguishable: while HCR
   seems to be absolutely specific for eIF-2, DAI can phosphorylate certain histones as well (although their target
   sites on eIF-2 are identical).

3. They posess different autokinase components.

4. Their modes of activation are completely different; DAI has
   an absolute requirement for dsRNA and ATP, and is unaffected
   by hemin whereas HCR activation does not require ATP and is
   prevented by hemin.

5. High levels of cAMP and other purines inhibit the activation
   of DAI more strongly than its activity. HCR activation is
   insensitive to these compounds, but its activity sensitive.

6. Antibodies which inhibit HCR do not affect DAI (Trachsel & London, Personal communication).

There may be other protein kinases in reticulocytes and other cells which can phosphorylate the α subunit of eIF-2. The inhibition caused by addition of GSSG or glucose-6-phosphate depletion may activate such a kinase (31), but at the time of writing the characterization of these conditions is incomplete. They may very well work through HCR via some alternative activation pathway.

Properties of HCR

1. HCR Assays

In order to discuss the properties and purification of HCR it is necessary to explain briefly how it can be assayed, since different assays detect different forms of this entity.

a. Inhibition of protein synthesis. Fractions of interest are mixed with a complete lysate and incubated under protein synthesis conditions for 30 - 40 minutes. HCR inhibits protein synthesis, but so do many other things; it is often quite difficult to find HCR at low levels in column effluents where salt gradients have been used for elution. In order to quantitate HCR it is necessary to make a series of dilutions. Plotting percent inhibition against the log of the dilution allows the estimation of that dilution of the HCR preparation which causes 50% inhibition of protein synthesis; one unit of HCR is the amount which causes this level of inhibition. These assays contain hemin, and are difficult to use for assaying reversible HCR, whose presence leads to 'kinky' incorporation kinetics. If 2mM GTP is present, neither reversible nor intermediate HCR are detected, and the assay becomes fairly specific for irreversible HCR. The assay can be made extremely sensitive by adding a large amount of the HCR solution, provided that this solution is compatible with protein synthesis conditions. For this reason, we normally try to get the HCR into a low salt buffer containing Mg^{2+} at pH 7.0 - 7.5.

b. eIF-2 kinase activity. In principle the eIF-2 kinase activity of HCR has great advantages over the above assay, but in practice there are drawbacks, chief of which is the need for large amounts of purified eIF-2 which is beyond our means to make (one should start with 2 litres of lysate, i.e. 50 - 100 anemic rabbits). In addition, it is always necessary to analyze the reactions on SDS-polyacrylamide gels because simple TCA

precipitation is not specific enough, especially with impure
preparations of HCR. Finally, one has to measure rates of eIF-2
phosphorylation in order to get an accurate quantitative measure of HCR, which means running and quantitating many gel
tracks, which though possible is not well suited to the needs
for a purification assay.

3. Measuring the autokinase band. An alternative to the
eIF-2 kinase assay is the HCR autokinase assay. If HCR is
incubated briefly with γ-^{32}P-ATP, it becomes rapidly labeled
with ^{32}P, and the extent of labeling is quite a good measure of
the amount of active HCR present (21, 41). It is still necessary
to analyze the reactions on an SDS-polyacrylamide gel, but
because this is an assay for extent rather than rate of phosphorylation, it is possible to analyze a useful number of samples
at once. Furthermore, the assay is very tolerant of salts and
other potentially troublesome presences. By using ATP at
specific activities of 10 - 50Ci/mM it is possible to detect
very small quantities of HCR in about 4 - 6 hrs from the
start of the incubation to the developed autoradiogram. A more
complete description of the autokinase, its relationship to
inhibitory and eIF-2 kinase activities and its possible significance is discussed below.

4. Assay for ProHCR. Purification of ProHCR is complicated
by it lack of activity in any of the assays mentioned above. In
fact, good ProHCR has no activity at all until after its activation. Activation of likely fractions can be done either by
incubating fractions for some time at 40°, or by adding N-ethyl
maleimide to a final concentration of about 0.5mM provided
there are no thiols present. Each activated fraction is then
assayed by one of the procedures described above.

2. The Size and Shape of HCR

The literature is very confusing about the molecular weight
of HCR. Values ranging from 60,000 to 500,000 exist depending
on the laboratory and the method of determination (32 - 36).
Such enormous discrepancy allow of several interpretations,
from downright incompetance, through the possibility that we
are all studying different things, to the interesting possibility that HCR may exist in a variety of molecular forms which
may be physiologically significant. Accordingly I recently
undertook a systematic study of HCR in various states of purity
and activation to try to discover the correct answer.

Gel filtration on AcA 34 (figure 1) yields a value of
320,000 for the molecular weight, in agreement with Gross and

Rabinovitz (34). No reproducible difference in elution position is found between ProHCR and HCR, and crude preparations elute in the same position as the most highly purified in agreement with Trachsel et al. (36).

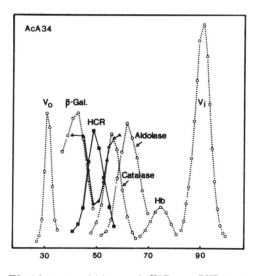

Figure 1. Elution position of HCR on LKB AcA 34. The column (1 x 40cm) was run in 25mM KCl, 10mM NaCl, 10mM HEPES pH 7.2, 1mM $MgCl_2$. Front marker is ribosomes, back marker $^{32}PO_4$ HCR was detected by assays 1 (▲——▲) and 3 (■——■).

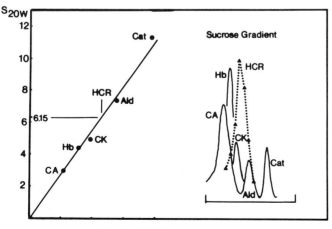

Figure 2. Sedimentation of HCR on 5 - 20% sucrose gradient run in the same buffer as AcA 34 column for 8h at 50,000 rpm.

By combining the data from gel filtration with data from
sucrose density gradient centrifugation it is possible to obtain estimates of the molecular weight and frictional ratio of
HCR (39). From the elution positions of catalase and hemoglobin
the effective pore size of AcA 34 can be estimated as 20nm (37)
which leads to a value of 6.15nm for the Stokes radius of HCR.
The S value of HCR obtained from the data in figure 2 is 6.15,
in good agreement with the value obtained by Ranu et al. (35)
and Trachsel et al. (36) (see Martin and Ames, ref. 38).
Combining these data according to the method of Siegel and
Monty (39) yields a molecular weight of 156,000 and a frictional
ratio of 1.73 for HCR. This corresponds to a very elongated
shape for HCR or else an unusually high degree of hydration, or
a combination of the two. It is clear from the fact that HCR
behaves as if it were larger than catalase on gel filtration,
but only half the size of catalase by sedimentation, using the
same buffers and the same preparation of HCR, that something
funny is going on, and the long-thin explanation seems quite
reasonable. However, there were grounds for suspicion, and it
seemed advisable to check that HCR was not dissociating on the
sucrose gradient, and also to obtain some independent estimate
of its subunit composition. I therefore labeled the autokinase
component of HCR with γ-^{32}P-ATP, sephadexed the labeled HCR
into triethanolamine hydrochloride buffer pH 8, and crosslinked
the preparation with 1mg/ml dimethylsuberimidate at room temperature, with the result shown below in figure 3.

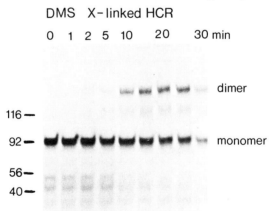

Figure 3. Crosslinking HCR with dimethylsuberimidate.
Experiment described in the text. The starting material
is shown in the first track of the autoradiograph; the
last track had less material loaded. Reaction stopped by
TCA precipitation, samples washed in acetone and taken
up directly in SDS gel sample buffer. This was a 5%
Weber and Osborn gel.

It is quite clear that there is one major, and one minor cross-linked product of the autokinase band (migrating with lacto-peroxidase (92,000 Mr)) on this gel. The major crosslinked band is exactly twice the molecular weight of the autokinase band, implying that HCR is a dimer. The minor band is probably the product of crosslinking between the major autokinase band and a minor band migrating at about 50,000 Mr, which appears to be a proteolytic degradation product of the major autokinase band, since its tryptic phosphopeptides are the same as those of the major band (my unpublished results). When this preparation was run on a 10% Weber & Osborn gel, the molecular weights of the monomer and dimer appeared to be 79,000 and 160,000 respectively. On Laemmli stacking gels, the monomer molecular weight is apparently 95,000 - 105,000, so there is still some anomaly which prevents perfect correspondance between all the molecular weight determinations, but I am inclined to think that the true structure of HCR is that of a homodimer of two 80,000 Mr sub-units. When I ran the crosslinked preparation on a sucrose gradient, the crosslinked dimer ran exactly with the monomer, showing that the 6.15S species is indeed a dimer, and not the dissociated monomer. The crosslinking data appear to exclude an $\alpha\beta$ structure for HCR, and the structure α_2 seems to accord better with the autokinase property, in that it is easier to imagine each active subunit phosphorylating the other, rather than an intrachain self-phosphorylation.

As reported by Trachsel et al. (36), there is no detectable difference in sedimentation between active and ProHCR. However, I did notice a slight increase in the S value of HCR when hemin was present, no more than 7%, and very hard to be sure of.

Activation of HCR

The classical way to activate HCR is to incubate a crude reticulocyte S100 for several hours at 40° in the absence of added hemin. Addition of 20 - 40µM hemin prevents or greatly delays the appearance of HCR activity. This type of activation still occurs in highly purified preparations of ProHCR, so there does not appear to be any absolute requirement for low molecular weight cofactors, although for reasons that will emerge, I would hesitate to be dogmatic on this point. What about oxygen, for example? The rate of activation of HCR is independent of the dilution of the extract, which argues that the rate limiting step is unimolecular. It seems reasonable to interpret these data to indicate that HCR is inherently unstable and undergoes a spontaneous alteration in its configuration unless hemin is present to stabilize it in its inactive state. Accordingly, activation of HCR should be seen as a kind of

partial denaturation of the protein. At higher temperatures,
between 40° and 45°, the activation is more rapid, and hemin is
ineffective at preventing it. However, the presence of high
levels (2mM) of GTP antagonizes the action of HCR activated at
these temperatures, and GTP can also promote the conversion of
HCR to ProHCR under other circumstances to be described below.
GTP is actually better than hemin at preventing the activation
of HCR, whereas ATP has no effect, except that it antagonizes
the action of GTP (20).

  b. Reaction with -SH reagents. An alternative way to activate ProHCR is to treat it with N-ethylmaleimide (NEM). The
activation occurs during a period of about 30 minutes at 30° in
crude lysate, but the concentration of NEM required for maximal
activation shows some curious features, which are illustrated
in figure 4.

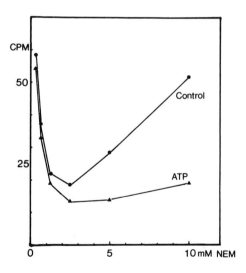

  Figure 4. Activation of gel-filtered S100 with increasing
concentrations of NEM in the presence or absence of 1mM ATP.
The presence of HCR was detected after neutralization of the
NEM with an equimolar amount of dithiothreitol, by adding 2μl
of the samples with 18μl of complete lysate protein synthesis
mix containing 2mM GTP. Counts are in thousands per 2μl sample
of this incubation. Both the preincubation with NEM and the
HCR assay were 30 minutes at 30°.

  This experiment shows that high concentrations of NEM fail
to activate HCR, and that the presence of ATP enhances the
activation of HCR by high concentrations of NEM. The reason

for this behavior seems to be that high concentrations of NEM
inactivate HCR, and that ATP protects against this inactivation.
As ProHCR is purified, less NEM is required for activation; 0.5
mM is very effective once the ProHCR is free of hemoglobin
provided no other thiols are present. Inactivation of HCR by
NEM occurs very rapidly at higher temperature (40°). The evidence suggests that HCR posesses two functionally significant
sites which can react with NEM: an activation site and an inactivation site, which may well be the active center, given that
its substrate, ATP, seems to offer some protection. A further
point of interest is that the inactivation of ProHCR by high
concentrations of NEM seems to occur much more rapidly than the
inactivation of HCR itself, suggesting that the accessibility
of the active site to NEM is different in the two states. However, these conjectures must await confirmation by labeling HCR
with ^{14}C NEM and analyzing the labeled peptides.

The activation of ProHCR by NEM may indicate that protein
-SH groups are involved in the activation process, in which
case one would expect that the presence of thiols like dithiothreitol would help to prevent activation. Gross (40) has reported that this is indeed the case, but my own experiences are
somewhat at variance with his, for I have repeatedly found that
high levels of DTT (5 - 10mM) tend to promote activation of
crude HCR. This is particularly vividly illustrated by some
attempts to purify HCR on thiopropyl sepharose 6B. ProHCR binds
quantitatively to this material, and can be eluted from the
column with DTT. However, the eluted material is fully active
HCR, which is puzzling, for the very fact that it can be recovered from the column must indicate that it has intact -SH
groups. I wonder whether HCR contains intrachain disulfide
bridges whose rupture or rearrangement may lead to activation,
as has been shown to occur in the case of papain (49).

   c. Activation of ProHCR at high pH. Further suspicions that
-SH groups may be important in the activation of ProHCR came
from studies of the effect of pH on activation.

Partially purified ProHCR was incubated at a range of pH
from 6.5 - 9.5 by addition of 1M PIPES, HEPES or Tris buffers
to give 0.1M buffer. The pH was measured at 20°, and overlapping
ranges of the different buffers were used to check that they
had no effect on activation apart from the pH of their solutions. The various samples were then incubated at 30° for 40 min
and assayed for HCR by their capacity to inhibit protein synthesis. Serial dilutions of each sample were assayed in order to
quantitate the amount of activation of HCR at each pH. The
results of one such experiment are shown in figure 5 overleaf.

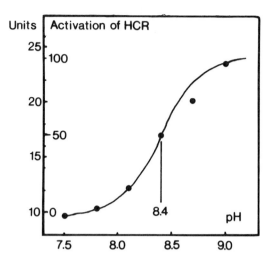

Figure 5. Activation of ProHCR by incubation at high pH.

It is clear that HCR is activated by raising the pH. At
pH 9.5 the activation was almost instantaneous, and seemed to
occur even at 0°; conversely, at pH 6.5 it was quite hard to
activate ProHCR by incubation at 30°. But the interesting
thing is that the apparent pK for this activation is in the
range 8.3 - 8.5, which corresponds to the pK of cysteine.

d. Activation of ProHCR at high hydrostatic pressure.
I stumbled across one final way to activate ProHCR while trying to make S100 from lysate by centrifugation in the SW 50.1
rotor. Despite great care with the temperature and the addition
of what I supposed to be optimal amounts of hemin, these S100s
contained surprisingly high levels of intermediate HCR. The
light dawned when I found that activation only occurred if the
rotor was spun above a certain speed. Accordingly I placed
some lysate in a sealed off plastic syringe in a french press,
pumped it up to 1kB, and found that it contained some form of
HCR. The presence of hemin made no difference, and all operations were done on ice or at 4°. The only effective counter
to this form of activation proved to be GTP, and GTP was also
capable of causing the reversion of the inhibitor back to Pro-
HCR, as can be seen in figure 6. Analysis of the inhibitor by
gel filtration showed that it eluted in the same position as
classically activated HCR, that pressurizing these fractions
also activated an inhibitor, and that increased autokinase activity was present in these fractions.

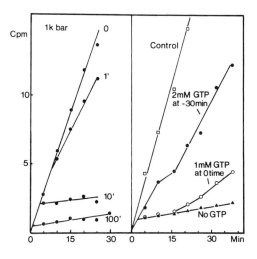

Figure 6. Activation of HCR by high hydrostatic pressure.
A lysate was subjected to 1000 atmospheres pressure for the
times indicated in the left panel, and then assayed for its
capacity to synthesize protein. The right hand panel shows the
result of mixing the 100 min pressed lysate with fresh lysate,
in the presence or absence of 1mM GTP in the final mixture.
Control indicates the synthesis obtained in unmixed fresh
lysate; 2mM GTP at -30 min means that the pressed lysate was
incubated for 30 min at 30° with 2mM GTP before mixing with
fresh untreated lysate (giving 1mM GTP in the subsequent incubation. All counts are thousands per 2μl samples.

I have been unable to detect the low molecular weight inhibitory species reported by Henderson and Hardesty (42) as being
formed under these circumstances despite many attempts, including samples kindly sent to me by them. There must be some
difference in our assays.

Presumably this pressure activation of HCR implies that
HCR has a smaller volume than ProHCR, but other interpretations
are doubtless possible. I was interested to read a recent (43)
report that high pressure causes partial oxidation of -SH
groups in lactate dehydrogenase. Can this be so in the case of
HCR? There does seem to be a remarkable body of evidence
beginning to accumulate which points to the involvement of -SH
groups in the activation process. The only trouble is that it
is all rather vague, merely suggestive stuff, which really
requires some good protein chemistry to confirm or refute it.

Purification of ProHCR

Most of the modes of activation of ProHCR described above were discovered as a result of attempts to purify ProHCR with a view to studying its activation. Investigation of each effect has led us off the path and we have not yet been able to obtain pure ProHCR. The following outline procedure yields fairly pure (estimated 25%) fairly non-inhibitory ProHCR as shown in figure 7. One starts by making S100 from lysate to which 2mM GTP has been added (to stop pressure activation). This is made 40% saturated ammonium sulfate, and the precipitate dissolved in the AcA 34 column buffer described in figure 1. It is fractionated on AcA 34 or Sepharose 6B, and peak fractions loaded onto DEAE-Sepharose CL 6B. The ProHCR is eluted with a gradient of KCl. It emerges at about 0.2M KCl. I have not yet found a good way of going on, having taken some terrible losses on phosphocellulose and hydroxyapatite, and getting activation on thiopropyl sepharose as previously mentioned. However, these ProHCR preparations do seem to be devoid of any other kinase activity, do have very low HCR activity, but can be strongly activated by incubation at 40° or by 0.1 - 0.5mM NEM at 30°.

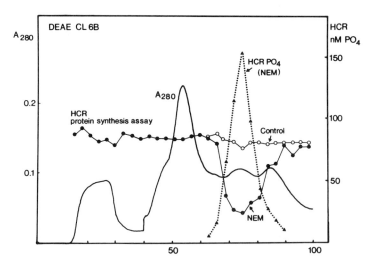

Figure 7. A representative elution profile of ProHCR from DEAE Sepharose CL6B. Buffer was 20mM PIPES pH 6.8, 0.1mM EDTA, 10% v/v glycerol, 50mM KCl to 300mM KCl (exponential gradient started at fraction 40). HCR was assayed before and after activation by 0.5mM NEM by protein synthesis and autokinase assays.

The Autokinase Reaction

All preparations of HCR contain a component which labels strongly and rapidly with γ-^{32}P-ATP (21, 41). As purification of HCR proceeds other labeled bands are lost, and this component, which comigrates precisely with both inhibitory (see fig. 7) and eIF-2 kinase activities is left. I have previously referred to this as a self-phosphorylated component, but it was not shown whether it represented a true, intramolecular autophosphorylation, an intermolecular reaction, or simply the phosphorylation of a contaminant which happens to comigrate with HCR. The last is unlikely, given the highly idiosyncratic behavior on sizing separations. In order to determine whether the labeling with ^{32}P was an intramolecular reaction or not, a series of reactions containing serial dilutions of HCR with contant ATP and salts were made up, and the rate of labeling of the 96,000 Mr (Laemmli gel) followed at 15° by sampling into gel sample buffer containing 10mM EDTA at intervals. Each sample was diluted with this buffer to give the same concentration of protein, analyzed on a 15% gel, autoradiographed and quantitated by densitometry of the film with the result shown in figure 8. A control reaction contained a mixture of beef heart cAMP-dependent protein kinase and calf thymus histones which was similarly diluted to give an idea of what a genuinely intermolecular reaction would look like.

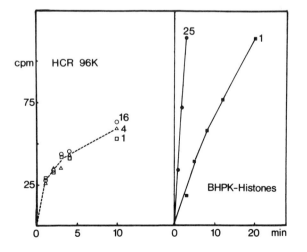

Figure 8. Rate of labeling of (left panel) HCR 96,000 Mr component and (right panel) histone H1 with beef heart protein kinase in the presence of 0.1mM γ-^{32}P-ATP at the relative concentrations indicated on the figure. Counts in thousands.

As the figure shows, the rate of labeling of the 96,000 Mr component was the same over a 16-fold concentration range expressed as the rate per μg of protein, whereas the rate of histone labelling declined with dilution in the control reaction (whereas the autokinase component of the beef heart protein kinas behaved like the HCR autokinase - data not shown). A further indication of the intramolecular nature of the reaction is shown in figure 9, which shows the relative rates of labeling of eIF-2 and the autokinase band in a preparation of ProHCR and HCR activated by incubating the ProHCR at 40° for 2hrs. While the labelling of eIF-2 proceeds linearly, the autokinase saturates after a minute or two. If there were competition between eIF-2 and the 96,000 Mr component, one should not observe such labeling kinetics. This experiment also demonstrates the excellent correlation between the autokinase level and the eIF-2 kinase activity, and allows a rough calculation of the rate of eIF-2 labeling at a given level of HCR: one mole of HCR phosphate labels one mole of eIF-2 with phosphate per minute at 30°. From data obtained in the experiments like that shown in fig. 7 it appears that 50% inhibition of protein synthesis is caused by around 2nMolar HCR phosphate (I emphasize that these are very approximate estimates, which could be wrong by a factor of about 50% owing to uncertainty about precise specific activity of the ^{32}P ATP and the counting efficiency in gel slices). Our lysates contain about 12.5 nMolar 40S subunits, which ought to have all their eIF-2 phosphorylated in 5 - 10 minutes if the in vitro estimates apply under the conditions of protein synthesis in the lysate. This seems a reasonable correspondance, that is, the added HCR's estimated eIF-2 kinase activity is of the right order of magnitude to account for the observed inhibition.

There is one very puzzling feature about the autokinase band, which is that I have never been able to detect a corresponding coomassie blue staining band. Either it contains a very large number of phosphorylatable sites, or it stains very poorly, or both. Some proteins, like casein and phosvitin do stain very feebly.

As figure 9 shows, ProHCR has much lower autokinase activity than activated HCR. The autokinase activity of ProHCR can be completely suppressed by hemin or GTP, whereas hemin has little or no effect on the labeling of the autokinase of activated HCR, in agreement with Trachsel et al (36) and Gross and Mendelewski (41). The correlation between autokinase activity and inhibitory activity is excellent for a given preparation of HCR, but varies somewhat between different preparations and different activation procedures. It is tempting to suggest that phosphorylation of HCR is necessary for its activation, but as Gross and Mendelewski point out, this idea seems to be inherently untestable, since ATP is the substrate for both the autokinase and the eIF-2 kinase. However, since the presence of ATP during activation of ProHCR by warming does not accelerate the process, some process other than phosphorylation appears to be the rate-limiting step in activation.

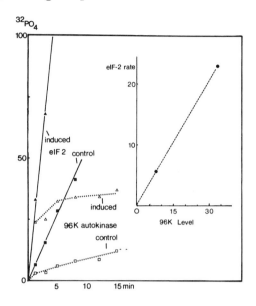

Figure 9. Labeling of the autokinase and the αsubunit of eIF-2 by γ-^{32}P-ATP at 30° in protein synthesis salts (125mM KCl 10mM NaCl, 10mM HEPES pH 7.2, 1mM MgCl$_2$, 0.5mM DTT). Rates of labeling were measured by densitometry of the autoradiograph of a 15% SDS-polyacrylamide gel on which reactions were analyzed.

When ^{32}P labeled HCR is added into a complete lysate under protein synthesis conditions, the label is completely stable no matter what additions (hemin, 2mM GTP, 5mM cAMP) are made. I have found it extremely difficult to discover whether HCR gets phsophorylated in the crude lysate incubated in the absence of hemin owing to the presence of a very strongly labeled polypeptide which migrates in almost the same position as HCR on polyacrylamide gels, but which can be shown to be unrelated to HCR by several criteria. This is, I think, an important question which remains open at present.

What Turns on HCR?

There are three explicit models to account for the activation of HCR in the literature. The oldest, and I think the truest, is due to Rabinovitz who simply suggested that HCR underwent a conformational change in the absence of hemin (32). Ranu and London speculated that a model of the kind:

$$RC \rightleftharpoons R + C$$

might be applicable (35), but no longer favor that hypothesis because of their later experiments (36). As I have already discussed, there is really no evidence for this class of model.

The third suggestion is due to Ochoa's group (44 - 46), who propose that HCR is activated by phosphorylation catalyzed by cAMP-dependent protein kinase (cAPK). Hemin would act by preventing the activation of cAPK by cAMP, according to this model. The evidence for this idea is rather indirect; on the one hand Datta et al. (44) showed that addition of the catalytic subunit of cAPK to a lysate inhibited protein synthesis, while it was also shown (45, 46) that hemin prevented the activation of cAPK by cAMP in purified preparations of the enzyme. They did not show that hemin inhibited the activation of the endogenous reticulocyte cAPK under conditions of protein synthesis in the crude lysate, or that during the inhibition of protein synthesis by added catalytic subunit of beef heart cAPK that HCR was activated.

Despite the attractiveness of the idea, which helps to explain the effects of ATP described by Balkow et al. (20), there are several serious abjections to it. First, it is quite clear that HCR does not need ATP for activation. Second, one might expect that cAMP would inhibit protein synthesis in the lysate, which it does not. Third, one finds inhibition by added beef heart cAPK in the presence of added cAMP in the presence of hemin, which ought to inhibit according to Datta et al. We decided to test the hypothesis as stringently as possible, first asking whether it could be shown that hemin did inhibit the activity of the endogenous reticulocyte cAPK under protein synthesising conditions. Accordingly, we ran incubations with fructose 1,6 bisphosphate as the energy source and as a means of generating γ-^{32}P-ATP of more or less constant specific activity during the incubation (47). We added beef heart cAPK, or cAMP, or hemin in all possible combinations, and analyzed the pattern of labeled proteins on SDS-polyacrylamide gels as shown in figure 10. The results are rather complicated but clearcut. The left panel of fig. 10 shows that both added beef heart cAPK (HOLO on the legend) and cAMP caused the labeling of new bands on the gel. This enhanced labeling was completely unaffected by the presence of added hemin, which did stimulate protein synthesis. Thus we conclude that levels of hemin which are adequate to stimulate protein synthesis in the lysate are inadequate to inhibit protein kinases under these conditions. A similar experiment in which histones were added as substrate gave the same result.

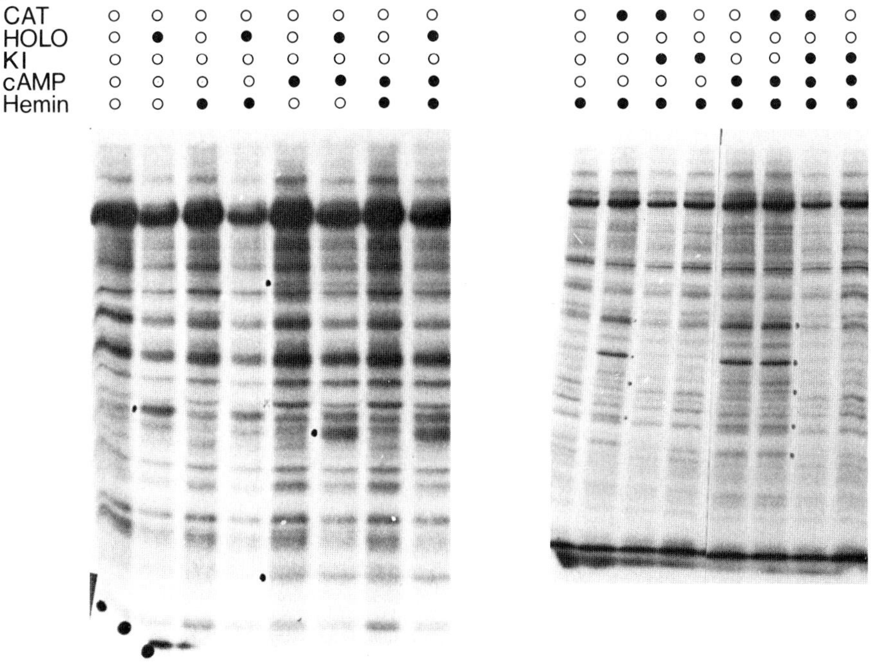

Figure 10. Patterns of protein phosphorylation in the reticulocyte lysate. Left panel was incubated for 20 min with the ATP generating system of Jackson & Hunt (47), right panel for 5 min with γ-^{32}P ATP in gel-filtered lysate. Key to caption: CAT = Catalytic subunit of rabbit muscle cAPK; HOLO = beef heart cAPK; KI = rabbit muscle heat stable kinase inhibitor (48); cAMP was 50μM, hemin, 25μM.

The right panel of fig. 10 shows a slightly different experiment, in which the effects of the heat stable kinase inhibitor was tested in lysate to see whether it would act like hemin. This gel shows clearly that addition of kinase inhibitor completely abolishes the stimulation of phosphorylation caused either by addition of cAMP (50μM) or skeletal muscle cAPK catalitic subunit. However, it had absolutely no effect on protein synthesis in the absence of hemin. If HCR were truly activated by cAPK, the kinase inhibitor should have blocked its activation.

A further interesting point emerged from these experiments, in that it proved that skeletal muscle catalytic subunit of cAPK did not inhibit protein synthesis, in contrast to the result reported by Datta et al. (44) using the beef heart enzyme. Comparison of the patterns of phosphorylation promoted by the two enzymes indeed shows that their substrate specificity is different. Addition of cAMP to the lysate gives a pattern of phosphorylation completely identical to that obtained when the muscle cAPK is added, but the addition of the beef heart enzyme causes enhanced labelling of additional bands on the gel. It is certainly true that beef heart cAPK inhibits protein synthesis in the presence of cAMP, though the inhibition is only seen at very high levels of added enzyme, and is still rather feeble. But it is not clear that it acts by activating HCR; in fact, I have tried to activate purified ProHCR in the presence of cAPK and cAMP with ATP, without success. When one tests whether HCR preparations contain a substrate for cAPK by adding γ-^{32}P-ATP, it turns out they do not. If cAPK does activate HCR, it must do so very indirectly. It occurred to me that the inhibition seen with added cAPK might be due to inhibition of protein phosphatases by activation of phosphatase inhibitor 1, the phosphorylatable phosphatase inhibitor that occurs in rabbit skeletal muscle (50). Preliminary attempts to detect such an entity in the lysate are encouraging but inconclusive, and addition of muscle phosphatase inhibitors had no effect on protein synthesis except at very high levels where one worried about contaminants. This may be a promising trail to follow.

III. CONCLUSIONS

This account of our present understanding of the control of protein synthesis in reticulocyte lysates should have made it abundantly clear how much remains to be understood. As I see it we do not know how phosphorylation affects the function of eIF-2. Neither is the activation of HCR explicable in molecular terms. I would not regard it as sure that this activation is the essentially passive process implied by the original Rabinovitz model, although I do believe that the experiments and arguments set forth above eliminate the particular 'active' model proposed by Datta et. al. (44 - 46).

Perhaps the blankest area of the map of this regulatory network is the part concerned with the reversal of inhibition.

In both intact reticulocytes and the lysate, addition of hemin to an inhibited system restores the initial rate of protein synthesis after a lag. Presumably this means that HCR is converted back to ProHCR, and phospho-eIF-2 is dephosphorylated. While it is quite easy to show that the lysate contains phosphatases which will dephosphorylate eIF-2, it has proven difficult to achieve significant purification of this enzyme or enzymes. I have also been unable to show that the activity of these phosphatases alters in response to supposedly physiologically significant environmental changes.

It seems to me that the provocative statement in Datta et al.'s paper, "The mechanism by which hemin prevents the formation of the inhibitor and maintains protein synthesis is thus explained" (45) was, on the whole, somewhat premature.

ACKNOWLEDGMENTS

I thank Cathy Lane for doing innumerable HCR assays as well as other expert technical assistance. Thanks also to Richard Jackson for invaluable advice and encouragment, to Jean Thomas for help with cross-linking, to Ora Rosen and Philip Cohen for advice and gifts of enzymes, and to Marty Gross, Boyd Hardesty Hans Trachsel, Vivian Ernst and Dan Levin for many discussions and communication of unpublished results.

REFERENCES

1. Kruh, J., and Borsook, H., J. Biol. Chem. 220, 905 (1956).
2. Bruns, G.P., and London, I.M., Biochem. Biophys. Res. Commun. 18, 236 (1965).
3. Grayzel, A.I., Horchner, P., and London, I.M. Proc. Nat. Acad. Sci. USA 55, 650 (1966).
4. Zucker, W.V., and Schulman, H.M., Proc. Nat. Acad. Sci. USA, 59, 582 (1968).
5. Adamson, S.D., Herbert, E., and Godschaux, W., Arch. Biochem. Biophys., 125, 671 (1968).
6. Maxwell, C.R., and Rabinovitz, M., Biochem. Biophys. Res. Commun. 35, 79 (1969).
7. Kosower, N.S., Vanderhoff, G.A., Benerofe, B., Hunt, T., and Kosower, E.M., Biochem. Biophys. Res. Commun., 45, 816 (1971).

8. Ehrenfeld, E., and Hunt, T., Proc. Nat. Acad. Sci. USA
    68, 1075 (1971).
9. Darnbrough, C.H., Hunt, T., and Jackson, R.J., Biochem.
    Biophys. Res. Commun. 48, 1556 (1972).
10. Legon, S., Jackson, R.J., and Hunt, T., Nature New Biol.
    241, 150 (1973).
11. Balkow, K., Miruno, S., Fisher, J.M., and Rabinovitz, M.,
    Biochim. Biophys. Acta 324, 397 (1973).
12. Mathews, M.B., Hunt, T., and Brayley, A., Nature New Biol.
    243, 230 (1973).
13. Beuzard, Y., Rodvien, R., and London, I.M., Proc. Nat.
    Acad. Sci. USA 70, 1022 (1973).
14. Lodish, H.F., and Desalu, O., J. Biol. Chem. 248, 3520
    (1973).
15. Kaempfer, R.O., Biochem. Biophys. Res. Commun. 61, 591
    (1975).
16. Clemens, M.J., Henshaw, E.C., Rahamimoff, H., and London,
    I.M., Proc. Nat. Acad. Sci. USA 71, 2946 (1974).
17. Ralston, R.O., Das, A., Dasgupta, A., Roy, R., Palmieri,
    S., and Gupta, N.K., Proc. Nat. Acad. Sci. USA 75, 4858
    (1978).
18. Legon, S., Brayley, A., Hunt, T., and Jackson, R.J.,
    Biochem. Biophys. Res. Commun. 56, 745 (1974).
19. Ernst, V., Levin, D.H., Ranu, R.S., and London, I.M.,
    Proc. Nat. Acad. Sci. USA 73, 1112 (1976).
20. Balkow, K., Hunt, T., and Jackson, R.J., Biochem. Biophys.
    Res. Commun. 67, 366 (1975).
21. Farrell, P.J., Balkow, K., Hunt, T., Jackson, R.J., and
    Trachsel, H., Cell 11, 187 (1977).
22. Kramer, G., Henderson, A.B., Pinphanichakarn, P., Wallis,
    M., and Hardesty, B., Proc. Nat. Acad. Sci. USA 74, 1445
    (1977).
23. Ranu, R.S., London, I.M., Das, A., Dasgupta, A., Majumdar,
    A., Ralston, R., Roy, R., and Gupta, N.K., Proc. Nat.
    Acad. Sci. USA 75, 745 (1978).
24. Trachsel, H., and Staehelin, T., Proc. Nat. Acad. Sci.
    USA 75, 204 (1978).
25. Safer, B., and Anderson, W.F., Critical Reviews in
    Biochemistry, December 1978 p. 261 (1978).
26. de Haro, C., and Ochoa, S., Proc. Nat. Acad. Sci. USA 75,
    (1978).
27. Dasgupta, A., Das, A., Roy, R., Ralston, R., Majumdar, A.,
    and Gupta, N.K., J. Biol. Chem. 253, 6054 (1978).
28. Roberts, W.K., Hovanessian, A., Brown, R.E., Clemens, M.J.
    and Kerr, I., Nature 264, 477 (1976).
29. Zilberstein, A., Kimchi, A., Schmidt, A., and Revel, M.,
    Proc. Nat. Acad. Sci. USA 75, 4734 (1978).

30. Lebleu, B., Sen, G.C., Cabrer, B., and Lengyel, P., Proc. Nat. Acad. Sci. USA 73, 3107 (1976).
31. Ernst, V., Levin, D.H., and London, I.M., Proc. Nat. Acad. Sci. USA 75, 4110 (1978).
32. Gross, M., and Rabinovitz, M., Biochim. Biophys. Acta 287, 340 (1972).
33. Adamson, S.D., Yau, P.M.-P., Herbert, E., and Zucker, W.V. J. Mol. Biol. 63, 247 (1972).
34. Gross, M., and Rabinovitz, M., Biochem. Biophys. Res. Commun. 50, 832 (1973).
35. Ranu, R.S., and London, I.M., Proc. Nat. Acad. Sci. USA 73, 4349 (1976).
36. Trachsel, H., Ranu, R.S., and London, I.M., Proc. Nat. Sci. USA 75, 3654 (1978).
37. Ackers, G.K., Biochemistry 3, 723 (1964).
38. Martiñ, R.G., and Ames, B.N., J. Biol. Chem 236, 1372 6 (1961).
39. Siegel, L.M., and Monty, K.J., Biochim. Biophys. Acta 112, 346 (1966).
40. Gross, M., Biochim. Biophys. Acta 520, 642 (1978).
41. Gross, M., and Mendelewski, J., Biochim. Biophys. Acta 520, 650 (1978).
42. Henderson, B.A., and Hardesty, B., Biochem. Biophys. Res. Commun. 83, 715 (1978).
43. Schmid, G., Ludemann, H.-D., and Jaenicke, R., Eur. J. Biochem 86, 219 (1978)
44. Datta, A., de Haro, C., Sierra, J.M., and Ochoa, S., Proc. Nat. Acad. Sci. USA 74, 1463 (1977)
45. Datta, A., de Haro, C., Sierra, J.M., and Ochoa, S., Proc. Nat. Acad. Sci. USA 74, 3326 (1977).
46. Datta, A., de Haro, C., and Ochoa, S., Proc. Nat. Acad. Sci. USA 75, 1148 (1978).
47. Jackson, R.J., and Hunt, T., FEBS Letters 93, 235 (1978).
48. Walsh, D.A., Ashby, C.D., Gonzalez, C., Calkins, D., Fischer, E.H., and Krebs, E.G., J. Biol. Chem. 246, 1977 (1971)
49. Brocklehurst, K., and Kierstan, M.P.J., Nature New Biol. 119, 167 (1973).
50. Nimmo, G.A., and Cohen, P., Eur. J. Biochem 87, 341 (1978)

DISCUSSION

J.E. DARNELL: Do you think that the inhibitor you are studying is the kind of inhibitor that accumulates in cells in general when initiation of inhibition has taken place. I am thinking, for example that inhibition occurs during mitosis and after heat shock. Do you think these events are mediated by the same kind of inhibition?

T. HUNT: I would like to think it is the same kind of thing but in fact there is no evidence of this. We have done some experiments with cell-free systems from other cells. It is certainly true that if you make inhibitor from cells that have been stressed in the kind of ways you mentioned, they are very inactive. But under those conditions there is no evidence that we could find the phosphorylation of eIF-2, so I am inclined to think that those inhibitions occur by different mechanisms. However, I should say that if you add HCR and phosphorylate the eIF-2 in such systems, you can show a very strong inhibition of initiation, so I think the question is really open at present.

E. KAMINSKAS: Is it not possible that it is not the phosphorylation of eIF-2 that results in inhibition of initiation factor formation, but rather the accumulation of ADP? I think it has been shown that ADP inhibits initiation factor complex formation and also that low adenylate energy charge does the same. Has anyone done the experiment of adding adenylate kinase and adenylate deaminase to the system to decrease ADP accumulation?

T. HUNT: These enzymes are already present in the lysate. Secondly, we looked very hard to see if there was any change in the concentration of ATP, ADP, GTP and GDP during these inhibitions and we could find no evidence for any changes.

V. ERNST: I would like to make one comment to complement this talk. You saw that when Tim added catalytic subunit, or cyclic AMP, to the reticulocyte lysate, there was a phosphoprotein labelled under these conditions. We have evidence that neither eIF-2 nor the eIF-2 kinase are labelled in the presence of cyclic AMP or the catalytic subunit of cAMP-dependent protein kinase under these conditions.

T. HUNT: Yes, those are our findings too.

DETERMINANTS IN PROTEIN TOPOLOGY

Gunter Blobel

The Rockefeller University
New York, NY   10021

Proteins are synthesized with specific structural infor-mation that determines their topology. Based in part on experimental evidence and in part on theoretical considera-tions one can distinguish four groups of protein sequences that are determinants for protein topology. These are: (1) signal sequences (2) stop-transfer sequences, (3) insertion sequences and (4) sorting sequences. Signal sequences trigger translocation of proteins across distinct cellular membranes. Stop-transfer sequences abrogate translocation. Both sequences interact with specific membrane proteins that function as translocators. Insertion sequences serve to integrate monotopic integral membrane proteins into the lipid bilayer wihout the help of translocators. Sorting sequences function to route protein traffic following translocation or integration into membranes.

An obligatory step in the intracellular pathway of numerous proteins is translocation across or insertion into a distinct cellular membrane. Recent in vitro studies have succeeded to establish some of the rules for these fundament-al cellular activities common to both prokaryotic and eukary-otic cells.

Translocation across membranes

The information for translocation is present in the polypeptide chain in the form of a "signal" sequence (1) that in many cases is short-lived, i.e., cleaved during or shortly after translocation. The signal sequence serves as an address. Its target is a specific transport system in the membrane. Direct as well as circumstantial evidence points to the existence in eukaryotic cells of at least six and maybe as many as eight distinct signals (see below) each addressed to its own specific transport system and each cleaved (if cleaved) by a specific signal peptidase.

347

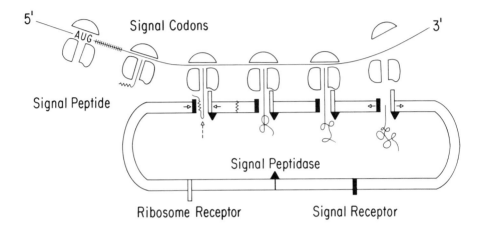

Figure 1. Hypothetical model for cotranslational
translocation across a membrane vesicle (taken from ref. 1).

        Various stages of the translation of a mRNA for a
secretory protein on a membrane-bound polysome are indicated.
The polysome contains six ribosomes. The first two ribosomes
near the 5' end of mRNA are not yet bound to the membrane.
The nascent chain is indicated to grow in a tunnel in the
large ribosomal subunit. The signal peptide portion is
indicated as a zig-zagged line at the amino-terminus of the
nascent chain (see second and third ribosome) or cleaved from
the nascent chain and located in the membrane (see membrane
between ribosome 3 and 4). The signal-peptidase site of the
nascent chain is indicated by an upward pointing arrow head
connected to a dashed line. Ribosome receptor and signal
receptor activity are represented by two different integral
membrane proteins.

        It should be noted that the signal peptide portion is
not always amino terminally located and not always cleaved
(1,4). Thus, in the case of an internal signal sequence,
there is ribosome binding and tunnel formation only after the
internal signal has been synthesized.

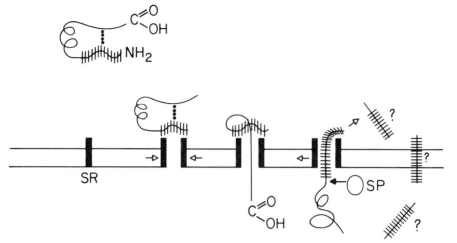

Figure 2. Hypothetical model for the posttranslational translocation of a protein across a membrane (taken from ref. 1).

The protein to be translocated contains a signal sequence, indicated here as a cross-barred region on the amino-terminal of the polypeptide chain. The protein is folded (indicated by dots) but the signal sequence is exposed and free to interact with the signal receptors (SR). This interaction is indicated to result in the aggregation of the signal receptors in the plane of the membrane to form a pore.

Although not indicated in this simple model, it is conceivable that the signal sequence for posttranslational translocation possesses two distinct domains. Both of these domains could be analogous to the two separate sites which have been invoked for cotranslational translocation: one, which could be the analogue of the translocation domain of the signal sequence and another one which could be the analogue of a site on the large ribosomal subunit. By further analogy to cotranslational translocation, the two putative domains of the signal sequence for posttranslational translocation could be recognized by two distinct receptors, each of them represented by a specific membrane protein.

Translocation is indicated arbitrarily to proceed carboxyl-terminal first. Signal peptidase (SP) would cleave when translocation is complete or nearly complete. Cleavage could be coupled to disassembly of the port by lateral diffusion of the receptors in the plane of the membrane. No preferences are expressed as to the location of the signal peptide.

The signal addressed to the rough endoplasmic reticulum (RER) has been determined for a number of secretory proteins (for summary see ref. 1), for a bitopic (see below) integral membrane protein (2) and for a lysosomal protein (3). Although the primary structure of these signal sequences is not identical, there is strong evidence from competition experiments (2,4,5) that they share common secondary structural features so that there is likely to be only one RER signal that is common to the multitude of secretory, lysosomal and bitopic integral membrane proteins. This RER signal is anywhere from 15-30 amino acids long (1). The middle portion is hydrophobic. The flanks contain charged or hydrophilic residues. The penultimate residue is either Gly, Ala, Ser, Cys or Thr, i.e., amino acids with short side chains. The RER signal is usually located on the amino terminus of the nascent chain and is cleaved from the nascent chain. However, recently an internal and uncleaved signal sequence has been found in ovalbumin (4).

The signal addressed to the prokaryotic plasma membrane (for summary see ref. 1) is structurally and functionally related to the RER signal.

The primary structure of the signal addressed to the chloroplast envelope has been determined so far only for one protein, the small subunit of ribulose-1,5-bisphosphate carboxylase (6). It is located in this case on the amino terminus, contains 44 amino acid residues, is not particularly hydrophobic, and is basic in nature (3 Arg, 3 Lys versus 1 Asp and 1 Glu).

No sequence information is available so far for the signals addressed to other membranes.

There are two distinct mechanisms by which signal sequences affect translocation. In one mechanism, referred to as "cotranslational" translocation (7-9) transfer of the polypeptide chain is strictly coupled to translation and is mediated by a ribosome-membrane junction (see Fig. 1). In another formula, referred to as "posttranslational" translocation (9-10), transfer of the polypeptide chain is independent of protein synthesis, i.e., it occurs after translation, and is not mediated by a ribosome-membrane junction (see Fig. 2).

The membrane's transport system, referred to as protein translocator, is envisioned to consist of at least two sub- units, both integral transmembrane proteins. The assembled translocator has been proposed to form a proteinaceous pore (1,7). Each of the contributing translocator subunits would contain a discrete monovalent "receptor" domain. Subunit assembly to form a pore would be triggered by the interaction of these receptor domains with a bivalent (or multivalent) ligand.

In the case of cotranslational translocation this "ligand" would be represented by a site on the large ribo- somal subunit and by the signal sequence of the nascent chain, whereas in posttranslational translocation the ligand would be represented by the signal sequence above. The bivalency of the ligand for posttranslational translocation therefore was proposed to reside in two discrete regions of the signal sequence, each of them addressed to a correspond- ing receptor domain of the translocator subunits (1).

The pore was proposed to be disassembled by disaggrega- tion of the translocator subunits in the plane of the membrane immediately following tranlocations (1). Reassembly and disassembly of the pore for the purpose of translocation across the membrane of a single polypeptide chain was deemed necessary in order to exclude bulk transport of other solutes and to maintain gradients that presumably exist between compartments separated by a membrane.

Signal-specific translocator subunits are presumably restricted in their subcellular distribution to a distinct cellular membrane. A similar restricted localization is likely for the corresponding signal peptidases which are generally situated in the membrane on a side opposite to that of the receptor domain of the translocator subunits.

Membranes equipped with cotranslational translocators are (see Table 1): the prokaryotic plasma membrane, the inner mitochondrial membrane, the thylakoid membrane and the RER membrane. It is likely that all these cotranslational translocators have a common evolutionary origin. Extensive structural and functional conservation of the traanslocator subunits, of the signal, and of signal peptidase has been demonstrated (11).

TABLE 1: TRANSPORT SYSTEMS FOR UNIDIRECTIONAL PROTEIN TRANSLOCATION ACROSS MEMBRANES

I. COTRANSLATIONAL TRANSLOCATION

   1. prokaryotic plasma membrane   —secretory, membrane proteins

   2. inner mitochondrial membrane   —membrane proteins

   3. thylakoid membrane   —membrane proteins

   4. rough endoplasmic reticulum   —secretory, lysosomal, membrane proteins

II. POSTTRANSLATIONAL TRANSLOCATION

   1. outer mitochondrial membrane   —intermembrane proteins (cytosol-synthesized)

   2. mitochondrial envelope   —matrix and membrane proteins (cytosol-synthesized)

   3. outer chloroplast membrane   —intermediate proteins (cytosol-synthesized)

   4. chloroplast envelope   —stroma and membrane proteins (cytosol-synthesized)

   5. peroxisomal membrane   —peroxisomal content proteins

Each of the listed membranes contains one kind of translocator (in multiple copies) that responds to a unique signal shared by the multitude of proteins to be translocated.

Membranes that contain posttranslational translocators for import of cytosol-synthesized proteins are (see Table 1): the mitochondrial envelope (12), the chloroplast envelope (13,14), and, surprisingly, also the peroxisomal membrane (15).

Both the mitochondrial as well as the chloroplast envelope are likely to contain two distinct posttranslational translocators each (1); one that imports proteins from the cytosol across only the outer membrane into the intermembrane space, and another one that transports proteins from the cytosol into the matrix across both the outer and inner membrane. Each of these two translocators is envisioned to respond to a different signal. Since plant cells contain both mitochondria and chloroplasts it is clear that the two signals for import into mitochondria (matrix and inter-membrane space) must be different from the two signals for import into the chloroplast (stroma and intermembrane space).

How could protein transport across two membranes be accomplished by one signal? One possibility which has been considered (1) is that transport proceeds via junctional complexes between the outer and inner membranes. Such junctional complexes should be constructed of at least four subunits, all bi- or polytopic (see below) integral membrane proteins. Each membrane would contain two of these subunits: those in the outer membrane would each contain a cytosol-accessible receptor domain (see above) and a "coupling" domain exposed at the intermembrane space; those in the inner membrane would each contain a coupling domain exposed to the intermembrane space and linked (transiently or permanently) to a matching coupling domain of the two subunits in the outer membrane. Thus, the cytosol-exposed monovalent receptor domains that are part of the translocator subunits in the outer membrane could be triggered to aggregate by the bivalent signal sequence (see above) and, via coupling to the translocator subunits in the inner membrane, a proteinaceous pore could be formed that would extend across both membranes.

It is likely that there exists yet another group of protein translocators that are mechanistically related to the posttranslational varieties but function in the import of heterotopically synthesized proteins across the plasma membrane into the cytosol. The existence of such translocators in the plasma membrane of both prokaryotic and eukaryotic cells is indicated by their ability to import certain toxins (e.g., diptheria toxin or the colicins) whereby a portion of the toxin may function as a signal sequence. Although it is possible that toxins would take advantage of the pathway which is normally used for the import of non-toxic, hetero-topically synthesized proteins, examples of the latter have not yet been identified.

Integration into membranes

Translocation of specific proteins across distinct cellular membranes is not a problem unique to soluble proteins. Many membrane proteins require translocation of one or of several distinct regions of the polypeptide chain in order to achieve their characteristic orientation with respect to the lipid bilayer. The intriguing question then arises of how selective translocation of a discrete segment(s) of the polypeptide chain is accomplished.

In considering this problem an attempt has been made (1) to classify integral membrane proteins (IMPs) based on the orientation of the polypeptide chain with respect to the hydrophobic core and the hydrophilic environment of the lipid bilayer. Accordingly, IMPs have been classified as monotopic, bitopic and polytopic (Fig. 3).

Monotopic IMPs are unilateral in nature, i.e., they ·contain hydrophilic domain(s) only one one side of the lipid bilayer and hydrophobic insertion domain that anchors the protein to the core of the lipid bilayer but that does not reach beyond the hydrophobic core into the hydrophilic environment on the opposite side of the lipid bilayer.

Bitopic and polytopic IMPs are bilateral in nature, i.e., they contain hydrophilic domains on both sides of the membrane. The polypeptide chain of bitopic IMPs traverses the lipid bilayer only once whereas that of polytopic IMPs traverses the lipid bilayer more than once i.e., polytopic IMPs possess mutliple hydrophilic domains on opposite sides of the membrane.

The insertion of monotopic IMPs into the lipid bilayer is proposed to proceed without the mediation of transloca-tors. However, a number of monotopic IMPs will have to be translocated across the membrane before they can be inserted into the lipid bilayer opposite from their side of synthesis. These monotopic IMPs would be synthesizd with a signal that

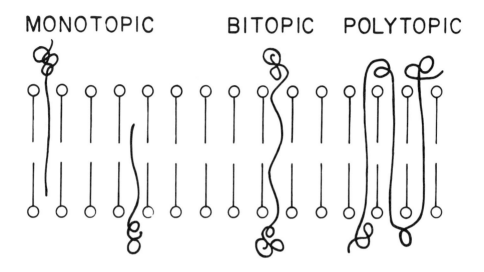

Figure 3.   Classification of integral membrane proteins into three groups: monotopic, bitopic and polytopic.   See text for detailed description.

is targeted to one of the protein translocators described above and insertion would proceed following translocation of the hydrophobic insertion domain.

Unlike monotopic IMPs, the integration into the membrane of bitopic and polytopic IMPs was proposed to be strictly coupled to translocation (1,2,16,17). It is envisioned that all members of this large group of proteins utilize the cotranslational or posttranslational translocators described above for soluble proteins. All members would therefore be synthesized with signal sequences (cleaved or uncleaved). However, unlike the case of soluble proteins, the translocation of the polypeptide chain of bitopic and polytopic IMPs would be interrupted (2,161,17). The information to abrogate translocation was proposed to reside in a specific segment of the polypeptide chain which has been named "stop-transfer" sequence (9). The stop-transfer sequence was proposed to interact with the assembled translocator to cause its disassembly into subunits, thereby preventing translocation of the remainder of the chain (1,2,17). Bitopic IMPs would require only one signal and one stop-transfer sequence. Polytopic IMPs may require multiple signal (uncleaved) and stop-transfer sequences, alternating with each other along the polypeptide chain (16).

Thus the complex orientation that is observed for membrane proteins could be accomplished by three types of sequences: first, by a signaal sequence that initiates translocation across a membrane; second, by a stop-transfer sequence that abrogates translocation; both signal and stop-transfer sequence interact with a specific translocator in the membrane; and third, by an insertion sequence which is able to penetrate into the hydrophobic core of a lipid bilayer without the help of a specific translocator.

Sorting

Following translocation across membranes by signal sequences and orientation within membranes by both signal sequences and stop-transfer sequences, many proteins have not yet completed their intracellular pathway. For example, both secretory and lysosomal proteins need to be separated within the lumen of the ER; many bitopic integral membrane proteins that are biosynthetically inserted into the RER need to be routed from there to various other cellular membranes. It has been proposed (1) that the information for sorting resides in specific "sorting" sequences. By analogy to

signal and stop-transfer sequences it is likely that there
are only a few group-specific sorting sequences which are
common to a large number of proteins. For example, all
bitopic integral membrane proteins of the eukaryotic plasma
membrane could share a common sorting sequence distinctly
different from that of their counterparts in the inner
nuclear envelope.

Sorting sequences could be transiently associated with
the protein and cleaved upon completion of the sorting
process. Alternatively, they could be permanent features,
analogous to the stop-transfer sequences. In this case they
could represent distinct sites for post-translational
modification. Physical sorting could be achieved if the
modifying enzymes were membrane-bound and had a distinct
localization along different membrane tracks, thereby acting
as shunts.

Evolution and protein topology

The mechanism by which a large number of different
proteins acquired structurally similar signals was most
likely by signal amplification and recombination. Subsequent
editing may have been achieved by RNA splicing. It will be
interesting to see whether these "topological" sequences for
translocation, for positioning within the membrane and for
subsequent sorting are still represented in the DNA as
spliced elements.

The proposal that many species of proteins could have
acquired during evolution all the topological information for
their intracellular pathway from a relatively small repertoire of specific sequences could be extended to explain the
"pleiotopic" intracellular localization of any one particular
species of protein. Thus, by gene duplication and recombination with any one of the "topological" sequences, one species
of protein could have acquired two or more distinct intracellular addresses. Examples for such pleiotopic proteins
may be species that are both cytosolic and membrane-bound
(actin? tubulin?) or that are both secretory and membrane-
bound (IgM).

This work was supported by Grant no. CA 12413 from the
National Cancer Institute and Grant no. NP-268 from the
American Cancer Society.

REFERENCES

1)   Blobel, G., Walter, P., Chang, C.N., Goldman, B.M., Erickson, A.H. and Lingappa, V.R. (1979). Soc. Exp. Biol. 33: 9-36.

2)   Lingappa, V.R., Katz, F.N., Lodish, H.F. and Blobel, G. (1978). J. Biol. Chem. 253, 8667-8670.

3)   Erickson, A.H. and Blobel, G. In preparation.

4)   Lingappa, V.R., Lingappa, J.R. and Blobel, G.   In preparation.

5)   Lingappa, V.R., Shields, D., Woo, S.L.C. and Blobel, G. (1978). J. Cell Biol. 79, 567-72.

6)   Schmidt, G., Devillers-Thiery, A., Desruisseaux, H., Blobel, G. and Chua, N.H. In preparation.

7)   Blobel, G. and Dobberstein, B. (1975). J. Cell Biol. 67, 835-51.

8)   Blobel, G. (1978). In FEBS 11th Meeting Copenhagen 1977, Vol. 43, ed. B.F.C. Clark, H. Klenow and J. Zeuthen, pp. 99-108.

10)   Dobberstein, B., Blobel, G. and Chua, N.H. (1977). Proc. Nat. Acad. Sci. USA, 74, 1082-5.

11)   Chang, C.N. and Blobel, G. In preparation.

12)   Maccecchini, M.L., Rudin, Y., Blobel, G. and Schatz, G. (1979). Proc. Nat. Acad. Sci. (USA), 76, 343-347.

13)   Chua, N.H. and Schmidt, G.W. (1978). Proc. Nat. Acad. Sci. (USA), 75, 6110-14.

14)   Highfield, P.E. and Ellis, R.J. (1978). Nature London, 271, 420-424.

15)   Goldman, B.M. and Blobel, G. (1978). Proc. Nat. Acad. Sci. (USA) 75, 5066-70.

16)   Bonatti, S., Cancedda, R. and Blobel, G. (1979). J. Cell Biol. 80, 219-224.

17)   Chang, C.N., Model, P. and Blobel, G. (1979). Proc. Nat. Acad. Sci. (USA) 76, 1251-1255.

DISCUSSION

R. PERRY: I almost hesitate to ask for more information
after your beautiful presentation, but you introduced the
problem of sorting mechanisms.   Are there any clues as to how
particular types of proteins are directed to the appropriate
membrane system?   Can you make any remarks at all on this
point?

G. BLOBEL: There is some very nice work going on with
lysosomal proteins.   These proteins contain phosphorylated
mannose residues.   It is conceivable that lysosomal proteins
contain a particular "sorting" sequence which is absent in
secretory glycoproteins.   This sequence could provide a
recognition site for a specific kinase that phosphorylates
mannose residues of glycoproteins.   A restricted location of
this kinase in a particular region of the smooth endoplasmic
reticulum or the Golgi complex would then act as a shunt to
separate secretory from lysosomal proteins within the lumen
of the endoplasmic reticulum.

B. RABIN: The situation is actually that we can show with one
particular reticular membrane protein, cytochrome P450, that
a large proportion of the newly synthesized membrane protein
is translocated into the vesicle.   We can show this in vivo,
in vitro and in situ and actually what we find is that the
newly biosynthesized protein plugs back into the membranes in
the Golgi.   Have you any thoughts about whether the Golgi
acts like a traffic policeman for the location of proteins
for export to membrane destinations?

G. BLOBEL: This is a good question.   In vitro translation
and translocation experiments might provide an answer to this
question.   Experiments of this sort involving P450 are in
progress in my laboratory.   [Answer revised by speaker from
the original.   Ed.]

H. LODISH: Gunter, since you drew on one of your diagrams a
"stop transfer" sequence for synthesis of integral membrane
proteins I wonder if I could make a brief comment and ask you
a question about it.   I think that a "stop transfer" sequence
is really an incorrect concept for the following reason - if
you look at the situation where the synthesis of either an
integral membrane protein or a secretory protein has just
been completed what you have is either protein transmembrane:
The NH_2-terminus of either protein faces the lumen, and the
30 amino acids at the COOH-terminus -- the last to be added

-- remain on the cytoplasmic side, since, at the moment
synthesis is completed, these are embedded in the large
ribosome subunit. Then, with secretory proteins and membrane
proteins, the ribosome falls off, leaving roughly 30 amino
acids on the cytoplasmic side. So as far as the integral
membrane proteins are concerned nothing more needs be done.
That is the way they wind up and I think the question is what
drives secreted proteins into the lumen.

G. BLOBEL: There could be bitopic integral membrane proteins
which have the bulk of their carboxy terminal portion on the
cytoplasmic side of the membrane. In this case a
"stop-transfer" sequence would be located near the amino
terminus of the nascent chain. Thus a stop-transfer sequence
could provide the means to orient the polypeptide chain of
bitopic or polytopic membrane proteins with respect to the
lipid bilayer.

PRECURSORS IN THE BIOSYNTHESIS OF INSULIN
AND OTHER PEPTIDE HORMONES

Shu J. Chan
Christoph Patzelt
John R. Duguid
Paul Quinn
Allen Labrecque
Barbara Noyes
Pamela Keim
Robert L. Heinrikson
Donald F. Steiner[1]

Department of Biochemistry
University of Chicago
Chicago, Illinois

SUMMARY

Recent studies have established that limited intracellular proteolysis may participate at several stages in the biosynthesis of many small polypeptide hormones, as exemplified by insulin. Conversion of proinsulin to insulin by trypsin- and carboxypeptidase B-like enzymes occurs in the pancreatic B cell after the peptide has been sequestered from the cytosolic compartment and transported to the Golgi area. An earlier role for proteolysis concerns the conversion of the initial translation product, preproinsulin, to proinsulin during its compartmentalization in the rough endoplasmic reticulum. Preproinsulin, a methionine-initiated peptide which contains a hydrophobic 24-residue NH_2-terminal extension, was originally identified as the product of the cell-free translation of insulin mRNA, but more recently it has also been detected in small amounts in intact rat islets.

[1]Supported by NIH grants AM 13914 and AM 20595 and grants from the Kroc Foundation and the Lolly Coustan Memorial Fund.

Its rapid formation and disappearance during pulse-chase studies are consistent with its role as a biosynthetic precursor and with the proposal that the prepeptide region serves as a leader sequence in the formation of the ribosome-membrane junction and in the vectorial discharge of the peptide. A model for segregation is presented, based on important structural features and physical properties of the prepeptides which may enable these to enter and span the microsomal membrane.

Studies on insulin mRNA obtained from transplantable islet cell tumors indicate that the major component contains approximately 600 nucleotides and possesses a 5' "cap" structure as well as a 3' polyadenylate "tail". A closer examination of the insulin tumor mRNA has also revealed two larger polyadenylated RNA fractions having apparent molecular weights of 280,000 and 360,000. These forms, which hybridize to the rat insulin genes and which were shown by ribonuclease T_1 digests to share nucleotide sequences with the major mRNA component, may represent precursors of the mature template. These results indicate that several sequential stages of macromolecular processing are required in the formation of the secreted hormone.

HUMAN PROINSULIN

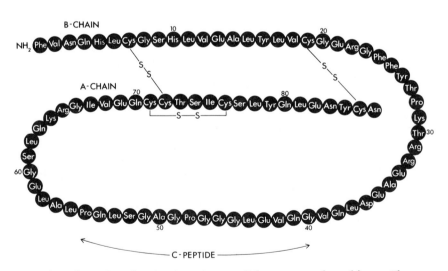

Fig. 1. *Covalent structure of human proinsulin. The region from positions 31 to 65 is removed to produce insulin from proinsulin. (Reproduced from Oyer, et al., J. Biol. Chem. 246, 1375, 1971.)*

I. INTRODUCTION

Post translational modification of proteins by limited intracellular proteolysis of larger precursors was first observed to occur in the biosynthesis of insulin. Proinsulin at first appeared to provide a unique and specialized solution for the rather unusual biosynthetic problems associated with the linkage of the separate chains of insulin via disulfide bonds (1). However subsequent studies in many laboratories have shown that proteolytic modification is an almost universally occurring mechanism in the synthesis of many secreted peptides and a variety of other proteins as well. It is increasingly evident that the "readout" of the linearly coded genetic messages in cells generally involves some processive "editing" of the coded information to subserve a variety of purposes. Thus selective cutting, and in some instances splicing as well, are natural and presumably necessary concommitants of the mechanism for information storage and retrieval that is utilized universally by living organisms.

Recent studies in our laboratory and a number of others on the biosynthesis of insulin and other proteins have considerably advanced our understanding of the roles played by protein precursors in biosynthesis, segregation and assembly. This report will briefly summarize recent progress in this general area. We also will present preliminary results indicating that post-transcriptional processing of the messenger RNA for preproinsulin may occur during the expression of the genes for insulin.

II. PROINSULIN AND THE PROPROTEINS

The primary structure of human proinsulin, shown in Fig. 1, is typical of the structure of proinsulins throughout the vertebrate kingdom. The characteristic disulfide arrangement first described by Ryle *et al*. (2) for bovine insulin is always present and the proteins are all similar in overall size. The gap spanned by the connecting segment in fully folded proinsulin is only 8-10 Ångstroms (3). Nonetheless, the connecting segment (C-peptide and 2 pairs of linking basic residues), though highly variable and seemingly without biological function, has retained a length of 26 to 35 residues

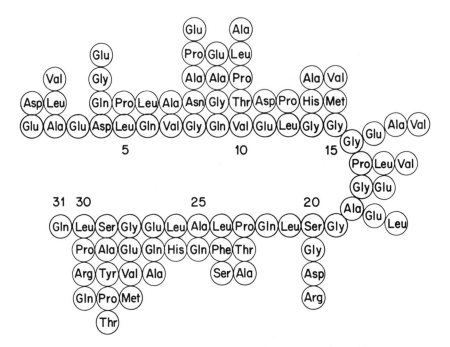

Fig. 2.   Amino acid sequence of human proinsulin C-peptide, with known substitutions occurring in nine other mammalian and one avian C-peptide shown alongside.   Deletions occur in the dog (residues 4-11), pig (residues 18 and 19), sheep and ox (residues 22-26), and guinea pig (residues 25 and 26).   (These sequences do not include the basic residues at either end, which link the C-peptide to the insulin chains.)   Reproduced in modified form from reference 4 (the rabbit C-peptide sequence has been added per Dr. R. Chance, unpublished data).

(Figure 2) throughtout an evolutionary span of almost 500 million years (4, 5). Moreover, it has been shown (6, 7, 8) that the function of the connecting peptide in promoting the correct oxidation of sulfhydrils in proinsulin can be accomplished by much smaller cross-linking reagents. Further strong evidence supporting the conclusion that the connecting segment could be shortened can be adduced from the elegant recent work of Rinderknecht and Humbel (9) on IGF, a circulating proinsulin like peptide which, interestingly, has a connecting peptide consisting of only 11 residues. Nonetheless this peptide organizes to form the typical insulin fold. Also Markussen (10) has successfully created and reoxidized a miniproinsulin in which the entire connecting peptide and one residue from the C-terminus of the B chain have been eliminated. We might therefore ask why the largely non-functional and highly variable C-peptide has been so well conserved in vertebrate evolution.

Although the persistence of the relatively long connecting peptide segment could indicate a lack of selective pressure to eliminate or shorten this region, it seems more likely that it serves some important function that is unrelated to the simple structure-making role that it has usually been assigned. One possibility, as suggested by Snell and Smyth (11), is that the redundant connecting segment covers and protects the active site of insulin, thus accounting for the low biological activity of the precursor. Obviously, if this were *not* the case there would be little need to convert pro-insulin to insulin, as for example is the case with IGF which is not converted to an insulin-like form (9). However, there seems to be no clear reason for maintaining an inactive or less active form of insulin during its biosynthesis, or elsewhere in the organism. Thus the simpler and more reasonable explanation can be proposed (12) that the C-peptide functions as a spacer which enlarges the proinsulin molecule sufficiently to span the ribosome-membrane junction. It has been estimated by Patzelt *et al.* (12) that 65-70 amino acid residues represents the minimum length of extended peptide chain which will permit successful vectorial discharge and sequestration within the cisternal spaces of the rough endoplasmic reticulum. The virtue of this explanation for the existence of the seemingly larger-than-necessary connecting sequent is that it is consistent with the structures of IGF as well as of the precursors of many other peptide hormones including parathyroid hormone, ACTH, MSH and endorphin, and probably of gastrin, glucagon and a variety of other small gastrointestinal and/or neuroendocrine peptides.

IGF is especially interesting in this context since it has a relatively shortened C-segment as compared with proinsulin. However, overall length in this case is maintained by the presence of additional residues at the C-terminus of the A chain (9). Thus this molecule has grown longer in one region while undergoing shortening in another, the overall effect being to conserve a proinsulin-like structure of similar overall size (13). But then, why hasn't this change occurred during the evolution of proinsulin? The important reason here could well be that the C-terminal Asn residue of the A chain is essential for insulin binding to its receptors. Removal or replacement of this amino acid, or addition of residues to the A chain, has been shown repeatedly to lower the biological activity markedly and hence could be expected to

exert a negative selective effect in evolution. This constraint does not apply to IGF which binds to a separate class
of receptors which strongly discriminate against insulin (14).
And, as discussed in more detail in a later section, the
presence of the presequence at the N-terminus, with its special functions and cleavage mechanisms, as well as the probable existence of nucleation sites within this region (12),
may prevent random growth of the proinsulin chain in its N-
terminal region where additions would be expected to be better tolerated from the standpoint of preserving the biological activity of the resultant insulin molecule (15). The
ultimate test of the validity of this "Minimum Length Hypothesis", of course, will lie in its success in predicting the
existence of precursors for many other small biologically
active peptides. Thus far no exceptions to this simple rule
have appeared, although relatively little is known about the
structure and size of many of these precursor forms.

III. PREPROINSULIN AND THE PRESECRETORY PROTEINS

Many recent studies, such as those by Milstein *et al.*
(16), Blobel and Dobberstein (17), Chan *et al.* (18), and
Kemper *et al.* (19), have shown that the *in vitro* translation
products of the mRNAs for various secretory proteins carry
a 20-to-30 residue N-terminal extension that contains a high
proportion of hydrophobic amino acids. The partial struc-

Bovine: X -Leu- X - X - X -Leu- X - X -Leu-Leu- X -Leu-Leu- X -Leu- X - X - X - X - X - X - X -Phe...

Rat: I.   (Met)Ala-Leu-Trp-Met-Arg-Phe-Leu-Pro-Leu-Leu-Ala-Leu-Leu-Val-Leu-Trp-Glu-Pro-Lys-Pro-Ala-Gln-Ala-Phe...
        II.   Ile   Phe Phe   Ile

Sea
raven: Met-Met- X - X -Leu- X - X - X - X - X - X - X -Leu-Leu- X -Leu-Leu- X -Leu- X - X - X - X - X - X - X - X...

Angler
fish: X -Ala-Leu- X -Leu- X - X -Phe- X -Leu-Leu-Val-Leu-Leu-Val-Val- X - X - X - X - X - X -Ala-Val...

Hagfish: X -Leu- X - X - X -Leu-Ala-Ala- X - X - X -Leu- X -Leu-Leu-Leu- X - X -Ala- X - X - X -Ala- X - X - Arg...

*Fig. 3. Partial amino acid sequences of the N-terminal
extensions of preproinsulins of several vertebrate species
(18, 40, 41, 45, 46, 66). The hagfish is a member of the
Cyclostomata, or jawless vertebrates, which are believed to
have diverged from the main gnathostomian line approximately
500 million years ago.*

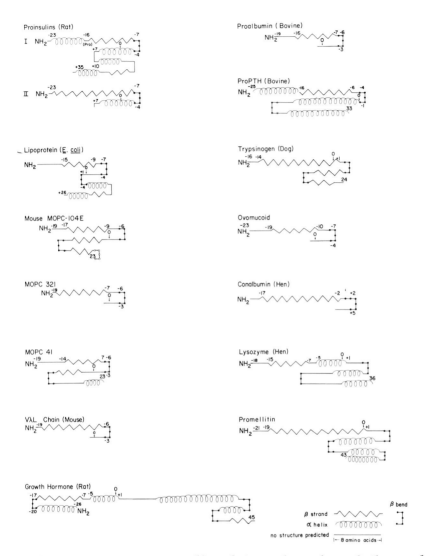

Fig. 4. Structures predicted from the rules of Chou and Fasman for the N-terminal regions of several presecretory proteins. (See text for details.)

tures of the N-terminal extensions of preproinsulins from several species is shown in Fig. 3. We have recently shown by means of incorporation studies with N-formylated initiator Met tRNA that rat preproinsulin I is initially synthesized with a residue of methionine at its N-terminus, which is subsequently rapidly removed (20). The prepeptide of preproin-

sulin appears to be well conserved throughout vertebrate evolution in terms of both length and hydrophobic character.
However, there is little or no evident homology among the
prepeptides of a variety of other presecretory proteins (21,
22, 23, 24). It has been proposed by Milstein et al. (16)
and Blobel and Dobberstein (17) that this region assists in
the formation of the ribosome-membrane junction leading to
the vectorial discharge and segregation of the nascent polypeptides. Although the details of this mechanism remain unclear, certain important structural features of the prepeptides can be defined by comparing the known sequences of a
wide variety of the prepeptides and applying the rules of
Chou and Fasman (25) for predicting their secondary structure.
As shown in Fig. 4 (preceding page), essentially all the prepeptides contain a region of six or more hydrophobic residues
arranged in an extended chain (or beta pleated sheet) structure. Following this is a section of 5 or 6 residues that
consists of amino acids with small aliphatic side chains
which might be expected to form a beta turn near the sites of
cleavage.

We believe that the lipophilic central region of the prepeptide interacts initially with the non-polar lipid bilayer
of the membrane of the rough endoplasmic reticulum in conjunction with additional, but still largely conjectural, recognition or orienting sites on the large ribosomal subunit
and the membrane (see Fig. 5). Perhaps these binding components are related to the "ribophorin" proteins described
recently by Kreibich and Sabatini (26). As a consequence of

POSSIBLE ROLE OF PRESEQUENCES IN TRANSMEMBRANE SEGREGATION OF SECRETORY PROTEINS

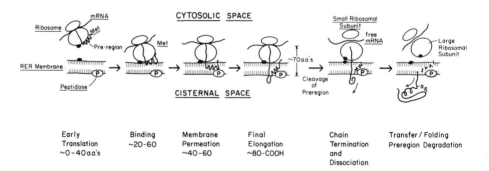

Fig. 5. Proposed mechanism for segregative transfer of
presecretory proteins across the membrane of the rough endoplasmic reticulum (Reproduced from ref. 67).

this initial interaction the hydrophobic region of the prepeptide dissolves in the lipid bilayer while its N-terminal
region, which is hydrophilic and more highly variable, remains outside the membrane. The nascent peptide then "loops
across" the membrane, as shown diagrammatically in Fig. 5, as
chain elongation proceeds allowing the peptide to diffuse
across the membrane. Once threaded through the lipid bilayer the nascent peptide chain will continue to diffuse across
the membrane at little energy cost unless a relatively large
region of hydrophobic amino acids appears within the sequence.
Such a cluster might bring diffusion to a halt and thereby
establish a permanent transmembrane orientation (insertion)
for the protein. In the case of most secretory proteins,
however, no such interuption in transfer is visualized and
the completed peptide chain is finally "pulled" entirely
through the membrane and into the cisternal space by the
folding of N-terminal portions of the product peptide which
has been freed from the presequence by a protease(s) believed
to be associated with the inner membrane surface of the rough
microsomes (27). In keeping with this model several studies
with reconstituted systems have provided evidence that the
signal sequences are rapidly cleaved from the nascent secretory product before polypeptide chain completion (28, 29),
thus accounting for the apparent difficulty in detecting the
presecretory forms in intact tissues (30, 31).

It should be pointed out that the model presented in
Fig. 5 differs significantly from the previously proposed
Signal Hypothesis of Blöbel and Dobberstein (28). In our
formulation the nascent protein chain is considered to pass
directly through the lipid bilayer, rather than through a
hydrophilic pore assembled from subunits in the membrane, and
also, the original N-terminus of the presecretory form is
visualized as remaining outside the microsomal membrane after
"looping over" occurs. The central hydrophobic region of the
presequence is long enough to span the hydrophobic core of
the lipid bilayer and this transmembrane orientation may
serve to appropriately direct the interaction of the presequence with the cleavage enzyme(s). Whether additional specific membrane "receptors" for the presequence are required
is a question which requires much further investigation.

We have recently demonstrated the biosynthesis of small
amounts of preproinsulin in intact isolated rat islets of
Langerhans (32). Preproinsulin only corresponded to a few
percent of the proinsulin peak, even after very short pulse
periods at lowered temperatures. The major proportion of the
preproinsulin rapidly disappeared during a 3 minute chase
period, while a minor fraction appeared to be turning over
more slowly (32). The low proportion of preproinsulin, and

its rapid turnover rate, is consistent with predictions of
the model discussed above. In cell free or reconstituted
systems the cleavage of the nascent peptide chain of preproteins has been estimated to occur after the polymerization of
at least 70-80 amino acids (17, 29). Both preproinsulin (109
residues) and preproparathyroid hormone (115 residues) are
both only slightly longer than this minimum length and therefore may undergo cleavage at about the time of chain completion. Recent studies by Mauer (33) have also shown a similar
small accumulation of preprolactin during pulse labeling of
pituitary slices. However, since this peptide is almost
twice as large as preproinsulin it seems likely that alternative explanations, such as heterotopic translation, warrant
further consideration and investigation.

Thus far only one secreted protein has been found which
lacks a presecretory form. This protein is ovalbumin, a protein secreted by cells in the oviduct. At first glance the
N-terminal sequence of ovalbumin reported by Palmiter *et al.*
(34) reveals none of the structural features described above
for the prepeptides (Fig. 4). Instead the N-terminal sequence
can be predicted to form two short amphipathic helices instead of the characteristic hydrophobic beta structure pre-
dictable for many of the presequences. Thus this region in
ovalbumin appears more likely to be part of the globular
structure of the protein itself. However, closer examination
reveals the presence of 8 hydrophobic residues at the N-terminus of ovalbumin. It is possible that this region allows
for initial binding as described in the model of Fig. 5. But
·the subsequent acetylation of its N-terminus may remove the
hydrophilic anchor necessary for orientation for cleavage.
This segment may then be pulled through the membrane after
synthesis as the protein folds.

In the case of cytoplasmically synthesized subunits of
the chloroplast enzyme ribulose diphosphate carboxylase (35),
as well as mitochondrial ATP'ase (36) and cytochrome oxidase
complex (37), evidence is accumulating which indicates that
these proteins or their subunits are synthesized as larger
precursors which are cleaved and transferred into the membrane from a cytoplasmic pool *after* synthesis is completed.
Perhaps the transfer of these proteins across organellar membranes is accomplished by a receptor mechanism analogous to
that which mediates the cellular uptake of diphtheria toxin
and other related toxic proteins (38). These proteins are
essentially precursors of enzyme subunits whose "extra" sequences may serve as recognition sites for membrane binding
and uptake, or as suggested earlier by Steiner *et al.* (39)
they could also serve to mask hydrophobic segments which will
interact with the lipid bilayer of membranes when exposed by

proteolysis at appropriate membrane surfaces. Clearly, a
variety of possible mechanisms exist for the insertion or
transfer of proteins into or across membranes in biological
systems, and much more detailed analysis in individual cases
will be necessary to fully define these processes.

IV. STUDIES ON INSULIN GENES

The messenger RNA for preproinsulin from normal rat and
fish islets and from ox pancreas has been studied by translation *in vitro* in cell free systems in several laboratories
(18, 40, 41, 42, 43). It is similar to many other eukaryotic
mRNAs in having a 3' polyadenylate tail and a 5' terminal
"cap" structure (12, 44). Ullrich *et al.* (45) have cloned
segments of the mRNAs for one or both rat preproinsulins
without prior purification of whole islet polyadenylated RNA
used to prime the reverse transcriptase reaction. They have
reported the nucleotide sequence for a cloned region of rat
preproinsulin I mRNA extending from residue 10 in the presequence to the C-terminus of the A chain region of proinsulin
and including the 57 nucleotide non-coding region which precedes the 3' poly A tract. More recently their results have
been extended by Villa Komaroff et al. (46) to include all
but the amino-terminal 3 residues of rat preproinsulin I.
We have purified and characterized preproinsulin mRNA
from rat insulinoma material. These tumors, which can be induced by streptozotocin or by X-irradiation, contain large
amounts of insulin and of preproinsulin mRNA. Preproinsulin
mRNA purified from an X-ray induced transplantable insulinoma
(47) had a similar specific translational activity to that of
purified globin mRNA, and on sucrose gradients had a sedimentation coefficient of 9.3 (48). It migrated as a single symmetrical band when electrophoresed on polyacrylamide gels in
the presence of formamide, with an observed molecular weight
of approximately 200,000. This size, which corresponds to a
length of 550-600 nucleotides, could code for only one preproinsulin, and therefore separate mRNAs must be presumed to
exist to code for each of the two non-allelic rat insulins
(49, 50).
Duguid and Steiner (51) have recently reported the identification of two larger polyadenylated RNAs that share sequence
homology with preproinsulin mRNA. These components were identified in poly A RNA extracted from small, non-necrotic tumors.
Their molecular weights were: Component III - 360,000; Component II - 280,000. Comparison of partial and complete RN'ase
T_1 digests of Components II and III with those of preproinsu-

lin mRNA (designated Component I) indicated significant primary and secondary structural similarities among the three
RNAs (51). Both Components II and III yielded additional
spots on two dimensional maps of complete T_1 digests.

We also investigated the hybridization of these three islet tumor RNAs with Eco R1 generated restriction fragments of
the total rat genome. Three DNA components which strongly
hybridize with preproinsulin mRNA have been detected, having
molecular weights of 6.0×10^6, 1.6×10^6 and 1.4×10^6.
Components II and III also hybridized with these same components, as demonstrated by further cleavage analysis using
several restriction enzymes but Component III appeared to
hybridize preferentially to the larger DNA fragment (51).

The simplest interpretation of these results is that the
two larger RNAs represent precursors to Component I, *i.e.*, of
one or both of the rat preproinsulin mRNAs. Several laboratories, including Ross (52), Curtis and Weissman (53), Kwan
et al. (54), Bastos and Aviv (55), Strair *et al.* (56) and
Weber *et al.* (57), have recently reported the characterization of larger precursors of the mRNAs of a number of proteins. Furthermore, Leder and coworkers (58, 59) have shown
that an inserted sequence in the coding region of the mouse
β-globin gene is represented in the 15S β-globin mRNA precursor. Thus the existence of larger RNA molecules sharing sequences with translatable insulin mRNA presents the possibility that the insulin genes may have a similar structural organization.

The relatively large abundance of Components II and III
contrasts with the abundance of the 15S precursor to globin
mRNA (less than 0.1%) (60). However, it may be speculated
that tumors such as the malignant X-ray induced tumors used
for these studies may have altered mechanisms for mRNA transcription or processing, just as many tumors have protein
precursor processing deficiencies (61). Whether such defects
as ectopic hormone synthesis (62) results from the translation of improperly processed or aberrant transcription products is an interesting question which deserves further investigation.

The rat genomic fragments discussed above clearly must
represent those regions corresponding to the genes for preproinsulins I and II. In similar collaborative studies with
Dr. Philip Leder and associates at the NIH we have identified
two fragments of the mouse genome which hybridize with cDNA
prepared from purified rat preproinsulin mRNA (12, 63). Further efforts are currently in progress to purify and clone
these genomic fragments in suitable λ phage systems (64, 65).

The isolation of the mRNA and ultimately of the gene for insulin seems to offer new possibilities for studies on the mechanisms which regulate the differentiation of β cells and the expression of insulin synthesis as a cellular function. It can be anticipated that studies of this kind will provide fresh insights into diseases such as diabetes where insulin production is deficient and perhaps also on others such as malignancy, where gene expression is altered either through mutational events or derangement of normal developmental processes.

ACKNOWLEDGMENTS

The authors wish to thank Drs. Howard S. Tager, Arthur H. Rubenstein, Sture Falkmer, Stefan Emdin, Jon Marsh and Mr. Randolph Hastings for their interest in and contributions to various aspects of the work discussed in this review. We are endebted to Dr. William Chick at the Joslin Research Laboratories (Boston) for his kindness in making the X-ray induced transplantable insulinoma from the New England Deaconess Hospital (46) available to this laboratory. We also wish to express our gratitude to Dr. Philip Leder and his associates at the NIH for their assistance and support in our initial efforts to undertake studies on the nature of the insulin genes. We also thank Ms. Sally Parks for invaluable assistance in the preparation of this manuscript.

REFERENCES

1. Steiner, D.F., and Clark, J.L., *Proc. Natl. Acad. Sci. USA 60*, 622 (1968).
2. Ryle, A.P., Sanger, F., Smith, L.F., and Kitai, R., *Biochem. J. 60*, 541 (1955).
3. Blundell, T., Dodson, G., Hodgkin, D., and Mercola, D., *Adv. in Protein Chem. 26*, 279 (1972).
4. Steiner, D.F., *Diabetes 27 (Suppl. 1)*, 145 (1978).
5. Steiner, D.F., Terris, S., Emdin, S.O., Peterson, J.D., and Falkmer, S., *in* "Early Diabetes in Early Life", p. 41. Academic Press, New York, (1975).
6. Brandenburg, D., and Wollmer, A., *Hoppe-Seyler's Z. Physiol. Chem. 354*, 613 (1973).
7. Busse, W.D., Hansen, S.R., and Carpenter, F.H., *J. Amer. Chem. Soc. 96*, 5949 (1974).
8. Steiner, D.F., *Diabetes 26*, 322 (1977).
9. Rinderknecht, E., and Humbel, R.E., *J. Biol. Chem. 253*, 2769 (1978).
10. Markussen, J. Presented at the Symposium on Proinsulin, Insulin, and C-Peptide, Tokushima, Japan, 1978, in press.
11. Snell, C.R., and Smyth, D.G., *J. Biol. Chem. 250*, 6291 (1975).
12. Patzelt, C., Chan, S.J., Duguid, J., Hortin, G., Keim, P., Heinrikson, R.L., and Steiner, D.F., *in* "Regulatory Proteolytic Enzymes and their Inhibitors, Proc. Vol. 47, Symp. A6" (S. Magnusson, *et al.*, eds.), p. 69. Pergamon Press, New York, (1978).
13. Blundell, T.L., Bedarkar, S., Rinderknecht, E., and Humbel, R.E., *Proc. Natl. Acad. Sci. USA 75*, 180 (1978).
14. Zapf, J., Schoenle, E., Froesch, E.R., *Eur. J. Biochem. 87*, 285 (1978).
15. Wood, S.P., Blundell, T.L., Wollmer, A., Lazarus, N.R., and Neville, R.W.J., *Eur. J. Biochem. 55*, 531 (1975).
16. Milstein, C., Brownlee, G.G., Harrison, T.M., and Mathews, M.B., *Nature New Biol. 239*, 117 (1972).
17. Blöbel, G., and Dobberstein, B., *J. Cell Biol. 67*, 835 (1975).
18. Chan, S.J., Keim, P., and Steiner, D.F., *Proc. Natl. Acad. Sci. USA 73*, 1964 (1976).
19. Kemper, B., Habener, J.F., Mulligan, R.C., Potts, J.T., Jr., and Rich. A., *Proc. Natl. Acad. Sci. USA 71*, 3731 (1974).
20. Chan, S.J., Ackerman, E., Quinn, P., Sigler, P., and Steiner, D.F. Manuscript in preparation.

21. Habener, J.F., Rosenblatt, M., Kemper, B., Kronenberg, H.M., Rich, A., and Potts, J.T., Jr., *Proc. Natl. Acad. Sci. USA 75*, 2616 (1978).
22. Schechter, I., Zemell, R., and Burstein, Y., *in* "FEBS Proceedings Vol. 47, Symp. A6", p. 103. Pergamon Press, New York, (1978).
23. Palmiter, R.D., Thibodeau, S.N., Gagnon, J., and Walsh, K.A., *in* "FEBS Proceedings Vol. 47, Symp. A6", p. 89. Pergamon Press, New York, (1978).
24. Tonegawa, J., Maxam, A.M., Tizard, R., Bernard, O., and Gilbert, W., *Proc. Natl. Acad. Sci. USA 75*, 1485 (1978).
25. Chou, P.Y., and Fasman, G.D., *Biochemistry 13*, 222 (1974).
26. Kreibich, G., Grebenau, R., Mok, W., Pereyra, B., Rodriguez-Boulan, E., Ulrich, B., and Sabatini, D.D., *Fed. Proc. 36*, 656 (1977).
27. Jackson, R.C., and Blöbel, G., *Proc. Natl. Acad. Sci. USA 74*, 5598 (1977).
28. Blobel, G., and Dobberstein, B., *J. Cell Biol. 67*, 852 (1975).
29. Boime, I., Szczesna, E., and Smith, D., *Eur. J. Biochem. 73*, 515 (1977).
30. Schmeckpeper, B.J., Adams, J.M., and Harris, A.W., *FEBS Letts. 53*, 95 (1975).
31. Sussman, P.L., Tushinski, R.J., and Bancroft, F.C., *Proc. Natl. Acad. Sci. USA 73*, 29 (1976).
32. Patzelt, C., Labrecque, A.D., Duguid, J.R., Carroll, R.J., Keim, P., Heinrikson, R.L., and Steiner, D.F., *Proc. Natl. Acad. Sci. USA 75*, 1260 (1978).
33. Mauer, R., and McKean, D.J., *J. Biol. Chem. 253*, 6315 (1978).
34. Palmiter, R.D., Gagnon, J., and Walsh, K.A., *Proc. Natl. Acad. Sci. USA 75*, 94 (1978).
35. Highfield, P.E., and Ellis, R.J., *Nature 271*, 420 (1978).
36. Schatz, G., personal communication (1978).
37. Poyton, R., personal communication (1978).
38. Pappenheimer, A.M., Jr., *Ann. Rev. Biochem. 46*, 69(1977).
39. Steiner, D.F., Kemmler, W., Clark, J.L., Oyer, P.E., and Rubenstein, A.H., *in* "Handbook of Physiology - Endocrinology I" (D.F. Steiner and N. Freinkel, eds.), p. 175. The Williams and Wilkins Company, Baltimore, (1972).
40. Lomedico, P.T., Chan, S.J., Steiner, D.F., and Saunders, G.F., *J. Biol. Chem. 252*, 7971 (1977).
41. Shields, D., and Blöbel, G., *Proc. Natl. Acad. Sci. USA 74*, 2059 (1977).

42. Permutt, M.A., Beisbroeck, J., Chyn, R., Boime, I., Szczesna, E., and McWilliams, D., *in* "Polypeptide Hormones: Molecular and Cellular Aspects, CIBA Fdn. Symp. 41", p. 97. Elsevier/Excerpta Medica, Amsterdam, (1976).
43. Okamoto, H. Presented at the Symposium on Proinsulin, Insulin, and C-Peptide, Tokushima, Japan, 1978, in press.
44. Duguid, J.R., Chan, S.J., Keim, P., Heinrikson, R., Hortin, G., Hofmann, C., Labrecque, A., Chick, W., and Steiner, D.F., *in* "Diabetes, International Congress Series # 413" (J.S. Bajaj, ed.), p. 9. Excerpta Medica, (1977).
45. Ullrich, A., Shine, J., Chirgwin, J., Pictet, R., Tischer, E., Rutter, W.K., and Goodman, H.M., *Science* 196, 1313 (1977).
46. Villa-Komaroff, L., Efstratiadis, A., Broome, S., Lomedico, P., Tizard, R., Naber, S.P., Chick, W.L., and Gilbert, W., *Proc. Natl. Acad. Sci. USA* 75, 3727 (1978).
47. Chick, W.L., Warren, S., Chute, R.N., Like, A.A., Lauris, V., and Kitchen, K.C., *Proc. Natl. Acad. Sci. USA* 74, 628 (1977).
48. Duguid, J., Steiner, D.F., and Chick, W.L., *Proc. Natl. Acad. Sci. USA* 73, 3539 (1976).
49. Smith, L.F., *Amer. J. Med.* 40, 662 (1966).
50. Clark, J.L., and Steiner, D.F., *Proc. Natl. Acad. Sci. USA* 62, 278 (1969).
51. Duguid, J.R., and Steiner, D.F., *Proc. Natl. Acad. Sci. USA* 75, 3249 (1978).
52. Ross, J., *J. Mol. Biol.* 106, 403 (1976).
53. Curtis, P.J., and Weissmann, C., *J. Mol. Biol.* 106, 1061 (1976).
54. Kwan, S., Wood, G.T., and Lingrel, J.B., *Proc. Natl. Acad. Sci. USA* 74, 178 (1977).
55. Bastos, R.N., and Aviv, H., *Cell* 11, 641 (1977).
56. Strair, R.K., Skoultchi, A.I., and Shafritz, D.A., *Cell* 12, 133 (1977).
57. Weber, J., Jelinek, W., and Darnell, J.E., Jr., *Cell* 10, 611 (1977).
58. Tilghman, S.M., Curtis, P.J., Teimeier, D.C., Leder, P., and Weissmann, C., *Proc. Natl. Acad. Sci. USA* 75, 1309 (1978).
59. Tilghman, S.M., Tiemeier, D.C., Seidman, J.G., Peterlin, B.M., Sullivan, M., Maizel, J.V., and Leder, P., *Proc. Natl. Acad. Sci. USA* 75, 725 (1978).
60. Ross, J., and Knecht, D.A., *J. Mol. Biol.* 119, 1 (1978).
61. Rubenstein, A.H., Steiner, D.F., Horwitz, D.L., Mako, M.E., Block, M.B., Starr, J.I., Kuzuya, H., and Melani, F., *Recent Prog. Horm. Res.* 33, 435 (1977).

62. Kahn, C.R., Rosen, S., Weintraub, B., Fajans, S., and Gorden, P., *N. Engl. J. Med. 297*, 565 (1977).
63. Duguid, J.R., Leder, P., Leder, A., and Steiner, D., unpublished data.
64. Leder, P., Tiemeier, D., and Enquist, L., *Science 196*, 175 (1977).
65. Benton, W.D., and Davis, R.W., *Science 196*, 180 (1977).
66. Chan, S., Emdin, S., Falkmer, S., and Steiner, D.F., unpublished data.
67. Duguid, J.R., Patzelt, C., Chan, S.J., Labrecque, A., Hastings, R., Quinn, P., and Steiner, D.F. Presented at the Symposium on Proinsulin, Insulin, and C-Peptide, Tokushima, Japan, 1978, in press.

DISCUSSION

G. BLOBEL: Is the small amount of preproinsulin which you detect in vivo, in tissue slices, segregated within vesicles? One way one could explain the existence of preproinsulin molecules is that they result from translation that was uncoupled from translocation across the endoplasmic reticulum membrane. In this case preproinsulin would be found in the cytosol and not within vesicles and could not serve as a "precursor" to proinsulin.

D.F. STEINER: Unfortunately, there is no clear answer on this question. What we found was that if we homogenize the pulsed islets and isolate fractions then we can no longer demonstrate any preproinsulin. Evidently it is degraded or "converted" during the manipulations. What we have done in the experiments is to preincubate the islets for at least 15 minutes and then to add the labelled amino acids so that the system isn't perturbed in any way at the time the incorporation of precursor amino acids is initiated. But even under those circumstances the amount of preproinsulin appears to be small. The fact that at very early times and lower temperatures there is relatively more of it is at least suggestive that it is not an artifact, although heterotopic translation (followed by degradation) is difficult to rule out. There are now reports in three systems, as mentioned in my lecture, of the transient production of preproteins in intact endocrine cells.

T. HUNT: I was very interested in your suggestion that the role of the N-terminus of ovalbumin is that of a signal sequence, but has that not been tested? If you produce the non-acetylated form in vitro in the presence of membrane vesicles, does it go across the vesicles or not?

D.F. STEINER: I believe Dr. Blobel has done such experiments.

G. BLOBEL: Yes, it can be segregated in vitro but we don't know whether the in vitro segregated form is acetylated.

D.F. STEINER: Actually in terms of the model that I was discussing, if the ovalbumin were not being acetylated one would not expect this to interfere with its segregation. It is that once segregated it might remain a membrane linked protein, whereas if it is acetylated then the end can slip out of the membrane after the bulk of the ovalbumin peptide has been transferred into the luminal space and is folding to its native conformation. Thus, acetylation may be related to non-cleavage because it may prevent the preregion from assuming the postulated trans-membrane orientation which may be necessary for cleavage.

T. HUNT: I thought you were making the point that acetylation would make the prepeptide more hydrophobic.

D.F. STEINER: It would, but the more important effect would be to allow the amino terminus to slip through the membrane and be "wrapped up" into the completed protein. Presumably it would still loop across initially as postulated in the model whether acetylated or not. It should be possible to examine these alternatives experimentally.

BIOSYNTHESIS AND PROCESSING OF GLYCOPROTEINS

E.C. HEATH, S.A. BRINKLEY, R.C. DAS, and T.H. HAUGEN
Department of Biochemistry
University of Iowa
Iowa City, IA

Considerable progress has been achieved in the past five years providing a clearer understanding of the mode of bio-synthesis and assembly of glycosylamine-linked carbohydrate sidechains of glycoproteins. The structural details of many carbohydrate sidechains of this type have now been charac-terized (1). The carbohydrate moiety is linked to the poly-peptide chain by way of a glycosylamine linkage utilizing the amide nitrogen of an Asn residue in the protein and the linkage region invariably consists of the trisaccharide Man-β1,4-GlcNAc-β1,4-GlcNAc-β-Asn. Beyond the invariant tri-saccharide linkage region of the sidechain, two or more branch structures usually occur, each of which contains Man, GlcNAc, Gal, and one of the sialic acids. In some instances, a fucose residue substitutes the trisaccharide linkage region, attached to the inner-most GlcNAc. On the basis of amino acid sequence determinations of many polypeptide chains in the vicinity of the carbohydrate sidechain, the tripeptide -Asn-[X]-Ser(Thr)- appears to be invariant at the site of glycosylation. It should be noted however, that while all carbohydrate sidechains of this class are attached to an Asn residue in this tripeptide sequence, there are many examples of proteins that possess this tripeptide sequence within their structure that is not glycosylated.

Regarding the biosynthesis of the carbohydrate side-chains of this type, it is now clear that two separate groups of enzymes participate in its biosynthesis, assembly, and attachment to the polypeptide chain. As shown in Figure 1, a group of glycosyl transferases, primarily localized in the Golgi apparatus appears to be solely responsible for the sequential addition of peripheral carbohydrate residues of

the oligosaccharide sidechain, including the NeuNAc-Gal-
GlcNAc trisaccharide at the reducing termini of completed
chains, as well as fucose residues that may be located in
terminal non-reducing positions elsewhere in the oligosaccharide.

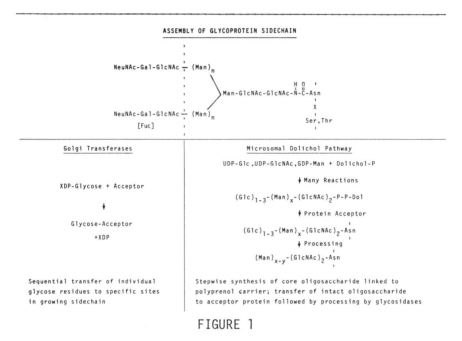

FIGURE 1

These enzymes characteristically catalyze the transfer of
single glycose residues from their corresponding nucleotide
derivatives to a non-reducing terminal acceptor position of
the growing oligosaccharide sidechain to increase the degree
of polymerization of the oligosaccharide by one carbohydrate
residue.

    Conversely, the Man- and GlcNAc- containing core struc-
ture of the oligosaccharide sidechain is assembled as an
intact oligosaccharide structure while linked to the polyprenol, dolichol, and after it has been assembled in this form,
is transferred en bloc to an appropriate Asn residue of the
polypeptide chain (2). The synthesis and assembly of the
lipid-linked core oligosaccharide structure initially
includes the invariant Man-GlcNAc-GlcNAc linkage region, a
variable number of additional mannose residues, as well as

one to three glucose residues in non-reducing terminal positions.   It now appears that, while the Glc-, Man-, and
GlcNAc-containing lipid-linked oligosaccharide is the primary
donor in the initiation of the carbohydrate sidechain on the
protein (3), this structure is rapidly processed by a series
of glycosidases that remove the glucose residues and a portion of the mannose residues shortly after the oligosaccharide is incorporated into protein (4).

Many individual enzymatic steps of the dolichol pathway
have now been studied in detail and the overall sequence of
reactions involved in this pathway are outlined in Figure 2.

Reactions of the Dolichol Pathway

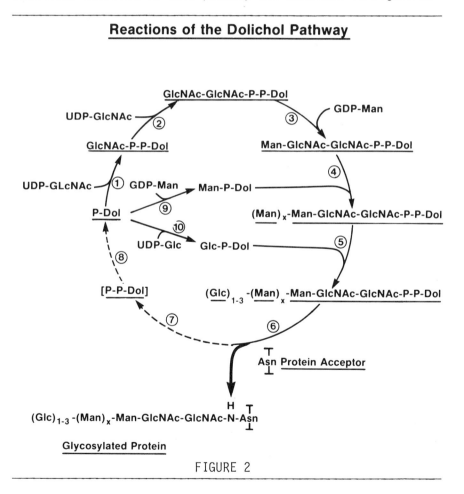

FIGURE 2

The assembly of the oligosaccharide is initiated by a phospho-
glycosyl transferase (Reaction 1) that transfer the GlcNAc-P
to dolichol monophosphate to establish a pyrophosphoryl linkage between GlcNAc and the polyprenol. Two glycosyl transferases (Reactions 2 and 3) sequentially transfer GlcNAc and
Man to the polyprenol derivative to produce the invariant
Man-β1,4-GlcNAc-β1,4-GlcNAc-trisaccharide derivative of dolichol. Dolichol monophosphate also serves as a glycose acceptor in two other reactions (Reactions 9 and 10) in which Man
(from GDP-Man) and Glc (from UDP-Glc) are transferred to the
polyprenol derivative to form the corresponding mannosyl
monophosphoryl dolichol (Man-P-Dol) and glucosyl monophosphoryl dolichol (Glc-P-Dol), respectively. Each of these
monoglycosyl derivatives of the polyprenol appear to be the
primary donors of additional Man and Glc residues that are
then transferred (Reactions 4 and 5, respectively) to the
growing oligosaccharide pyrophosphoryl dolichol derivative,
ultimately resulting in the formation of the complex Glc-,
Man-, and GlcNAc-containing oligosaccharide derivative of the
polyprenol. The intact complex oligosaccharide is then transferred (Reaction 6) to an appropriate Asn residue of a protein acceptor to establish the glycosylamine bond between
protein and the complex oligosaccharide (5). Although it has
not yet been demonstrated, it is possible that dolichol pyrophosphate is a product of the oligosaccharide transferase
(Reaction 7) and if so, may be converted to dolichol monophosphate (Reaction 8) to regenerate the monophosphorylated
derivative so that dolichol monophosphate may serve in a
catalytic fashion in the assembly of the oligosaccharide
sidechain and glycosylation of protein. However, it is
important to note that Reactions 7 and 8 are hypothetical
reactions that may be involved if the dolichol pathway functions in a cyclic manner.

It now appears quite clear that the dolichol pathway
plays a major role in establishing the covalent linkage of
carbohydrate to protein, the details of which are illustrated
in Figure 3. As indicated above, the structure of the trisaccharide linkage region formed in this reaction appears to
be identical to the comparable invariant trisaccharide structure found in glycoproteins of this class and the specificity of the oligosaccharide transferase for the appropriate
-Asn-[X]-Ser(Thr)- amino acid sequence of the polypeptide
chain has been firmly established (5).

INITIATION OF OLIGOSACCHARIDE
SIDECHAIN IN GLYCOPROTEIN BIOSYNTHESIS

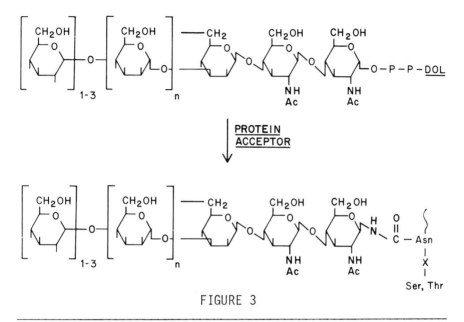

FIGURE 3

It appears that the dolichol pathway is primarily
localized in the rough endoplasmic reticulum and that the
introduction of the core oligosaccharide structure onto the
polypeptide chain occurs at the nascent peptide stage of
protein synthesis, prior to completion of the translation
process (6), as shown in Figure 4. Thus, secretory glycoproteins are synthesized with a leader sequence, or signal
peptide, at their N-terminus, traverse the microsomal membrane and are finally deposited in the lumen of the
endoplasmic reticulum after the leader sequence has been
removed by proteolytic processing (7). During transit of
the growing nascent peptide through the microsomal membrane
the complex oligosaccharide assembled as the polyprenol-
linked derivative in the membrane is transferred to an
appropriate Asn residue of a polypeptide chain resulting in
the compartmentalization of the newly synthesized, processed
protein that bears the complex Glc-, Man-, and GlcNAc-
containing oligosaccharide structure. Shortly after deposi-
tion of the partially glycosylated protein in the lumen, the
oligosaccharide structure undergoes several processing steps
by specific glycosidases that result in the removal of a

portion of the mannose residues and all of the glucose residues in the oligosaccharide structure.

Biosynthesis and Processing of Secretory Glycoproteins

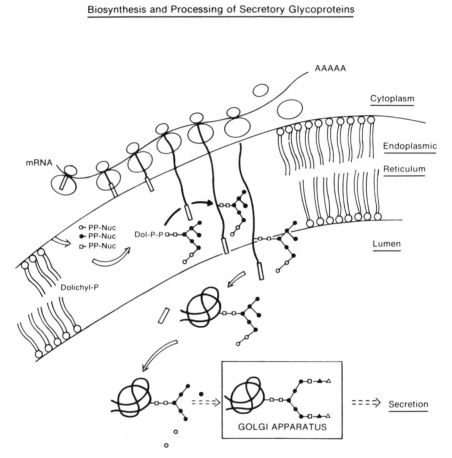

Figure 4. Symbols used: 0, glucose; □ , N-acetylglucosamine; ●, mannose; △ , sialic acid; ▲ , galactose;
▭ , signal peptide.

These processing steps may in fact occur just prior to, or
after, the protein has reached the Golgi apparatus where the
glycosyl transferases complete the peripheral regions of the
carbohydrate sidechain.   Since the introduction of the carbohydrate sidechain into the nascent peptide appears to be a
co-translational event, further information concerning the
sequence of events of these processes and to relate these

processes to the mobilization of the newly synthesized glycoprotein it would be necessary to establish an in vitro translation system that would permit the de novo biosynthesis of a
secretory glycoprotein.

Since most of our previous work relating to the enzymatic
steps of the dolichol pathway were conducted with a mouse
myeloma tumor (MOPC-46B) that produces a glycosylated kappa-
type immunoglobulin light chain, we chose to develop this
system for these studies. Messenger RNA was isolated from
myeloma cells by extraction in the presence of guanidine
thiocyanate followed by ethanol precipitation (8). Some of
our studies were conducted with the polyadenylated fraction
of mRNA from these cells, although we have found more recently that the crude preparations of mRNA obtained after repeated
ethanol precipitation are quite satisfactory for most purposes.
Using ^{35}S-Met as a tracer in a rabbit reticulocyte cell-free
translation system (9), the synthesis of protein was totally
dependent upon the addition of mRNA and was linear for
periods up to 60-90 min. The radioactive translation products formed in this system were analyzed by polyacrylamide
gel electrophoresis in the presence of sodium dodecyl sulfate
(PAGE-SDS) as shown in Figure 5.

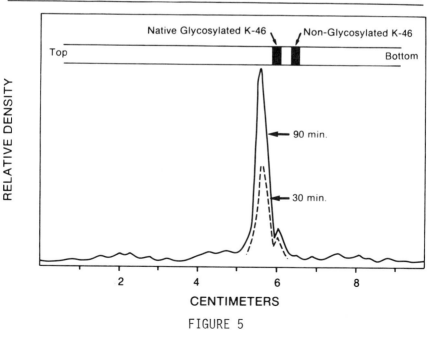

SDS - GEL ELECTROPHORESIS OF TRANSLATION PRODUCTS

FIGURE 5

The major translation product formed in this system corresponded to neither native glycosylated kappa-46 nor the non-
glycosylated form of this protein, but rather migrated
slightly slower than the native glycosylated form of the
protein. In order to determine if this product of translation is related to the immunoglobulin light chain, cell-free
translation systems were analyzed by a double antibody precipitation technique using anti-sera prepared against native
kappa-46 light chain as shown in Figure 6.

SDS - ELECTROPHORESIS OF ANTIBODY PRECIPITABLE TRANSLATION PRODUCTS

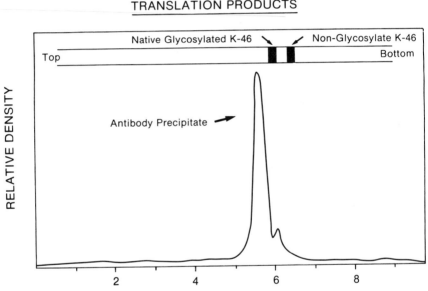

CENTIMETERS

FIGURE 6

Again, the major product of translation that migrates slower
than native kappa-46 was specifically precipitated with the
antibody directed against the native protein and most of the
other minor products of translation were eliminated by this
technique. On the basis of these results, we tentatively
concluded that the major product of translation of the myeloma mRNA preparation is a higher molecular weight form of the
kappa-46 light chain. Since this protein is a secretory protein, it appeared likely that the higher molecular weight

form of the protein produced in cell-free translation systems
may represent a precursor form of the protein with an additional leader sequence attached to the N-terminus of the
polypeptide chain.

In order to verify this tentative conclusion, ^{3}H-Leu-
labeled preparations of native kappa-46 and the antibody precipitable cell-free translation product were isolated and
subjected to sequence analysis as shown in Figure 7.

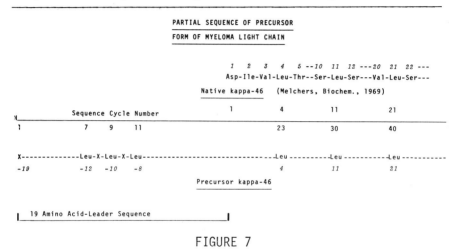

FIGURE 7

The results of this experiment indicated that native kappa-46
contains leucine residues at positions 4, 11 and 21 whereas
the precursor form of kappa-46 contains three additional
leucine residues in an N-terminal sequence consisting of 19
amino acids. This additional N-terminal sequence precisely
accounts for the difference in molecular weight of the cell-
free translation product and the native protein as judged by
their migration in calibrated PAGE-SDS systems.

Attempts to conduct the translation of kappa-46 mRNA in
the presence of dog pancreas microsomal membranes, in an
effort to demonstrate the coupling of the translation system
with the membrane associated proteolytic processing step (6),
resulted in the formation of two major translation products
corresponding to the non-glycosylated polypeptide chain of
kappa-46 and the non-glycosylated precursor form of the protein as shown in Figure 8.

TWO DIMENSIONAL GEL ELECTROPHORESIS OF MYELOMA RNA DIRECTED PROTEINS

TRANSLATED IN THE PRESENCE OF DOG PANCREAS MEMBRANES

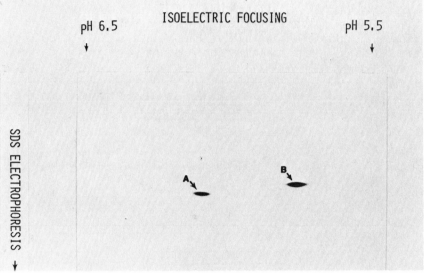

Figure 8. The positions of migration of standard samples of non-glocosylated kappa-46 (A) and non-glycosylated precursor kappa-46 (B) are indicated by the arrows.

Reticulocyte cell-free translation systems were supplemented with dog pancreas microsomes and after incubation, the mixture was treated with Triton X-100 and the anti-kappa-46 precipitable material was analyzed by two-dimensional PAGE. Initially, isoelectric focusing was conducted in the horizontal direction between pH 5.5 and pH 6.5, followed by PAGE-SDS in the vertical direction. ^{35}S-Met-labeled proteins were then visualized by autoradiography and revealed the presence of two components, A, co-migrating with non-glycosylated kappa chain and B, co-migrating with the non-glycosylated precursor form of the protein. Further analysis of the translation products produced in the presence of dog pancreas membranes indicated that only a small proportion of the newly synthesized protein is glycosylated as well as proteolytically processed in this system. In an effort to identify microsomal membrane preparations that would carry out the glycosylation step more efficiently, various membrane preparations (rat liver, rat pituitary, hen oviduct, and mouse myeloma) were analyzed for their efficiency in carrying out the glycosylation step as well as the proteolytic processing step and it was found that hen oviduct membranes were most effective, as shown in Figure 9.

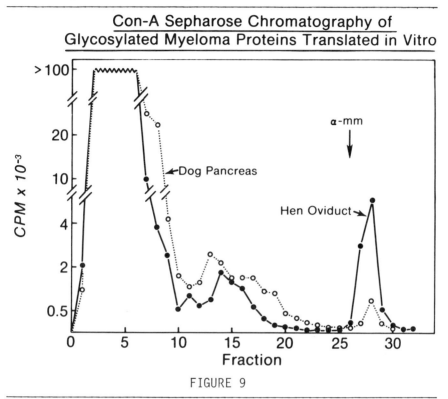

Con-A Sepharose Chromatography of Glycosylated Myeloma Proteins Translated in Vitro

FIGURE 9

A comparison of the amount of glycosylated translation product obtained in the presence of hen oviduct membranes with the quantity obtained with a similar amount of dog pancreas membranes indicates that the oviduct membranes are at least an order of magnitude more effective in catalyzing this process. In this experiment cell-free translation systems were, incubated in the presence of either preparation of microsomal membrane, the mixture was dissolved in Triton X-100 and then applied to a concanavalin-A derivatized Sepharose column. After extensive washing of the column with detergent-containing buffer glycoproteins were eluted from the resin in a buffer containing 1% α-methyl mannoside.

SDS GEL ELECTROPHORESIS OF MYELOMA RNA TRANSLATION PRODUCTS

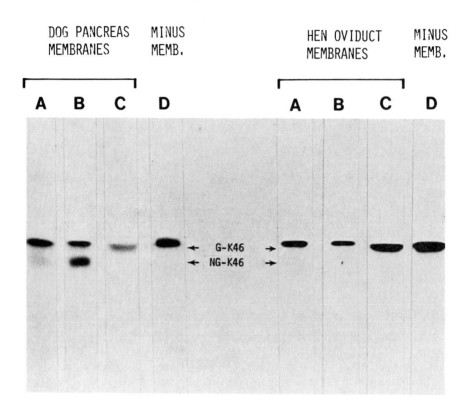

A, TOTAL;   B, AB PPT'D;   C, CON-A ELUATE-AB PPT'D;   D, PRE-K46

Figure 10. The positions of migration of standard samples of glycosylated native kappa-46 (G-46) and the non-glycosylated form of this protein (NG-46) are indicated by arrows.

As shown in Figure 10, analysis of this product by PAGE-SDS indicated that the concanavalin-A adsorbed protein is a partially glycosylated form of kappa-46 light chain. These results clearly indicate that microsomal membranes from either dog pancreas or hen oviduct have the capacity to carry out the proteolytic as well as the glycosylation steps of co-translational processing of the newly synthesized protein.

However, it is important to note that glycosylation and proteolysis of the precursor protein do not appear to be closely
coupled in the dog pancreas membrane system since a relatively large quantity of non-glycosylated, but protease processed
form of the protein was produced whereas only a small proportion of the partially glycosylated native polypeptide was
formed. In contrast, with hen oviduct membranes present during translation, no detectable amount of non-glycosylated
native polypeptide was formed (Lane B) whereas a large proportion of the protein synthesized is glycosylated and
processed by proteolysis under these conditions. Therefore
it appears that in the hen oviduct membrane, the glycosylation system and the protease are closely coupled since
glycosylation and proteolysis of the newly synthesized protein occurred together and there was no evidence to indicate
that proteolysis occurred in the absence of glycosylation, as
was the case with dog pancreas membranes.

In order to further analyze the co-translational processing systems, we felt that it was essential to develop an assay
procedure for glycosylation that would provide quicker and
more reliable quantitative data regarding this process.

CON-A SEPHAROSE FILTER ASSAY

1. ALIQUOT OF TRANSLATION MIXTURE DILUTED IN COLD Met-CONTAINING BUFFER
    AND MIXED WITH CON-A SEPHAROSE SLURRY. TO SAMPLES INCUBATED IN THE
    ABSENCE OF MEMBRANES, AN IDENTICAL QUANTITY OF MEMBRANES WAS ADDED
    IMMEDIATELY PRIOR TO ASSAY.

2. APPLIED SLURRY TO NUFLOW FILTERS AND WASHED REPEATEDLY (15 X 2 ML)
    WITH ORIGINAL BUFFER CONTAINING 0.1% NP40. CONTROL SAMPLES WERE
    TREATED IDENTICALLY EXCEPT THE WASH SOLUTION CONTAINED 1% ALPHA-
    METHYL MANNOSIDE.

3. MOIST FILTERS COUNTED IN SOLUENE:TOLUENE SCINTILLATION SYSTEM.

Example: Translation of myeloma proteins in presence of hen oviduct membranes

Assay:	Sample	CPM on Filter
	Complete	29,940
	Complete + α-MM	16,240
	Minus Memb.	15,015

FIGURE 11

As shown in Figure 11, we have recently developed an assay
for glycosylated protein, utilizing the specific binding of
the glycosylated forms of the protein to concanavalin-A
derivatized Sepharose. An aliquot of the translation mixture is diluted in unlabeled methionine-containing buffer and
mixed with a slurry of concanavalin-A Sepharose. In order to
control the system for the possible interference by endogenous membrane glycoproteins, samples that were incubated in
the absence of membranes are supplemented with an identical
quantity of membranes immediately prior to assay. The slurry
of concanavalin-A Sepharose is collected on a filter and
washed exhaustively with detergent-containing buffer and
duplicate aliquots are treated identically except that the
wash solution contains α-methyl mannoside. Finally, the
radioactive content of the concanavalin-A Sepharose is determined by a scintillation counting system. The results
indicate that the assay procedure is relatively reliable in
spite of the blank values obtained; we have verified that the
radioactivity present in the blank samples is relatively
constant ($\pm$10%), non-proteinaceous, and most likely represents non-specific binding of residual ^{35}S-Met to the gel.
Further, we have confirmed that the radioactive material
bound to the incubated samples may be fully recovered as
partially glycosylated kappa-46; no detectable quantity of
non-glycosylated forms of the protein were detected in
material eluted from the concanavalin-A Sepharose with α-
methyl mannoside or with 1% SDS.

Using this assay system, we have attempted to determine
optimal conditions for glycosylation in the cell-free translation system supplemented with microsomal membranes. The
addition of membrane markedly inhibits the cell-free protein
translation system as shown in Figure 12; at a concentration
of 20-25 OD_{260} units of membrane per ml. the incorporation
of ^{35}S-Met into protein is inhibited approximately 50%. The
basis for this inhibition is not well understood but in a
complex cell-free protein synthesizing system any one of a
variety of explanations may be possible.

Inhibition of Translation of Myeloma mRNA in the Presence of Membrane

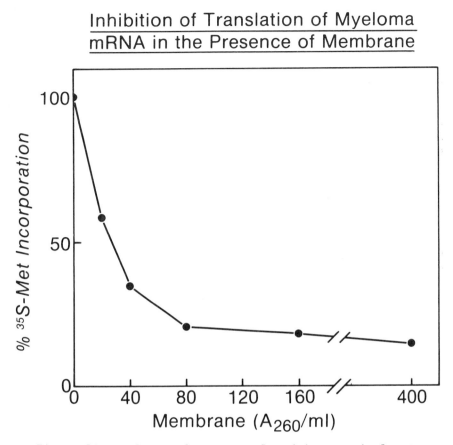

Figure 12. Myeloma RNA was translated in a reticulocyte cell-free system containing hen oviduct microsomal membrane as indicated.

Since the co-translational processing steps require interaction of the polysome, the nascent peptide chain, and the membrane, it is possible that partial disruption of membrane integrity with detergent may facilitate these interactions, and consequently, the processing steps. Initially, we determined the effect of added detergent on the translation system in the absence of membranes and these results are shown in Figure 13. It is obvious that relatively high concentrations of a non-ionic detergent such as NP-40 have relatively modest effect on protein synthesis per se. There is a rather abrupt increase in the level of inhibition to approximately 30% at low concentrations of detergent that is not greatly increased at concentrations up to 0.5% detergent.

Myeloma mRNA Translation in the Presence of Detergent

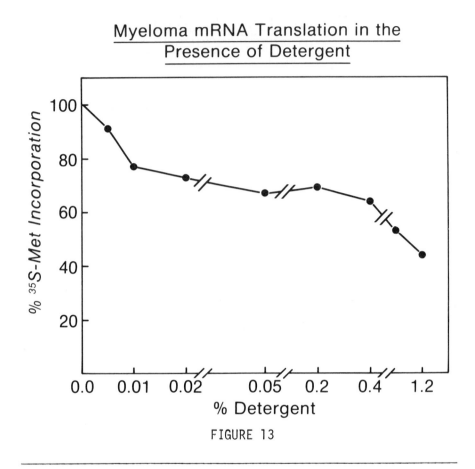

FIGURE 13

On this basis, we examined the relative effect of detergent on protein synthesis in the presence of increasing concentrations of membrane as shown in Figure 14.

Effect of Detergent/Membrane
on Protein Synthesis

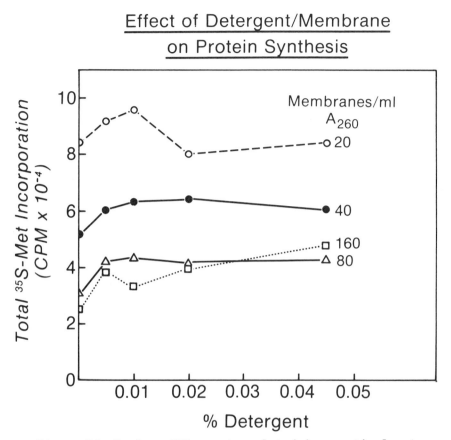

Figure 14. Myeloma RNA was translated in a reticulocyte cell-free system containing hen oviduct microsomal membrane and NP-40 as indicated.

Again, it was noted that increasing membrane concentrations markedly inhibited protein synthesis in the absence of detergent and each of these values reproducibly showed a slight increase at low concentrations (0.005 to 0.01%) of detergent and in most instances, the values were relatively constant as the detergent concentration was increased several fold. However, examination of the efficiency of glycosylation of newly synthesized protein under these conditions indicated that the detergent was exerting a much more significant effect on this process (Figure 15).

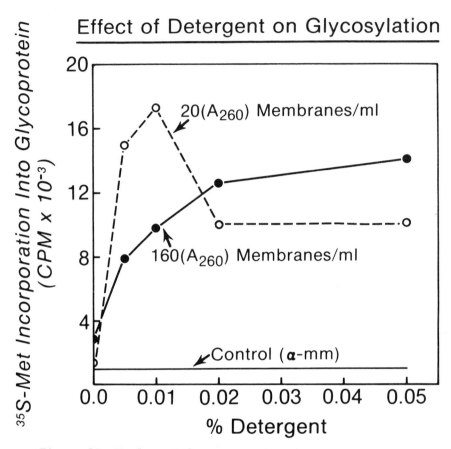

Figure 15. Myeloma RNA was translated in a reticulocyte cell-free system containing hen oviduct microsomal membrane and NP-40 as indicated. Assays were conducted as described in Figure 11.

Using the concanavalin-A filter assay, cell-free translation systems containing 20 or 160 OD_{260} units of membrane per ml. were assayed at increasing concentrations of detergent within the range that modestly stimulated overall synthesis of protein. It can be seen that at low concentrations of membrane (20 OD_{260} per ml.) the formation of glycosylated translation product was greatly enhanced. At higher membrane concentrations (160 OD_{260} per ml.), the maximum stimulation of glycosylation was observed at a somewhat higher detergent concentration (0.02-0.04%). In each instance the overall stimulation of glycosylation was 5- to 10-fold and appeared to be comparable regardless of the membrane concentration in the

incubation mixture. However, since the addition of increasing
concentrations of membrane markedly inhibits the overall incorporation of ^{35}S-Met into protein, a more striking effect of
detergent on the efficiency of glycosylation at higher membrane concentrations was apparent (Figure 16).

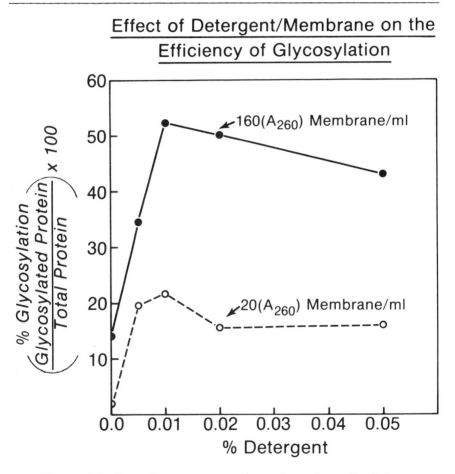

Effect of Detergent/Membrane on the Efficiency of Glycosylation

Figure 16. Experiment was conducted as described in
legend to Figure 15.

In this case the effect of increasing concentrations of detergent is plotted against per cent glycosylation; that is, the
per cent of the total protein synthesized that is synthesized
in a glycosylated form under these conditions. It is obvious
that low concentrations of detergent markedly enhance the

efficiency of glycosylating newly synthesized protein particularly at the higher membrane concentrations. In the
experiment containing 160 OD_{260} per ml., the per cent of the
total protein synthesized in the glycosylated form increased
from 14% in the absence of detergent to approximately 52% of
the protein synthesized in the glycosylated form at 0.01%
NP-40. In contrast, while the extent of stimulation of glycosylation was comparable at lower membrane concentrations,
because the efficiency of glycosylation is low (approximately
3%) at this membrane concentration, the maximum efficiency of
glycosylation at .01% NP-40 reached only about 22%. Thus, it
appears that partial disruption of the integrity of microsomal membrane vesicles permits more efficient utilization of
the membrane-associated processing steps involved in co-
translational modification of the newly synthesized protein.
The observation that the membrane-associated translation
systems will not only tolerate substantial concentrations of
detergent but, in fact, are enhanced under these conditions,
offers the possibility of further evaluation of the dolichol
glycosylation pathway by supplementation of translation
systems with exogenously supplied dolichol monophosphate,
sugar nucleotides, as well as polyprenol-linked monosaccharides and oligosaccharides.

     Using similar techniques, we have studied the de novo
biosynthesis of bovine pancreatic ribonuclease that may provide additional information concerning membrane-associated
co-translational processing of secretory glycoproteins. This
enzyme also is a secreted protein and is an example of a
protein that possesses the tripeptide sequence -Asn-Leu-Thr-
and yet bovine ribonuclease usually occurs approximately 5%
in the glycosylated form and approximately 95% in the non-
glycosylated form as it is isolated from pancreatic fluid.
Isolation of a cell-free translation system that would permit
the de novo biosynthesis of ribonuclease may provide the
possibility to determine the basis for the inefficiency of
the glycosylation system with this protein. Secondly, the
ribonuclease system also would permit evaluation of a fundamental question regarding the compartmentalization of
secretory proteins in the lumen of the endoplasmic reticulum;
that is, if a hydrolytic enzyme such as ribonuclease were
translated in a precursor form and if, by chance, the translation occurred on non-membrane-bound ribosomes, would the
precursor form of the enzyme then be in a position to destroy
cytoplasmic mRNA? It is also possible that a precursor form
of the enzyme may be biologically inactive and consequently

its accidental translation in the cytoplasm would have no
serious consequences on the integrity of the cell. In this
case, the active enzyme could be found only compartmentalized
in the lumen of the endoplasmic reticulum after the leader
sequence of the precursor form of the protein had been
removed by proteolytic processing.

   To approach these questions mRNA was isolated from bovine
pancreas by the guanidine thiocyanate procedure previously
described and, as shown in Figure 17, the translation system
was totally dependent upon mRNA producing a variety of proteins including a species approximately 2,000 greater in
molecular weight than native non-glycosylated RNase. Precipitation of the translation mixtures with antibody directed
against native RNase resulted in the isolation of the 16,000
molecular weight species that appeared to be a precursor
form of RNase. This tentative conclusion was further supported when the translation system was supplemented with microsomal membranes as shown in Figure 18. Two-dimensional PAGE
analysis of the products revealed that both hen oviduct membranes and dog pancreas membranes provide the capacity to
convert a portion of the precursor form of ribonuclease to a
species that comigrates with native non-glycosylated ribonuclease in this system. Again, hen oviduct membranes were
strikingly more efficient than were dog pancreas membranes in
their capacity for proteolytic processing of the precursor
form of the protein. Closer examination of the translation
products (Figure 19) in the presence of dog pancreas membranes
indicated that a small proportion of the processed protein is
glycosylated. Antibody precipitation of a translation mixture containing dog pancreas membranes revealed the presence
of both glycosylated and non-glycosylated RNase but it is
clear after analysis of this material on concanavalin-A
Sepharose that a portion of the product, coincidentally
corresponding to the precursor form of ribonuclease, is the
glycosylated form of the enzyme. It appears therefore that
ribonuclease is synthesized in a cell-free system in a precursor form with 15-20 additional amino acids probably
occurring as a leader sequence at the N-terminus of the protein, and secondly, that for an as yet unknown reason, the
-Asn-Leu-Thr- sequence of ribonuclease does not provide a
site that may be efficiently glycosylated by the membrane-
associated glycosylation system.

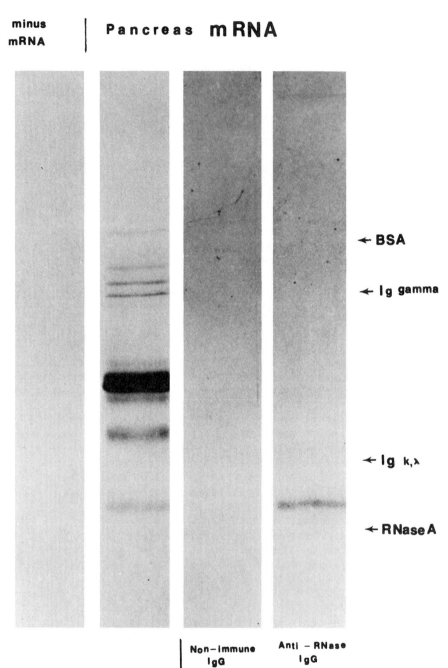

minus
mRNA

Pancreas **mRNA**

← **BSA**

← **Ig gamma**

← **Ig k,λ**

← **RNase A**

Non-immune
IgG

Anti – RNase
IgG

**Total
Protein**

Immuno-precipitate

Figure 17. Pancreas RNA translation products were analyzed
by PAGE-SDS as described in the text.

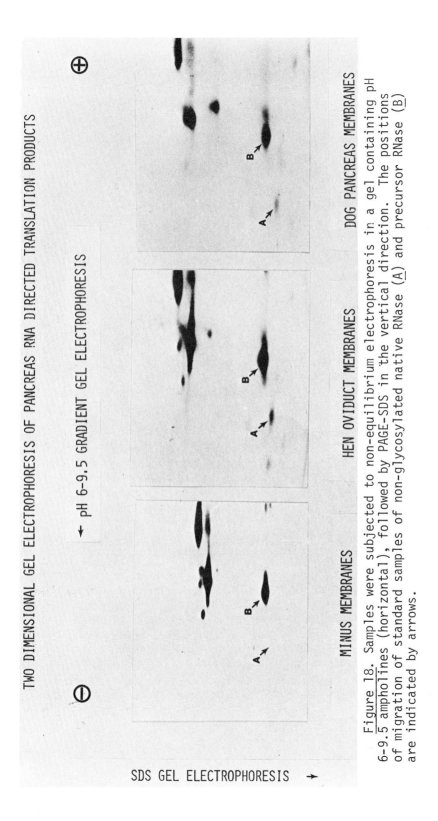

TWO DIMENSIONAL GEL ELECTROPHORESIS OF PANCREAS RNA DIRECTED TRANSLATION PRODUCTS

← pH 6-9.5 GRADIENT GEL ELECTROPHORESIS

⊖   ⊕

SDS GEL ELECTROPHORESIS →

MINUS MEMBRANES   HEN OVIDUCT MEMBRANES   DOG PANCREAS MEMBRANES

Figure 18. Samples were subjected to non-equilibrium electrophoresis in a gel containing pH 6-9.5 ampholines (horizontal), followed by PAGE-SDS in the vertical direction. The positions of migration of standard samples of non-glycosylated native RNase (A) and precursor RNase (B) are indicated by arrows.

SDS GEL ELECTROPHORESIS OF PANCREAS RNA TRANSLATION
PRODUCTS FORMED IN PRESENCE OF DOG PANCREAS MEMBRANES

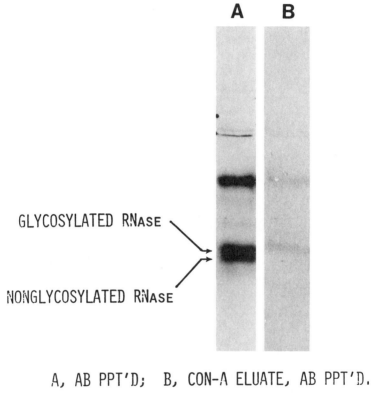

A   B

GLYCOSYLATED RNASE

NONGLYCOSYLATED RNASE

A, AB PPT'D;   B, CON-A ELUATE, AB PPT'D.
FIGURE 19

To address the question regarding the potential biological
activity of the precursor form of ribonuclease, attempts were
made to assay the biological activity of the pre-ribonuclease
synthesized in the cell-free translation system in the absence of membrane. As shown in Figure 20, an assay procedure
was developed that permitted the isolation and analysis of
the precursor form of the protein.

ASSAY OF DE NOVO SYNTHESIZED PRE-RNase.

1. ISOLATE PRE-RNase BY ANTIBODY PRECIPITATION.

2. DISSOCIATE AND DENATURE ANTIGEN-ANTIBODY COMPLEX
   WITH 8 M UREA AND 20 mM DTT.

3. ALLOW REDUCED PROTEINS TO AIR OXIDIZE.

4. ASSAY REOXIDIZED PROTEIN FOR ABILITY TO
   HYDROLYZE POLY-C.

FIGURE 20

Initially, the precursor protein was isolated as an antigen-antibody precipitate which was then dissociated under denaturing conditions. The denatured protein was then allowed to air-oxidize and finally was assayed for its ability to hydrolyze polycytidylic acid. The results of this experiment are outlined in Figure 21 and indicate that the precursor form of the protein is quite capable of catalyzing the hydrolysis of the polynucleotide substrate. Estimations of the specific activity of the precursor ribonuclease indicated that it is identical (within experimental error) to the specific activity of crystalline commercial bovine ribonuclease. We have concluded therefore, that the precursor form of secretory proteins may have the potential for full biological activity although, with the reservation that the conditions used to demonstrate the potential biological activity of the precursor protein in our experiments may not be reproduced in vivo.

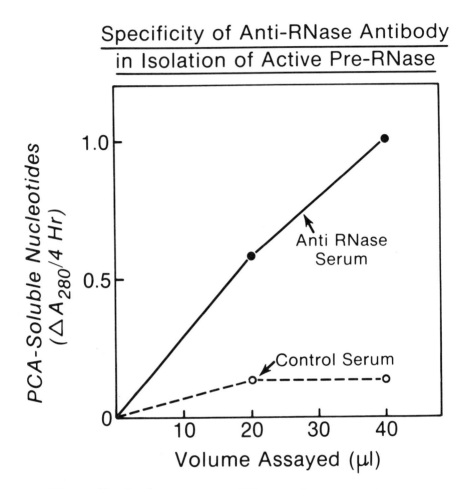

Figure 21. Bovine pancreas RNA translation products were
assayed for RNase activity as described in Figure 20.
Mixtures were incubated for 4 hrs and 5% perchloric acid-
soluble nucleotides were determined by measuring the
absorbancy at 280 nm.

REFERENCES

(1) Kornfeld, R., and Kornfeld, S., Ann. Rev. Biochem., 45,
     217 (1976).

(2) Waechter, C.J., and Lennarz, W.J., Ann. Rev. Biochem.,
     45, 95 (1976).

(3) Robbins, P.W., Hubbard, S.C., Turco, S.J., and Wirth,
     D.F., Cell, 12, 893 (1977).

(4)   Kornfeld, S., Li, E., and Tabas, I., J. Biol. Chem., 253, 7771 (1978).

(5)   Struck, D.K., Lennarz, W.J., and Brew, K., J. Biol. Chem., 253, 5786 (1978).

(6)   Rothman, J.E., and Lenard, J., Science 195, 743 (1977).

(7)   Blobel, G., and Dobberstein, B., J. Cell Biol., 67, 835 (1975).

(8)   Chirgwin, J., Przybyla, A., and Rutter, W.J., In Press.

(9)   Pelham, H.R.B., and Jackson, R.J., Eur. J. Biochem., 67, 247 (1976).

DISCUSSION

H. LODISH: You said at the beginning that you thought the sugar dolichol was assembled on the cytoplasm surface of the endoplasmic reticulum but the sugar nucleotide should not be able to cross the membrane. Isn't it true that in the Golgi the sugar nucleotides are obviously on the inside of the membrane?

E. HEATH: Yes, that is a problem. I meant to emphasize in that discussion that the sugar nucleotide, the primary source of generating sugar nucleotides, is in the cytoplasmic enzyme systems that synthesize sugar nucleotides and that they were being delivered to the dolichol pathway or to the Golgi transferases from cytoplasmic sources. The only thing that I am saying is that the sugar nucleotides come from the cyto-plasm and whether they are transported through the microsomal membrane or delivered directly to the dolichol pathway is yet to be established.

A. HASILIK: I would like to know about the precursor RNase – whether its specific activity on poly(C) is comparable to that of the purified enzyme.

E. HEATH: Yes, the specific activity of the precursor RNase is identical to that of pure RNase A with polycytidylic acid as substrate. The enzyme was recovered from the antigen-antibody precipitate generally by the procedure which I described to you. In many controls, we added picogram quan-tities of RNase and showed that we were able to quantitative-ly recover all of the activity we added, plus the activity that was synthesized. The best analysis of the situation was when we synthesized the precursor protein in the presence of

carrier free methionine and determined that the specific
enzymatic activity of the precursor is exactly the same as
pure RNase.

A. HASILIK: After you recovered the precursor by this
procedure did it still run on gels as the precursor form of
RNase?

E. HEATH: Yes, the material we recovered runs on a 2D gel
exactly as the precursor form of protein. We do not convert
it chemically by any manipulations to the active form. It is
still the precursor form of the protein and as best as we can
determine its specific activity on poly C is exactly the same
as RNase.

B. HARTLEY: I would like to congratulate you on another
beautiful experiment which shows that even the precursor form
of RNase is active after formation of the disulfide bridges.
This provides evidence that those of us who have been trying
to persuade people like Chris Hampton and David Phillips that
the formation of disulfide bridges is the most important
approach to translational modification that you can have
because in the absence of the disulfide bridges, the protein
is not refolded, the protein simply does not exist. Clearly,
the RNase is an example of this. You can see that the
nuclease is indeed inactive when the disulfide bridges are
reduced. This is of very great importance to the cell as it
would be indeed to proteolytic enzymes because any error that
is made would be a complete disaster to the cell. Therefore
the evolution of disulfide bridges is the key step to pro-
tecting the cell against folding inside whereas the concentration of glutathione and the reducing potential prevents
the folding.   I think it is a very valuable experiment.

E. HEATH: Yes, I think one must have a reservation about its
potential enzymatic activity. I certainly agree with you that
the likelihood of its recovering activity in the cytosol is
probably very small.

THE ENZYMATIC CONVERSION OF MEMBRANE
AND SECRETORY PROTEINS TO GLYCOPROTEINS

William J. Lennarz

Department of Physiological Chemistry
The Johns Hopkins University School of Medicine
Baltimore, Maryland

The oligosaccharide chain of N-glycosidically linked
glycoproteins is invariably found attached to an Asn
that is part of the sequence -Asn-X-Ser(Thr)-, where X
may be one of the 20 amino acids. Using crude or purified RER preparations from hen oviduct it was found that
proteins containing unglycosylated sequences could be
glycosylated via lipid-linked saccharides only after they
were denatured and S-alkylated. With denatured α-
lactalbumin (α-LA) as a model substrate the minimal peptide sequence required for glycosylation was studied.
The smallest peptide fragment isolated from α-LA that
was active as an acceptor of the oligosaccharide chain
was a heptapeptide. To more precisely define the minimal
acceptor requirements a series of tripeptides were chemically synthesized. The results indicate that simple tripeptides of the type Asn-X-Ser(or Thr) serve as substrates, providing that their amino and carboxy termini
are blocked.

To study the possible role of glycosylation of proteins in secretion we have investigated the effect of
tunicamycin, an inhibitor of saccharide-lipid synthesis
and hence protein glycosylation, using hepatocytes.
These studies revealed that inhibition of the glycosylation of several secretory proteins has no inhibitory
effect on their secretion.

Finally, using α-LA bearing a glucose-containing
oligosaccharide chain as a substrate, we have undertaken
a search for glycosidases that act on the oligosaccharide
chain of glycoproteins. A glucosidase that excises Glc

from the oligosaccharide chain of α-LA has been detected
in crude oviduct membranes. Currently we are investigating the subcellular localization and the properties
of this neutral glucosidase, as well as its role in
processing of the oligosaccharide chain of glycoproteins.

INTRODUCTION

In the mid 1960's it became apparent that a variety of
complex glycans associated with the cell membrane of prokaryotes were synthesized by a novel mechanism involving
lipid-linked sugars (1). These compounds were of two types,
those in which sugars were attached via a phosphodiester
bridge to the long chain polyisoprenoid, undecaprenol, and
those in which the sugar moiety was attached via a pyrophosphate bridge to this lipid. It is now clear that analogous
compounds participate in the synthesis of N-glycosidically
linked glycoproteins in a wide variety of eukaryotes (2).   The
structures of the two types of lipids are shown in Fig. 1.

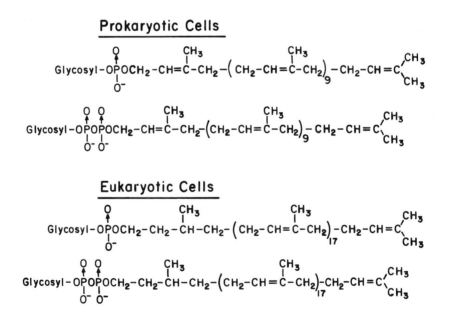

Figure 1.   Structures of polyisoprenoid-linked saccharides
found in prokaryotes and eukaryotes.

The saccharide lipids found in the prokaryotes and the eukaryotes are similar in two respects: in both cases the sugars can be linked by either a phosphodiester or a pyrophosphate bridge and the lipid carrier is a long chain polyisoprenoid. However, there are two distinct differences in the eukaryotic and prokaryotic lipid carriers. The prokaryotic lipids are generally C_{55} compounds consisting of 11 isoprenoid units, whereas the eukaryotic lipids, termed dolichols, are much longer and contain 17 to 20 isoprenoid units. The other distinct difference between the prokaryotic and eukaryotic lipids is in the former all the isoprene residues are unsaturated, whereas in the latter the α-isoprene residue is saturated. In the context of the function of polyisoprenoid phosphates of the eukaryotic type it is of interest to consider the length of dolichol in respect to its orientation in membranes. Examination of molecular models of dolichol phosphate indicate that in an extended form it is approximately 100 Å in length, which is significantly greater than the length of the lipid bilayer. In view of this it will be of particular interest to gain an understanding of the organization of this molecule in the membrane in which it functions (see below).

EARLY STUDIES ON THE BIOSYNTHETIC REACTIONS INVOLVED IN ASSEMBLY OF OLIGOSACCHARIDE CHAINS

Our studies on the participation of saccharide lipids in assembly of N-glycosidically linked glycoproteins have utilized, for the most part, oviduct tissue from laying hens, since this tissue is highly active in synthesis of a variety of glycoproteins, most notably ovalbumin. The overall sequence of reactions documented in these studies, using crude membranes from hen oviduct, is outlined in Fig. 2. Initially, it was possible to demonstrate the formation of mannosylphosphoryldolichol from GDP-mannose and the formation of N,N'-diacetylchitobiosylpyrophosphoryldolichol from UDP-N-acetylglucosamine. As shown, after formation of the mannosyl lipid and the N,N'-diacetylchitobiosyl lipid, further steps involve addition of a mannosyl unit to the disaccharide lipid directly from GDP-mannose, followed by further elongation by the addition of mannose units from mannosylphosphoryldolichol. The resulting oligosaccharide lipid, containing 7-9 hexose units, is subsequently transferred from its lipid carrier to endogenous membrane proteins (3-7). Based on a variety of experiments comparing the oligosaccharide attached to lipid with

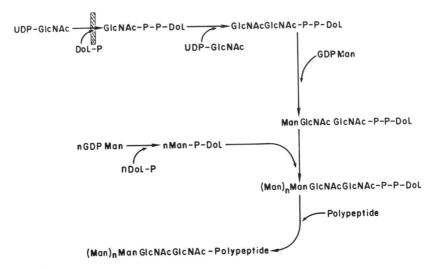

Figure 2.   Sequence of reactions involved in assembly of the oligosaccharide chain of N-glycosidically linked protein. The cross-hatched bar indicates the site of inhibition by tunicamycin.

that subsequently transferred to protein it was concluded that this transfer occurs en bloc, without any alteration or loss of saccharide units from the saccharide chain. In order to ascertain the relationship between the oligosaccharide attached to protein via this pathway, and the oligosaccharide chains found in many secretory proteins, structural studies on both the oligosaccharide attached to the endogenous membrane proteins and that attached to the lipid were undertaken. By use of both enzymatic and chemical methods it was possible to establish (8) that the core region of this oligosaccharide has the structure shown in Fig. 3. This structure is, in fact, that found in a wide number of secretory proteins, thus suggesting that the lipid linked pathway is responsible for the synthesis of the oligosaccharide chain of not only membrane glycoproteins, but secretory glycoproteins as well. Subsequent studies, to be discussed below, strongly support this idea.

$$(O-\alpha-Man)_n-O-\beta-Man\,(1\to4)-O-\beta-GlcNAc\,(1\to4)-N-GlcNAc-Asn$$

CORE OLIGOSACCHARIDE

Figure 3.   Structure of the core region of the oligosaccharide chain of oligosaccharide-lipid.

EXOGENOUS PROTEINS AND PEPTIDES AS ACCEPTORS

Because the endogenous proteins labeled in crude oviduct membrane preparations were ill-defined, except in terms of apparent size (6), we undertook to use soluble exogenous proteins as acceptors for the oligosaccharide chain of oligosaccharide-lipid. As noted by Marshall (9), all N-linked glycoproteins that had been sequenced were found to have the carbohydrate attached to an Asn in the sequence Asn-X-Ser or Asn-X-Thr, where X represents any one of the 20 amino acids. Our attempts to use native proteins containing the sequence -Asn-X-Ser(or Thr)- as exogenous substrates for the oviduct membrane transferase were unsuccessful. However, it was found that if such proteins were first denatured, reduced and alkylated, the resulting unfolded proteins were active as acceptors of the oligosaccharide chain of oligosaccharide-lipid (10-11). This is illustrated in the case of sulfitolyzed ribonuclease A in Fig. 4. It is apparent that in the absence of exogenous

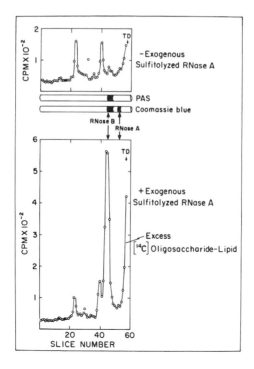

Figure 4.   SDS-PAGE analysis of enzymatically glycosylated ribonuclease.

protein acceptor only low levels of glycosylated membrane proteins are formed. However, in the presence of exogenous, unfolded ribonuclease A a new glycosylated polypeptide with the mobility of the glycosylated counterpart of ribonuclease A, ribonuclease B, is formed.

These findings suggested that perhaps only a limited region of the polypeptide was necessary to interact with the oligosaccharide transferase. To test this idea we turned to another protein that contains potential acceptor sites, α-lactalbumin (12). In collaboration with Dr. Keith Brew at the University of Miami, reduced and alkylated α-lactalbumin was subjected to the step-wise fragmentation shown in Fig. 5. Each of the fragments obtained was then tested as an acceptor of the oligosaccharide-lipid. All fragments containing the tripeptide sequence -Asn-Gln-Ser- at positions 45-47 were found to be active as acceptors except the hexapeptide formed from the heptapeptide by Edman degradation. The inactivity of the hexapeptide was shown not to be because the addition of the carbohydrate chains actually occurs at Asn_{44} in the heptapeptide. This was accomplished by subjecting the glycosylated glycopeptide formed from the heptapeptide to sequential Edman degradation. The results of this experiment clearly showed that the labeled carbohydrate chain was attached to Asn_{45}.

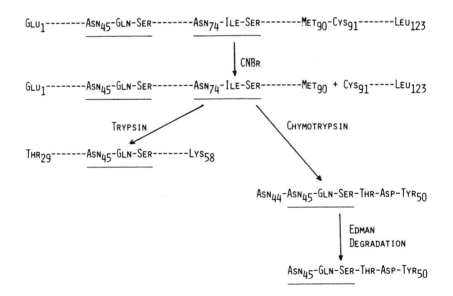

Figure 5. Scheme for the step-wise degradation of α-lactalbumin.

On the basis of these experiments it was clear that little, if any, additional sequence information was necessary for a peptide to serve as substrate. To document this we altered our approach and used synthetic tri- and tetrapeptides prepared by Dr. Brew, as well as several small peptides isolated from ribonuclease A by Drs. Greg Grant and Ralph Bradshaw at Washington University. Each of the peptides was tested as a substrate for the oligosaccharide transferase. As shown in Table I, it is clear that simple tripeptides of the type Asn-X-Ser or Asn-X-Thr are active as acceptors when both the carboxy and the amino termini are blocked. It is apparent, however, that this blockage need not be accomplished by addition of another amino acid residue at either end, but can also be effected by N-acetylation (in the case of the amino terminus) or by formation of the methyl ester or methyl amide (in the case of the carboxy terminus). These results, coupled with the comparison of the relative activity of these peptides with denatured protein substrates, indicate that the primary, and perhaps exclusive site of interaction of the oligosaccharide transferase is with the tripeptide, and that the remaining portions of the polypeptide play little or no role in

Table I. RELATIVE ACCEPTOR ACTIVITIES FOR THE GLYCOSYLATION OF PEPTIDES BY HEN OVIDUCT MICROSOMES

Active Acceptors	Rel. Acceptor Activity Per nmole of Peptide[*]	Inactive Acceptors	Rel. Acceptor Activity Per nmole of Peptide[*]
Ac-Asn-Leu-Thr-Lys	8.8	[+]Asn-Leu-Thr-Lys	0.03
Ac-Asn-Leu-Thr-N-CH$_3$ (H)	14.3	Asn-Leu-Thr	< .01
Ac-Asn-Leu-Thr-O-CH$_3$	4.1	[+]Ac-Asn-Leu-Thr	0.16
Ser-Arg-Asn-Leu-Thr-Lys	3.6		
CM-RNase A (-Asn$_{34}$-Leu$_{35}$-Thr$_{36}$-)	3.0		
Ac-Asn-Leu-Ser-Leu	2.1	Asn-Leu-Ser-Leu	< .01
Gly-Asn-Leu-Ser-Leu	1.0		
Ac-Asn-Gly-Ser-GlyOET	2.9	Asn-Gly-Ser	< .01
Ac-Asn-Gly-Ser-N-CH$_3$ (H)	0.7	Ac-Asn-Gly-Ser	< .01
[+]Ac-Asn-Gly-Ser-O-CH$_3$	0.2	Ac-Asn-Gly	< .01
CM-α-Lac (Bovine) (-Asn$_{45}$-Gln$_{46}$-Ser$_{47}$-)	12.0		

[*]Activity relative to Gly-Asn-Leu-Ser-Leu with the same microsomal preparation.
[+]Observable activity at very high concentrations.

defining substrate specificity in glycosylation. However, it
is apparent from the results discussed earlier that the sequence of the polypeptide chain, and hence its secondary
structure, may determine the accessibility of a given -Asn-X-
Ser(Thr)- site to the oligosaccharide transferase.

THE ROLE OF GLYCOSYLATION IN SECRETION
OF GLYCOPROTEINS

As the present time there is considerable interest in the
biological functions of the oligosaccharide chains. One of
the widely held views has been that the oligosaccharide chains
play an essential role in the secretion of glycoproteins.
With the availability of tunicamycin, a drug that specifically
blocks the formation of N-acetylglucosaminylpyrophosphoryl-
dolichol, the first intermediate in synthesis of the oligosaccharide chain (see Fig. 1), it became feasible to test
this requirement of glycosylation for secretion. Initially,
it was shown that oviduct slices continued to synthesize the
polypeptide chain of ovalbumin in the presence of tunicamycin,
but that it was not glycosylated (13). These results clearly
established that this secretory glycoprotein was synthesized
via lipid intermediates. However, because these experiments
were carried out in tissue slices, it was not possible to
definitively determine whether or not the unglycosylated
ovalbumin had been secreted. To accomplish this objective we
turned to studies with primary cultures of rat liver and
chick liver hepatocytes (14). (Experiments in the latter system were carried out in collaboration with Mrs. P. Siuta
Mangano and Dr. M.D. Lane at Johns Hopkins.) In the rat hepatocyte system we focused our attention on the synthesis and
secretion of the glycoprotein, transferrin, and as a control,
the synthesis and secretion of the non-glycoprotein, serum
albumin. Similarly, with chick hepatocytes we examined synthesis and secretion of the apo B chain of VLDL, a glycoprotein, and the light chain of VLDL, an unglycosylated protein.
Using labeled amino acids and labeled glucosamine, cells were
incubated in the absence or presence of tunicamycin and the
effects of the drug on synthesis of the polypeptide chain, as
well as on its glycosylation and secretion, were monitored.
The polypeptides of interest were measured by antibody precipitation and by SDS-polyacrylamide gel electrophoresis. The
results of this study, summarized in Table II, indicate that
tunicamycin had a very limited inhibitory effect on polypeptide synthesis. In contrast, a marked inhibition of glycosylation, as assessed by the incorporation of N-acetylglucosa-

TABLE II. EFFECT OF INHIBITION OF GLYCOSYLATION ON SECRETION OF GLYCOPROTEINS

| System Utilized | Glycoproteins | Proteins | Effect of Tunicamycin on: | | |
			Polypeptide Synthesis	Glycosylation % of Control	Secretion
Rat Hepatocytes {	Transferrin		80%	< 20%	78%
		Serum Albumin	80%	-	75%
Chick Hepatocytes {	Apo-B of VLDL		80%	< 10%	83%
		Light Chain of VLDL	80%	-	85%

mine, was observed. However, despite this severe inhibition
of glycosylation, the extent of secretion was only slightly
below that observed in the absence of tunicamycin, and did
not differ between the two secreted proteins that are normally
glycosylated (i.e., transferrin and the apo B chain of VLDL)
and the two non-glycoproteins. Thus, although in some cases
protein glycosylation may well be a prerequisite to secretion,
on the basis of these findings we conclude that this is certainly not an absolute requirement.

SUBCELLULAR DISTRIBUTION OF THE ENZYMES
INVOLVED IN ASSEMBLY OF OLIGOSACCHARIDE
LIPID AND TRANSFER TO PROTEINS

     All of the enzymatic studies described above on the
assembly of oligosaccharide-lipid and on transfer of the
oligosaccharide chain to protein were carried out with crude
membrane preparations. It was of interest to determine the
subcellular localization of this enzyme system; therefore a
scheme for the fractionation of subcellular components of
oviduct was developed (15). This fractionation scheme yielded
highly purified preparations of rough endoplasmic reticulum
and a smooth membrane fraction composed of the smooth endoplasmic reticulum and Golgi vesicles. Chemical and ultra-
structural studies with each of these preparations indicated
that the smooth membrane fraction was devoid of nucleic acids
and consisted of smooth membranes and vesicles free of ribosomes. Conversely, the rough membrane fraction was rich in
nucleic acid and consisted of sealed vesicles studded with
ribosomes. As shown in Table III, the enzymes involved in
saccharide-lipid synthesis and lipid-mediated glycosylation
were highly enriched in the rough endoplasmic reticulum
fraction as compared to the smooth membrane fraction, with

WILLIAM J. LENNARZ

TABLE III. RELATIVE GLYCOSYLTRANSFERASE ACTIVITIES OF ROUGH ENDOPLASMIC RETICULUM
(RER) AND SMOOTH MEMBRANE FRACTION (SMF)

Component Labeled	Membrane Fraction		RER/SMF
	RER	SMF	
	cpm		
Experiment I[*]			
Using GDP-[^{14}C]Man:			
Man-P-Dol	49,820	2,240	22.3
Oligosaccharide-P-P-Dol	8,039	638	12.5
Membrane Protein	8,325	293	28.4
Using UDP-[^{3}H]Gal:			
Membrane Protein	160	2,345	0.048
Experiment II[*]			
Using GDP-[^{14}C]Man:			
Man-P-Dol	34,200	2,450	14.0
Oligosaccharide-P-P-Dol	1,850	270	6.85
Exogenous Protein	11,600	750	15.5
Using UDP-[^{3}H]Gal:			
Gal-Lipid	19	22	Nil
Oligosaccharide-Lipid	33	92	Nil
Exogenous Protein	230	4,800	0.048

[*]In Experiment I incorporation was measured in the absence of exogenous protein
acceptors. In Experiment II CM-α-lactalbumin was added in the mannose incorporation
studies and agalactoorosomucoid was added in the galactose incorporation studies.

the enrichment values ranging from 12 to 28 fold. Conversely,
galactosyl transferase, an enzyme marker for Golgi membranes,
was enriched 20 fold in the smooth membrane fraction.

These findings are consistent with the earlier observations of Kiely et al. (16), who showed that nascent ovalbumin
still attached to tRNA was already glycosylated. Thus it is
clear that the translation and glycosylation are temporally
closely related events and that both occur in the rough endoplasmic reticulum. This conclusion provides a rational
explanation for the findings discussed above on the requirements for unfolded substrates in glycosylation.

These observations, as well as recent studies from several
laboratories (17-18) on the disposition of newly translated
membrane and secretory glycoproteins in the endoplasmic reticulum, led us to propose the model shown in Fig. 6 for the participation of saccharide lipids in protein glycosylation in
the rough endoplasmic reticulum. As suggested in this model,
sugar nucleotides in the cytoplasm react with dolichol phosphate derivatives at the external face of the endoplasmic reticulum membrane to form the oligosaccharide-lipid. As shown,
oligosaccharide-lipid undergoes a translocation to the inner
face, where the transfer of the oligosaccharide chain to

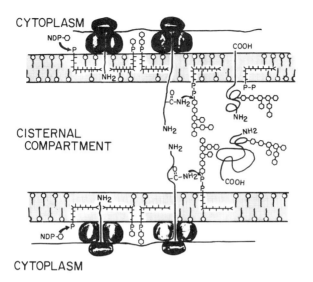

Figure 6.   Model proposed for the participation of
saccharide-lipids in protein glycosylation in the rough
endoplasmic reticulum.

protein occurs. However at the present time the alternative
possibility, i.e., that transfer of the oligosaccharide to the
protein occurs at the external face of the membrane, and that
as translation proceeds the polypeptide is inserted through
the membrane with its carbohydrate unit already attached, is
equally tenable.

Current studies in our laboratory are being directed toward the development of topological probes to determine the
sidedness of the saccharide-lipid intermediates and the site
of oligosaccharide transfer. Since NEM was found to inhibit
synthesis of the disaccharide-lipid, N,N'-diacetylchitobiosyl-
pyrophosphoryldolichol, we prepared dextran-NEM. This im-
permeable reagent also inhibited N,N'-diacetylchitobiosylpy-
rophosphoryldolichol synthesis, suggesting that at least one
of the enzymes involved in synthesis of this intermediate was
exposed on the cytoplasmic face of the rough endoplasmic reticulum. At present we are using galactosyl transferase to
study the orientation of the disaccharide-lipid in membranes.
Using unilamellar phosphatidylcholine liposomes prepared in

the presence of labeled N,N'-diacetylchitobiosylpyrophospho-
ryldolichol it was found that galactosyl transferase could
serve as a probe for the orientation of N,N'-diacetylchito-
biosylpyrophosphoryldolichol in bilayers. In the presence of
UDP-Gal and this soluble transferase only the externally oriented disaccharide units of N,N'-diacetylchitobiosylpyrophos-
phoryldolichol in the liposomes were galactosylated. Preliminary experiments using this probe to study the orientation of
saccharide-lipids in the rough endoplasmic reticulum indicate
that 50-70% of the newly synthesized disaccharide-lipid is
accessible on the cytoplasmic face of the rough endoplasmic
reticulum. Providing that appropriate control experiments
establish that the rough endoplasmic reticulum remains sealed
under the various conditions used in these experiments, these
results suggest that assembly of oligosaccharide-lipid and
possibly transfer of the oligosaccharide chains to the proteins being translated may, in fact, occur on the cytoplasmic
side of the rough endoplasmic reticulum.

SYNTHESIS AND PROCESSING OF GLUCOSE-CONTAINING OLIGOSACCHARIDE CHAINS

    Although Leloir et al. in their early studies on the
involvement of saccharide-lipids in synthesis of glycoproteins
reported the occurrence of glucose-containing oligosaccharide-
lipids (19), it was only following the observations of Spiro
et al. (20) and Turco et al. (21) that it became clear that
these larger oligosaccharides linked to dolichol might be the
naturally occurring intermediate and donors of their oligosaccharide chains to glycoproteins. These observations
prompted us to investigate the synthesis of a glucose-containing oligosaccharide-lipid in oviduct membranes (22). These
studies revealed that indeed cell-free preparations of oviduct
do synthesize a glucose-containing oligosaccharide-lipid.
Moreover, the oligosaccharide chain of the oligosaccharide-
lipid synthesized in the cell-free membrane preparation contains 12 to 14 hexose units, and appears to be similar if not
identical to that found in the oligosaccharide-lipid isolated
from oviduct tissue by Spiro et al. (20). Using this glucose-
containing oligosaccharide-lipid as a substrate in membrane
preparations it was possible to demonstrate that the intact
oligosaccharide chain, still containing glucosyl residues, was
transferred from the lipid to both endogenous membrane proteins and soluble, exogenous proteins such as S-carboxymethy-
lated α-lactalbumin. To investigate the possible processing
of the saccharide units on this larger glucose-containing

oligosaccharide chain, glycosylated α-lactalbumin was synthesized in vitro, isolated and purified. Then, as shown in
Fig. 7, the glycosylated α-lactalbumin was used as a substrate
to search for potential processing glycosidases that could
modify the oligosaccharide chain after it was attached to protein. Using doubly labeled glycosylated α-lactalbumin containing radioactive glucose and N-acetylglucosamine residues
it was found that oviduct membranes contain an enzyme that
catalyzes the excision of the glucose residues, while leaving
the remainder of the glucosamine labeled oligosaccharide chain
intact (23). Thus, as shown in Fig. 8, upon incubation of
doubly labeled glycosylated α-lactalbumin with oviduct membranes virtually all of the glucosyl residues are released
from the oligosaccharide chain whereas the N-acetylglucosa-
minyl units, and therefore the core oligosaccharide, remains
attached to the polypeptide. We have solubilized and partially purified the glucosidase responsible for this enzymatic
excision of glucose from the oligosaccharide chain, and are
currently studying its properties and specificity. Whether or

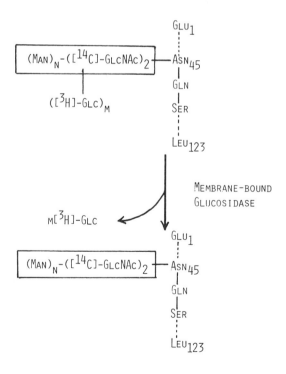

Figure 7. Use of glycosylated α-lactalbumin as a
substrate for a membrane-bound glucosidase.

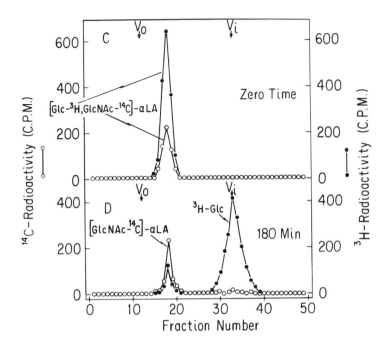

Figure 8.   Selective excision of glucosyl residues from
the doubly labeled oligosaccharide chain of glycosylated α-
lactalbumin.

not other processing glycosidases, such as a mannosidase of
the type reported by Tabas and Kornfeld (24), are also
present in oviduct membrane preparations remains to be established.

ACKNOWLEDGMENTS

    The work from the author's laboratory was supported by
grants from the National Science Foundation (PCM 74-11986) and
the National Institutes of Health (GM 21451 and GM 24281).   I
am deeply indebted to the following coworkers who, during the
past seven years, have worked so diligently and enthusiasti-
cally on the biochemistry of N-linked glycoproteins: W.W.
Chen, U. Czichi, J.J. Elting, J.A. Hanover, G.W. Hart, K.E.
Kronquist, J.J. Lucas, D.D. Pless, D.K. Struck and C.J.
Waechter. It is a pleasure to acknowledge Ms. A. Fuhr for
her expert preparation of this manuscript.

REFERENCES

1. Scher, M., and Lennarz, W. J., Biochim. Biophys. Acta
    Rev. 265, 417 (1972).
2. Waechter, C. J., and Lennarz, W. J., Ann. Rev. Biochem.
    45, 95 (1976).
3. Waechter, C. J., Lucas, J. J., and Lennarz, W. J.,
    J. Biol. Chem. 248, 7570 (1973).
4. Lucas, J. J., Waechter, C. J., and Lennarz, W. J.,
    J. Biol. Chem. 250, 1992 (1975).
5. Chen, W. W., and Lennarz, W. J., J. Biol. Chem. 251,
    7802 (1976).
6. Pless, D. D., and Lennarz, W. J., J. Biol. Chem. 250,
    7014 (1975).
7. Chen, W. W., and Lennarz, W. J., J. Biol. Chem. 252,
    3473 (1977).
8. Chen, W. W., Lennarz, W. J., Tarentino, A. L., and Maley,
    F., J. Biol. Chem. 250, 7006 (1975).
9. Marshall, R. D., Biochem. Soc. Symp. 40, 17 (1974).
10. Pless, D. D., and Lennarz, W. J., Proc. Natl. Acad. Sci.
    U.S.A. 74, 134 (1977).
11. Kronquist, K. E., and Lennarz, W. J., J. Supramolec.
    Struct. 8, 51 (1978).
12. Struck, D. K., Lennarz, W. J., and Brew, K., J. Biol.
    Chem. 253, 5786 (1978).
13. Struck, D. K., and Lennarz, W. J., J. Biol. Chem. 252,
    1007 (1977).
14. Struck, D. K., Siuta, P. B., Lane, M. D., and Lennarz,
    W. J., J. Biol. Chem. 253, 5332 (1978).
15. Czichi, U., and Lennarz, W. J., J. Biol. Chem. 252, 7901
    (1977).
16. Kiely, M. L., McKnight, G. S., and Schimke, R. T.,
    J. Biol. Chem. 251, 5490 (1976).
17. Katz, F. N., Rothman, J. E., Lingappa, V. R., Blobel, G.,
    and Lodish, H. F., Proc. Natl. Acad. Sci. U.S.A. 74,
    3278 (1977).
18. Lingappa, V. R., Lingappa, J. R., Prasad, R., Ebner, K.
    E., and Blobel, G., Proc. Natl. Acad. Sci. U.S.A. 75,
    2338 (1978).
19. Behrens, N. H., Carminatti, H., Staneloni, R. J., Leloir,
    L. F., and Cantarella, A. I., Proc. Natl. Acad. Sci.
    U.S.A. 70, 3390 (1973).
20. Spiro, M. J., Spiro, R. G., and Bhoyroo, V. D., J. Biol.
    Chem. 251, 6420 (1976).
21. Turco, S. J., Stetson, B., and Robbins, P. W., Proc.
    Natl. Acad. Sci. U.S.A. 74, 4411 (1977).

22. Chen, W. W., and Lennarz, W. J., J. Biol. Chem. 253,
     5774 (1978).
23. Chen, W. W., and Lennarz, W. J., J. Biol. Chem. 253,
     5780 (1978).
24. Tabas, I., and Kornfeld, S., J. Biol. Chem. 253, 7779
     (1978).

DISCUSSION

E.G. KREBS: Do you have anything akin to kinetic parameters
with the peptides. That is, yes they are substrates, but are
they good substrates?

W.J. LENNARZ: Both the proteins and the synthetic peptides are
poor substrates in one sense, since we obtain only a low
percentage of glycosylation. All one can say is that the small
peptides are comparable in acceptor activity to the denatured
polypeptides. One could argue that the reason the extent of
glycosylation is very poor is that glycosylation occurs at the
inner face of the E.R. membrane and that only a small fraction
of the acceptors are accessible to the lumen. However if you
purposefully make the E.R. leaky you don't enhance the extent of
conversion.

S.KORNFELD: Another way you could look at the orientation of
oligosaccharide-P-P-dolichol would be to treat the RER with
endoglycosidase and see if all the oligosaccharide was
released. I wonder if you have done that.

W.J. LENNARZ: No. That experiment is on the drawing board.

R.K. KELLER: Did you prelabel the membranes before you did the
transfer in the last experiment to see where exogenous α-
lactalbumin was glycosylated?

W.J. LENNARZ: We have done it by prelabelling, but more
frequently simply by adding sugar nucleotide at the same time as
we add the potential acceptor protein. After the incubation, we
separated the membranes from the supernatant and asked where
the glycosylated α-lactalbumin was. 100% of it was found in the
supernatant. Now that could mean that a small fraction of the
membranes burst open and that population of membranes is doing
the glycosylation. Or the protein acceptor is dipping in the
membrane and bouncing back out again - which I find rather
unlikely. Or the glycosylation goes on at the cytoplasmic face
of sealed membranes.

E.C. HEATH: I would like to hear Bill's comments on this. With the magnificent demonstration that the glycosylation really only needs the -Asn-[X]-Ser (Thr)-tripeptide, we are really faced with a serious dilemma with regard to bovine RNase and alpha lactalbumin. Prolactin has a signal sequence and I don't believe it is ever glycosylated. There is no evidence that I know of that this occurs and yet it has glycosylatable polypeptides. This can be rationalized if there were some secondary or tertiary structure of the protein that might mask the site, but with a protein like alpha lactalbumin or RNase, where it is glycosylated in a proportion of molecules, I find it much more difficult to rationalize.

W.J. LENNARZ: My thoughts on that have been that in order for 100% of the polypeptide to be glycosylated, then one has to have the rate of assembly and transfer of oligosaccharide chains to the protein, either equal to or greater than translation, and maybe that situation doesn't exist in the pancreas with regard to ribonuclease.   So if one could, for instance, slow down translation or up glycosylation then perhaps you can get 100%. I think this is a sloppy coupled system to have, but that is just the way I would think about it.

E.C. HEATH: Yes, but if you look at the pituitary there are lots of other secretory proteins that would be 100% glycosylated and therefore to justify the lack of, or partial, glycosylation on kinetic grounds appears difficult.

W.J. LENNARZ: There is of course the other issue is that we don't really know when folding starts and what the molecular architecture of the signal peptide sequence has to be in terms of its three dimensional structure to interact with the transferase. Possibly nearby amino acids may impart the wrong conformation to the tripeptide so that it will never get glycosylated. At present, I cannot tell which of these factors is important.

OLIGOSACCHARIDE PROCESSING DURING GLYCOPROTEIN BIOSYNTHESIS

Ira Tabas, Ellen Li, Mark Michael, and Stuart Kornfeld

Departments of Medicine and Biochemistry
Washington University School of Medicine
St. Louis, Missouri

Many soluble and membrane-bound glycoproteins contain asparagine-linked oligosaccharide units which can be divided into two basic categories, termed simple (or high mannose) and complex (1). Typical examples of these two oligosaccharides are shown in Figure 1. The oligosaccharides share a common core structure but differ in their outer branches. The work of many investigators has established that the biosynthesis of asparagine-linked oligosaccharide units is initiated by the "en bloc" transfer of a preformed oligosaccharide from an oligosaccharide pyrophosphoryldolichol intermediate to an asparagine residue of the nascent polypeptide chain (2). Only recently, however, has it been demonstrated that the synthesis of complex-type oligosaccharides involves the transfer of a glucose-containing high mannose-type oligosaccharide from the lipid carrier to the nascent polypeptide and that following transfer to the protein the oligosaccharide is "processed" by the removal of glucose and mannose residues (3-6). We have studied oligosaccharide processing in vesicular stomatitis virus-infected Chinese Hamster Ovary cells (4,7-9). This virus contains a single glycoprotein, termed G, which has two complex-type oligosaccharides (10). VSV-infected Chinese Hamster Ovary cells synthesize a lipid-linked oligosaccharide which has the structure shown in Figure 2 (7). This oligosaccharide has three branches, one of which contains 3 glucose residues. The glucose residues probably function as a signal for transfer of the oligosaccharide to the nascent protein (11). The sequence of oligosaccharide processing on the viral G protein was determined using the following approach. Virus-infected cells were incubated for 3 to 5 minutes with either [^{3}H]mannose, [^{3}H]glucosamine, or [^{3}H]galactose to label the mannose, N-acetylglucosamine and glucose residues, respectively, of the lipid-linked oligosaccharide. The cells were then suspended in fresh medium and chased for up to two hours. Glycopeptides were prepared from the cells following various chase intervals and treated with C. perfringens endo-β-N-

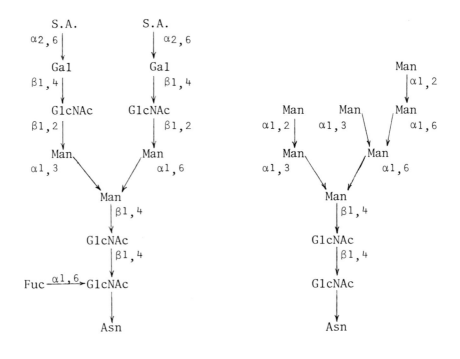

Complex-Type   High Mannose-Type

Figure 1 - Typical structures of asparagine-linked
           oligosaccharides

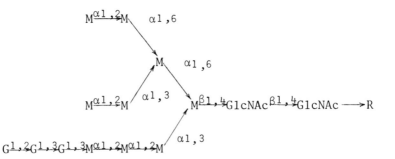

Figure 2 - Proposed structure of the lipid-linked oligo-
           saccharide (7)

R = P⟶P⟶Dolichol

acetylglucosaminidase C_{II}. This enzyme cleaves the di-N-acetylchitobiose unit of glycopeptides with the structure (Man)$_{>3}$(GlcNAc)$_2$-Asn but will not act on (Man)$_3$(GlcNAc)$_2$-Asn or glycopeptides with complex-type oligosaccharides (12). Consequently it can be used to distinguish these two classes of glycopeptides. The released oligosaccharides were then separated by paper chromatography and their structures determined (8). With this information the sequence of sugar removal could be deduced. The results of these studies are summarized in Figure 3. Within a few minutes after transfer to the protein, processing is initiated by the removal of two of the three glucose residues. At 20 to 30 min the last glucose residue is removed and processing proceeds rapidly with the ultimate removal of 6 of the 9 mannose residues and the addition of the outer branch sugars. The sequence of mannose removal is very precise. Thus only one Man$_8$ and one Man$_7$ isomer were detected among the processing intermediates, indicating that mannose removal occurs in a defined sequence. After the four mannose residues linked α1→2 have been removed, an N-acetylglucosamine residue is transferred to the mannose residue linked α1→3 to the β-linked core mannose. The transfer of this N-acetylglucosamine residue then signals a specific α-mannosidase to cleave the 2 mannose residues linked to the mannose which is linked α1→6 to the β-linked core mannose. A second N-acetylglucosaminyltransferase then adds another outer N-acetylglucosamine residue to the mannose residue exposed by the removal of the two mannose residues. That the (Man)$_5$(GlcNAc)$_2$-Asn structure is the physiologic ·substrate for the initial N-acetylglucosaminyltransferase is best demonstrated by the finding that a mutant line of CHO cells (termed 15B) which is deficient in UDP-GlcNAc: glycoprotein N-acetylglucosaminyltransferase activity accumulates this intermediate (13). Narasimhan, et al. (14) have shown that Chinese Hamster Ovary cells contain two UDP-GlcNAc: glycoprotein N-acetylglucosaminyltransferases (termed GlcNAc-transferase I and II) with distinct substrate specificities and that GlcNAc-transferase I can use (Man)$_5$(GlcNAc)$_2$-Asn as an acceptor. They reported that a variant line, PHA^{R}I, which appears to be identical to clone 15B, has a selective deficiency of GlcNAc-transferase I. These data indicate that the first N-acetylglucosamine residue is transferred to the (Man)$_5$(GlcNAc)$_2$-Asn processing intermediate by GlcNAc-transferase I while the second N-acetylglucosamine residue is most likely transferred by GlcNAc-transferase II. Following the addition of the outer N-acetylglucosamine residues, the synthesis of the complex-type oligosaccharide is completed by the addition of galactose, fucose and sialic acid residues.

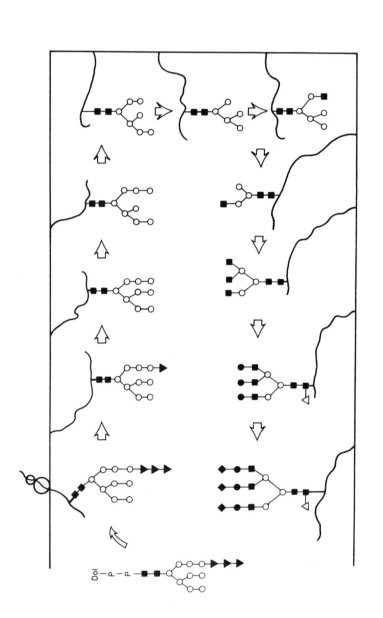

Figure 3 – Proposed sequence for the synthesis of complex-type oligosaccharides. Dol=dolichol. The symbols are ■, N-acetylglucosamine; ○, mannose; ●, glucose; ▼, galactose; ◆, fucose; △, sialic acid.

We have begun to characterize some of the glycosidases
involved in oligosaccharide processing. A partially purified
glucosidase preparation has been obtained from calf liver
which converts $(Glc)_2(Man)_9(GlcNAc)_2$ to $(Man)_9(GlcNAc)_2$ but
is totally inactive toward $(Glc)_3(Man)_9(GlcNAc)_2$. This indicates that at least two glucosidases participate in the removal of the glucose residues from the oligosaccharide precursor. Since the glucosidase preparation is only 100-fold
purified, it is quite possible that it contains two different
glucosidases, one that converts $(Glc)_2(Man)_9(GlcNAc)_2$ to
$(Glc)_1(Man)_9(GlcNAc)_2$ and another that converts $(Glc)_1(Man)_9$
$(GlcNAc)_2$ to $(Man)_9(GlcNAc)_2$. A Golgi α-mannosidase has
been purified which converts $(Man)_9(GlcNAc)_2$ to $(Man)_5(GlcNAc)_2$,
indicating that it has a strict specificity for mannose residues linked $\alpha1\rightarrow2$ (see Figure 2). This enzyme is distinct
from the α-mannosidase which is involved in the late stage of
processing and which cleaves mannose residues linked $\alpha1\rightarrow3$ and
$\alpha1\rightarrow6$. Interestingly, both of these enzymes appear to be different from the Golgi α-mannosidase characterized by Touster's
group (15,16).

The observations concerning the biosynthesis of complex-
type oligosaccharides and the finding that only a single,
very large lipid-linked oligosaccharide is present in several
tissues and cell lines (3,17) has led to the concept that
high mannose and complex-type oligosaccharides are derived
from the same lipid-linked oligosaccharide. To test this
concept we elucidated the structures of the asparagine-linked
high mannose-type oligosaccharides isolated from CHO membrane
glycoproteins and compared them with the structures of the
major lipid-linked oligosaccharide and the peptide-bound processing intermediates in the biosynthesis of complex-type
chains. The structures of the $(Man)_9(GlcNAc)_2$, $(Man)_8$
$(GlcNAc)_2$ and $(Man)_5(GlcNAc)_2$ oligosaccharides from the membrane glycoproteins were identical to the structures of the
processing intermediate of the same size (18). There was too
little of the $(Man)_6(GlcNAc)_2$ processing intermediate to
characterize but this intermediate is probably identical to
the $(Man)_6(GlcNAc)_2$ membrane glycoprotein unit based on the
structures worked out for the processing intermediates containing 7 and 5 mannose residues. A $(Man)_7(GlcNAc)_2$ unit was
not isolated from the membrane glycoproteins. This structural identity between the high mannose-type oligosaccharides
on membrane glycoproteins and the processing intermediates
provides support for the concept that high mannose-type oligosaccharides arise by "incomplete" processing of the protein bound oligosaccharide along the same pathway which leads
ultimately to the formation of a complex-type oligosaccharide.
The occurrence in ovalbumin of oligosaccharides with struc-

tures that are a hybrid of high mannose and complex-type chains is also compatible with this concept (19,20). It will be of great interest to elucidate the factors which determine whether the precursor oligosaccharide is partially processed to give rise to a high mannose-type unit or completely processed to a complex-type unit.

REFERENCES

1. Kornfeld, R. and Kornfeld, S., Annu. Rev. Biochem. 41, 217 (1976).
2. Waechter, C. and Lennarz, W., Annu. Rev. Biochem. 41, 95 (1976).
3. Robbins, P.W., Hubbard, S., Turco, S., and Wirth, D., Cell 12, 289 (1977).
4. Tabas, I., Schlesinger, S., and Kornfeld, S., J. Biol. Chem. 253, 716 (1978).
5. Hunt, L., Etchison, J., and Summers, D., Proc. Natl. Acad. Sci. 75, 754 (1978).
6. Staneloni, R. and LeLoir, L., Proc. Natl. Acad. Sci. 75, 1162 (1978).
7. Li, E., Tabas, I., and Kornfeld, S., J. Biol. Chem. 253, 7762 (1978).
8. Kornfeld, S., Li, E., and Tabas, I., J. Biol. Chem. 253, 7771 (1978).
9. Tabas, I., and Kornfeld, S., J. Biol. Chem. 253, 7779 (1978).
10. Reading, C., Penhoet, E., and Ballou, C., J. Biol. Chem. 253, 5600 (1978).
11. Turco, S., Stetson, B., and Robbins, P.W., Proc. Natl. Acad. Sci. 74, 4411 (1977).
12. Ito, S., Muramatsu, R., and Kobata, A., Arch. Biochem. Biophys. 171, 78 (1975).
13. Li, E. and Kornfeld, S., J. Biol. Chem. 253, 6426 (1978).
14. Narasimhan, S., Stanley, P., and Schachter, H., J. Biol. Chem. 252, 3926 (1977).
15. Dewald, B. and Touster, O., J. Biol. Chem. 248, 7223 (1973).
16. Tulsiani, D., Opheim, D., and Touster, O., J. Biol. Chem. 252, 3227 (1977).
17. Spiro, M., Spiro, R., and Bhoyroo, V., J. Biol. Chem. 251, 6420 (1976).
18. Li, E. and Kornfeld, S., J. Biol. Chem., in press (1979).
19. Tai, T., Yamashita, K., Ito, S., and Kobata, A., J. Biol. Chem. 252, 6687 (1977).

20. Yamashita, K., Tachibana, Y., and Kobata, A., J. Biol.
     Chem. 253, 3862 (1978).

                          DISCUSSION

B. HARTLEY: The glycosyl group has a modulated effect on the
secretion and establishment in the plasma membrane but the
full biological significance must be in its final target and
apt to increase the circulation. Is that your view?

S. KORNFELD: I think that ultimately the carbohydrate units
of glycoprotein will turn out to have several roles. One
will be to influence the physical properties of the protein
backbone while another will be to serve as recognition
signals in biologic interactions. Both of these functions
will be of significance.

A. HASILIK: I have a question about your experiments on
altering the amino acid sequences. Do you think that the
change in the composition of the oligosaccharide is really
due to a recognition of the changed amino acid sequences? It
could equally be a consequence of possible changes in the
dwell-in time of the altered polypeptides in proximity of the
glycosyl transferases. Do you know anything about the timing
of the transfer of these products through the intercellular
membrane system?

S. KORNFELD: We have not yet examined the rate of movement
of the modified viral glycoprotein through the intracellular
membranous systems. All we know is that if you incorporate
the amino acid analogs into the viral G protein and examine
the final product many hours later, it contains a totally
different oligosaccharide unit.

J. LEIBOWITZ: Dr. Kornfeld, correct me if I am wrong, but I
think that in one of your papers on oligosaccharides of IgE,
you showed a linkage region of mannose-N-acetyl glucosamine-
mannose-N-acetylglucosamine. Can you comment on this?

S. KORNFELD: There are a number of reports describing
oligosaccharide structures that have core regions which could
not have been described from the lipid-linked oligosaccharide
that I have been discussing. Either the structures are
wrong, or else there must be other minor pathways for glycosylation which lead to these unusual oligosaccharides. Since
we know there is at least a second pathway for glycosylation,

it is certainly possible to conceive of other, minor pathways
which lead to the formation of specialized oligosaccharide
structures.

B. HARTLEY: Do you think that this is all these enzymes are
doing? Is the mutant sick?

S. KORNFELD: The reason the mutant was detected in the first
place by Trowbridge and Hyman was because it does not express
a cell surface glycoprotein termed thy-1. Trowbridge and
Hyman also showed that the glycoprotein does get glycosylated, but with an altered sugar chain. Since our data with
the VSV G protein indicated that the folding of the nascent
protein is greatly influenced by the oligosaccharide unit, it
seems to us that the presence of an altered oligosaccharide
on a nascent protein could lead to abnormal folding. We
postulate therefore that the reason the thy-1 antigen is not
expressed in these cells is that it does not have the right
oligosaccharide. Since the majority of the glycoproteins
appear to be functional and the cells grow alright, it seems
that most of the glycoproteins in these cells do not have an
absolute requirement for correct glycosylation.

STRUCTURAL BASIS OF THE REGULATION
OF GALACTOSYLTRANSFERASE

Keith Brew
Richard H. Richardson
Sudhir K. Sinha

Department of Biochemistry
University of Miami School of Medicine
Miami, Florida

I.   INTRODUCTION

Of the glycosyltransferases that are involved in the assembly of the oligosaccharide moieties of N-glycosidically linked glycoproteins, the most extensively characterized enzyme is the galactosyltransferase that catalyzes the transfer of galactose from UDP-galactose into a β1-4 linkage with non-reducing terminal N-acetylglucosamine (GlcNAc).This reaction, like a number of other terminal proteins processing events occurs in the Golgi apparatus; the galactosyltransferase is an intrinsic component of these membranes in many mammalian cells (1-3).

Galactosyltransferase catalyzes the transfer of galactose to a number of other substrates besides glycoproteins with terminal GlcNAc residues, including the free monosaccharide GlcNAc (4) and GlcNAc oligosaccharides such as N, N diacetylchitobiose (5). Free glucose can also serve as a substrate, to give lactose as a product, although the Km for glucose of about 2M is so high as to preclude this as a physiologically significant reaction. However, in the presence of the modifier protein, α-lactalbumin, the Km for glucose is reduced by 10^3, so that lactose synthesis is effectively catalyzed at physiological concentrations of glucose (see 6). Together, galactosyltransferase and α-lactalbumin form the lactose synthase enzyme system which catalyzes the biosynthesis of lactose in the lactating mammary gland. Because of the localization of galactosyltransferase in the Golgi membranes, the biosynthesis of lactose occurs within the lumen of the Golgi apparatus, and lactose is secreted from the cell by the same mechanism as the secretion of proteins: namely by the formation of secretory vesicles, and the expulsion of their contents from the cell by reverse pinocytosis. The modifier protein, α-lactalbumin, is secreted at the same time, a phenomenon of considerable regulatory significance that is discussed in detail elsewhere (6). The production of α-lactalbumin in significant amounts only in the lactating mammary gland ensures that galactosyltransferase functions in lactose synthesis only in the lactating mammary gland, but in glycoprotein synthesis in other tissues (6).

α-Lactalbumin exerts regulatory effects on all reactions catalyzed by galactosyltransferase. These effects can be categorized into three groups:

(1) an enhancement of monosaccharide binding, as indicated by reduced Km values, observed with good substrates (GlcNAc), marginal substrates (glucose) and very poor substrates (xylose, 2-deoxyglucose). Other monosaccharides, such as 4-deoxyglucose, become inhibitors only in the presence of α-lactalbumin, indicating that their binding is also increased by α-lactalbumin.

(2) a reduction of product release rates (lowered Vm values) with GlcNAc, but not with poor monosaccharide substrates such as glucose.

(3) competitive inhibition with glycoprotein or oligosaccharide substrates, indicative of mutually exclusive binding. This is observed with substrates as small as the disaccharide diacetylchitobiose (5).

The protein that evokes these effects with galactosyltransferase, α-lactalbumin, provides one of the most interesting ramifications of the lactose synthase system, as it is strikingly homologous in primary structure with lysozymes of the chicken egg white type, a group that includes the mammalian lysozymes (7). When the sequences of α-lactalbumins and avian or mammalian lysozymes are aligned, approximately 40% of the amino acids in corresponding positions are identical, and many more are similar in chemical nature. The arrangements of the four disulfide bonds of α-lactalbumin and lysozyme are also identical, and a variety of circumstantial and direct evidence indicates that the conformations of the two groups of proteins are closely similar (8). It is clear that the structural genes for the α-lactalbumins and lysozymes have been derived from a common ancestor through a process of gene duplication and mutational change. Present evidence suggests that the ancestral protein was a lysozyme in function, and that the gene duplication event occurred at some early time during the evolutionary development of the mammals (see 6). Although the two groups of proteins are so similar in structure, in functional terms they are separated by a profound gap. One group, the lysozymes, has a catalytic function in the hydrolysis of microbial cell wall polysaccharide, whereas the α-lactalbumins are effectors of monosaccharide binding to galactosyltransferase but have no catalytic activity or ability, in isolation, to bind monosaccharides (9). Investigations of the molecular basis of the regulation of galactosyltransferase by α-lactalbumin are therefore relevant to the broad question of how protein structural changes during evolution can lead to the development of completely new functions and biochemical capacities in organisms. There is also an intrinsic interest in attempting to understand the basis of the unique regulatory properties of lactose synthase, which are mediated through reversible protein-protein interactions.

II. STRUCTURE OF GALACTOSYLTRANSFERASE

Although, at its intracellular site of action, galactosyltransferase is membrane-bound, soluble forms are also present in secretions such as milk, colostrum and serum (4, 10, 11). The soluble enzymes are readily purified by affinity chromatography procedures and have been used for most studies of the properties of galactosyltransferase. The enzyme from bovine colostrum, used in the studies described here, is a glycoprotein of molecular weight, 50,000 (10) whereas the enzyme from milk contains a number of components differing in molecular weight the largest of which is similar to the colostrum enzyme. These appear to be the products of proteolytic action (10). Recently, the enzyme from Golgi membranes prepared from lactating sheep mammary gland has been solubilized and purified. The main component appears to be a glycoprotein of molecular weight 65,000 to 70,000, considerably larger than the soluble enzymes hitherto studied. However, the enzymic and regulatory properties were found to be essentially identical with those of the soluble galactosyltransferase from colostrum (12). Obviously the larger membrane-derived enzyme is of considerable interest with respect to its mode of insertion in Golgi membranes, but it is difficult to prepare in adequate quantities for detailed study. For the present it appears justified to study the regulatory aspects of galactosyltransferase using the soluble enzyme form.

III. GALACTOSYLTRANSFERASE-α-LACTALBUMIN INTERACTIONS AND REGULATION

The interaction of α-lactalbumin with galactosyltransferase is unlike that between subunits of most multimeric enzymes, being reversible and highly ligand dependent. From crosslinking and various binding studies, it is clear that only 1:1 complexes are formed betwen the two proteins (6, 13). Table 1 summarizes the dissociation constants of a number of complexes of galactosyltransferase with α-lactalbumin, and acceptor substrates (14). They show clearly the enhancement of the association of the two proteins obtained by the successive addition of the ligands Mn^{2+}, UDP-hexose and monosaccharide. The strongest association (Kd of 0.17 μM) is achieved in the presence of Mn^{2+}, UDP-glucose and monosaccharide, a dead-end inhibitory complex resembling a central complex in the catalytic cycle of galactosyltransferase. In steady state kinetic studies, the effects of α-lactalbumin in the promotion of monosaccharide binding, or as a competitive inhibitor of oligosaccharide or glycoprotein binding, are characterized by an association constant closely similar to that for α-lactalbumin binding with a galactosyltransferase. Mn^{2+} UDP-galactose complex. Inhibition patterns support the view that the association of α-lactalbumin with this complex leads to the inhibitory effects of α-lactalbumin on the transfer of galactose to glycoproteins or oligosaccharides. Another important feature of the association is the mutual enhancement, or synergism, in the binding of α-lactalbumin and monosaccharides.

TABLE 1. Dissociation Constants of Ligands from Complexes with Galactosyltransferase.

Ligand	Complex	K_d	
α-Lactalbumin	E.α-lac	∞	
	E.Mn.α-lac	213	μM
	E.Mn.UDP-gal.α-lac	9.1	μM
	E.Mn.UDP-glc.α-lac	12.5	μM
	E.Mn.GlcNAc.α-lac	1.4	μM
	E.Mn.UDP-glc.Glc.α-lac	0.2	μM
GlcNAc	E.Mn.GlcNAc	23	mM
	E.Mn.α-lac.GlcNAc	0.2	mM
Glucose	E.Mn.UDP-glc.Glc	455	mM
	E.Mn.UDP-glc.α-lac.Glc	8.3	mM

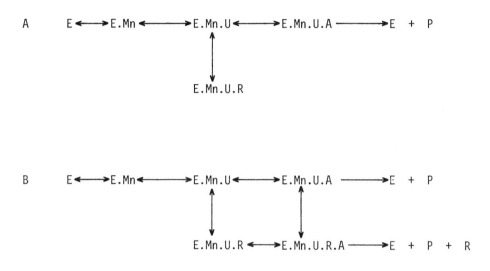

Fig. 1 a + b present general mechanistic schemes based on steady state kinetic and binding studies (14) which can be used to account for the regulatory effects of α-lactalbumin on reactions involving oligosaccharide or monosaccharide substrates. In both cases, α-lactalbumin and acceptor bind to a complex of galactosyltransferase with Mn^{2+} and UDP-galactose (E. Mn^{2+}. UDP-gal). For a glycoprotein or oligosaccharide, the binding is mutually exclusive, leading to competitive inhibition by α-lactalbumin (1b). With a monosaccharide substrate, random, synergistic binding of α-lactalbumin and monosaccharide occurs to give two types of enzyme complex that can give rise to products (E.Mn^{2+}.UDP-gal. A and E. Mn. UDP-gal. A.R), as shown in Fig. 1b. Thus at low, unsaturating concentrations of acceptor, a higher proportion of the enzyme is pulled into complexes that can give rise to products. Rate equations derived for this scheme predict, reasonably well, the observed kinetic effects with glucose as substrate. In the case of GlcNAc as substrate, the kinetic effects of α-lactalbumin can be explained by this scheme if it is assumed that the central complex containing α-lactalbumin (E. Mn^{2+}. U.A.R) gives rise to products at a lower rate than the central complex that does not contain α-lactalbumin (E.Mn.U.A).

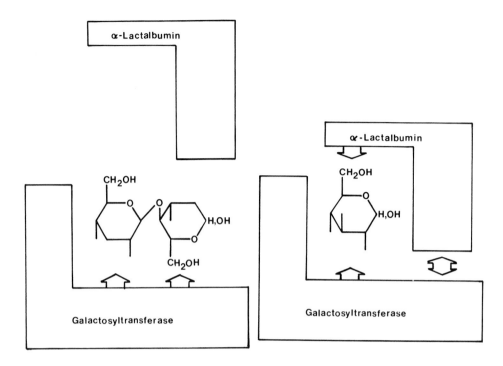

Fig. 2 presents a model that provides a conceptual basis for understanding the effects of α-lactalbumin on the binding of acceptors by galactosyltransferase as well as the enhancement of α-lactalbumin binding by monosaccharides. Three assumptions underlie this model: (1) the effects of α-lactalbumin are "direct" in nature, i.e. they are not mediated by the induction of a conformational change in galactosyltransferase as a result of binding at a site distant from the catalytic site of the enzyme. (2) α-lactalbumin plays some direct physical role in the binding of monosaccharides (3) the acceptor binding site of galactosyltransferase contains more than one subsite for monosaccharide binding. The former is reasonable in that it avoids unnecessarily complex explanations of the effects of α-lactalbumin, and that the kinetic effects of α-lactalbumin appear to be mediated by rapid equilibrium binding whereas the induction of a conformational change is likely to be a slow process in relation to the catalytic cycle of galactosyltransferase. The second assumption is based on the view that the homology of α-lactalbumin with lysozyme must be associated with a functional similarity at the molecular level, while the third assumption is supported by the observation that β-1-4 linked GlcNAc oligomers or glycopeptides have lower Km values than the free monosaccharide (5).

In the model, α-lactalbumin binds on the surface of galactosyltransferase at a site composed at least partly of a secondary subsite for monosaccharides (i.e. that to which the reducing terminal GlcNAc of diacetyl chitobiose would bind). This is consistant with the mutually exclusive binding of oligosaccharides and α-lactalbumin with the enzyme. We propose that the mode of binding brings a region of α-lactalbumin that is capable of forming direct interactions with monosaccharides into proximity with the acceptor monosaccharide binding site of galactosyltransferase. This results in the formation of a hybrid binding site for monosaccharides consisting of part of the α-lactalbumin surface and part of the galactosyltransferase binding site. A higher affinity binding site for monosaccharides is thereby formed and the mutual enhancement of binding by monosaccharides and α-lactalbumin becomes understandable. Although this model is plausible, it should be emphasized that it was initially formulated in the absence of any direct evidence as to the nature of functional regions in α-lactalbumin. Its relationship to recent evidence on the nature of such areas is discussed below.

IV. NATURE OF INTERACTION SITE

The model proposed above for the regulatory action of α-lactalbumin suggests that the α-lactalbumin molecule should possess proximal areas for interaction with galactosyltransferase, and for interaction with monosaccharides when in a complex with the enzyme. Unfortunately little specific information has been obtained hitherto on the nature of such functional areas in α-lactalbumin (see 6). The elucidation of these functional regions is also important in developing an understanding of the nature of structural changes associated with the functional difference between α-lactalbumin and lysozyme.

To attempt to map the regions of α-lactalbumin that are proximal to the binding site for galactosyltransferase, we have investigated changes in the reactivities of the 13 amino groups of α-lactalbumin that occur on complex formation with galactosyltransferase. These groups are a reasonable choice for study as they are widely distributed over the surface of the α-lactalbumin model, and can be specifically modified using acid anhydrides. A procedure similar to that originally developed by Hartley and coworkers (15) and termed "competitive labelling" was used. In this, parallel solutions of α-lactalbumin and a 1:1 molar combination of α-lactalbumin and galactosyltransferase, in the presence of appropriate ligands to induce complex formation, were reacted with trace amounts of high specific-activity [^{3}H]-acetic anhydride. Under these conditions, minimal acetylation of the protein amino groups occurs, the extent of modification of each group reflecting its microenvironment in the protein and the pK of the amino group. Changes in reactivity in the presence of galactosyltransferase will indicate groups that are perturbed on complex formation. By minimising the extent of labelling (in this case to about 0.2 acetyl groups per protein molecule), a major pitfall of the conventional protective labelling approach is avoided, namely the induction of conformational changes in proteins by extensive chemical modification. The precise conditions of the labelling experiments are shown in Fig. 3. The two solutions each contained the same amount of bovine α-lactalbumin, and the same concentrations of ligands (Mn^{2+}, UDP-glucose, GlcNAc), but only one contained galactosyltransferase. Under these conditions α-lactalbumin and galactosyltransferase form the most tightly-associating complex known (Kd < 0.2μM), so that approximately 80% of the proteins will be in a complex. As the labelling is kinetic, the fact that α-lactalbumin is continuously associating and dissociating from the complex does not invalidate this approach, although it would if conventional modification procedures were used. Labelling was acheived by the addition of 1.51 μmoles (8 mc) of acetic anhydride dissolved in 110 μl of dry acetonitrile and the solutions left for 4 h. so that the labelling reagent was completely hydrolyzed or reacted. The reaction mixtures were separated by gel filtration with columns of Biogel P60 to give α-lactalbumin from one experiment, and both α-lactalbumin and galactosyl-transferase from the other. The bulk of the radioactive reagent (97-99%) reacted with the solvent in both cases, only 0.3% being incorporated into α-lactalbumin molecules (< 0.17 acetyl groups per molecule)

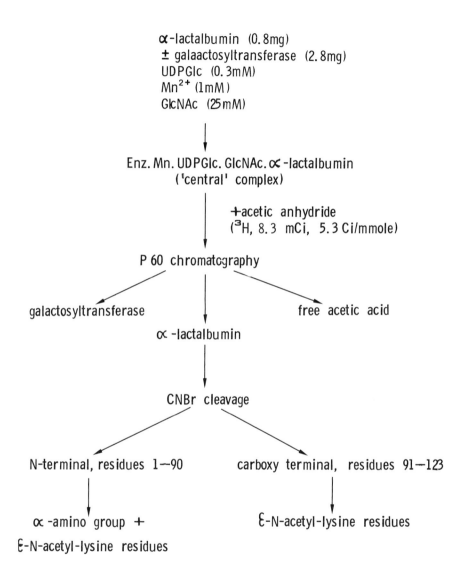

Fig 3. Scheme for the differential labelling experiment, see text for details.

The α-lactalbumin recovered from the modification performed in the presence of galactosyltransferase contained less (74%) radioactivity than that modified in free solution, indicating that some degree of protection had been acheived. To determine the distribution of label in the individual amino groups, α-lactalbumin from each experiment was completely acetylated with nonradioactive acetic anhydride under denaturing conditions, to produce a chemically uniform product, and was mixed with a sample of α-lactalbumin that had been completely acetylated with $[^{14}C]$-acetic anhydride. Each sample was reduced, carboxymethylated and then fragmented as indicated in Fig. 3. Peptides were separated by a variety of column procedures,to give eventually small peptides (thermolysin or chymotryptic peptides) each containing a single residue of lysine (or two in one case) whose location in the polypeptide chain was identifiable. In this way, the specific activity of ^3H (measured as a $^3H/^{14}C$ ratio) in each of the amino groups of the two α-lactalbumin samples were determined.

A comparison of the results for the two samples showed little change in the reactivities of most amino groups, with protection factors ($^3H/^{14}C$ for control $^3H/^{14}C$ for complex), of 1.0 to 1.35. The only two exceptions were Lysine 5, a highly reactive residue in the free protein which showed a three-fold drop in reactivity in the complex (protection factor 3.) and Lysine 114 a much less reactive residue which increased two fold in reactivity (protection factor of 0.5) in the complex.

V.   LOCATION OF THE BINDING SITE
ON α-LACTALBUMIN

The relationship of these results to the conformation of α-lactalbumin is illustrated in Fig. 4. This shows a schematic polypeptide chain structure, following the conclusions of Browne and coworkers (8) and Warme and coworkers (16), obtained by modifying the polypeptide backbone of lysozyme to accomodate structural differences in α-lactalbumin. In the absence of a high resolution structure for α-lactalbumin, this model is a reasonable basis for interpreting our results, although it will undoubtedly prove to be inaccurate in detail. Examination of this structure shows that while the amino groups are widely dispersed over the surface of the molecule (the α-carbons of the appropriate residues being numbered in Figure 4), the two that are affected by complex formation are localized in one region, the lower right hand side, when the molecule is viewed towards the region corresponding to the lysozyme active site cleft region with the amino terminus at the lower end. The changed reactivities should not be construed as conclusive proof that this region of the molecule is directly involved in binding to galactosyltransferase, as it is possible that some conformational change could accompany complex formation that would lead to altered reactivities. In particular, the increased reactivity of the E-NH$_2$ group of Lysine 114 could reflect a greater exposure of this group to the solvent following a conformational alteration. Alternatively, a proximity of this group to a positively charged group on galactosyltransferase in the complex, could also lead to enhanced reactivity. However the very specific nature of the results, the localized region in which these two side chains are situated, and the proximity of this region to the area corresponding to the active site region of lysozyme add credence to the view that the residues may be close to the functional areas of α-lactalbumin. In addition, previous studies have shown that modification of the imidazole group of histidine 32 in α-lactalbumin by ethoxyformylation, leads to a major reduction in activity (17). The side chain of this residue is proximal to that of Lysine 4 (Fig. 4). Inactivation of α-lactalbumin has also been shown to result from sulfenylation of Tryptophan 118, which also lies in this area (18).

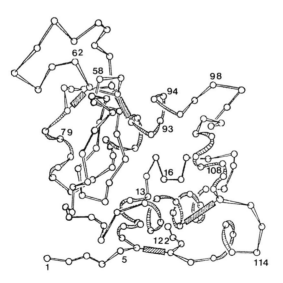

Fig. 4. A putative structure for α-lactalbumin based on the structure of chicken egg white lysozyme elucidated from X-ray analysis (8, 16). The structure has been redrawn to accommodate the necessary deletions with minimal disturbance of secondary structure.

An examination of the nature of the surface of the α-lactalbumin molecule in this area, which adjoins and overlaps with the monosaccharide binding subsite designated F in lysozyme (8) suggests that it has a somewhat hydrophobic surface, as a result of sequence changes from lysozyme including the substitution of a residue of leucine (110 in α-lactalbumin) for arginine (114 in lysozyme). A hydrophobic character for this region is also suggested by energy-minimization structural predictions for α-lactalbumin (15). Evidence that the association of α-lactalbumin with galactosyltransferase involves hydrophobic interactions comes from the observation that the association is enhanced by high ionic strength (13), and is disrupted by organic solvents such as dimethylformamide and dimethylsulfoxide at concentrations that do not affect the conformations of either of the two proteins (9).

These results provide some support for the proposed mode of action of α-lactalbumin depicted in Fig. 3, and a possible basis for the functional divergence of α-lactalbumin and lysozyme. Thus, structural modifications in the area neighbouring subsite F in lysozyme resulting from a limited number of sequence changes appear to have modified this in α-lactalbumin from a monosaccharide binding region to a hydrophobic binding site for galactosyltransferase, while neighbouring areas (possibly corresponding to subsites D or E) have retained an ability to contribute to monosaccharide binding in a complex with galactosyltransferase. Further studies of the association sites of α-lactalbumin and glactosyltransferase are currently in progress.

Acknowledgements

This work was supported by a grant (GM-20363) from the National Institute of General Medical Sciences. K.B. is also the recipient of a Research Career Development Award (K04-00147) from NIGMS.

REFERENCES

1. Schachter, H., Jabbal, I., Hudgin, R.L., Pinteric, L., McGuire, E.J. and Roseman, S. (1970) J. Biol. Chem. <u>245</u>, 1090-1100.

2. Keenan, T.W., Morre, D.J. and Cheethan, R.D. (1970) Nature <u>228</u>, 1105-1106.

3. Fleischer, B., Fleischer, S. and Ozawa, H. (1969) J. Cell. Biol. <u>43</u>, 59-79.

4. Brew, K., Vanaman, T.C. and Hill, R.L. (1968) Proc. Nat. Acad. Sci., Wash. 59, 491-497.

5. Bell, J.E., Beyer, T.A. and Hill, R.L. (1976) J. Biol. Chem. 251, 303-3013.

6. Hill, R.L. and Brew, K. (1975) Adv. in Enzymology and Rel. Areas of Mol. Biol. 43, 411-489.

7. Brew, K., Castellino, F.J., Vanaman, T.C. and Hill, R.L. (1970) J. Biol. Chem. 245, 4750.

8. Brown, W.J., North, A.C.T., Phillips, D.C., Brew, K., Vanaman, T.C. and Hill, R.L. (1969) J. Mol. Biol. 65-86.

9. Brew, K., Richardson, R. and Sinha, S.K. (1978) Proceedings of 12th FEBS Meeting, Dresden, in press.

10. Powell, J.T. and Brew, K. (1974) Eur. J. Biochem. 48, 217-228.

11. Turco, S.J. and Heath, E.C. (1976) Archives Biochem. Biophys. 176, 352-357.

12. Smith, C.A. and Brew, K. (1977) J. Biol. Chem. 252, 7294-7299.

13. Powell, J.T. and Brew, K. (1975) J. Biol. Chem. 250, 6337-6343.

14. Powell, J.T. and Brew, K. (1976) J. Biol. Chem. 251, 3653-3663.

15. Kaplan, H. Stevenson, K.J. and Hartley, B.S. (1971) Biochem. J. 124, 289-299.

16. Warme, P.K., Momany, F.A., Rumball, S.V., Tuttle, R.W. and Scheraga, H.A. (1974) Biochemistry 13, 768-781.

17. Schindler, M., Sharon, N. and Prieels, J-P. (1976) Biochem. Biophys. Res. Commun. 69, 167-173.

18. Schechter, Y., Patchornik, A. and Burnstein, Y. (1974) J. Biol. Chem. 249, 413-419.

DISCUSSION

B.S. HARTLEY: I would like to ask one question.   Do these sugars bind to α-lactalbumin on their own?

K. BREW: We have carried out binding experiments by the method of Hummel and Dreyer, in which a gel filtration column was equilibrated with radioactively labelled monosaccharides (glucose or N-acetylglucosamine) and a sample of α-lactalbumin in the same solution is passed down it. If binding occurs, a peak of radioactivity under the protein should be found, followed by a trough. We observed the reverse: a trough under the protein, followed by a peak, and attribute this to the fact that monosaccharides are excluded from the volume of solution occupied by the α-lactalbumin molecules. From this I would say that α-lactalbumin does not bind monosaccharides significantly under these conditions.

B.S. HARTLEY: But you postulated a conserved region in the α-lactalbumin molecule as being a sugar binding site.

K. BREW: I would only suggest that it can bind sugars when it is in a complex with galactosyltransferase.   The environment then would be very different from that in free solution.

B.S. HARTLEY: Does lysozyme bind monosaccharides in that site alone?

K. BREW: In subsite E it does bind them, although fairly weakly.

S. KORNFELD: Does   lysozyme   interact   with   galactosyl-transferase?

K. BREW: No, lysozyme has no effect on the activity of galactosyltransferase.

B.S. HARTLEY: Is the area on the surface of α-lactalbumin around lysine 5 considerably different between lysozyme and α-lactalbumin?

K. BREW: Yes, there are some quite interesting substitutions, one of which I mentioned in my talk. This involves arginine 114 in lysozyme which directly interacts with monosaccharides in subsite F, and is changed to a leucine in α-lactalbumin, thereby removing the possibility of this interaction as it involves the guanidino group of the arginine. It appears in general that this area may be more hydrophobic in α-lactalbumin.

G. BLOBEL: Does galactosyltransferase have any peripheral sugars?

K. BREW: Other groups, including Dr. K. Ebner's have performed carbohydrate analyses of galactosyltransferase. I believe that it contains galactose and sialic acid.

R.K. KELLER: Is it possible that the oligosaccharide of α-lactalbumin sits in the disaccharide binding site of galactosyltransferase?

K. BREW: Only a small proportion, about 10%, of bovine α-lactalbumin molecules are glycosylated. The unglycosylated and glycosylated forms are equally active, so the carbohydrate can play no role in its function. Other species variants of α-lactalbumin (human and guinea-pig) have no glycosylation 'signal sequence' and are completely devoid of carbohydrate.

E.G. KREBS: Is this system unique or are there many other examples in which an interacting protein changes the specificity of an enzyme?

K. BREW: The analogy that one usually draws is with tryptophan synthase which also consist of two distinct proteins and undergoes activity changes through their interactions, but is in many ways very different from lactose synthase. The closest thing that I can think of as an analogy with the way α-lactalbumin functions is that of metal activators in enzymes, where you get a metal-enzyme-substrate bridge system.   α-Lactalbumin appears to act as a co-factor for the binding of the substrate.   I don't know of any other protein that has this function.

SPECIFICITY CONSIDERATIONS RELEVANT
TO PROTEIN KINASE ACTIVATION AND FUNCTION

Tung-Shiuh Huang
James R. Feramisco[1]
David B. Glass[2]
Edwin G. Krebs

Howard Hughes Medical Institute
Laboratory at the University of Washington
Department of Pharmacology
Seattle, Washington

I. INTRODUCTION

The phosphorylation and dephosphorylation of proteins
catalyzed by protein kinases and phosphoprotein phosphatases
was recognized as a process approximately 25 years ago (1, 2)
and served to explain the mechanisms of interconversion of
the two forms of glycogen phosphorylase (3). Since that time
more than twenty other examples of phospho-dephospho enzymes
have been discovered (for a review see ref. 4). The rate of
growth of this field since 1954-55 is illustrated in Fig. 1.
Nonenzymic proteins, as well as enzymes, undergo phosphorylation-dephosphorylation, and, although in these instances the
function(s) served are less well understood than for enzymes,
the potential importance of these reactions as applied to
this latter set of proteins is very great. The concept is
underscored by the existence of phosphoproteins in essentially all cellular compartments and structures and the
probability that protein phosphorylation-dephosphorylation
reactions always serve a useful purpose.
Steps involving the phosphorylation and dephosphorylation
of enzymes occur in a number of regulatory cascade systems,
and it is clear that the protein kinase-catalyzed steps in
these processes are carefully controlled. It is probable

---

[1]Present address: Cold Spring Harbor Laboratory, Cold
Spring Harbor, New York.
[2]Present address: Department of Pharmacology, Emory
University School of Medicine, Atlanta, Georgia.

449

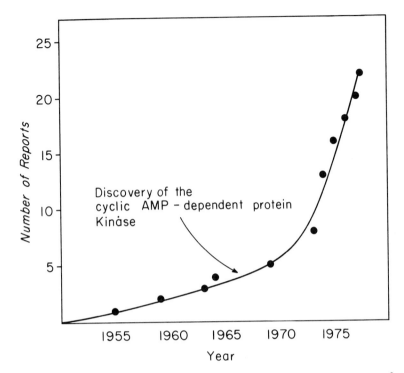

FIGURE 1.   The cumulative number of enzymes reported to
undergo phosphorylation–dephosphorylation since 1950.

that the phosphoprotein phosphatases are also regulated, but
this class of enzymes has not been studied as extensively as
the kinases.   For the protein kinases to function in a
meaningful manner it is essential that they respond to specific signals that impinge on them and that in turn they mani-
fest selectivity in the reactions that they catalyze.   In the
present paper both of these aspects of specificity will be
examined briefly.   New data with respect to the specificity
of action of the cyclic AMP-dependent protein kinase will be
presented.

II.   CLASSIFICATION OF PROTEIN KINASES BASED
ON THEIR RECOGNITION OF SPECIFIC SIGNALS

The ability of protein kinases to respond to specific signals impinging on them is recognized explicitly by the existence of a classification scheme that is based on the nature
of the signal.   This system of nomenclature is shown in

Table I. The first protein kinase to be named according to
the scheme presented in Table I was the cyclic AMP-dependent
protein kinase (5) followed shortly by the cyclic GMP-depen-
dent protein kinase (6). In 1973, after it became known that
phosphorylase kinase is a potentially multifunctional enzyme
that catalyzes the phosphorylation of protein other than
phosphorylase $\underline{b}$, it was suggested (7) that this kinase be
designated as a Ca^{2+}-dependent protein kinase rather than
being referred to by its more restrictive title. The enzyme
was known to require Ca^{2+} from earlier work (9, 10). In 1974
a different Ca^{2+}-dependent protein kinase, the myosin light
chain kinase, was described (11). Thus it would appear that
several enzymes differing in their substrate specificity may
belong to this set (see below). The most recent protein kinase shown to be regulated by a specific effector compound in
the double-stranded RNA-dependent protein kinase found in
cells treated with interferon (for a minireview, see ref. 8).
The final category of protein kinase referred to in Table I,
i.e., the nonspecified protein kinases, might also be thought
of as "messenger-independent" protein kinases, although such
a designation could prove to be misleading for several reasons. First, it is possible that specific messengers interacting directly with members of the group may be discovered
at some later date. Second, some of the nonspecified protein
kinases can be regulated by substrate level control, i.e.,
through the allosteric interaction of metabolites with the
phosphorylatable protein substrate. This latter type of regulation, which can also be highly specific, plays a prominant
role in the regulation of phosphoprotein phosphatases (4).

TABLE I. Classification of Protein Kinases Based on
Their Regulatory Agents

Category	Name	References
1	Cyclic AMP-dependent protein kinase(s)	5
2	Cyclic GMP-dependent protein kinase(s)	6
3	Ca^{2+}-dependent protein kinase(s)	7
4	Double-stranded RNA-dependent protein kinases	(see 8 for minireview)
5	Nonspecified protein kinases	—

A further point should be made in reference to the nomenclature of the Ca^{2+}-dependent protein kinases. It has been
shown (12, 13) that myosin light chain kinase acquires its
Ca^{2+}-sensitivity through its interaction with the Ca^{2+}-depen-
dent modulator (CDR, calmodulin) of cyclic nucleotide phosphodiesterase (14). Waisman et al. (15), noting that the myosin
light chain kinase appears to be multifunctional, recommended
that it be referred to as "modulator-dependent" protein kinase,
a name that might seemingly also be appropriate for phosphorylase kinase (16). In our opinion, however, such a designation
is not in keeping with the system employed in Table I, which
emphasizes the nature of the regulatory signal, i.e., variations in the concentration of cyclic AMP, cyclic GMP, Ca^{2+}, or
double-stranded RNA. The Ca^{2+}-dependent modulator does not
regulate enzymes by virtue of variation in its cellular concentration insofar as is known.

III.   AMINO ACID SEQUENCES OF PHOSPHORYLATED SITES
IN NATURAL PROTEIN SUBSTRATES OF CYCLIC AMP-DEPENDENT
PROTEIN KINASE

A.   Phosphorylation Site in Glycogen Synthase

In an earlier report (17) the amino acid sequence of a
phosphorylated tryptic peptide obtained from rabbit muscle
glycogen synthetase phosphorylated by the cyclic AMP-dependent
protein kinase was shown to be Ser-Asx-Ser(P)-Val-Asp-Thr-Ser-
Ser-Leu-Ser-Pro-Pro-Thr-Glu-Ser-Leu-Ser-Ser-Ala-Pro-Leu-Gly-
Glu-Gln-Asp-Arg. Proud et al. (18) who also determined a
part of this same sequence, designated this site as phosphorylation "site-1" in the synthetase. It was of interest to
know more of the structure on the amino terminal end of the
phosphorylation site. Accordingly, the present work was undertaken.

Rabbit skeletal muscle glycogen synthetase I, 2.4 mg/ml,
was phosphorylated in a reaction mixture containing 12 µg/ml
cyclic AMP-dependent protein kinase catalytic subunit, 0.25
mM [γ-^{32}P]ATP, 10 mM magnesium acetate, 50 mM Tris-HCl (pH
7.5), 45 mM β-mercaptoethanol, 5% sucrose and 100 mM NaF.
Incubation was carried out at 30°. The time course of phosphorylation is shown in Fig. 2. Enzyme phosphorylated to the
extent of 1.7 mol ^{32}P/90,000 g glycogen synthetase, i.e.,
equivalent to enzyme removed at 120 min in the experiment of
Fig. 2, was used for the sequence determination.

[^{32}P]glycogen synthetase D (160 mg) was digested with subtilisin at a ratio of 100:1 (w/w) at 30° for 7 min in the
presence of 50 mM Tris-HCl, pH 7.5, 2 mM EDTA, 45 mM β-

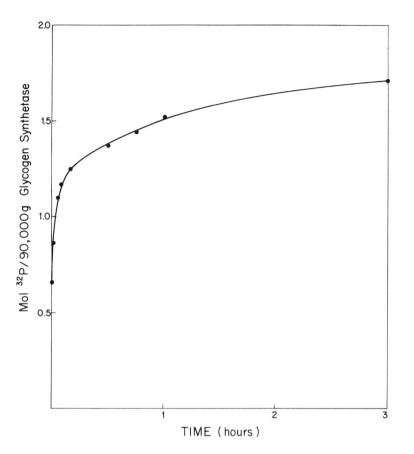

FIGURE 2. Phosphorylation of glycogen synthetase I by cyclic AMP-dependent protein kinase catalytic subunit in the presence of [γ-^{32}P]ATP and Mg^{2+}.

mercaptoethanol, and 5% sucrose. The ^{32}P-labeled peptides were fractionated on a Sephadex G-50 column (Fig. 3). Fraction IV was further purified by high voltage electrophoresis and sequenced by conventional procedures. The sequence of the isolated peptide was found to be Ser-Ser-Gly-Gly-Ser-Lys-Arg-Ser-Asn-Ser(P)-Val-Asp-Thr-Ser. This peptide, which overlaps the previously determined tryptic peptide by nine amino acid residues, provides seven additional residues on the amino terminal side of the phosphorylated serine.

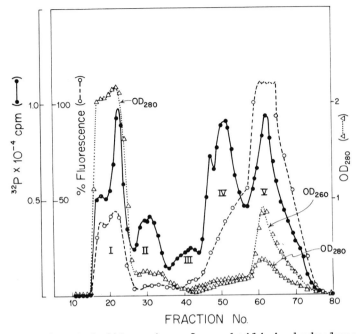

FIGURE 3.   Gel filtration of a subtilisin hydrolysate of
32P-labeled glycogen synthetase A.   Sephadex G-50 column (1.9
x 150 cm) equilibrated with 30% acetic acid and eluted with
the same buffer.   Fractions of 5.6 ml were collected, and 5
µl from each fraction was used to measure radioactivity (●)
and relative fluorescence (o).   Absorbance was measured at 280
and 260 µM (Δ).

B.   Phosphorylation Site in Bovine Heart Type II
Regulatory Subunit of Cyclic AMP-Dependent
Protein Kinase

Bovine heart Type II regulatory subunit–cyclic AMP complex
from the cyclic AMP-dependent protein kinase, 3.8 mg/ml, was
phosphorylated in a reaction mixture containing 12 µg/ml
cyclic AMP-dependent protein kinase catalytic subunit, 125 µM
[γ32-P]ATP, 10 mM magnesium acetate, 5 mM MOPS buffer, pH 7.0,
100 mM NaCl, 0.5 mM EDTA, and 15 mM β-mercaptoethanol.   Incubation was at 30°.   The time course of phosphorylation is
shown in Fig. 4.   A 90 mg sample was used for determination of
the sequence of amino acids at (one of) the phosphorylated
site(s).

The 32P-labeled regulatory subunit was partially digested
with subtilisin at a substrate to subtilisin ratio of 100:1
(w/w) at 30° for 5 min in 5 mM MOPS, pH 7.0, 100 mM NaCl, 0.5

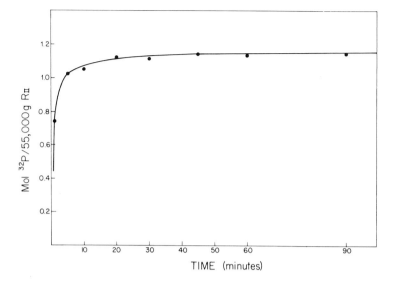

FIGURE 4.   Phosphorylation of type II regulatory subunit
of cAMP-dependent protein kinase by its catalytic subunit.

mM EDTA, and 15 mM β-mercaptoethanol.   The ^{32}P-peptides were
fractionated by gel filtration on Sephadex G-50 (Fig. 5).
Fractions 41 through 49 from the G-50 column (Peak I) were
pooled and further purified by ion exchange chromatography
(Fig. 6).   A major peptide (I) was obtained in the pure form
and an unsuccessful attempt was made to sequence it using the
sequinator due to blockage of the third step of degradation.
It was then digested further with chymotrypsin, following
which it was possible to isolate a major ^{32}-P-labeled chymotryptic peptide using Sephadex G-25, SP-Sephadex chromatography, and high voltage electrophoresis at pH 1.9.   The
amino acid sequence of this peptide was determined to be Asp-
Arg-Arg-Val-Ser(P)-Val using the sequinator in the presence
of cytochrome C as a carrier to prevent washing out from the
spinning cup during the run.

IV.   DISCUSSION

    Protein kinases are highly selective in the reaction that
they catalyze.   Thus most of the cellular proteins do not
undergo phosphorylation-dephosphorylation, at least at significant rates.   In a study designed to examine this point Cohen
et al. (19) examined seventeen of the glycolytic and related

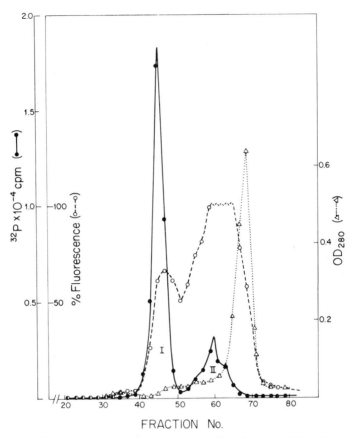

FIGURE 5.   Fractionation of subtilisin peptides by gel filtration. A subtilisin hydrolysate of ^{32}P-labeled Type II regulatory subunit of cAMP-dependent protein kinase was applied to a Sephadex-G-50 column (1.9 x 150 cm), previously equilibrated with 30% acetic acid and eluted with the same buffer. 5.6 ml/fractions were collected, 5 µl and 10 µl from each fraction was used to determine radioactivity (•) and relative fluorescence (o). Absorbance was measured at 280 µM (Δ).

enzymes in order to determine whether or not they would serve as substrates for pure phosphorylase kinase, but the only enzyme found to undergo phosphorylation was glycogen phosphorylase. Other investigators have had similar experiences when testing available proteins picked at random as potential substrates for this and other protein kinases. When crude tissue extracts are treated with [γ-^{32}P]ATP, or when intact

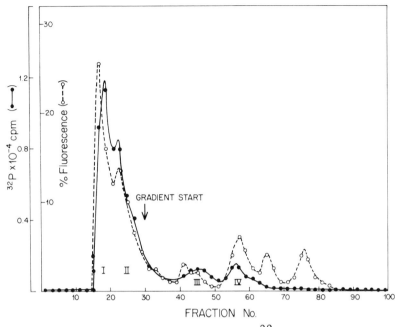

FIGURE 6.   Further purification of ^{32}P-labeled subtilisin peptide from Sephadex-G-50 fraction I over SP-Sephadex column (0.9 x 45 cm).   The column was previously equilibrated with 0.2 M pyridine acetate buffer pH 3.1 and eluted with gradient established by 75 ml each of 0.2 pH 3.1 and 2M pH 5 pyridine acetate buffer.   Fractions of 1.5 ml were collected, and 5 µl from each fraction was used to determine radioactivity (•) and relative fluorescence(o).

cells are incubated with ^{32}Pi, the great majority of the proteins present are not phosphorylated.

What factors determine the specificity of protein kinases? Two early studies showed that peptides derived from protein substrates by partial proteolysis could be phosphorylated by phosvitin kinase (20) or by phosphorylase kinase (21), indicating that for these enzymes it was not essential that the native structure of the substrate be present.   Notwithstanding the existence of these reports, it was at first believed that the cyclic AMP-dependent protein kinase required that its substrates possess some unique native configuration.   This concept had arisen because the first phosphorylated site sequences examined had shown no obvious structural similarity and attention was thus drawn away from primary structure.   (In these early studies only tryptic peptides lacking basic residues were sequenced.)   Later, however, when it was shown that

the cyclic AMP-dependent protein kinase catalyzed the phosphorylation of small peptides (22) or denatured proteins (23,
24) interest was focused on primary structural determinants.
It became clear that relatively short amino acid sequences on
either side of the susceptible serine or threonine residue are
sufficient to provide most of the information needed for this
kinase to act. Inspection of the amino acid sequences at the
phosphorylation sites determined in the present study, together with those elucidated heretofor for other physiologically
significant substrates of the cyclic AMP-dependent protein
kinase, shows that they fall into two categories (Table II).
In the first group an Arg-Arg-sequence is separated from the
phosphorylated serine by a single amino acid residue. In the
second group a Lys-Arg-sequence is separated from the serine
by two amino acid residues.

TABLE II. Sites Phosphorylated by Cyclic AMP-Dependent
Protein Kinase

Substrate	Sequence	Ref.
Type II regulatory subunit of cAMP-dependent protein kinase	Asp-<u>Arg-Arg</u>-Val-Ser(P)-Val-	this work
Liver pyruvate kinase	Leu-<u>Arg-Arg</u>-Ala-Ser(P)-Val-	(25)
Rabbit skeletal muscle phosphorylase kinase α-subunit	Phe-<u>Arg-Arg</u>-Leu-Ser(P)-Ile-	(26)
Rabbit skeletal muscle glycogen synthetase	<u>Lys-Arg</u>-Ser-Asn-Ser(P)-Val-	this work
Rabbit skeletal muscle phosphorylase kinase β-subunit	<u>Lys-Arg</u>-Ser-Gly-Ser(P)-$^{Val}_{Ile}$	(26)
Histone H1	<u>Lys-Arg</u>-Lys-Ala-Ser(P)-Gly	(27)

The fact that primary structural information by itself
allows one to predict that a given peptide or sequence of
amino acids in a protein will be phosphorylated by a given
protein kinase does not mean that higher orders of protein
structure are unimportant with respect to protein kinase
specificity. Obviously, a given primary structure determines
the secondary and tertiary structure of proteins, and even for
small peptides particular configurations are favored depending
on the sequence and presence of certain amino acid residues.

Attention has been given to the possibility that β-turns may be important in determining protein kinase specificity (28). Other features of secondary structure and possibly some aspects of tertiary structure may also be important. It is apparent that the exposure of a potential phosphorylation site would be critical in determining whether or not phosphorylation can occur.

REFERENCES

1. Fischer, E. H., Krebs, E. G., J. Biol. Chem. 216, 121-132 (1955).
2. Sutherland, E. W., Wosilait, W. D., Nature 175, 169-171 (1955).
3. Cori, G. T., Green, A. A., J. Biol. Chem. 151, 31-38 (1943).
4. Krebs, E. G., Beavo, J. A., Ann. Rev. Biochem. (1979) In Press.
5. Walsh, D. A., Perkins, J. P., Krebs, E. G., J. Biol. Chem. 243, 3763-3765 (1968).
6. Kuo, J. F., Greengard, P., J. Biol. Chem. 245, 2493-2498 (1970).
7. Krebs, E. G., Stull, J. T., England, P. J., Huang, T. S., Brostrom, C. O., Vandenheede, J. R., In "Protein Phosphorylation in Control Mechanisms," Miami Winter Symp., 5, 31-45 (1973).
8. Hunt, T., Nature, 273, 97-98 (1978).
9. Meyer, W. L., Fischer, E. H., Krebs, E. G., Biochemistry, 3, 1033-1039 (1964).
10. Ozawa, E., Hosoi, K., Ebashi, S., J. Biochem., 61, 531-533 (1967).
11. Pires, E., Perry, S. V., Thomas, M. A. W., FEBS Lett., 41, 292-296 (1974).
12. Dabrowska, D., Sherry, J. M. F., Aramatorio, D. K., Hartshorne, D. J., Biochemistry, 17, 253-258 (1978).
13. Yagi, K., Yazawa, M., Kakiuchi, S., Ohshima, M., Uenishi, K., J. Biol. Chem., 253, 1338-1340 (1978).
14. Cheung, W. Y., Biochem. Biophys. Res. Commun., 29, 478-482 (1967).
15. Waisman, D. M., Singh, T. J., Wang, J. H., J. Biol. Chem. 253, 3387-3390 (1978).
16. Cohen, P., Burchell, A., Foulkes, J. G., Cohen, P. T. W., Vanaman, T. C., Nairn, A. C., FEBS Lett., 92, 287-293 (1978).
17. Huang, T. S., Krebs, E. G., Biochem. Biophys. Res. Commun. 75, 643-650 (1977).

18. Proud, C. G., Rylatt, D. B., Yeaman, S. J., Cohen, P., FEBS Lett., 80, 435-442 (1977).
19. Cohen, P., Antoniw, J. F., Davison, M., and Tyler, C., In Metabolic Interconversion of Enzymes 1973, eds. E. H. Fischer, E. G. Krebs, H. Neurath, E. R. Stadtman, Springer-Verlag, Berlin, pp. 33-42 (1974).
20. Rabinowitz, M., Lipmann, F., J. Biol. Chem., 235, 1043-1050 (1960).
21. Nolan, C., Novoa, W. B., Krebs, E. G., Fischer, E. H., Biochemistry, 3, 542-551 (1964).
22. Daile, P., Carnegie, P. R., Biochem. Biophys. Res. Commun., 61, 852-858 (1974).
23. Bylund, D. B., Krebs, E. G., J. Biol. Chem., 250, 6355-6361 (1975).
24. Humble, E., Berglund, L., Titanji, V., Ljungström, O., Edlund, B., Zetterqvist, Ö., Engström, L., Biochem. Biophys. Res. Commun, 66, 614-621 (1975).
25. Edlund, B., Andersson, J., Titanji, V., Dahlqvist, U., Ekman, P., Zetterqvist, Ö., and Engström, L., Biochem. Biophys. Res. Commun., 67, 1516-1521 (1975).
26. Yeaman, S. J., Cohen, P., Watson, D. C., Dixon, G. H., Biochem. J., 162, 411-421 (1977).
27. Langan, T. A., Ann. N. Y. Acad. Sci., 185, 166-180 (1971).
28. Small, D., Chou, P. Y., Fasman, G. D., Biochem. Biophys. Res. Commun., 79, 341-346 (1977).

DISCUSSION

R. SCOTT: Most of the comments you have made refer to soluble kinase. Do you think that a similar classification of substrate specificity is going to apply to membrane bound kinases?

E.G. KREBS: The membrane bound protein kinases are more difficult to study and are not as well understood as the soluble protein kinases. In so far as is known the cyclic AMP-dependent protein kinases of membranes seem to show the same specificity as their soluble counterparts. It is not known with certainty whether identical proteins are involved.

D.F. STEINER: How hard and fast is the rule of paired basic residues in terms of the catalytic unit of cyclic AMP dependent protein kinase?

E.G. KREBS: Not   hard   and   fast   insofar   as   <u>in</u>   <u>vitro</u>
phosphorylations are concerned. The phosphorylation of serine
50 in denatured lysozyme is a puzzle to us because the only
arginine that is present is about 5 or 6 residues away.   If
there are no basic residues at all, we have not found any
significant rate of phosphorylation.

THE ROLE OF PROTEIN PHOSPHORYLATION IN THE
COORDINATED CONTROL OF INTERMEDIARY METABOLISM

P. Cohen[1], N. Embi, G. Foulkes, G. Hardie,
G. Nimmo, D. Hylatt, and S. Shenolikar

Department of Biochemistry
University of Dundee
Dundee
Scotland, U.K.

I. PROTEIN PHOSPHORYLATION AS A REGULATORY DEVICE

Glycogen phosphorylase was the first enzyme whose
activity was shown to be regulated by a phosphorylation-
dephosphorylation mechanism (1), and two other enzymes
involved in glycogen metabolism, phosphorylase kinase (2) and
glycogen synthase (3), were the second and third recorded
examples of this phenomenon. Over the past few years,
however, the number of proteins whose activities have been
shown to be regulated in this manner has increased
dramatically (4), and it seems likely that protein phosphorylation is the major general mechanism by which metabolic
events in eukaryotic cells are controlled by external
physiological stimuli.

In the first part of this article, recent progress in
understanding the regulation of glycogen metabolism in
skeletal muscle by phosphorylation-dephosphorylation will be
summarized. In the second part, evidence will be presented
that several of the proteins in glycogen metabolism are also
involved in the regulation of other metabolic processes.
These findings have raised the exciting possibility that there
are a relatively simple network of regulatory pathways which
allow the activities of diverse but functionally related
cellular processes to be controlled in a synchronous manner.

Supported by grants from the Medical Research Council, London,
the British Diabetic Association and the Wellcome Trust.

II. IDENTIFICATION OF CALMODULIN AS A SUBUNIT OF MUSCLE PHOSPHORYLASE KINASE

Phosphorylase kinase is the enzyme at which the neural and hormonal mechanisms for stimulating glycogenolysis meet in a single protein, since its activity is dependent on calcium ions and is stimulated by cyclic AMP dependent protein kinase (reviewed in 5). Several years ago we reported that phosphorylase kinase possessed the structure $(\alpha\beta\gamma)_4$, where the molecular weights of the α, β and γ-subunits were 145,000, 128,000 and 45,000 respectively (6). The α and β-subunits were found to be the components phosphorylated by cyclic AMP dependent protein kinase (6-10) and this raised the question of which subunit(s) bound the calcium ions, that were essential for activity.

Recently, phosphorylase kinase has been shown to contain a fourth component, termed the δ-subunit (11). This subunit has a molecular weight of 17,000, and is in equimolar concentration with the other subunits. Phosphorylase kinase therefore has the structure $(\alpha\beta\gamma\delta)_4$. The δ-subunit was missed for several years since it migrated as a faintly staining band with the bromophenol blue dye front, on the 5% polyacrylamide gels previously used to resolve the α, β and γ-subunits.

The δ-subunit has been demonstrated to be identical to a calcium binding protein termed calmodulin. This protein was first identified as a factor which stimulated the activity of the high K_m cyclic nucleotide phosphodiesterase of brain tissue (12-14), but it has subsequently been implicated in the control of a number of reactions which are stimulated by calcium ions. These include a calcium stimulated adenylate cyclase in brain (15), the calcium ATPase of erythrocyte membranes, which is involved in calcium transport (16,17), the depolymerisation of microtubules (18), myosin light chain kinase in skeletal muscle (19) and smooth muscle (20), and NAD kinase in higher plants (21).

The amino acid sequence of calmodulin shows a 50% identity with troponin C, the protein which confers calcium sensitivity to actomyosin ATPase in the muscle contractile apparatus (22-25). Both proteins bind four molecules of calcium per mole with high affinity (23). Calmodulin can substitute for troponin C in restoring calcium sensitivity to actomyosin ATPase in reconstitution experiments (26), but troponin C is at least 500-fold less effective than calmodulin in the activation of cyclic nucleotide phosphodiesterase (27).

Troponin C appears to be a specialized form of calmodulin which functions only in the contractile apparatus, while calmodulin itself seems to have a variety of functions, and may well be the major calcium receptor of eukaryotic cells (see section V).

Phosphorylase kinase has been reported to possess 16 binding sites for calcium ions per $(\alpha\beta\gamma\delta)_4$ with affinities in the micromolar range (28). Since calmodulin can bind four molecules of calcium per mole, the calcium binding measurements are consistent with the view that all the high affinity binding sites are located on the δ-subunit of phosphorylase kinase.

Electrical excitation of muscle causes calcium ions to be released from the sarcoplasmic reticulum, and the rise in the concentration of calcium ions in the muscle sarcoplasm is believed to represent the signal which synchronises glyco-genolysis and muscle contraction. The finding that calcium ions activate both processes by binding to extremely similar calcium binding proteins (calmodulin and troponin C) puts this concept on a much firmer molecular basis.

Calmodulin is bound very tightly to phosphorylase kinase and does not dissociate even in the presence of 8M urea, provided that calcium ions are present (A.C. Niarn, unpublished work). In this respect it resembles the inter-action between troponin C and troponin I (26). However the addition of excess calmodulin activates purified phosphory-lase kinase considerably. The activation varies from preparation to preparation and ranges from 2-fold to 7-fold. It appears to result from the binding of a second molecule of calmodulin, since the calmodulin stimulated activity can be blocked by the addition of the drug trifluoperazine (29) or by troponin I, whereas the calcium dependent activity in the absence of excess calmodulin is unaffected by these compounds. Furthermore, phosphorylase kinase preparations always contain stoichiometry quantities of calmodulin, irrespective of the extent of stimulation by excess calmodulin. The physiological relevance of the activation by excess calmodulin is not yet clear (S. Shenolikar and P. Cohen, unpublished work).

III. THE REGULATION OF GLYCOGEN SYNTHASE BY
THREE PROTEIN KINASES

It is well established that cyclic AMP dependent protein kinase catalyses the phosphorylation of glycogen synthase, thereby converting the enzyme to a form which is more dependent on the allosteric activator glucose-6P (30,31). The phosphorylation of phosphorylase kinase and glycogen synthase by cyclic AMP dependent protein kinase take place at very similar rates <u>in vitro</u> (32) and these phosphorylations almost certainly underlie the synchronous activation of glyco-genolysis and inhibition of glycogen synthesis, which occurs in skeletal muscle in response to adrenalin (reviewed 5,6).

However, the activity of phosphorylase kinase and the rate of glycogenolysis also depend on calcium ions, and this raises the question of whether the activity of glycogen synthase is also controlled by this divalent cation. Several years ago, we reported that purified preparations of glycogen synthase contained traces of a protein kinase distinct from cyclic AMP dependent protein kinase that could phosphorylate glycogen synthase. This activity, which was termed glycogen synthase kinase-2, was unaffected by cyclic AMP or by the specific protein inhibitor of cyclic AMP dependent protein kinase. Moreover it had a different K_m for ATP (0.4 mM vs 0.02 mM) and it could use GTP as a phosphoryl donor ($K_m = 5$ mM) whereas cyclic AMP dependent protein kinase could not (33). This activity has been shown to phosphorylate a serine residue (site-2) distinct from either of the two sites phos-phorylated by cyclic AMP dependent protein kinase (sites 1a and 1b), and to convert glycogen synthase to a form which is more dependent on glucose-6P (34,35, Table 1).

The activity of the endogenous glycogen synthase kinase-2 was found to be unaffected by the presence or absence of calcium ions (33), but was stimulated by calmodulin in the presence of calcium ions (35). The stimulation varied from preparation to preparation and ranged from 1.5-fold to more than 10-fold. The same serine residue.(site-2) was phos-phorylated either in the presence or absence of calmodulin and calcium ions, and this suggested that both phosphorylations were catalysed by the same enzyme (35).

Roach and Larner (36) recently reported that phosphory-lase kinase catalysed the phosphorylation of glycogen synthase. These results and the activation of glycogen synthase kinase-2 by calmodulin therefore raised the question of whether

TABLE I. Phosphorylation of Glycogen Synthase by Three Protein Kinases

Protein kinase	Site	Sequence	Inactivation		Reference
Cyclic AMP dependent protein kinase	1a	arg–ala–ser(P)–	Yes	(++)	32
	1b	ser–asn–ser(P)–val–asp–thr–	No	(–)	32
Glycogen synthase kinase–2 (phosphorylase kinase)	2	thr–leu–ser(P)–val–ser–ser–	Yes	(+)	34, 35
Glycogen synthase kinase–3	3	?	Yes	(+++)	see text

TABLE II. Amino Terminal Sequences of Rabbit Muscle Glycogen
Phosphorylase and Glycogen Synthase Containing the Sites of
Phosphorylation by Phosphorylase Kinase (34, 39-41).

Phosphorylase

$$
\begin{array}{c}
\text{P} \\
10 \qquad\qquad\qquad 1 \quad 15 \qquad\qquad\qquad\qquad\qquad 20
\end{array}
$$

glu-lys-arg-lys-gln-ile-ser-val-arg-gly-leu-ala-gly-val-glu

Synthase

$$
\begin{array}{c}
\text{P} \\
1 \qquad\qquad 5 \qquad\qquad 1 \qquad\qquad 10 \qquad\qquad\qquad 15
\end{array}
$$

pro-leu-ser-arg-thr-leu-ser-val-ser-ser-leu-pro-gly-leu-glu

Identical residues or conservative differences are underlined.
The numbers indicate distances from the N-terminus of each
enzyme.

phosphorylase kinase and glycogen synthase kinase-2 could be
the same enzyme. The activity of the endogenous glycogen
synthase kinase-2 in purified preparations of glycogen
synthase is unaffected by calcium ions in the absence of added
calmodulin (33) whereas the activity of purified phosphorylase
kinase is almost completely dependent on calcium ions (37),
and this observation originally led us to conclude that the
two enzymes were distinct (33). However a more detailed
analysis has shown that the procedure used for the purification of glycogen synthase (38) causes phosphorylase kinase to
completely lose its requirement for calcium ions. In addition
we have confirmed the results of Roach and Larner, and shown
further that the phosphorylation of glycogen synthase by
purified phosphorylase kinase occurs exclusively at site-2
(Table I). This suggests that glycogen synthase kinase-2 and
phosphorylase kinase may indeed be the same enzyme (D.B.
Rylatt, N. Embi and P. Cohen, unpublished work).

Site-2 is located only seven amino acids from the N-
terminus of glycogen synthase (34) and the phosphoserine in
phosphorylase a is 14 amino acids from the N-terminus (39,40).
Thus phosphorylase kinase phosphorylates analogous regions of
phosphorylase and glycogen synthase, and it is therefore of
interest that the amino acid sequences surrounding these two
serine residues are similar (Table II). It is not yet known
whether site-2 on glycogen synthase becomes phosphorylated
in vivo. It is however tempting to speculate that this

phosphorylation can regulate glycogen synthesis during muscle contraction.

Following our original report that glycogen synthase was phosphorylated by a protein kinase whose activity was unaffected by cyclic AMP or calcium ions (33), we partially purified such an enzyme from skeletal muscle extracts (42). This enzyme is now known to be distinct from both cyclic AMP dependent protein kinase and phosphorylase kinase and has been termed glycogen synthase kinase-3. This enzyme inactivates glycogen synthase to a greater extent than cyclic AMP dependent protein kinase and phosphorylase kinase, for a given amount of phosphate incorporated, and preliminary peptide mapping studies indicate that it phosphorylates a serine residue (site-3) distinct from sites 1a, 1b and 2 (Table I) (D.B. Rylatt and P. Cohen, unpublished results). The physiological role of this enzyme is unknown, but it is possible that it is involved in the regulation of glycogen synthase by insulin (reviewed 5,6).

IV. THE REGULATION OF PROTEIN PHOSPHATASE-1 BY INHIBITOR-1

A single enzyme in mammalian skeletal muscle, which has been termed protein phosphatase-1 (6), catalyses the dephosphorylation of phosphorylase, phosphorylase kinase and glycogen synthase, and therefore carries out each of the dephosphorylations which inhibit glycogenolysis or activate glycogen synthesis. The evidence supporting this statement has been reviewed (5,6). This finding has emphasized the potential importance of protein dephosphorylation in the control of glycogen metabolism since the activity of protein phosphatase-1 would be expected to influence the state of phosphorylation of several interconverting enzymes.

Huang and Glinsmann demonstrated the existence of two heat stable proteins in skeletal muscle, termed inhibitor-1 and inhibitor-2, which inhibited phosphorylase phosphatase. They also made the key observation that inhibitor-1 was only an inhibitor after it had been phosphorylated by cyclic AMP dependent protein kinase (43,44). Inhibitor-1 has been purified to homogeneity in this laboratory and characterized extensively (45-48). It is phosphorylated by cyclic AMP dependent protein kinase <u>in vitro</u> at similar rates to phosphorylase kinase and glycogen synthase, and the phosphorylation takes place on a unique threonine residue (45). The molar concentration of inhibitor-1 (1.5 µM) is comparable to phosphorylase kinase $(2.5 \text{ } \mu\text{M})$ and glycogen synthase $(2.8 \text{ } \mu\text{M})$ and is higher than the concentration of protein phosphatase-1 (46). Inhibitor-1 inhibits the phosphorylase phosphatase, phosphorylase kinase phosphatase and glycogen synthase phosphatase activities of protein phosphatase-1 in an identical manner $(K_i = 1.6 \text{ nM})$ and is only an inhibitor in its phosphorylated form (47). All of these properties are consistent with the view that inhibitor-1 is important in the control of protein phosphatase-1, and we have recently demonstrated that inhibitor-1 is phosphorylated in skeletal muscle in vivo. In normal fed animals the basal level of phosphorylation was $31 \pm 7\%$ and it increased to $70 \pm 12\%$ following an intravenous injection of adrenaline (48). These results indicate that the classical glycogenolytic cascade involving the activation of phosphorylase kinase by cyclic AMP dependent protein kinase, and the activation of phosphorylase by phosphorylase kinase, represents only "half the story". The phosphorylation of inhibitor-1 by cyclic AMP dependent protein kinase, and the inactivation of protein phosphatase-1 by inhibitor-1, may be equally as important in elevating the level of phosphorylase a.

Protein phosphatase-1 may also be the enzyme which dephosphorylates inhibitor-1 in vivo. The reason for making this surprising statement is that inhibitor-1 does not inhibit its own dephosphorylation, and the kinetic constants for the dephosphorylation of inhibitor-1 are similar to other substrates of protein phosphatase-1 (47). This remarkable property might be related to the fact that inhibitor-1 is phosphorylated on a threonine residue, whereas all other substrates for cyclic AMP dependent protein kinase and all other phosphoproteins in muscle are phosphorylated on serine residues (50).

Inhibitor-1 is a novel protein in metabolic regulation which mediates the control of a protein phosphatase by a protein kinase. It not only represents a mechanism for amplifying the effect of cyclic AMP, but also for amplifying changes in the activity of protein phosphatase-1 (provided that protein phosphatase-1 is the enzyme which dephosphorylates inhibitor-1 in vivo).

V. ROLE OF THE REGULATORY COMPONENTS OF GLYCOGEN METABOLISM IN THE CONTROL OF OTHER METABOLIC PROCESSES

A list of enzymes found in the cytoplasm of mammalian cells whose activities appear to be modulated by phosphorylation-dephosphorylation is given in Table III. The general feature that seems to be emerging is that enzymes which have a biodegradative function are activated by phosphorylation, whereas enzymes which have a biosynthetic function are inactivated by phosphorylation. This suggests that common interconverting enzymes, or common regulator molecules acting on different interconverting enzymes, could be used to synchronize the activities of different metabolic pathways in response to neural and hormonal stimuli. The evidence which implicates several of the protein components involved in glycogen metabolism in the regulation of other metabolic events will be reviewed briefly in the following section.

A. Cyclic AMP Dependent Protein Kinase and Calmodulin

It is now well established that cyclic AMP dependent protein kinase does not merely function in the regulation of glycogen metabolism, but that it mediates most, if not all, of the actions of hormones which work through cyclic AMP (58). The regulatory subunit of cyclic AMP dependent protein kinase seems to be the major receptor for cyclic AMP in mammalian cells, and its interaction with cyclic AMP causes it to dissociate from the catalytic subunit, thereby activating the enzyme (59). Some enzymes which are regulated by cyclic AMP dependent protein kinase are shown in Table III.

It seems likely that calmodulin is the major calcium receptor in the cytoplasm of mammalian cells (section II), and that it plays a role almost perfectly analogous to the regulatory subunit of cyclic AMP dependent protein kinase. In the case of skeletal muscle, nervous stimulation causes calcium ions to pass from the sarcoplasmic reticulum to the cytoplasm. The binding of calcium to calmodulin (or troponin C, in the case of the muscle contractile apparatus) then triggers the activation of a number of enzymes, including protein kinases such as phosphorylase kinase and myosin light chain kinase. This concept is illustrated in Figure 1. It is also important to emphasize that these pathways could be interchangeable in different tissues and may be closely interlinked. Thus electrical excitation of brain tissue causes an elevation of cyclic AMP (60), while the interaction of

TABLE III. Enzymes Found in the Cytoplasm of Mammalian Cells that are Regulated by Phosphorylation

	Type of protein kinase			
	cAMP	Calmodulin	Other	Biodegradative pathway
Activation by Phosphorylation				
Glycogen phosphorylase (37)	−	+	−	glycogenolysis
Myosin (19)	−	+	−	ATP hydrolysis
Phosphorylase kinase (51)	+	+*	−	glycogenolysis
Triglyceride lipase (52)	+	−	−	triglyceride breakdown
Cholesterol esterase (53)	+	−	−	cholesterol ester hydrolysis
				Biosynthetic pathway
Inactivation by Phosphorylation				
Glycogen synthase (5,6)	+	+	+	glycogen synthesis
Acetyl CoA carboxylase (54)	+	+	+	fatty acid synthesis
Glycerol phosphate acyl transferase (55)	+	−	−	triglyceride synthesis
HMG CoA reductase (56)	−	−	+	cholesterol synthesis
Initiation factor eIF2 (57)	−	−	+	protein synthesis
L-type pyruvate kinase (58)	+	−	−	gluconeogenesis

*Phosphorylase kinase phosphorylates itself

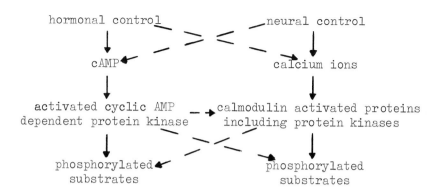

FIGURE 1. Role of cyclic AMP dependent protein kinase and calmodulin in the neural and hormonal control of intermediary metabolism.

adrenaline with its α-receptors in the liver leads to an elevation of the cytoplasmic calcium ion concentration (61). Calmodulin dependent protein kinases and cyclic AMP dependent protein kinase might also frequently phosphorylate the same enzyme (as in the case of glycogen synthase) while cyclic AMP dependent protein kinase may regulate calmodulin dependent protein kinases, as in the case of phosphorylase kinase and perhaps smooth muscle myosin light chain kinase (62).

B. Protein Phosphatase-1

Protein phosphatase-1 catalyses each of the dephosphorylations which inhibit glycogenolysis or activate glycogen synthesis, and therefore dephosphorylates sites phosphorylated by both cyclic AMP dependent protein kinase and other types of protein kinase. The ability of protein phosphatase-1 to catalyse a number of functionally related dephosphorylations raises the question of whether the specificity of the enzyme is even broader, and whether it might carry out many of the dephosphorylations that inhibit biodegradative pathways or stimulate biosynthetic pathways (Table III).

One of the first pieces of evidence which implicated cyclic AMP dependent protein kinase in the control of metabolic processes other than glycogen metabolism was the finding that its activity was very similar in all mammalian tissues, and was high even in tissues where glycogen metabolising enzymes were almost non-existent. We therefore

made a similar study of the distribution of protein phospha-
tase-1, by measuring the phosphorylase phosphatase activity of
a variety of tissue extracts, as well as the susceptability of
the phosphorylase phosphatase activity to the two proteins,
inhibitor-1 and inhibitor-2, which appear to inhibit
protein phosphatase-1 specifically.   These experiments
indicated that protein phosphatase-1 or a closely analogous
protein was present at very similar concentrations in all
mammalian tissues tested; e.g. the specific activity of
phosphorylase phosphatase in adipose tissue extracts was only
two-fold lower than in skeletal muscle, whereas the phosphorylase and phosphorylase kinase activities were several hundred
fold lower (63).   These experiments support the idea that
protein phosphatase-1 is involved in the regulation of
metabolic processes other than glycogen metabolism.

Severson et al (64) have stated that a highly purified
preparation of phosphorylase phosphatase from rabbit liver was
able to inactivate hormone sensitive triglyceride lipase from
chicken adipose tissue, which had been activated by cyclic AMP
dependent protein kinase.   Ingebritsen et al (56) have also
reported that liver hydroxymethylglutaryl CoA reductase, which
had been inactivated by incubation with ATP and magnesium ions,
could be reactivated by a protein phosphatase preparation
which appeared to copurify with liver phosphorylase phosphatase.   As discussed in the following section, protein phos-
phatase-1 has also been implicated in the regulation of acetyl
CoA carboxylase activity (Table III).   These observations
strengthen the view that protein phosphatase-1 may have a
number of physiological substrates.

VI. REGULATION OF FATTY ACID SYNTHESIS BY PHOSPHORYLATION-DEPHOSPHORYLATION

The rate at which cytoplasmic acetyl CoA is converted to
long chain fatty acids is stimulated by insulin in fat cells
(65) and inhibited by glucagon in hepatocytes (66).   The
influence of hormones on this pathway is therefore analogous
to their effects on glycogen synthesis.

The rate limiting enzyme in fatty acid synthesis is
acetyl CoA carboxylase.   It is completely dependent on the
allosteric activator citrate for activity, and inhibited by a
very low concentration of palmityl CoA (67).   Following
reports that acetyl CoA carboxylase might be inactivated by a
phosphorylation mechanism in the liver (68-70), we initiated a

study of the regulation of fatty acid synthesis in lactating mammary gland. Simple procedures were developed for the isolation of large quantities of acetyl CoA carboxylase and fatty acid synthetase, and measurements of the alkali labile phosphate bound covalently to the purified enzymes showed that the carboxylase contained 3.2 ± 0.5 molecules of phosphate per subunit whereas the synthase contained only 0.2 molecules per subunit (54,71).

The purified acetyl CoA carboxylase was shown to contain traces of three different protein kinases that could phosphorylate the enzyme. One was cyclic AMP dependent protein kinase which phosphorylates acetyl CoA carboxylase at a similar rate to the enzymes of glycogen metabolism (54). Peptide mapping studies of tryptic phosphopeptides suggest that two sites (site-1 and site-2) are phosphorylated. A second endogenous protein kinase was unaffected by cyclic AMP or the specific protein kinase inhibitor (54), and is specific for the phosphorylation of site-2. If calmodulin and calcium ions are included in incubations, the phosphorylation of acetyl CoA carboxylase is usually stimulated about 1.5-fold and the additional phosphate is incorporated into a distinct tryptic phosphopeptide (site-3). This suggests that a third protein kinase is involved in this reaction (D.G. Hardie, unpublished work).

Each of the phosphorylations described above can be reversed by protein phosphatase-1, which dephosphorylates acetyl CoA carboxylase and glycogen synthase at very similar rates. The activity ratio phosphorylase phosphatase/acetyl CoA carboxylase phosphatase is identical in extracts of skeletal muscle or mammary gland, and both activities are inhibited in an identical manner by inhibitor-1 and inhibitor-2. These experiments suggest that protein phosphatase-1 is the major acetyl CoA carboxylase phosphatase activity in mammary gland (J.G. Foulkes, D.G. Hardie and P. Cohen, unpublished work).

The phosphorylation of purified acetyl CoA carboxylase by endogenous protein kinases does not appear to influence the V_{max} of the enzyme, the K_a for citrate, or the K_m for acetyl CoA. However, if the enzyme is incubated with protein phosphatase-1, the activity assayed at suboptimal concentrations of citrate is increased 2.5-fold. The activation, which can be prevented by inhibitor-2, is accompanied by the loss of about one molecule of alkali labile phosphate per subunit. It therefore appears that acetyl CoA carboxylase, as isolated, is in an inactivated state (D.G. Hardie, unpublished work).

Although these investigations are still at an early stage,
there is clearly a striking resemblance between the regulation
of glycogen synthase and acetyl CoA carboxylase.   The two
enzymes may well be phosphorylated by identical interconverting enzymes, and the results obtained so far, strongly
support the idea that hormones control the pathways of
glycogen and fatty acid synthesis in a coordinated manner,
through very similar protein phosphorylation and dephosphorylation reactions.

REFERENCES

1. Krebs, E.G. and Fischer, E.H., Biochim. et Biophys. Acta,
    $\underline{20}$, 150-157 (1956).
2. Krebs, E.G., Graves, D.J. and Fischer, E.H., J. Biol.
    Chem., $\underline{234}$, 2867 (1959).
3. Friedman, D.L. and Larner, J., Biochemistry, $\underline{2}$, 669-675,
    (1963).
4. Nimmo, H.G. and Cohen, P., Adv. Cyc. Nuc. Res., $\underline{8}$, 145-
    266 (1977).
5. Cohen, P., Curr. Top. Cell. Reg., $\underline{14}$, 117-196 (1978).
6. Cohen, P., Eur. J. Biochem., $\underline{34}$, 1-14 (1973).
7. Hayakawa, T., Perkins, J.P. and Krebs, E.G., Biochemistry,
    $\underline{12}$, 574-580 (1973).
8. Cohen, P., Watson, D.C. and Dixon, G.H., Eur. J. Biochem.,
    $\underline{51}$, 79-92 (1975).
9. Yeaman, S.J. and Cohen, P., Eur. J. Biochem., $\underline{51}$, 93-
    104 (1975).
10. Antoniw, J.F. and Cohen, P., Eur. J. Biochem., $\underline{68}$, 45-
    54 (1976).
11. Cohen, P., Burchell, A., Foulkes, J.G., Cohen, P.T.W.,
    Vanaman, T.C. and Nairn, A.C., FEBS Lett., $\underline{92}$, 287-293,
    (1978).
12. Kakiuchi, S., Yamazaki, R. and Nakajima, H., Proc. Japan
    Acad., $\underline{46}$, 589-592 (1970).
13. Cheung, W.Y., Biochem. Biophys. Res. Commun., $\underline{38}$, 533-
    538 (1970).
14. Cheung, W.Y., J. Biol. Chem., $\underline{246}$, 2859-2869 (1971).
15. Brostrom, C.O., Huang, Y.C., Breckenridge, B.M. and Wolff,
    D.J., Proc. Natl. Acad. Sci. USA, $\underline{72}$, 64-68 (1975).
16. Gopinath, R.M. and Vicenzi, F.F., Biochem. Biophys. Res.
    Commun., $\underline{77}$, 1203-1209 (1977).
17. Jarrett H.W. and Penniston, J.J., Biochem. Biophys. Res.
    Commun., $\underline{77}$, 1210-1216 (1977).
18. Marcum, J.M., Dedman, J.R., Brinkley, B.R. and Means,
    A.R., Proc. Natl. Acad. Sci. USA, $\underline{75}$, 3771-3775 (1978).

19. Yagi, K., Yazawa, M., Kakuichi, S., Oshimo, M. and Uenishi, K., J. Biol. Chem., 253, 1338-1340 (1978).
20. Drabowska, R., Aromatoril, O.D., Sherry, J.M.F. and Hartshorne, D.J., Biochem. Biophys. Res. Commun., 78, 1263-1272 (1977).
21. Anderson, J.M. and Cormier, M.J., Biochem. Biophys. Res. Commun., 84, 595-602 (1978).
22. Vanaman, T.C., Sharief, F. and Watterson, D.M., in "Calcium binding proteins and calcium function" (Wasserman et al, eds.), p. 107-116, Elsevier/North Holland, Amsterdam/New York (1977).
23. Dedman, J.R., Jackson, R.L., Schreiber, W.E. and Means, A.R., J. Biol. Chem., 253, 343-346 (1978).
24. Grand, R.J.A. and Perry, S.V., FEBS Lett., 92, 137-142, (1978).
25. Vanaman, T.C., Sharief, F., Awramik, J.L., Mendel, P.A. and Watterson, D.M., in "Contractile systems in non-muscle tissues" (S.V. Perry et al, eds.), p. 165-174, Elsevier/North Holland, Amsterdam/New York (1976).
26. Amphlett, G.W., Vanaman, T.C. and Perry, S.V., FEBS Lett., 72, 163-167 (1976).
27. Dedman, J.R., Potter, J.D. and Means, A.R., J. Biol. Chem., 252, 2437-2440 (1977).
28. Kilimann, M.W. and Heilmeyer, L.M.G., Eur. J. Biochem., 73, 191-197 (1977).
29. Levin, R.M. and Weiss, B., Molec. Pharmacol., 13, 690-697 (1977).
30. Soderling, T.R., Hickenbottom, J.P., Reimann, E.M., Hunkeler, F.L., Walsh, D.A. and Krebs, E.G., J. Biol. Chem., 245, 6617-6628 (1970).
31. Schlender, K.K., Wei, S.H. and Villar-Palasi, C., Biochim. Biophys. Acta, 191, 272-278 (1969).
32. Proud, C.G., Rylatt, D.B., Yeaman, S.J. and Cohen, P., FEBS Lett., 80, 435-442 (1977).
33. Nimmo, H.G. and Cohen, P., FEBS Lett., 47, 162-167 (1974).
34. Rylatt, D.B. and Cohen, P., FEBS Lett., in the press (1979).
35. Rylatt, D.B., Embi, N. and Cohen, P., FEBS Lett., in the press (1979).
36. Roach, P.J., Delpoli-Roach, A.A. and Larner, J., J. Cyc. Nuc. Res., 4, 245-257 (1978).
37. Brostrom, C.O., Hunkeler, F.L. and Krebs, E.G., J. Biol. Chem., 246, 1961-1967 (1971).
38. Nimmo, H.G., Proud, C.G. and Cohen, P., Eur. J. Biochem., 68, 21-30 (1976).
39. Titani, K., Cohen, P., Walsh, K.A. and Neurath, H., FEBS Lett., 55, 120-123 (1975).

40. Titani, K., Koide, A., Hermann, J., Ericsson, L.H., Kumor, S., Wade, R.D., Walsh, K.A., Neurath, H. and Fischer, E.H., Proc. Natl. Acad. Sci. USA, <u>74</u>, 4762-4766 (1977).

41. Huang, T.S. and Krebs, E.G., FEBS Lett., in the press (1979).

42. Nimmo, H.G., Proud, C.G. and Cohen, P., Eur. J. Biochem., <u>68</u>, 31-44.

43. Huang, F.L. and Glinsmann, W.H., FEBS Lett., <u>62</u>, 326-329 (1976).

44. Huang, F.L. and Glinsmann, W.H., Eur. J. Biochem., <u>70</u>, 419-426 (1976).

45. Cohen, P., Rylatt, D.B. and Nimmo, G.A., FEBS Lett., <u>76</u>, 182-186 (1977).

46. Nimmo, G.A. and Cohen, P., Eur. J. Biochem., <u>87</u>, 341-351 (1978).

47. Nimmo, G.A. and Cohen, P., Eur. J. Biochem., <u>87</u>, 353-365 (1978).

48. Shenolikar, S., Foulkes, J.G. and Cohen, P., Biochem. Soc. Trans., <u>6</u>, 935-937 (1978).

49. Foulkes, J.G. and Cohen, P., in "Hormones and cell regulation", <u>3</u>, in the press (H.G. van der Molen, J.E. Dumont and J. Menez, eds.), Elsevier/North Holland, Amsterdam/New York (1979).

50. Shenolikar, S. and Cohen, P., FEBS Lett., <u>86</u>, 92-98 (1978).

51. Walsh, D.A., Perkins, J.P., Brostrom, C.O., Ho, E.S. and Krebs, E.G., J. Biol. Chem., <u>246</u>, 1968-1976 (1971).

52. Steinberg, D., Adv. Cyc. Nuc. Res., <u>7</u>, 157-198 (1976).

53. Beckett, G.J. and Boyd, G.S., Biochem. Soc. Trans., <u>3</u>, 892-894 (1975).

54. Hardie, D.G. and Cohen, P., FEBS Lett., <u>91</u>, 1-7 (1978).

55. Nimmo, H.G. and Houston, B., Biochem. J., <u>176</u>, 607-610 (1978).

56. Ingebritsen, T.S., Lee, H.S., Parker, R.A. and Gibson, D.M., Biochem. Biophys. Res. Commun., <u>81</u>, 1268-1277 (1978).

57. Farrell, P.J., Balkow, K., Hunt, T. and Jackson, R.J., Cell, <u>11</u>, 187-200 (1977).

58. Engstrom, L., Curr. Top. Cell. Reg., <u>13</u>, 29-51 (1978).

59. Krebs, E.G., Curr. Top. Cell. Reg., <u>5</u>, 99-133 (1972).

60. Kebabian, J.W., Adv. Cyc. Nuc. Res., <u>8</u>, 421-508 (1977).

61. Blackmore, P.F., Brumley, F.T., Marks, J.L. and Exton, J.H., J. Biol. Chem., <u>253</u>, 4851-4858 (1978).

62. Adelstein, R.S., Conti, M.A., Hathaway, D.R. and Klee, C.B., J. Biol. Chem., in the press (1978).

63. Burchell, A., Foulkes, J.G., Cohen, P.T.W., Condon, G. and Cohen, P., FEBS Lett., <u>92</u>, 68-72 (1978).

64. Severson, D.L., Khoo, J.C. and Steinberg, D., J. Biol. Chem., 252, 1484-1489 (1977).
65. Halestrap, A.P. and Denton, R.M., Biochem. J., 132, 509-517 (1973).
66. Cook, G.A., Nielsen, R.C., Hawkins, R.A., Mehlman, M.A., Lakshmanan, M.R. and Veech, R.C., J. Biol. Chem., 252, 4421-4424 (1977).
67. Bloch, K. and Vance, D., Ann. Rev. Biochem., 46, 263-298 (1977).
68. Carlson, C.A. and Kim, K., J. Biol. Chem., 248, 378-380 (1973).
69. Smith, S. and Williamson, I.P., Biochem. Soc. Trans., 5, 737-739 (1977).
70. Carlson, C.A. and Kim, K., Arch. Biochem. Biophys., 164, 478-489 (1974).
71. Hardie, D.G. and Cohen, P., Eur. J. Biochem., in the press (1979).

DISCUSSION

H. SEGAL: My question is in regard to the state of phosphorylation of glycogen synthetase. You mentioned that there were separate sites of phosphorylation for cyclic AMP dependent protein kinase and for the calcium dependent kinase. If I understand your experiments what you do is prepare an unphosphorylated synthetase and then phosphorylate it specifically with one kinase or another to study the dephosphorylation. My question is - in vivo - in a starved animal, for example, from which one can prepare a synthetase b, what is the state of phosphorylation of the enzyme obtained under those conditions?

P. COHEN: That is a very good question. What Dr. Segal said is correct. In our experiments on phosphatase specificity we have only used substrates phosphorylated with either cyclic AMP dependent protein kinase or glycogen synthetase kinase-3. If one combined both of these kinases one might produce a substrate that would require a different protein phosphatase. Protein phosphatase-I can dephosphorylate the ^{32}p-labelled synthase substrates we have used, but this does not rule out the possibility that the generation of a different substrate may require a different enzyme for the dephosphorylation of that substrate.

Now in regard to the state of phosphorylation of glycogen synthetase in vivo, our experiments so far indicate the

following: in normal fed animals glycogen synthetase contains
2 moles of phosphate per subunit and when one injects adrenaline the phosphate content goes up to 3 molecules per subunit. Now the question is, on what sites are those
phosphates? We have some evidence that in the normal-fed
state, most of the phosphate is not in sites labeled by
cyclic AMP dependent Protein kinase and that these sites only
become highly phosphorylated when adrenaline is injected. In
other words the increase from 2 moles of phosphate to 3 moles
of phosphate involves the phosphorylation of the sites
labeled by the cyclic AMP dependent protein kinase, as might
be expected. However, it is not clear whether glycogen synthetase kinase-2 (phosphorylase kinase) or glycogen synthetase kinase-3, or even conceivably some other enzyme, is
involved in the phosphorylation of the other sites observed
in the absence of adrenalin.

E. KREBS: I was not convinced that the evidence that phosphorylase kinase-catalyzed phosphorylation of glycogen
synthetase will prove to be physiologically significant. You
mentioned that the rate is very slow. I did not follow your
reasoning that due to the fact that the glycogen synthetase
concentration is less than that of phosphorylase, maybe one
only needs a slow rate. This could even make matters worse
because it would mean that with a very low concentration of
glycogen synthetase the reaction would be even slower because
you would not saturate the kinase with substrate.

P. COHEN: I think that is a valid point. It is clearly
important to carry out detailed comparative studies of the
kinetics of phosphorylation of glycogen synthetase and
phosphorylase kinase. However, at physiological concentrations of glycogen synthase (0.3 mg/ml) and phosphorylase (8
mg/ml) the relative rates of phosphorylation synthase/phos-
phorylase is about 1/20. However, the half time for the
phosphorylation of glycogen synthase is only 2-fold slower
than phosphorylase. This suggests that the phosphorylation
of glycogen synthase by phosphorylase kinase may be physiologically significant. Of course the crucial question is
whether electrical stimulation or adrenaline administration
promotes the phosphorylation of serine-7 in vivo. The
location of this site so close to the N-terminus of the
enzyme should make this analysis quite simple.

M. SINGH: My question has to do with your work on the Acetyl
CoA carboxylase. I understand that the enzyme is phosphorylated already and I also see that phosphorylation is supposed

to ultimately inactivate the enzyme. In terms of regulation, does it mean that there is a critical number of PO_4 groups to be put on before the enzyme is inactivated?

P. COHEN: The project is still at a preliminary stage, but the information we have at present is the following. If the purified enzyme which contains 3.5 molecules of phosphate per subunit is treated with protein phosphatase-1, the activity goes up 2.5-3 fold and the phosphate content decreases to about 2.5 molecules of phosphate per subunit.

As I mentioned, if you take the molecule with 3.5 molecules of phosphate per subunit, you can demonstrate the presence of 3 different protein kinase activities which phosphorylate the enzyme, but these additional phosphorylations do not seem to affect the activity directly. At the moment, we would like to think in terms of those phosphorylations controlling the rate of phosphorylation and dephosphorylation of the other sites, but I must emphasize that the work is still at an early stage and these ideas remain to be tested.

F. AHMAD: I am glad to note that your results on rabbit mammary gland acetyl CoA carboxylase and fatty acid synthetase are very similar to those reported by us on the carboxylase and synthetase which are purified from rat mammary gland. We found approximately 6 moles of phosphate per carboxylase protomer (M_r 240,000) and still these preparations had a specific activity of approximately 15 units/mg of protein. On the other hand you found approximately 3-4 moles of phosphate in rabbit mammary gland carboxylase but these preparations had specific activity of approximately 4 units/mg of protein which could be increased about 3 fold upon dephosphorylation. Have you been able to inactivate the dephosphorylated enzyme by phosphorylation?

P. COHEN: There are several points there. First, when I said the 3.5 molecules of phosphate per subunit, only the alkali labile phosphate bound covalently to the enzyme was measured, which is an important point. Second, the activation of acetyl CoA carboxylase which occurred upon the addition of protein phosphatase-1 was completely blocked by the specific protein phosphatase inhibitor, termed inhibitor-2, and the activation was accompanied by a decrease in the alkali labile phosphate content of 1 mole per subunit. This provides very strong evidence that this is indeed a dephosphorylation reaction. In terms of what enzyme(s) would rephosphorylate the activated enzyme, I would like to reserve judgement on that for the moment.

PROTEIN PHOSPHATASE C: PROPERTIES, SPECIFICITY
AND STRUCTURAL RELATIONSHIP TO A LARGER HOLOENZYME

Ernest Y.C. Lee,[1] James H. Aylward,[2] Ronald L. Mellgren[3]
and S. Derek Killilea[4]

Department of Biochemistry
University of Miami School of Medicine
Miami, Florida

INTRODUCTION

The interconversion of proteins between phosphorylated and
dephosphorylated forms has become recognized as an important
regulatory mechanism.   The concepts that have emerged have
developed largely from extensive studies of the enzymes of
glycogen metabolism, particularly those from rabbit skeletal
muscle (for reviews, see Rubin and Rosen, 1975; Nimmo and Cohen,
1978; Larner and Villar-Palasi, 1971; Roach and Larner, 1976;
Walsh and Krebs, 1973). While the focus of much of the research
has been on the phosphorylating enzymes, the protein kinases,
rather less is known of the properties and regulation of the
protein phosphatases.   The enzymology of the protein
phosphatase(s) which serve to dephosphorylate phosphorylase

[1]Studies presented in this paper were supported by NIH grant
AM18512, and in part by AM12532, and a grant from the American
Heart Association, Florida Affiliate Inc.   E.Y.C. Lee is an
Established Investigator of the American Heart Association.

[2]Present Address: Department of Physiology, Vanderbilt
University, Nashville, Tennessee

[3]Present Address: Department of Pharmacology, Medical College
of Ohio, Toledo, Ohio

[4]Present Address: Department of Biochemistry, North Dakota
State University, Fargo, North Dakota.

kinase, glycogen synthase and phosphorylase is still poorly understood. The major question which is raised is whether a single enzyme serves to dephosphorylate these phosphorylated enzymes, or whether discrete, specific phosphatases exist. The study of the protein phosphatase activities has been difficult, owing to difficulties of isolation and the occurrence of a variety of multiple enzyme forms. (for reviews, see Lee et al. 1976, Lee et al. 1978). In addition, the presence of multiple phosphorylation sites in glycogen synthase and phosphorylase kinase further complicates the issue.

Nevertheless, there bas been considerable progress in the study of the protein phosphatases during the past 5 years. These advances have largely concerned the fact that a number of protein phosphatase preparations have been unequivocally shown to act on several phosphoprotein substrates, and that extensive progress has been made in the isolation of at least one of these protein phosphatase activities. Our laboratory has been interested in the isolation and characterization of phosphorylase phosphatase per se, and in this paper we will review the information developed on this system, with an emphasis on a low molecular weight enzyme form which we have named protein phosphatase C.

Phosphorylase Phosphatase/Protein Phosphatase C.

Phosphorylase phosphatase was discovered as an enzymatic activity in rabbit muscle extracts which catalysed the conversion of phosphorylase a to phosphorylase b (Cori and Green, 1943). The nature of the interconversion was only later shown to involve phosphorylation and dephosphorylation of phosphorylase (Krebs and Fischer, 1956; Wosilait and Sutherland, 1956). Our initial attempts were directed at the isolation of the enzyme from rabbit liver. These early attempts were frustrated until it was discovered that treatment of rabbit liver extracts with high concentrations of ethanol at room-temperature led to the activation of phosphorylase phosphatase activity and its conversion to a discrete lower molecular weight form of M_r 35,000 (Brandt et al. 1974). The experimental demonstration of this effect is shown in Fig. 1. This low molecular weight form was readily amenable to further purification and a procedure was developed which allowed the separation of the enzyme in a homogeneous form from rabbit liver (Brandt et al. 1975a). The enzyme was characterized as having a molecular weight of ca. 35,000 by gel filtration and by SDS disc gel electrophoresis. The procedure which was developed also allowed the purification of the enzyme from rabbit skeletal muscle (Brandt et al. 1975a) and from bovine heart (Killilea et al. 1978). This M_r 35,000 phosphatase appears to be distributed

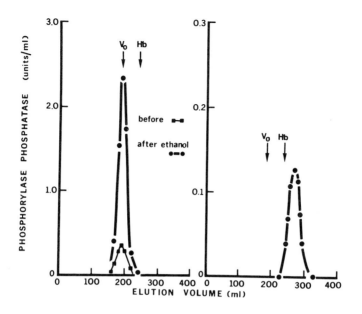

Fig. 1. The Effects of Ethanol Treatment on the Behavior
of Rat Liver Phosphorylase Phosphatase Activity on Sephadex G-
75 Chromatography.

Left Panel: A rat liver extract was chromatographed on a
Sephadex G-75 column. Fractions were assyed for phosphorylase
phosphatase activity (squares). Each fraction was subjected to
treatment with 5 volumes of ethanol and then assayed for
phosphorylase phosphatase activity (circles).

Right Panel: The void volume fractions (left panel) were pooled
and treated with ethanol, and then re-chromatographed and the
fractions assayed for phosphorylase phosphatase activity.

in both liver and muscle tissues, and it is also present in
tissues from different mammalian species. We have been able to
demonstrate the presence of the M_r 35,000 enzyme form in the
liver, muscle and heart extracts of rat tissues after ethanol
treatment and believe it likely that it will prove to be widely
distributed in mammalian tissues. These studies provided the
first example of a protein phosphatase involved in glycogen
metabolism which has been isolated (i.e. purified to apparent
homogeneity). We have named the enzyme protein phosphatase C

since, as discussed later, it is not specific for phosphorylase
a and because it appears to be the catalytic subunit of a larger
holoenzyme. Some properties of the enzyme found in this
laboratory are reported in Table I.

Table I

Properties of Phosphorylase Phosphatase Isolated as the Low

Molecular Weight Form

Tissue	Specific Activity	Molecular Weight	Reference
Rabbit liver	2,100[a]	35,000	Brandt et al. (1975a)
Rabbit muscle	755[a]	35,000	Brandt et al. (1975a)
Bovine heart	7,900[a]	35,000	Killilea et al. (1978)
Rabbit liver	5,130[b] 504[b]	34,000 30,500	Khandelwal et al. (1976)
Rabbit muscle	1,500[b]	33,000	Gratecos et al. (1977)

[a]Expressed as nmoles phosphorylase a dimer converted per minute
per mg protein

[b]Expressed as nmoles Pi released per minute per mg protein

Two other procedures have been reported for the isolation
of phosphorylase phosphatase as a form of M_r ca. 35,000 (Table
I). Khandelwal et al. (1976) reported the isolation of two
forms, of M_r 34,000 and 30,500 respectively. The larger form
resembles in most respects the enzyme isolated by Brandt et al.
(1975a) and is probably identical. The smaller form had a lower
specific activity toward phosphorylase a than the M_r 35,000
form. Gratecos et al. (1977) have reported a procedure for the
purification of rabbit muscle phosphorylase phosphatase as a
form of M_r ca. 33,000.

The studies from our laboratory (Killilea et al. 1976a 1976b) and from Krebs and co-workers (Khandelwal et al. 1976) of rabbit liver phosphorylase phosphatase allowed the first definitive demonstration using purified enzymes that the enzyme had an intrinsic activity toward other phosphoproteins, notably glycogen synthase and phosphorylase kinase. Thus far there have been no exceptions in our experience to the ability of protein phosphatase C to dephosphorylate proteins which have been phosphorylated by either phosphorylase kinase or cAMP-dependent protein kinase (Lee et al. 1978). A list of phosphoproteins which are substrates for liver phosphorylase phosphatase (protein phosphatase C) is shown in Table II. These studies established rigorously the enzymatic basis for the concept that a single protein phosphatase might mediate the dephosphorylation of the phosphoenzymes of glycogen metabolism.

Table II

Substrates for Liver Protein Phosphatase C

Phosphorylase	A,B
Phosphorylase kinase	A,B
R-subunit of cAMP-dependent protein kinase	A
Glycogen synthase	A,B
Hormone sensitive lipase	C
Troponin I	B
Pyruvate kinase	D
HMG-CoA reductase	E
Histone	A,B
Casein	B

References. A, Lee et al. 1976; B, Khandelwal et al. 1976;

C, Severson et al. 1977; D, Titanji, 1977;

E, Ingebritsen et al. 1978; Ingebritsen et al., this volume.

It should be noted that similar evidence had been developed
earlier with partially purified protein phosphatase
preparations (for reviews, see Lee et al. 1976, 1978). These
include studies of protein phosphatase activities which were
examined originally as synthase phosphatases (Kato and Bishop,
1972; Nakai and Thomas, 1974; Zieve and Glinsmann, 1973) and
which provided the first evidence for protein phosphatases of
broad specificity. Thus, it appears that phosphorylase
phosphatase, identified as the form we describe as protein
phosphatase C, is a relatively "non-specific"
phosphomonoesterase. This apparent lack of specificity,
however, must be viewed within the perspective which has been
developed by studies of the cAMP-dependent protein kinase
(Krebs et al., this volume). These studies have shown that the
ability of cAMP-dependent protein kinase to phosphorylate a
variety of unrelated proteins is not due to a lack of
specificity but can now be interpreted in some detail in terms
of specificity toward a limited peptide sequence. Indeed, from
this point of view one would not expect otherwise than that a
protein phosphatase which dephosphorylates a particular
phosphoprotein would also dephosphorylate other proteins
phosphorylated by the same protein kinase, i.e. the protein
phosphatases might display specificity toward a similar peptide
sequence. At the present time no such information is available
regarding protein phosphatase C. However, since it
dephosphorylates proteins which are substrates of both
phosphorylase kinase and cAMP-dependent protein kinase it is
still open to question whether it possesses some peptide site
specificity or whether it is indeed relatively nonspecific.

The enzymes of glycogen metabolism have been most
extensively characterized from rabbit muscle, and most of our
detailed knowledge of their enzymology is derived from this
tissue, with the exception of the protein phosphatases. It may
be noted that the specificity of rabbit muscle protein
phosphatase C has not yet been examined, although the
specificity of a partially purified preparation named protein
phosphatase III has been investigated (Antoniw et al. 1977;
Nimmo and Cohen, 1978).

The ability of protein phosphatase C and of a variety of
less well defined protein phosphatase preparations to act on
more than one of the phospho-enzymes of the glycogen system
raises the question of whether a single "multifunctional"
phosphatase exists. We have pointed out that the elucidation of
the specificity of protein phosphatase C provides the enzymic
basis for this possibility (Lee et al. 1978), but this question
must remain open at the present time due to our insufficient

knowledge of the enzymology of these enzymes. There is considerable uncertainty as to the nature and number of protein phosphatase enzymes in any one tissue that has been studied, and a number of preparations of rather different apparent physical properties have been described (Lee et al. 1978), all of which display activity toward several phosphoprotein substrates. Also, there are some reports which indicate that the major liver protein phosphatase activities toward glycogen synthase and phosphorylase may be separated (Tan and Nuttall, 1978; Kikuchi et al. 1977). Thus, unless we can assess that the enzyme which has been isolated a) represents the native enzyme and b) represents the major or sole enzymatic activity toward all its potential substrates, one cannot directly extend the in vitro findings to a physiological role. This is also true for the enzyme form we have described as protein phosphatase C, as will be discussed in more detail below. A final comment on the potential role of protein phosphatase C is that it may have broader functions than the control of glycogen metabolism, in terms of its capacity to reverse the actions of cAMP-dependent protein kinase (Killilea et al. 1976c).

On the Absence of Protein Phosphatase C in Tissue Extracts

As has been noted above (Fig. 1) the ethanol treatment which was used during the purification of protein phosphatase C leads to the dissociation of the enzyme to the M_r 35,000 form. In addition, treatment with trypsin, urea (Killilea et al. 1979) or a freeze-thaw in 0.2 M mercaptoethanol (Kato et al. 1974) were also found to bring about the same result. However, it should be noted that only the M_r 35,000 form derived after ethanol treatment has been isolated, and there is no certainty that this is in fact identical to the forms derived after either urea, trypsin or mercaptoethanol treatment. The study of this phenomena of dissociation (accompanied by large increases in activity) prompted us as well as Kato and co-workers (Brandt et al. 1974; Kato and Sato, 1974) to propose that phosphorylase phosphatase might exist as a complex of which the M_r 35,000 form is the catalytic subunit. The increase in activity associated with the reduction in molecular weight during these treatments suggested the removal of an inhibitory subunit. These considerations led to the discovery of heat-stable protein inhibitors of phosphorylase phosphatase activity (Brandt et al. 1975b).

---

[5]These protein inhibitors have since been extensively studied in rabbit muscle (Huang and Glinsmann 1976; Nimmo and Cohen, 1977; Cohen, this volume). Ironically there is no evidence as yet that these protein inhibitors are in fact associated as components of the holoenzyme form of phosphorylase phosphatase.

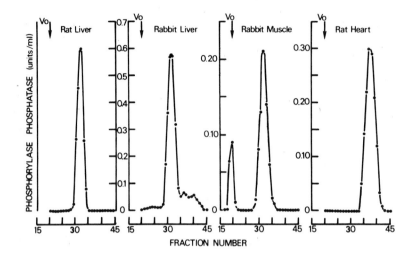

Fig. 2. Gel Filtration Behavior of Phosphorylase Phosphatase Activities in Various Tissue Extracts.

Liver and muscle extracts were obtained as 100,000 x g supernatants and were chromatographed on Bio-Gel A-0.5m (exclusion limit = 500,000). The rabbit muscle extract was a 20,000 x g supernatant. Fractions were assayed for phosphorylase phosphatase activity as by Brandt et al. (1975a).

These experimental findings were consistent with the hypothesis that phosphorylase phosphatase originated as a "holoenzyme" consisting of catalytic and inhibitory subunits. This hypothesis requires that phosphorylase phosphatase should be found to exist in a discrete high molecular weight form. Studies of the gel filtration behavior of phosphorylase phosphatase activity in crude tissue extracts revealed the existence of such a form (Lee et al. 1976; Lee et al. 1978; Killilea et al. 1979). A representative result is shown in Fig. 2. In liver, skeletal muscle or heart muscle, a single major molecular weight form of phosphorylase phosphatase activity was detected. In all tissues examined this form eluted with an apparent molecular weight of ca. 260,000, and could be converted to the M_r 35,000 form by ethanol treatment. What is also striking about these experimental findings (Fig.2) is the absence of significant amounts of lower molecular weight forms.

This is in contrast to the experience of most other Laboratories. We can rationalize this in terms of the hypothesis that the holoenzyme contains both non-catalytic and catalytic subunits, and that partial proteolysis of the non-catalytic subunits may give rise to enzyme species intermediate between the M_r 260,000 form and the M_r 35,000 form. Evidence for this has been found for both liver (Killilea et al. 1979) and rabbit skeletal muscle phosphatase (Mellgren et al., 1979). In liver extracts we have observed that vigorous homogenization of the liver leads to the appearance of multiple molecular weight enzyme forms, presumably because of the action of lysosomal proteases. In rabbit muscle extracts we have been able to show a stepwise conversion of the native enzyme to smaller forms by the action of neutral Ca^{++}-dependent proteases.

All the studies reported above are consistent with the holoenzyme hypothesis but do not constitute rigorous evidence. Nevertheless, these studies point inescapably to the existence of a native enzyme form which is larger than any of the enzyme forms which have been studied to date, although this was clearly recognized in our original report on the properties of protein phosphatase C (Brandt et al. 1975a).

Structural Relationship of Protein Phosphatase C to a Larger Holoenzyme

The hypothesis which we have advanced for the relationship of protein phosphatase C to the larger native enzyme from which it is derived is shown diagrammatically in Fig. 3. This hypothesis also provides an explanation for much of the phenomenology associated with the behavior of phosphorylase phosphatase. This can be explained by assuming a greater susceptibility of the putative inhibitor subunit to denaturants or to proteolysis. Final evidence for this must come from isolation of the native holoenzyme, which we have named "protein phosphatase H".

Attempts to isolate "protein phosphatase H" as a high molecular weight form of phosphorylase phosphatase have only been partially successful (Lee et al. 1978). A high molecular weight enzyme was extensively purified from rat liver. Despite the fact that the preparation contained a single molecular weight form of activity, multiple forms were observed by both ion-exchange chromatography and by disc gel electrophoresis. The latter procedure showed three major protein bands, all of which were associated with enzyme activity. On SDS disc gel electrophoresis three major polypeptide bands were observed, of M_r 65,000, 55,000 and 35,000 respectively (Fig. 4). The

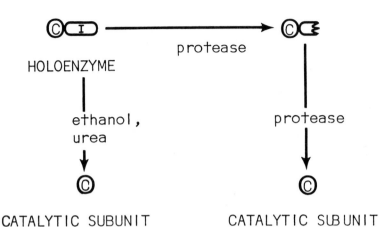

Fig. 3. Scheme Showing the Proposed Relationship Between Protein Phosphatase C and Its Putative Precursor.

In this scheme the holoenzyme is represented as a complex containing dissimilar subunits (here simply represented as consisting of one of each type). The "C" subunit is the catalytic subunit of M_r 35,000 which has been isolated as "protein phosphatase C". "I" is the putative regulatory subunit. In this scheme the "C" subunit is released by treatments with ethanol or urea by selective denaturation of the "I" subunit. Proteolysis of the holoenzyme leads to partial breakdown of the "I" subunit, giving rise to intermediate enzyme species and finally the free "C" subunit.

relative amount of the M_r 55,000 component appeared to be variable whereas that of the M_r 65,000 and 35,000 bands were present in a relatively constant ratio. This ratio approximated a 1:2 ratio of M_r 65,000 and M_r 35,000 polypeptides. We have speculated that the holoenzyme consists of one M_r 65,000 subunit and two of the M_r 35,000 subunit (Lee et al. 1978).

On this basis the holoenzyme would have a molecular weight of 135,000. The purified enzyme had an estimated molecular weight of 130,000 (the apparent molecular weight on gel filtration was 225,000). These results (Lee et al. 1978) can only be regarded as preliminary evidence. However, these

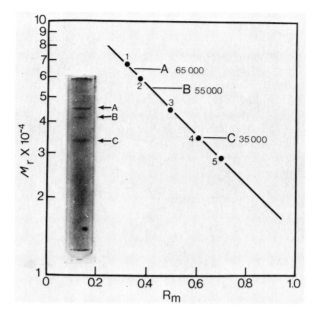

*Fig. 4.   SDS Disc Gel Electrophoresis of Rat Liver Protein
Phosphatase*

*A purified preparation of rat liver phosphorylase phosphatase
was subjected to SDS disc gel electrophoresis. The bands marked
A, B, and C respectively were found to have molecular weights as
indicated. The standards used were 1) bovine serum albumin, 2)
catalase, 3) ovalbumin, 4) lactate dehydrogenase, 5) carbonic
anhydrase.*

studies do show that phosphorylase phosphatase can be
extensively purified in a high molecular weight form, and that
the M_r 35,000 enzyme form which we had previously characterized
is physically present as a component of the larger enzyme.

We have also attempted to isolate a high molecular weight
protein phosphatase from rabbit muscle using phosphorylated
histone as a primary substrate (Aylward et al. 1978). The final
preparation which was obtained was active toward both
phosphorylase a and histone. The preparation contained one
major band on disc gel electrophoresis which was associated
with phosphatase activity toward histone and phosphorylase a.

On gel filtration the activity behaved with an apparent molecular weight of 250,000 (Stokes radius 54.5A). On sucrose density ultracentrifugation a sedimentation coefficient of 6.9 was obtained. The estimated molecular weight based on these data was 156,000. On SDS disc gel electrophoresis three polypeptide components corresponding to M_r 65,000, 55,000 and 35,000 were obtained(Fig. 5). The qualitative data are remarkably similar to those obtained with the rat liver preparations. Again, the evidence allows only the conclusion that the M_r 35,000 catalytic species is physically present in a larger enzyme form, and the final elucidation of the exact nature and stoichiometry of the subunits of native protein phosphatase H will depend on its rigorous isolation.

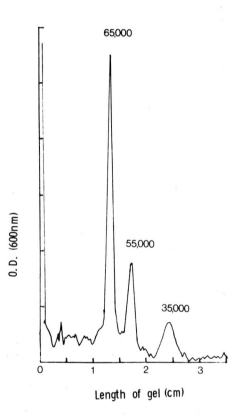

Length of gel (cm)

Fig. 5.   Densitometric Scan of Rabbit Muscle Protein Phosphatase After SDS Disc Gel Electrophoresis.

A sample of purified rabbit muscle protein phosphatase was run on SDS disc gel electrophoresis. The gel was scanned and the relative molecular weights of the three major peaks determined with appropriate standards.

Summary

There is now considerable evidence that both liver and muscle contain a protein phosphatase of discrete properties which can be isolated as a form of M_r 35,000, termed protein phosphatase C. While this enzyme has been studied largely by its action on phosphorylase a, it is able to dephosphorylate a variety of phosphoprotein substrates. It has been found that phosphorylase phosphatase exists as a high molecular weight form which contains protein phosphatase C as a catalytic subunit. Our current hypothesis is that the native holoenzyme also contains subunit(s) which have a regulatory or inhibitory function. The view that protein phosphatase C may serve a multifunctional role, and may be released from a less-active holoenzyme are shown diagrammatically in Fig. 6. It is purely a speculative scheme, since a physiologically relevant dissociation of protein phosphatase H has yet to be demonstrated and its enzymology is also not well-defined. Nevertheless, this scheme illustrates the concept of a single protein phosphatase which would be a focal point for the control of the dephosphorylation processes of glycogen metabolism, and also indicates that the role of the protein phosphatase may be seen as that of an "antagonist" to protein kinase with potentially wider functions than those restricted to glycogen metabolism (Killilea et al. 1976c).

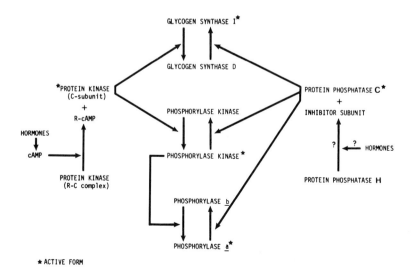

Fig. 6.   Scheme of Enzyme Interconversions in Glycogen Metabolism.

REFERENCES

Antoniw, J.F., Nimmo, H.G., Yeaman, S.J. and Cohen, P. (1977) Biochem. J. 162, 423-433.

Aylward, J.H., Mellgren, R.L. and Killilea, S.D. (1978) Fed. Proc. 37, 1808.

Brandt, H., Killilea, S.D. and Lee, E.Y.C. (1974) Biochem. Biophys. Res. Commun. 61, 598-604.

Brandt, H., Capulong, Z.L. and Lee, E.Y.C. (1975a) J. Biol. Chem. 250, 8038-8044.

Brandt, H., Lee, E.Y.C. and Killilea, S.D. (1975b) Biochem. Biophys. Res. Commun. 63, 950-956.

Cori, G.T. and Green, A.A. (1943) J. Biol. Chem. 151, 31-38.

Gratecos, D., Detwiler, T.C., Hurd, S. and Fischer, E.M. (1977) Biochemistry, 16, 4812-4917.

Huang, F.L. and Glinsmann, W.H. (1976) Eur. J. Biochem. 70, 419-426.

Ingebritsen, T.S., Lee, H.S., Parker, R.A. and Gibson, D.M. (1978) Biochem. Biophys. Res. Commun. 81, 1268-1277.

Kato, K. and Bischop, J.S. (1972) J. Biol. Chem. 247, 7420-7429.

Kato, K. and Sato, S. (1974) Biochim. Biophys. Acta 358, 299-307.

Kikuchi, K., Tamura, S., Hiraga, A. and Tsuiki, S. (1977) Biochem. Biophys. Res. Commun. 75, 29-37.

Killilea, S.D., Brandt, H., Lee, E.Y.C. and Whelan, W.J. (1976a) J. Biol. Chem. 251, 2363-2368.

Killilea, S.D. and Whelan, W.J. (1976b) Biochemistry, 15, 1349-1356.

Killilea, S.D., Brandt, H. and Lee, E.Y.C. (1976c) Trends Biochem. Sci. 1, 30-33.

Killilea, S.D., Aylward, J.H., Mellgren, R.L. and Lee, E.Y.C. (1978) Arch. Biochem. Biophys. 191, 638-646.

Killilea, S.D., Mellgren, R.L., Aylward, J.H. and Lee, E.Y.C. (1979) Arch. Biochem. Biophys. 193, 130–139.

Krebs, E.G.and Fischer, E.H. (1956) Biochim. Biophys. Acta, 20, 150–157.

Larner, J. and Villar-Palasi, C. (1971) Curr. Top. Cellu. Regul. 3, 195–236.

Lee, E.Y.C., Brandt,H., Capulong, Z.L. and Killilea, S.D. (1976) Adv. Enzyme Regul. 14, 467–490.

Lee, E.Y.C., Mellgren, R.L., Killilea, S.D. and Aylward, J.H. (1978) In "Regulatory Mechanisms of Carbohydrate Metabolism" (V. Esmann, ed.), FEBS Symposia 42, 327–346.

Mellgren, R.L., Aylward, J.H., Killilea, S.D. and Lee, E.Y.C. (1979) J. Biol. Chem. 254, 648–652.

Nakai, C. and Thomas, J.A. (1974) J. Biol. Chem. 249, 6459–6467.

Nimmo, H.G. and Cohen, P. (1977) Adv. Cyclic Nucl. Res. 8, 145–266.

Rubin, C.S. and Rosen, O.M. (1975) Annu. Rev. Biochem. 44, 831–887.

Severson, D.L., Khoo, J.C. and Steinberg, D. (1977) J. Biol. Chem. 252, 1484–1489.

Tan, A.W.H. and Nuttall, F.Q. (1978) Biochim. Biophys. Acta 522, 139–150.

Titanji, V.P.K. (1977) Biochim. Biophys. Acta 481, 140–151.

Wosilait, W.D. and Sutherland, E.W. (1956) J. Biol. Chem. 218, 469–481.

Walsh, D.A. and Krebs, E.G. (1973) The Enzymes, 8, 555–581.

Zieve, F.J. and Glinsmann, W.H. (1973) Biochem. Biophys. Res. Commun. 50, 872–878.

DISCUSSION

H.L. SEGAL: When you reported on the phosphorylase phosphatase activity in tissue extracts, I presume you were referring to high speed supernatant fractions from liver. Now as you know, there are phosphatase activies associated with the glycogen pellet. I wonder whether you think that it is the same enzyme that is distributed between the soluble phase and the pellet or do you think the pellet-associated activities are a different enzyme entirely?

E.Y.C. LEE: I believe that the enzyme activities we generally work with are the cytosolic ones. The gel filtration profiles which I showed are of 100,000 xg supernatants. There is a possibility that in the glycogen pellet there may be protein phosphatase activities which differ and I believe that Dr. Cohen has generally worked with the glycogen pellet enzymes.

T. HUNT: You sounded as if you are skeptical about these inhibitor proteins that Dr. Cohen discussed. Have you actually tested either of them on the holoenzyme?

E.Y.C. LEE: No, I don't think that I would say that I am skeptical. What I wanted to do was to prevent any misconceptions. Our discovery of the heat stable protein inhibitors originally gave us a basis for suggesting a high molecular weight enzyme, but in fact, there isn't any direct evidence that they are integral components of the "holoenzyme". Perhaps Dr. Cohen would comment on this.

P. COHEN: In muscle extracts, inhibitor-1, the phosphorylated inhibitor, is entirely present in its dephosphorylated form unless precautions are taken. In its dephosphorylated form it does not interact at all with the phosphatase. It can be resolved by DEAE-cellulose and it does not centrifuge down with the glycogen pellet whereas the phosphatase does. So I would not expect inhibitor-1 to be a component of an enzyme one isolates in the way Dr. Lee describes. As regards inhibitor-2, I also agree that this is not a component of the most highly purified phosphatase preparations that we have. In fact, we now have inhibitor-2 pure and it is present in extremely small amounts. We needed another 10-fold purification over that which was required for the purification of inhibitor-1, about 40,000-fold purification was required. In molar terms it must be insufficient to be able to form stoichiometric complexes with a phosphatase in muscle extracts. However, either inhibitor-2 or an analagous protein does seem to be present in higher amounts in other tissues.

E.Y.C. LEE: We have not tested whether the holoenzyme is
inhibited by the protein inhibitors but Dr. Cohen probably has.

P. COHEN: In tissue extracts one can titrate the phosphorylase
phosphatase activity with those inhibitors in a very similar
way to the purified enzyme.   If the phosphatase is in a high
molecular weight form in these tissue extracts this would imply
that the high molecular weight enzyme can be inhibited by these
inhibitor proteins.

REGULATION OF PYRUVATE DEHYDROGENASE BY
PHOSPHORYLATION/DEPHOSPHORYLATION[1]

Philip J. Randle
Nancy J. Hutson[2]
Alan L. Kerbey
Peter H. Sugden[3]

Nuffield Department of Clinical Biochemistry
University of Oxford
Oxford, U.K.

I. INTRODUCTION

In animals glucose is oxidised by glycolysis to pyruvate
in the cytosol followed by intramitochondrial oxidation of
pyruvate to acetyl CoA catalysed by the pyruvate
dehydrogenase complex (PDH complex). In most tissues the
acetyl CoA so formed is oxidised in the tricarboxylate cycle.
In liver, and in some species adipose tissue, the acetyl CoA

---

[1]Work supported by Medical Research Council (UK) and by
the British Diabetic Association.
[2]U.S. Public Health Service Postdoctoral Fellow, Present
address: Department of Physiology, Milton S. Hershey
Medical Center, Hershey, Pa, 17033.
[3]Wellcome Fellow in Basic Medical Sciences.

so formed may give rise to cytosolic acetyl CoA which is synthesised into long-chain fatty acids. The PDH reaction (equation 1) is the major rate-determining reaction in

$$\text{pyruvate} + \text{CoA} + \text{NAD}^+ \longrightarrow \text{acetyl CoA} + CO_2 + \text{NADH} + H^+ \qquad (1)$$

glucose oxidation although it is not the only regulated reaction in the pathway.

Flux through the PDH reaction is decreased by starvation and increased by glucose feeding; this has been shown in vivo and in tissues in vitro. Flux is also decreased in severely diabetic animals (eg in alloxan-diabetes) and this change in flux is reversed by insulin treatment of the diabetic animal. In most in vitro tissue preparations insulin has no effect or only small effects on flux through the PDH reaction. The exception is rat adipose tissue in which flux is markedly increased by insulin in vitro. Flux through the PDH reaction is decreased in vitro by oxidation of fatty acids (heart and skeletal muscles, liver, kidney, adipocytes) and ketone bodies (heart and skeletal muscles, kidney, brain). This is a competition for oxygen consumption between alternative fuels (glucose and lipid fuels) in which oxidation of lipid fuels inhibits flux through the PDH reaction. There is evidence that effects of diabetes and of starvation on flux through the PDH reaction are the result of mobilisation and oxidation of lipid fuels. Insulin which increases flux, decreases release of fatty acids from adipocytes and lipolytic hormones decrease flux through the PDH reaction. This background information has recently been reviewed (Randle et al, 1979).

This paper is concerned with mechanisms which regulate the PDH complex and hence flux through the PDH reaction. Knowledge of mechanisms began with the observation that the PDH reaction is inhibited by acetyl CoA (competitive with CoA) and NADH (competitive with NAD$^+$). Acetyl CoA and NADH are common end products of the oxidation of pyruvate and of oxidation of fatty acids and ketone bodies. This finding and the observation that diabetes or in vitro oxidation of lipid fuels increases the concentration ratio of acetyl CoA/CoA suggested a mechanism for regulation of flux through the PDH reaction (Garland and Randle, 1964a,b).

Linn et al (1969a,b) observed that bovine PDH complexes are phosphorylated and inactivated with MgATP by a kinase (PDH kinase) intrinsic to the complex. Dephosphorylation and reactivation was effected by a mitochondrial pyruvate dehydrogenase phosphate phosphatase (PDHP phosphatase). These

reactions are shown in equations 2 and 3. These reactions
have been shown in PDH complexes from ox kidney and heart, pig

PDH complex (active) + nMgATP
$$\longrightarrow \text{PDHP complex (inactive) + nMgADP} \qquad (2)$$

PDHP complex (inactive)
$$\longrightarrow \text{PDH complex (active) + nPi} \qquad (3)$$

heart, pigeon breast muscle, rat heart, rat liver and rat
epididymal adipose tissue and human heart muscle. The
reactions have also been shown in PDH complexes in rat heart,
liver, kidney and adipose tissue mitochondria. In
mitochondria reactions shown in equations 2 and 3 operate
simultaneously and their relative rates determine the
proportions of active and inactive complexes. In rat tissues
the proportion of active complex is decreased by starvation
and by alloxan diabetes. Effects of starvation are reversed
by feeding and those of diabetes by insulin injections
(Wieland et al, 1971a). In perfused rat heart oxidation of
fatty acids or ketone bodies in vitro decreases the proportion
of active PDH complex (Wieland et al, 1971b). In rat adipose
tissue in vitro insulin increases, and lipolytic hormones
decrease the proportion of active complex (Jungas, 1971;
Coore et al, 1971). These various changes in the proportion
of active complex were not associated with any alteration in
the total concentration of PDH complex (sum of active and
inactive complex). The mechanisms must therefore involve
regulation of the PDH kinase reaction (equation 2) or of the
PDHP phosphatase reaction (equation 3) or both. A more
detailed bibliography to this work is given by Randle et al
(1979). Before considering mechanisms regulating
phosphorylation/dephosphorylation we shall review the
chemistry of the PDH complex and of the PDH kinase and PDHP
phosphatase reactions.

II. THE PYRUVATE DEHYDROGENASE COMPLEX

A. Component Enzymes and Elementary Reactions

The PDH complex of animal tissues contain three component
enzymes, pyruvate decarboxylase (PDC), dihydrolipoyl
acetyltransferase (LAT) and dihydrolipoyl dehydrogenase (DLD).
They catalyse the elementary reactions shown in equations 4-7

(TPP is thiamin pyrophosphate; lip is lipoyl) (for reviews see Reed, 1960). The overall reaction (equation 1) is

$$\text{TPP.PDC} + \text{pyruvate} \longrightarrow \text{hydroxyethyl TPP}^-.\text{PDC} + CO_2 \qquad (4)$$

$$\text{hydroxyethyl TPP}^-.\text{PDC} + (\text{S.S})\text{lip.LAT}$$
$$\rightleftharpoons \text{TPP.PDC} + (\text{acetyl S HS})\text{lip.LAT} \qquad (5)$$

$$(\text{acetyl S HS})\text{lip.LAT} + \text{CoA}$$
$$\rightleftharpoons (\text{HS HS})\text{lip.LAT} + \text{acetyl CoA} \qquad (6)$$

$$(\text{HS HS})\text{lip.LAT} + \text{NAD}^+ \rightleftharpoons (\text{SS})\text{lip.LAT} + \text{NADH} + H^+ \qquad (7)$$
$$\text{DLD}$$

essentially non-reversible (K_{eq}, 10^8) on account of the PDC reaction (equation 4).

B. Subunit Composition of the Complex.

SDS polyacrylamide disc gel electrophoresis of bovine kidney or heart or pig heart PDH complexes shows four subunits. These are α and β subunits of pyruvate decarboxylase (mol. wts. 40,600 and 35,100); subunit of dihydrolipoyl acetyltransferase (mol. wt. 76,100); and subunit of dihydrolipoyl dehydrogenase (mol. wt. 58,200). The molecular weights are for the pig heart complex (Sugden & Randle, 1978 but see also Hamada et al, 1975, 1976). The molecular weights of the component pig heart enzymes indicate that the decarboxylase is a tetramer ($\alpha_2\beta_2$); that dihydrolipoyl acetyltransferase contains 24 subunits; and that dihydrolipoyl dehydrogenase is a dimer (Hayakawa et al, 1969; Hamada et al, 1975, 1976). The subunit stoichiometry of the pig heart complex has recently been determined by amidination of free amino groups with methyl 1-^{14}C acetimidate by the method of Bates et al (1975). The ratios were 4 PDC α chains: 4 PDC β chains : 2 LAT subunits : 1 DLD subunit (Sugden and Randle, 1978), consistent with data based on separation of component enzymes (Hamada et al, 1975,1976). It is suggested that the overall composition of the pig heart complex is

$$24 \text{ PDC } (\alpha_2\beta_2) : 24 \text{ LAT} : 6 \text{ DLD dimer} \qquad (8)$$

This is based on a molecular weight of 1.8×10^6 for the PDH holocomplex (Hayakawa et al, 1969). Electron microscopy of bovine kidney and pig complexes and of dihydrolipoyl acetyltransferase core, however, suggests icosohedral (5 3 2) symmetry which cannot easily be reconciled with the formula given in (8) (Ishikawa et al, 1966; Reed and Oliver, 1968; Hayakawa et al, 1969).

The molecular weights of the bovine kidney and heart subunits as determined by polyacrylamide disc gel electrophoresis are very similar to those of the pig heart subunits given above (Barrera et al, 1972). It has been suggested that the bovine PDH complex is composed of 60 PDC (αβ) : 60 LAT : 15 DLD dimer (Barrera et al, 1972). This is consistent with 5 3 2 symmetry but it assumes that PDC is present as αβ (difficult to reconcile with the stoicheiometry of phosphorylation); that the subunit molecular weight of the LAT subunit is 55,000 (based on sedimentation equilibrium in guanidinium hydrochloride and difficult to reconcile with SDS polyacrylamide disc gel electrophoresis); that DLD is lost on purification; and higher molecular weights for the PDH holocomplex and for the LAT holoenzyme (Barrera et al, 1972). The question of symmetry requires further study as does the low content of dihydrolipoyl dehydrogenase. Although lipoamide dehydrogenase is removed during purification of the PDH complex, the concentration in preparations of purified complex is rather constant and lipoamide dehydrogenase is a component of 2-oxoglutarate dehydrogenase removed during purification.

III. PHOSPHORYLATIONS AND DEPHOSPHORYLATIONS

A. Stoichiometry of Phosphorylations

SDS polyacrylamide disc gel electrophoresis of PDHP complexes from bovine kidney and heart, pig heart and pigeon breast muscle has shown that phosphorylations in the complex are confined to the α-chain of PDC (Barrera et al, 1972; Severin and Feigina, 1977; Sugden and Randle, 1978). Initial studies suggested incorporation of one mol. of phosphate per mol. of PDC α-chain (Linn et al, 1969a). However subsequent amino acid sequencing indicated more than one site of phosphorylation and suggested that only one site of phosphorylation is involved in inactivation (Reed et al, 1974). This led to the suggestion of half-site phosphorylation in the PDC tetramer (i.e. αP.αβ$_2$) for inactivation.

We have used three techniques in studying phosphorylations
in pig heart PDH complex. The first technique employs
stepwise inactivation and phosphorylation by repetitive
additions of limiting amounts of ATP (increments of between
2.5 and 10μM)(Sugden et al, 1978). Inactivation proceeded
rapidly until about 2% of active complex remained and at this
point incorporation was approximately 0.45 nmol P/unit of
complex inactivated (1 unit forms 1μmol NADH at 30°C).
Phosphorylation continued to a limit of approximately 1.3 nmol
P/unit of complex. The stoicheiometry of phosphorylation of
PDH complex phosphorylated to approximately 97.5% inactivation
by this technique has been estimated by radioamidination of
the ^{32}P PDHP complex. After separation of the α-chains of
PDC by polyacrylamide disc gel electrophoresis, dual isotope
counting permitted calculation of the ratio of mols P : mols
α-chain. This ratio was 1:2 (Sugden and Randle, 1978). These
results indicated that the inactivating and additional
phosphorylations are

$$(\alpha_2\beta_2)_n + nMgATP \longrightarrow (\alpha P.\alpha\beta_2)_n + nMgADP \qquad (9)$$

$$(\alpha P.\alpha\beta_2)_n + 2nMgATP \longrightarrow (\alpha_2 P_3.\beta_2)_n + 2nMgADP \qquad (10)$$

The second technique employed phosphorylation at a saturating
ATP concentration. With this technique inactivation and
phosphate incorporation were first order until 80-85% of the
complex was inactivated; incorporation was approximately
0.5 nmol P/unit of complex inactivated. Phosphorylation and
inactivation then continued to the limit value of
approximately 1.3 nmol P/unit. This technique was routinely
used to prepare fully phosphorylated complex. In the third
technique inactivation by the first technique was accomplished
with non-radioactive ATP. This preparation could then be used
to study the additional phosphorylations with $\gamma-^{32}P$ ATP.This
preparation incorporated approximately 0.8 nmol P/unit of
complex.

B. Amino Acid Sequence around Phosphorylation Sites

The amino acid sequences of phosphopeptides purified from
tryptic digests of preparations of pig heart PDHP complexes
made with $\gamma-^{32}P$ ATP have recently been determined by
automated Edman degradation in a Beckman 890C sequencer by
Catherine Waller and Dr. K.B. Reid of the Department of
Biochemistry, University of Oxford (Sugden et al, 1979). In

the fully phosphorylated PDHP complex prepared by the second
technique three sites of phosphorylation were identified in
two phosphopeptides (A,B) whose sequences are shown below.

Peptide A

                   Site 1
Tyr. His. Gly. His. Ser(P). Met. Ser. Asp. Pro. Gly. -
 1   2   3   4   5   6   7   8   9   10
          Site 2
- Val. Ser(P). Tyr. Arg.
  11   12   13   14

Peptide B

                     Site 3
Tyr. Gly. Met. Gly. Thr. Ser(P). Val. Glu. Arg.
 1   2   3   4   5   6   7   8   9

The location of phosphoserine A12 was confirmed by partial
acid hydrolysis which yielded Val. Ser(P) and Val. Ser(P). Tyr.
The phosphorylation site in Peptide B was confirmed by acid
hydrolysis and demonstration of the presence of Ser(P) and
absence of Thr(P). The sequences of phosphopeptides A and B
correspond to those of tryptic phosphopeptides from fully
phosphorylated bovine kidney and heart PDHP complexes (Yeaman
et al, 1978) except that the A8 aspartate in the pig complex
is replaced by asparagine in the bovine complexes. The
conditions used in the studies of Sugden et al (1979) would
not be expected to result in conversion of asparagine to
aspartate.

     The position of the inactivating site was identified as A5
serine in PDHP complex phosphorylated to 91% inactivation by
the incremental method (technique 1). The complex was cleaved
by cyanogen bromide followed by trypsin and yielded a
phosphopeptide with amino acid composition (Tyr:2His:1Gly:1Ser:
1P:1 homoserine); and a phosphopeptide with the amino acid
composition of peptide A but containing only one mol.
phosphate. It is concluded that partially phosphorylated PDHP
complexes may yield a third tryptic phosphopeptide C with the
same amino acid sequence as peptide A but with Ser 12. The
location of the inactivating phosphate in pig heart PDH
complex is identical with the location in tryptic
phosphopeptides from bovine complexes (Yeaman et al, 1978).

     High voltage electrophoresis of tryptic digests of pig
heart PDHP complexes using conditions described by Davis et al
(1979) separates phosphopeptides A,B and C with mobilities
relative to a DNP-lysine marker of 0.99, 0.61 and 1.40. Use
of this method of analysis showed that PDH complex
phosphorylated to 85-90% inactivation by the incremental
method (technique 1) contained 91% of ^{32}P in site 1 and 9% of
^{32}P in site 2; PDH complex during the first order phase of

phosphorylation by technique 2 contained 84% of ^{32}P in site 1, 15% in site 2 and 1% in site 3. Fully phosphorylated PDHP complex showed equal distribution of ^{32}P between sites 1,2 and 3. These results show that site 1 phosphorylation is much more rapid than site 2 phosphorylation which is more rapid than site 3 phosphorylation. This has also been shown with the bovine complex (Davis et al, 1977; Yeaman et al, 1978). This conclusion is confirmed by the observation that the rate of inactivation of pig heart PDH complex is approximately thirteen times greater than the rate of incorporation of ^{32}P into PDHP complex (αP.$\alpha\beta_2$).

These results may suggest that phosphorylation of sites 1, 2 and 3 is sequential as shown in one possible model depicted in Fig. 1. The evidence that phosphorylation is sequential is not entirely unequivocal and is principally that a phosphopeptide with the amino acid composition of A containing Ser 5 and Ser(P) 12 has not been identified. The results of sequence determinations (this study; Yeaman et al, 1978) show that sites 1 and 2 are on the same PDC α-chain; whether site 3 is on the same α-chain has yet to be demonstrated unequivocally. Because site 2 phosphorylation begins before site 1 phosphorylation is completed there is the possibility of competition for phosphorylation between sites 1 and 2 in a single complex. The PDH kinase content of the PDH complex is not known but it is certainly much less in molar terms than that of its substrate PDC. It is not known whether dephosphorylations are random or sequential and it has yet to be determined unequivocally whether a single phosphatase is involved in all three dephosphorylations.

IV. REGULATION OF PHOSPHORYLATION/DEPHOSPHORYLATION

Pyruvate or dichloroacetate are inhibitors of the PDH kinase reactions (Linn et al, 1969b; Whitehouse et al, 1974); these compounds have no effect on the PDHP phosphatase reaction. These compounds effect conversion of PDHP complexes into PDH complex in isolated mitochondria and in tissues in vivo or in vitro. These observations indicate that the PDH kinase and PDHP phosphatase reactions may operate simultaneously in vivo and thus constitute a cycle. The proportions of active and inactive complexes are therefore determined by the relative rates of the opposing kinase and phosphatase reactions.

In discussing the regulation of PDH kinase and PDHP

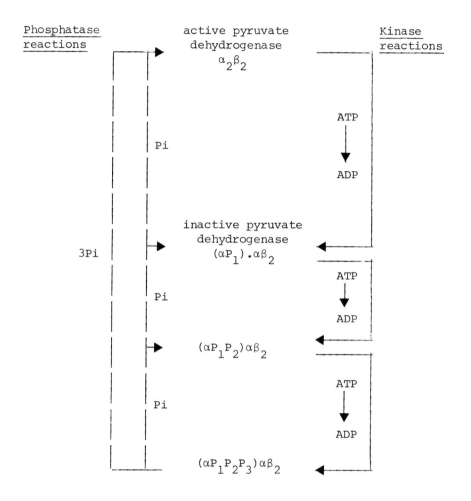

FIGURE 1. One model for phosphorylation-dephosphorylation
cycles in the pig heart pyruvate dehydrogenase complex. In
this model the three phosphorylations catalysed by the kinase
are shown in sequence (site 1, site 2, site 3), although
this has not been established unequivocally. It is not
known whether phosphatase reactions are random or sequential.

phosphatase reactions we are particularly concerned with mechanisms that may explain effects of starvation, diabetes, insulin and lipolytic hormones (see introduction). The general approach has been to define mechanisms of regulation in the purified complex; and to ascertain the extent to which they explain in vivo regulation by experiments with isolated mitochondria and isolated tissues. The PDH complex is intramitochondrial and it uses intramitochondrial substrates and is regulated only by intramitochondrial metabolites, so far as is known. In the purified PDH complex the kinase reactions can readily be studied in the absence of PDHP phosphatase. The kinase is tightly bound to the complex and remains with it during fractionation. The phosphatase remains in the supernatant during sedimentaiton of the complex at 150,000g and is readily separated. The phosphatase reactions may be studied by addition of phosphatase to PDHP complexes.

A. Regulation of Kinase and Phosphatase Reactions by Metabolites

The kinase reaction in purified bovine and porcine complexes is modulated by a number of metabolites. Those currently of principle biochemical interest are acetyl CoA and NADH (activators) and CoA, NAD^+ and pyruvate (inhibitors) (Linn et al, 1969b; Cooper et al, 1974; Pettit et al, 1975; Cooper et al, 1975). Pyruvate is non-competitive or uncompetitive against ATP whereas ADP is competitive. The kinetics of activation by acetyl CoA and NADH and of inhibition by NAD^+ and CoA have yet to be fully evaluated. These findings suggested regulation of the proportion of active PDH complex by mitochondrial concentration ratios of ATP/ADP, acetyl CoA/CoA and NADH/NAD^+. It was predicted that increasing the concentration ratios of these metabolite pairs would decrease the proportion of active PDH complex and vice versa. This prediction was borne out in experiments with rat heart mitochondria in which each concentration ratio was varied independently (Hansford, 1976; Kerbey et al, 1977).

The effect of these metabolites on the inactivating kinase reaction (equation 9) and the additional kinase reactions (equation 10) has recently been evaluated (Kerbey et al, 1979). In this study PDH complex activity was used to follow the inactivating phosphorylation. Incorporation of ^{32}P into PDHP $(\alpha P.\alpha\beta_2)$ (prepared with non-radioactive ATP) was used to follow the additional phosphorylations (predominantly site 2 phosphorylation). The results of this study, summarised in

TABLE I.   Metabolites on PDH Kinase Reactions

Metabolite	Parameter	Value for phosphorylation	
		Inactivating	Additional
MgATP	K_m (μM)	25.5	9.7
ADP	K_1 (μM) [a]	69.8	31.5
Pyruvate	K_2 (mM) [b]	2.8	1.1
Acetyl CoA/CoA	50% activation	0.23	0.16
NADH/NAD$^+$	50% activation	0.4	0.1

[a]Competitive inhibition.   [b]Uncompetitive inhibition.

Table I, show that both inactivating and additional
phosphorylation reactions are inhibited by ADP and pyruvate
and activated by increasing concentration ratio of acetyl
CoA/CoA and NADH/NAD$^+$.   The effects of these ratios on
additional phosphorylation reactions was complex as there was
evidence for some inhibition at ratios above 1 to 2.
    The phosphatase reaction with bovine or porcine PDHP
complexes requires Mg^{2+} (K_m approximately 1mM) and is
activated by Ca^{2+} (K_a approximately 0.1μM) (Linn et al, 1969a;
Denton et al, 1972).   The requirement for Mg^{2+} and Ca^{2+} has
been shown in mitochondria (Severson et al, 1974).   The
phosphatase reaction in the bovine complex is inhibited by
acetyl CoA and NADH (Pettit et al, 1975).   Inhibitory effects
in the porcine complex are small (15% or less).
    Systematic studies of the effects of the phosphatase on
inactivating and additional sites of phosphorylation have yet
to be reported.   Preliminary studies suggest that Mg^{2+} and
Ca^{2+} are required for both and that both are inhibited (c.15%)
by acetyl CoA and NADH.   The initial rate of reactivation of
fully phosphorylated porcine PDHP complex ($\alpha_2P_3\beta_2$) by pig
heart or ox heart phosphatase is approximately one-third the
rate of reactivation of PDHP complex ($\alpha P.\alpha\beta_2$) (Sugden et al,
1978).   Multi-site phosphorylation therefore inhibits reactivation of the complex.   The initial rate of release of ^{32}Pi
from fully phosphorylated complex by phosphatase preparations
is approximately 2.3 x faster than with PDHP complex ($\alpha P.\alpha\beta_2$)
(Sugden et al, 1978). The mechanism of these differences is
not known; one model is shown in equations (11) and (12) (only
$\alpha_2\beta_2$ is active).

$$3(\alpha P.\alpha\beta_2) \longrightarrow 3(\alpha_2\beta_2) + 3Pi \qquad\qquad (11)$$

$$3(\alpha_2P_3\beta_2) \longrightarrow (\alpha_2\beta_2) + 2(\alpha P.\alpha\beta_2) + 7Pi \qquad (12)$$

V. MECHANISMS MEDIATING EFFECTS OF INSULIN
DEFICIENCY AND OF INSULIN

Control of the PDH kinase reaction by the mitochondrial
concentration ratios of ATP/ADP, acetyl CoA/CoA and NADH/NAD$^+$
helps to explain some aspects of the physiological regulation
of the proportion of active complex in heart muscle. For
example, in rat heart muscle oxidation of fatty acids or
ketone bodies increases the concentration ratio of acetyl CoA/
CoA from approximately 0.005 (glucose) to 0.5-1 (glucose +
fatty acid or ketone body) (see Garland and Randle, 1964b;
Randle et al, 1970; Williamson et al, 1976). Through
activation of PDH kinase this change in ratio may account for
the fall in the proportion of active PDH complex induced by
oxidation of lipid fuels in the heart. The concentration ratio
of acetyl CoA/CoA is also increased in heart muscle by insulin
deficiency (Garland & Randle, 1964b); in rat epididymal adipose
tissue, insulin action in vitro may decrease the mitochondrial
ratio of acetyl CoA/CoA (Paetzke-Brunner et al, 1978). Thus
changes in the concentration ratio of acetyl CoA/CoA may be one
factor regulating the phosphorylation state of the PDH complex
in diabetes and in the action of insulin on adipocytes.
However, for reasons now to be detailed, other mechanisms may
also be involved which in insulin deficiency may be
quantitively more important.

A. Starvation and Alloxan Diabetes

When hearts of non-diabetic rats oxidise fatty acids the
proportion of active PDH complex is reduced from 25% (glucose
perfusion) to 7% (glucose + fatty acid). In hearts of alloxan-
diabetic rats the value is near zero (glucose perfusion)
(Kerbey et al, 1976). The concentration ratio of acetyl CoA/
CoA is no greater in diabetic hearts than in non-diabetic
hearts oxidising fatty acids (Garland & Randle 1964b) and
ratios of ATP/ADP and NADH/NAD$^+$ are unlikely to be relatively
increased in the diabetic tissue. Moreover it was known that
flux through the PDH reaction in rat heart with 4mM-pyruvate
as substrate is reduced to 10% of the control by alloxan
diabetes (Garland et al, 1964). This suggested that the action
of pyruvate to effect conversion of inactive PDHP complex to
active PDH complex is inhibited (relative to the control) in
hearts of diabetic rats. It was then shown that the proportion
of active PDH complex in hearts of alloxan-diabetic rats
perfused with medium containing 5 or 25mM-pyruvate was much

lower than in non-diabetic controls (Kerbey et al, 1976).

It was shown further that this resistance to the action of pyruvate on the interconversion of PDH and PDHP complexes persists into mitochondria prepared from rat heart (Kerbey et al, 1976). The proportion of active complex in mitochondria from hearts of diabetic rats with pyruvate was lower than in mitochondria from non-diabetic controls and this difference could not be explained by the concentrations of known metabolite effectors of the PDH kinase reaction (Kerbey et al, 1977). Starvation of the rat for 48h induced changes comparable to those seen in diabetes (Hutson et al, 1978). It was concluded from these studies that starvation and alloxan-diabetes must either render the kinase reactions in the PDH complex insensitive to pyruvate inhibition, or inhibit the PDHP phosphatase reaction or both. Further, the mechanism must be sufficiently stable to persist through isolation and incubation of rat heart heart mitochondria. In the event both mechanisms appear to operate, the kinase reaction is activated and relatively insensitive to pyruvate inhibition; and reactivation of PDHP complex by the phosphatase reactions is inhibited.

1. PDH Kinase Reaction in Starvation and Diabetes. The rate of the kinase reactions in PDH complex can be followed in extracts of rat heart mitochondria with $\gamma-^{32}P$ ATP by monitoring the disappearance of active complex and the formation of protein-bound ^{32}P provided that precautions are taken to minimise ATPase and proteolytic activities. Mitochondria are incubated without substrate, prior to extraction, to effect conversion of PDHP complex into PDH complex (Hutson and Randle (1978)). Representative data given in Table 2 show that the rate of phosphorylation and inactivation of PDH complex was accelerated in alloxan-diabetes and that this persisted in the presence of pyruvate. Qualitatively similar changes were seen in extracts of mitochondria prepared from hearts of rats starved for 48h (not shown). The increased intrinsic kinase activity of PDH complex in extracts of heart mitochondria of starved or diabetic animals is unlikely to be due to altered concentrations of known metabolite effectors of the kinase reaction (see Hutson and Randle, 1978). The mechanism of this increased kinase activity in starvation and diabetes is not known; the stability of the change suggests either an increased concentration of PDH kinase in the complex or some other stable form of modification e.g. covalent modification. An increase in intrinsic kinase activity in the PDH complex during starvation has also been seen in lactating mammary gland (Baxter and Coore, 1978).

TABLE 2.   Kinase Reaction in mitochondrial extracts[a]

Minutes	PDH Activity (% of Initial)[b] Mean ± S.E.M.			
	Non-diabetic		Diabetic[c]	
	Control	Pyruvate[d]	Control	Pyruvate[d]
0	100 ± 7.6	100 ± 7.0	100 ± 8.3	100 ± 8.5
1	63 ± 4.2	66 ± 7.3	20 ± 1.3	39 ± 2.8
2	40 ± 2.6	59 ± 6.5	8 ± 0.6	27 ± 3.0
3	22 ± 1.6	52 ± 5.8	4 ± 0.5	19 ± 2.9
6	8 ± 0.9	44 ± 4.9	3 ± 0.6	12 ± 2.2
9	4 ± 0.3	31 ± 5.5	3 ± 0.3	8 ± 1.5
	^{32}P incorporation (pmol/mg Mitochondrial Protein)			
1	23 ± 2.7	20 ± 0.8	49 ± 5.1	59 ± 7.2
2	39 ± 4.7	21 ± 3.0	68 ± 7.1	50 ± 9.3
3	49 ± 6.0	24 ± 3.8	79 ± 7.9	62 ±11.8
6	65 ± 6.8	29 ± 4.5	93 ± 8.3	76 ±11.7
9	70 ± 7.7	35 ± 6.0	101 ± 9.6	78 ±13.2

[a]Extracts of rat heart mitochondria incubated at 30°C with
0.3mM- γ-^{32}P ATP. [b]Initial activities (m-units PDH/mg
protein), non-diabetic 101 to 108, diabetic 88-91. [c]Diabetes
induced with alloxan. There were 9 observations in control
and 4 in pyruvate groups. [d]2.75mM-pyruvate.

    2. PDHP Phosphatase Reaction in Starvation and Diabetes.
Evidence for inhibition of the PDHP phosphatase reaction in
heart mitochondria of diabetic rats was obtained as follows
(Hutson et al, 1978). The PDH kinase reaction in mitochondria
oxidising succinate can be interrupted by addition of
oligomycin and an uncoupler of respiratory chain
phosphorylation. The rate of the phosphatase reaction can be
followed by monitoring the rate of increase in the
concentration of active PDH complex; this rate is reduced by
alloxan-diabetes or starvation. The rate of the phosphatase
reaction can likewise be followed in extracts prepared from
mitochondria in which conversion of PDH to PDHP has been
effected by incubation for 1 min with respiratory substrate.
With this technique the concentrations of divalent metal ion
activitors of the phosphatase can be equalised in different
mitochondrial extracts. The rate of the phosphatase reaction
was reduced by alloxan-diabetes or starvation. The rate of

reactivation of, or ^{32}P release from pig heart PDHP complex by
extracts of rat heart mitochondria was not influenced by
alloxan-diabetes or starvation (Kerbey et al, 1976; Hutson et
al, 1978). This suggested that some stable change in the PDHP
complex in mitochondria of starved or diabetic rats inhibited
its conversion into active PDH complex. This conclusion was
borne out by experiments with PDHP complexes partially
purified from hearts of non-diabetic and diabetic rats and
freed of PDHP phosphatase. The rate of reactivation of PDHP
complex from hearts of diabetic rats by PDHP phosphatase from
ox heart or rat heart was slower than with PDHP complex from
non-diabetic rats. The mechanism has yet to be established
but inhibition of the phosphatase reaction by multisite
phosphorylations in the PDHP complex is one interesting
possibility.

Baxter and Coore (1978) have shown inhibition of the PDHP
phosphatase reaction in extracts of lactating mammary gland
mitochondria following starvation of the rat. An important
point of difference from our studies was retention of this
difference with added pig heart PDHP complex suggesting
inhibition of PDHP phosphatase.

3. General conclusions. It would appear, currently, that
enhanced activity of PDH kinase may be primarily responsible
for the decreased proportion of active PDH complex in tissues
of starved and alloxan-diabetic rats (this increase has
presently been shown only in heart and in lactating mammary
gland). As a working hypothesis it is suggested that this may
facilitate multisite phosphorylation and inhibit conversion of
PDHP complex into PDH complex. The resistance to pyruvate may
thus arise primarily from the accelerated kinase reaction with
inhibition of the phosphatase as a secondary feature. Until
the mechanism of activation of the kinase reaction and of
inhibition of the phosphatase reaction have been established
unequivocally it is difficult to approach more fundamental
questions. These include the question as to whether effects
of insulin deficiency in starvation or diabetes on
interconversions in the PDH complex are initiated by
mobilisation and oxidation of lipid fuels; or whether they
arise by interruption of more direct effects of the hormone.

B. INSULIN AND LIPOLYTIC HORMONES

In rat adipose tissue insulin (which is antilipolytic)
increases and lipolytic hormones decrease the proportion of
active PDH complex (see introduction). It has recently been

suggested that this action of insulin may result from a lowering of the mitochondrial ratio of acetyl CoA/CoA (Paetzke-Brunner et al, 1978), perhaps as a result of activation of acetyl CoA carboxylase. While it is likely that such an effect of insulin may contribute to interconversions in the PDH complex the hypothesis, like an earlier one (Wieland et al, 1974) overlooks important observations. These include evidence that the PDHP phosphatase reaction is stimulated in mitochondria from adipose tissue exposed to insulin in vitro (Hughes and Denton, 1976); and that the proportion of active PDH complex is retained during incubation of mitochondria from adipose tissue exposed to insulin. These changes are not explicable in terms of mitochondrial ratios of ATP/ADP or NADH/NAD$^+$ or acetyl CoA/CoA (Denton et al, 1977). The approaches that we have used in exploring effects of insulin deficiency are applicable to in vitro actions of insulin.

REFERENCES

Barrera,C.R., Namihara, G., Hamilton, L., Munk, P., Eley, M.H., Linn, T.C. and Reed, L.J. (1972). Arch. Biochem. Biophys. 143, 343.

Bates, D.L., Harrison, R.A., and Perham, R.N. (1975). FEBS Lett. 60, 427.

Baxter, M.R., and Coore, H.G. (1978). Biochem. J., 174, 553.

Cooper, R.H., Randle, P.J. and Denton, R.M. (1974). Biochem. J. 143, 625.

Cooper, R.H., Randle, P.J., and Denton, R.M. (1975). Nature (London), 257, 808.

Coore, H.G., Denton, R.M., Martin, B.R., and Randle, P.J. (1971). Biochem. J. 125, 115.

Davis, P.F., Pettit, F.H. and Reed, L.J. (1977). Biochem. Biophys.Res. Commun. 75, 541.

Denton, R.M., Randle, P.J., and Martin, B.R. (1972). Biochem. J. 128, 161.

Denton, R.M., Bridges, B.J., Brownsey, R.W., Evans, G.L., Hughes, W.A., and McCormack, J. (1977). Proc. 11th FEBS Congress, 46, 81.

Garland, P.B., and Randle, P.J. (1964a). Biochem. J. 91, 6C.

Garland, P.B., and Randle, P.J. (1964b). Biochem. J. 93, 678.

Garland, P.B., Newsholme, E.A., and Randle, P.J. (1964). Biochem. J. 93, 665.

Hamada, M., Otsuka, K., Tanaka, N., Ogasahara, K., Koike, K., Hiroaka, T., and Koike, M. (1975). J. Biochem. (Tokyo). 78, 187.

Hamada, M., Hiraoka, T., Koike, K., Ogasahara, K., Kanzaki, T., and Koike, M. (1976). J. Biochem.(Tokyo). 79, 1273.

Hansford, R.G.(1976). J. Biol. Chem., 251 5483.

Hayakawa, T., Kanazaki, T., Kitamura, T., Fukuyoshi, Y., Sakwai, Y., Koike, K., Suematsu, T., and Koike, M. (1969). J. Biol. Chem. 244, 3660.

Hughes, W.A. and Denton, R.M. (1976). Nature (London). 26, 471.

Hutson, N.J. and Randle, P.J. (1978). FEBS Lett. 92, 73.

Hutson, N.J., Kerbey, A.L., Randle, P.J. and Sugden, P.H. (1978). Biochem. J., 173, 669.

Ishikawa, E., Oliver, R.M. and Reed, L.J. (1966). Proc. Natl. Acad. Sci. U.S.A. 56, 534.

Jungas, R.L. (1971). Metab. 20, 43.

Kerbey, A.L., Randle, P.J., Cooper, R.H., Whitehouse, S., Pask, H.T., and Denton, R.M. (1976). Biochem. J. 154, 327.

Kerbey, A.L., Radcliffe, P.M., and Randle, P.J. (1977). Biochem. J. 164, 509.

Kerbey, A.L., Radcliffe, P.M., Randle, P.J., and Sugden, P.H. (1979). Biochem. J. In Press.

Linn, T.C., Pettit, F.H. and Reed, L.J. (1969a). Proc. Natl. Acad. Sci. U.S.A. 62, 234.

Linn, T.C., Pettit, F.H., Hucho, F., and Reed, L.J. (1969b). Proc. Natl. Acad. Sci. U.S.A. 64, 227.

Paetzke-Brunner, I., Schon, H., and Wieland, O.H. (1978). FEBS Lett. 93, 307.

Pettit, F.H., Pelley, J.W., and Reed, L.J. (1975). Biochem. Biophys. Res. Commun. 65, 575.

Randle, P.J., England, P.J., and Denton, R.M. (1970). Biochem. J. 117, 677.

Randle, P.J., Sugden, P.H., Kerbey, A.L., Radcliffe, P.M. and Hutson, N.J. (1979). Biochem. Soc. Symp., 43. In Press.

Reed, L.J. (1960). Enzymes 2nd Ed. 3, 195.

Reed, L.J., and Oliver, R.M. (1968). Brookhaven Symp. Biol. 21, 397.

Reed, L.J., Pettit, F.H., Roche, T.E., Butterworth, P.J., Barrera, C.R., and Tsai, C.S. (1974). In "Metabolic Interconversion of Enzymes 1973" (E.H. Fischer, E.G. Krebs, H. Neurath, E.R. Stadtman, eds.) p.97. Springer-Verlag. Berlin-Heidelberg.

Severin, S.E., and Feigina, M.M. (1977). Adv. Enzyme Regul. 15, 1.

Severson, D.L., Denton, R.M., Pask, H.T., and Randle, P.J. (1974). Biochem. J. 140, 225.

Sugden, P.H., and Randle, P.J. (1978). Biochem. J. 173, 659.

Sugden, P.H., Hutson, N.J., Kerbey, A.L., and Randle, P.J.
    (1978). Biochem. J. 169, 433.
Sugden, P.H., Kerbey, A.L., Randle, P.J., Waller, C.A., and
    Reid, K.B.M. (1979). Biochem. J. In Press.
Whitehouse, S., Cooper, R.H. and Randle, P.J.(1974).
    Biochem. J. 141, 761.
Wieland, O.H., von Funcke, H., and Loffler, G. (1971a).
    FEBS Lett. 15, 295.
Wieland, O.H., Siess, E.A., Schulze-Wethman, F.H., von Funcke,
    H., and Winton, B. (1971b). Arch. Biochem. Biophys. 143
    593.
Wieland, O.H., Weiss, L., Loffler, G., Brunner, I., and
    Bard, S. (1974). In "Metabolic Interconversion of Enzymes
    1973" (E.H. Fischer, E.G. Krebs, H. Neurath and
    E.R. Stadtman, eds) p.117 Springer-Verlag, Berlin-
    Heidelberg.
Williamson, J.R., Ford, C., Illingworth, J., and Safer, B.
    (1976). Circ. Res. 38, 1.39.
Yeaman, S.J., Hutcheson, E.T., Roche, T.E., Pettit, F.H.,
    Brown, J.R., Reed, L.J., Watson, D.C., and Dixon, G.H.
    (1978). Biochemistry, 17, 2364.

DISCUSSION

E. KREBS: Has any work been done on the phosphorylation of simple peptides to see whether there are any effects of the regulatory compounds in this situation? This would be helpful in deciding whether metabolites were acting at the substrate level or whether they were acting on the kinase itself.

P. RANDLE: Not very much is known in relation to inhibition, so far. The original comments were that no peptides were very well phosphorylated, but there is one important difference that makes me a little bit skeptical about short-chain peptide studies. If you take the pyruvate decarboxylase component and put it with purified kinase, then it is a very poor substrate for kinase, and it is a much better substrate when the kinase is attached to the acetyl complex. The short peptide substrate is a very good substrate for the kinase. Since there is this difference between the holo enzyme as a substrate in relation to its state in the complex and the short chain peptides, I am not too convinced of the relevance of doing studies with short chain peptides or of any conclusions about what might happen in the complex.

P. COHEN: Would the following experiments be feasible?
Phosphorylate the first site with $-^{32}$P[ATP] and
phosphorylate the second and third sites with ATP [S] to
produce a derivative which is not susceptible to dephosphorylation.   The first site could then be dephosphorylated
leaving a complex which is phosphorylated specifically in
sites 2 and 3.   You might then be able to investigate whether
the rate of phosphorylation of that material differs from the
dephosphorylated enzyme.

P. RANDLE: I do not know anything about phosphorylation with
analogs of ATP. I think it is feasible.   It is clear at the
moment that the first two phosphorylation sites are on the
same peptide, and we are trying to establish that the third
is on the same peptide as well.   We would like to do that
before we start on the phosphatase.   I think there is an
important point there because one might get fooled by two
separate peptides running with the same electrophoretic
mobility.

E. KREBS: Are there any other phosphorylatable enzymes or
identified proteins in the mitochondria?

P. RANDLE: Well there are two potential candidates.   We have
reported on the inactivation of the branched chain complex in
heart muscle.   This is an analogous complex which attacks the
2-oxo acids of leucine, isoleucine and valine and which is
subject to comparable end-product inhibition.   That is inactive in heart mitochondria as you isolate them and is
activated by incubating the mitochondria without substrate or
with uncoupler.   It can also be reacted with the relevant
branch-chain alpha keto acids; it can also be reactivated.
If you activate the complex in mitochondria and then make
extracts and incubate with ATP it is inactivated, and the
inactivation is inhibited by the branch chain alpha keto
acid.   We do not know whether that is a phosphorylation.   It
might be an adenylation, for example, because there is only a
very small amount there.   The concentration is about 2-3%
that of the pyruvate dehydrogenase complex.   Now this phenomenon is not seen in liver or kidney mitochondria, and we
think that it is seen in skeletal muscle mitochondria.   I
think the branch chain complex is likely to be a candidate at
least in muscle tissue.   The second report was that there may
be a phosphorylation in the respiratory chain in relation to
the action of insulin on gluconeogenesis, which may be
brought about by cytoplasmic kinase acting on the outside of
mitochondria as opposed to the kinases on the inside.

E. KREBS: I was wondering whether there might be some sort of a concerted phosphorylation seen within the mitochondria similar to that which Phil Cohen was stressing would occur in the cytosol?

P. RANDLE: Well at the moment, I think one is faced with only one kinase in the mitochondria. Whether there are more complexities than there may be in phosphatase regulation I do not know.

ALLOSTERIC CONTROL OF E. COLI GLUTAMINE SYNTHETASE
IS MEDIATED BY A BICYCLIC NUCLEOTIDYLATION CASCADE SYSTEM

E. R. Stadtman
P. B. Chock
S. G. Rhee

Laboratory of Biochemistry
National Heart, Lung and Blood Institute
National Institutes of Health
Bethesda, MD   20014

I.   INTRODUCTION

Glutamine is a key intermediate in nitrogen metabolism. It serves directly as a building block in the biosynthesis of most proteins, and in addition, serves as a source of nitrogen atoms in the biosynthesis of all purine and pyrimidine nucleotides, of all amino acids, of glucosamine-6-P, p-aminobenzoic acid and of nicotinamide derivatives. Glutamine synthetase is therefore an enzyme of singular importance because it links the assimilation of NH_3, derived from catabolic processes, with biosynthetic pathways leading to the formation of protein, nucleic acids, complex poly-saccharides and various coenzymes (1). In addition, in some organisms glutamine synthetase itself appears to regulate the expression of structural genes concerned with the bio-synthesis of other enzymes in nitrogen metabolism (2).

Because it occupies a central position in nitrogen metabolism glutamine synthetase is obviously a strategic target for rigorous cellular control. To be effective, however, the system that regulates this enzyme must be able to sense simultaneous changes in the concentrations of many different metabolites and to integrate their combined effects in order that the level and catalytic activity of glutamine synthetase can be adjusted continuously to meet the ever changing demand to glutamine. Investigations in our laboratory in Bethesda and that of Holzer and his

associates in Freiburg, carried out over a period of nearly
15 years, have gone far in elucidating the highly complex but
elegant system that controls glutamine synthetase activity
in E. coli and other gram negative bacteria (3). This system
utilizes several different basic principles of cellular
regulation, including: (i) repression and derepression of
glutamine synthetase formation in response to availability
of ammonia nitrogen, (ii) cumulative feedback inhibition by
multiple end products of glutamine metabolism, (iii) divalent
cation mediated conformational changes in glutamine synthetase structure, and (iv), metabolic interconversion of
glutamine synthetase between covalently modified (inactive)
and unmodified (active) forms. Together these diverse
mechanisms provide the flexibility needed for effective
regulation of the enzyme. There is insufficient space to
summarize here the results of investigations on all four of
the regulatory mechanisms. In keeping with the theme of
this symposium, we shall summarize the results of our
studies on the allosteric regulation of glutamine synthetase
via the interconvertible enzyme cascade.

II. NUCLEOTIDYLATION CYCLES INVOLVED IN
GLUTAMINE SYNTHETASE REGULATION

The glutamine synthetase cascade is comprised of two
linked nucleotidylation cycles. One involves the cyclic
adenylylation and deadenylylation of glutamine synthetase;
the other involves the cyclic uridylylation and deuridylylation of Shapiro's regulatory protein, P_{II}.

A. The Adenylylation Cycle

As shown in Fig. 1A, the adenylylation of glutamine
synthetase (GS) involves the transfer of an adenylyl group
from ATP to each subunit of glutamine synthetase (4, 5). The
adenylyl group is attached to the enzyme in phosphodiester
linkage to the hydroxyl group of just one tyrosyl residue
in each subunit (6). Because glutamine synthetase is a
dodecamer composed of identical subunits (7, 8), up to 12
adenylyl groups can be attached to each enzyme molecule.
Moreover, because adenylylated subunits are catalytically
inactive under most physiological conditions, the specific
activity of the enzyme is inversely proportional to the
average number of adenylylated subunits per enzyme molecule.

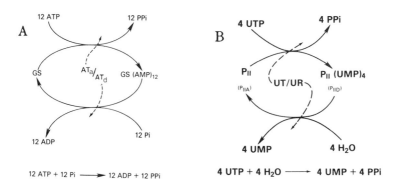

Figure 1. (A) *Glutamine Synthetase Adenylylation Cycle.* (B) P_{II} *Uridylylation Cycle.*

The adenylylation of glutamine synthetase is opposed by enzyme catalyzed phosphorolysis of the adenylyl-tyrosyl bonds to regenerate unmodified (active) GS and produce ADP (9). Because the deadenylylation and adenylylation reactions are catalyzed at separate sites designated AT_d and AT_a, respectively, on the same adenylyl transferase (AT), it follows that the two processes are intimately coupled; moreover, unless the capacity of AT to catalyze one reaction with respect to the other is strictly regulated, GS will undergo senseless cycling between adenylylated and unadenylylated states, the net result of which is simply phosphoryolysis of ATP to form ADP and PPi. Such futile cycling is prevented by interaction of the adenylylation cascade cycle with another nucleotidylation cycle that involves the uridylylation and deuridylylation of Shapiro's P_{II} protein (10).

B. The Uridylylation Cycle

As shown in Fig. 1B, a specific uridylyltransferase (UT) catalyzes transfer of a uridylyl group from UTP to each subunit of P_{II} (11, 12). Because P_{II} protein is composed of 4 identical subunits, up to 4 uridylyl groups can be bound per P_{II} molecule (13). As in the case of GS adenylylation, the uridylyl groups are attahced to P_{II} through phosphodiester linkage to the hydroxyl group of a tyrosyl residue in each subunit (13). The uridylylation

reaction is opposed by the action of a separate uridylyl
removing enzyme (UR) activity of the UT/UR complex. This
catalyzes hydrolysis of the uridylyl-tyrosyl bond to form
UMP and unmodified P_{II}. Because the uridylylation and
deuridylylation steps are catalyzed by separate catalytic
centers on the same enzyme or enzyme complex (UR/UT) (14),
the two steps are coupled to yield cyclic interconversion
of the P_{II} protein between modified and unmodified states.
The net result of this interconversion cycle is the hydrolysis of UTP to UMP and PPi.

C. The Bicyclic Cascade System

Linkage of the GS adenylylation cascade cycle with the
P_{II} uridylylation cascade cycle obtains from the fact that
the unmodified form of P_{II} (sometimes referred to as P_{IIA})
stimulates the capacity of adenylyl transferase to catalyze
the adenylylation of GS, whereas the uridylylated form of
P_{II}, (referred to as P_{IID}) is required to activate the
deadenylylation activity of adenylyltransferase (11).
Ultimately, however, the cyclic interconversions of P_{II} and
GS are regulated by the concentrations of various metabolites
that influence the activities of the converter enzymes;
i.e., AT_a, AT_d, UR and UT. Whereas at least 40 different
metabolites have been shown to effect one or more of these
enzymes (15), two metabolites, α-ketoglutarate (α-KG) and
L-glutamine (Gln), play a dominant role in GS regulation.
As shown in Fig. 2, these effectors exhibit reciprocal
effects on the activities of the converter enzymes.
Glutamine stimulates enormously adenylylation of GS (and
possibly the deuridylylation of P_{II}UMP) whereas it inhibits
the deadenylylation of $GS(AMP)_n$ and the uridylylation of
P_{II}. Conversely, α-ketoglutarate inhibits the adenylylation
of GS but it stimulates the deadenylylation of $GS(AMP)_n$ and
the uridylylation of P_{II}. It is evident from Fig. 2 that
if the ratio of Gln to α-KG is very high, P_{II} will be
converted almost entirely to P_{IIA} and the GS will become
fully adenylylated. But if the ratio of Gln to α-KG is
very low, the P_{II} will become mostly uridylylated and GS
will exist almost completely in the unmodified form. In
other words, a shift from a very low to a very high ratio
of Gln to α-KG will result in nearly complete inhibition of
GS, whereas a shift from a very high to a very low ratio
results in nearly complete activation. This and similar
considerations of other interconvertible enzyme systems
has led to the suggestion that such enzymes represent

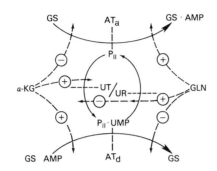

Figure 2. Metabolite Control of the Nucleotidylation Cycles by Gln and α-Kg. + indicates activation; – indicates inhibition.

metabolic switches that can be used to turn ON or OFF critical enzyme activities in response to metabolic demands. However, this concept is untenable because it implies that the converter enzyme that catalyzes the modification and demodification of an interconvertible enzyme must undergo reciprocal "all or none" changes in activity, or, more accurately, that the concentrations of positive and negative allosteric effectors that regulate the converter enzymes will vary reciprocally in an "all or none" manner. This view is unrealistic and is contraindicated by studies with glutamine synthetase (16) and pyruvate dehydrogenase (17) showing that the ratio of modified and unmodified enzyme forms vary over a wide range, depending on the nutritional state of organisms. It therefore appears, that covalent modification of an interconvertible enzyme is a dynamic process by means of which the specific activity of the enzymes can be varied progressively by shifts in the steady state distribution of modified and unmodified enzyme forms – shifts which are produced by changes in the concentrations of allosteric effectors that modulate the activities of the converter enzymes.

III.   THEORETICAL STEADY STATE ANALYSIS OF THE GLUTAMINE SYNTHETASE CASCADE

A better understanding of the interrelationship between metabolite regulation of the converter enzymes and the steady state level of GS adenylylation was obtained from a theoretical analysis of the bicyclic cascade model illustrated in Fig. 3.

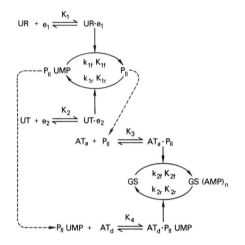

Figure 3. The Glutamine Synthetase Bicyclic Cascade.

For simplicity it is assumed that the cascade is initiated by the binding of an allosteric effector, e_1, with the UR enzyme to form the active $UR \cdot e_1$ complex which is needed to catalyze the conversion of $P_{II}UMP$ to P_{II}, and that this conversion is opposed by the binding of another effector e_2 to UT to form the $UT \cdot e_2$ complex which catalyzes the conversion of P_{II} back to $P_{II}UMP$. The assumed roles of effectors e_1 and e_2 therefore correspond to the physiologically important effectors α-ketoglutarate and glutamine, respectively. Fig. 3 shows also that the interconversion of P_{II} is coupled to the adenylylation and deadenylylation of GS because the binding of P_{II} to AT_a generates the $AT_a \cdot P_{II}$ complex that is required for catalysis of the adenylylation of GS, whereas the binding of $P_{II}UMP$ to the AT_d site produces the active complex, $AT_d \cdot P_{II}UMP$, that catalyzes the deadenylylation of $GS(AMP)\bar{n}$.

With reasonable simplifying assumptions that have been described elsewhere (15, 18, 19), it can be shown that in the steady state, the average number of adenylylated subunits per GS molecule, $\bar{n}$, is given by the expression:

$$\bar{n} = 12\left[\left(\frac{k_{1r}}{k_{1f}}\right)\left(\frac{k_{2r}}{k_{2f}}\right)\left(\frac{K_{1r}}{K_{1f}}\right)\left(\frac{K_{2r}}{K_{2f}}\right)\left(\frac{K_4}{K_3}\right)\left(\frac{K_2}{K_1}\right)\left(\frac{[UT]}{[UR]}\right)\left(\frac{[AT_d]}{[AT_a]}\right)\left(\frac{[e_2]}{[e_1]}\right) \times \right.$$

$$\left.\left(\frac{1 + K_1[e_1]}{1 + K_2[e_2]}\right) + 1\right]^{-1} \quad (1)$$

in which K_1, K_2, K_3 and K_4 are the association constants for the reactions leading to the activated converter enzyme complexes as shown in Fig. 3; k_{1f}, k_{1r}, k_{2f} and k_{2r} are specific rate constants for interconversions in the first and second cycles, respectively, as shown in Fig. 3; K_{1f}, K_{1r} and K_{2f}, K_{2r} are the association constants for the interactions of the activated converter enzyme complexes with their interconvertible protein substrates; e.g., K_{1f} = association constant for the reaction

$$UR \cdot e_1 + P_{II}UMP \xrightleftharpoons{} (UR \cdot E) \cdot (P_{II}UMP), \text{ etc.}$$

It is evident from Eq. 1 that the state of adenylylation, $\bar{n}$, is a multiplicative function of 9 different parameter ratios; i.e., the ratio of each parameter in the regeneration cascade to the corresponding parameter in the forward cascade.

A. Signal Amplification and Amplitude Effects.

For simplicity, only two allosteric effectors (e_1 and e_2) are included in the formal analysis shown here; however, any or all of the other 16 parameters in the steady state equation can be altered by single or multiple interaction(s) of one or more of the cascade enzymes with various metabolites. Therefore, in addition to the ratio $[e_2]/[e_1]$, the other 8 parameter ratios that govern the value of $\bar{n}$ can be varied by allosteric or substrate interactions. Furthermore, because of the multiplier effect of parameter ratios on the value of $\bar{n}$, the state of adenylylation of GS is extremely sensitive to multiple allosteric interactions. This is illustrated by the data in Fig. 4A which show how the value of $\bar{n}$ varies as a function of e_1 concentration when, one by one each parameter ratio (viz k_{1r}/k_{1f}, k_{2r}/k_{2f}, ... etc) is

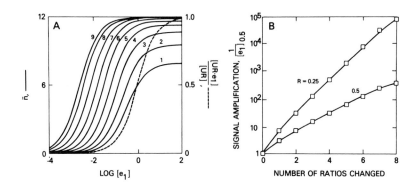

Figure 4. Computer simulated curves showing (A) The relationship between the number of parameter ratios decreased by one-half and the response of $\bar{n}$ to primary effector (e_1) concentration. (B) The relationship between the signal amplification and the number of parameter ratios (R) decreased from 1.0 to 0.5 or 0.25 (as indicated).

decreased stepwise by a factor of only one-half. It should be emphasized that in every case e_1 reacts directly with the UR enzyme only. As a point of reference the broken line in Fig. 4A is presented to show how the fractional saturation of UR with e_1 varies as a function of e_1 concentration when $K_{1} = 1.0$. The other curves in Fig. 4A show how the value of $\bar{n}$ responds to the binding of e_1 to UR. These curves illustrate two important features of the GS cascade system: (i) The maximal value of $\bar{n}$ that is obtainable with saturating levels of e_1 can vary depending upon the overall parameter ratio contribution. In other words, the amplitude (15) of the adenylylation reaction is variable. (ii) The concentration of e_1 required to achieve a given level of adenylylation varies with the overall parameter ratio, and can be much lower than the concentration of e_1 required to obtain comparable activation of the UR enzyme with which it reacts directly. For example, Curve 1 in Fig. 4A shows that when all parameter ratios in Eq. 1 have a value of 1.0, the concentration of e_1 required to obtain 50% adenylylation of GS ($\bar{n} = 6.0$) is the same ($[e_1] = 1.0$) as that required to obtain 50% saturation of the converter enzyme UR; but even with saturating levels of e_1 only 67% of GS can be adenylylated. The amplitude of the adenylylation reaction is therefore 8.0. However, as one by one

each parameter ratio is decreased to 0.5, the other curves
in Fig. 4A show that there is progressively a decrease in
the concentration of e_1 required to achieve a given state
of adenylylation and progressively an increase in the
amplitude of the adenylylation reaction (from $\bar{n}$ = 8.0 to
$\bar{n}$ = 12). Note that when all parameter ratios are assumed to
be 0.5, a concentration of e_1 that produces only 1-2%
activation of UR is sufficient to produce 50% adenylylation
of GS. In other words the bicyclic cascade system is
endowed with a large signal amplification potential with
respect to primary allosteric stimuli, e_1. For purposes of
comparison this signal amplification is defined (15) as the
ratio of the concentration of effector required to produce
50% activation of the converter enzyme (UR) to that required
to obtain 50% activation of the interconvertible enzyme
(GS). More explicitly, in the present case the signal
amplification is equal to the ratio $(1/K_1)/[e_1]_{0.5}$, where
$1/K_1$ is the apparent K_m for the interaction of e_1 with UR,
and $[e_1]_{0.5}$ is the concentration of e_1 required to produce
50% adenylylation of GS. Fig. 4B shows how the signal
amplification changes as one by one the 8 different parameter ratios are changed from either 1 to 0.5 or 1 to 0.25.
It can be seen that two-fold and four-fold changes in each
of all 8 ratios can lead to about 400-fold and 10,000-fold
increases, respectively, in the signal amplification
potential.

The nearly exponential increase in signal amplification
produced by simultaneous changes in the magnitude of all
the parameter ratios is readily apparent if for simplicity
it is assumed that all parameter ratios (R) except
$[e_2]/[e_1]$ in Eq. 1 are identical, i.e. $k_{1r}/k_{1f} = k_{2r}/k_{2f}$...
etc. = R. Then for the situation in which $[e_2] = 1.0$
and K_1 = 1.0, Equation 1, reduces to Eq. 2

$$\bar{n} = 12\left[R^8 \; \frac{1}{[e_1]} \; \frac{(1 + [e_1])}{(1 + R)} + 1 \right]^{-1} \qquad (2)$$

which shows that the value of $\bar{n}$ varies as a function of the
parameter ratio R to the 8th power. As a consequence,
relatively small changes in the value of the parameter
ratio, R, will result in an enormous shift in the range of
e_1 concentrations required to produce a given change in
the adenylylation of GS, as shown in Fig. 5A. Fig. 5B
shows that by decreasing the parameter ratio from 1 to 0.1,
it is theoretically possible to achieve a 10^8-fold increase

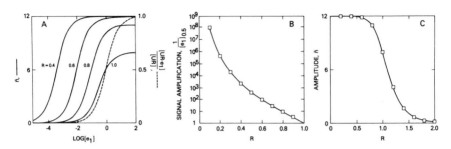

Figure 5. Computer simulated curves showing the relationship between the parameter ratio (R) and, (A) the value of $\bar{n}$ as a function of primary effector $[e_1]$ concentration, (B) the signal amplification, (C) the amplitude of the adenylylation reaction (see text for details).

in signal amplification, and Fig. 5C shows that the amplitude of the adenylylation reaction in the steady state can vary from 12 (complete adenylylation) to almost 0, as the parameter ratio is varied from 0 to 2.0.

It should be pointed out that signal amplification as defined here is a time independent variable. If applied to unidirectional (non-cyclic) phenomena, the definition is meaningless because in theory any amount of enzyme no matter how small, if given sufficient time, can catalyze the conversion of any amount of substrate to products. By analogy it might appear that any minimal fractional activation of a converter enzyme by its positive allosteric effector could yield sufficient activity to catalyze the covalent modification of any amount of protein substrate. Indeed the signal amplification should then be infinite. However, interconvertible enzymes systems are unique because the modification and demodification reactions oppose one another to yield steady states that are specified by the relative values of the reaction constants for the converter, enzymes that catalyze the forward and regeneration steps. The signal amplification as defined here is therefore a unique function of the activities of the opposing converter enzymes. Nevertheless, since the value of $\bar{n}$ and signal amplification are both time independent variables, it is evident that if signal amplification is to be of practical significance, attainment of the steady state must be achieved within a reasonable time period. In this respect it is pertinent to note that a theoretical analysis of

cyclic cascade systems by P. B. Chock (see (20) for details),
has shown the rate of covalent modification of an interconvertible enzyme in a multistep cascade is a multiplicative
function of the specific rate constants of all steps in the
cascade. Accordingly, the establishment of a steady state
in a bicyclic cascade can be achieved in the millisecond
time range.

B. Apparent Cooperativity

In the above analysis it was assumed that the only role
of effector e_1 is to activate the UR enzyme. However,
because the bicyclic cascade is composed of four different
converter enzyme activities and two interconvertible
proteins, there are at least 4 different steps in the
cascade that are susceptible to allosteric regulation.
Therefore, with any given effector multiple patterns of
regulation are possible depending upon its ability to serve
as a positive or negative modifier at different steps in
the cascade. It is therefore noteworthy that α-ketoglutarate
and glutamine can each serve as positive and negative
modifiers in at least two different steps in the GS cascade
(See Fig. 2). A theoretical analysis of such regulatory
patterns has already been published (15, 18, 19) and because
of space limitation, it cannot be described here in detail.
Essentially, the analysis shows that when the same effector
serves as a positive modifier in two forward steps in a
bicyclic cascade, or when the same effector is a positive
modifier of a forward step and also a negative modifier of
a regeneration step, then the fractional modification of the
interconvertible enzyme in the second cycle is a sigmoidal
function of the effector concentration. The cooperative-
type behavior is illustrated by the computer simulated curves
(solid lines) in Fig. 6 which show how the steady state value
of $\bar{n}$ varies as a function of e_1 concentration, at 3 different
parameter ratios, when it is assumed that e_1 activates the
UR enzyme and also the $AT_a.P_{II}$ complex. The dashed line
describes the fractional saturation of the converter enzyme
as a function of e_1 concentration when the association
constant is equal to 1.0. In contrast to the curves in
Fig. 5A, the slopes of the almost linear portions of the $\bar{n}$
versus log $[e_1]$ curves in Fig. 6 are considerably steeper
than that of the dashed line which describes a normal non-
cooperative saturation function. The steeper slope is
indicative of sigmoidicity in the response of $\bar{n}$ to e_1 concentration. From Hill-type plots the cooperativity coefficient

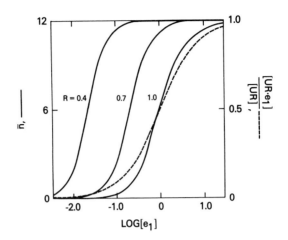

Figure 6. *Apparent cooperativity in the response of*
$\bar{n}$ *to primary effector* $[e_1]$ *concentration when this effector
activates both UR and* AT_a^1 *activities.*

(Hill number) of 1.7 is obtained. This corresponds to the
more accurately defined <u>sensitivity</u> <u>index</u> (15) of 2.7. It
follows, then, that in addition to the amplifier and amplitude
effects described above, the GS cascade is also capable of
generating cooperative-type responses to increasing concentrations of the allosteric effector glutamine.

IV.   VALIDITY OF THE CASCADE MODEL

A.   Metabolite Control of Steady State Levels
of Adenylylation

In the cascade model described here, it is assumed that
for any metabolic condition a steady state will be established in which the rates of adenylylation and deadenylylation of GS are equal. The assumption is supported by the
studies of Segal, <u>et</u> <u>al</u>. (16) showing that when a purified
preparation of GS is incubated in a mixture containing all
of the essential components of the cascade system, the
fraction of adenylylated subunits present changes with time
until a unique steady state value is reached which is dependent upon the concentrations of each one of the compounds in

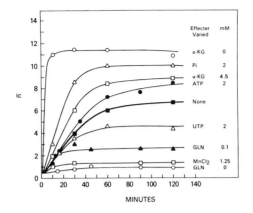

Figure 7. Effect of metabolite concentrations on the steady state level of adenylylated subunits. The heavy line (closed squares) shows the change in n̄ with time when 95 μg of GS was incubated in a mixture containing 20 mM MgCl$_2$, 20 mM Pi, 1 mM ATP, 1 mM UTP, 15 mM α-ketoglutarate, 0.3 mM glutamine and partially purified preparations of P$_{II}$, AT (containing also UR, UT) as previously described (16). The other curves illustrate the effect of changing the concentration of only one metabolite in the mixture, as indicated. The curves are derived from data in reference (16).

the mixture. The results of one such experiment are summarized in Fig. 7. The heavy line, designated by closed squares, shows that if unadenylylated GS is incubated in a mixture containing AT$_a$, AT$_d$, UR, UT, P$_{IIA}$, P$_{IID}$, and the effectors UTP, ATP, Pi, α-ketoglutarate, glutamine and Mg^{2+} or Mg^{2+} + Mn^{2+}, then the value of n̄ increases with time until a steady state value of 6.0 is reached. In.other words for these arbitrary conditions, the specific activity of GS is stabilized at one-half of its maximal value. It follows from the theoretical analysis (Eq. 1) that this steady state is determined by the relative magnitude of the products of 8 different reaction constants favoring the deadenylylation reaction and the products of the corresponding constants that favor the adenylylation reaction. Moreover, because one or more of the constants can be altered by changes in the concentrations of any one of the allosteric or substrate effectors in the incubation mixture, the steady state function predicts that a change in the concentration of any one of these effectors will cause a shift in the steady

state level of $\bar{n}$. This prediction is supported by the other
curves in Fig. 7, which show that when all other components
of the reaction mixture are held constant, a change in the
concentration of any one effector results in a shift in
the steady state value of $\bar{n}$.

B.   Signal Amplification and Sigmoidal Kinetics

The data in Fig. 7 confirm in principle the thesis that
the GS cascade is not merely a metabolic switch, to turn *ON*
or OFF GS activity, but rather it serves as a dynamic processing unit that is capable of integrating large amounts
of metabolic information and can adjust the specific
activity of GS in a smooth and continuous manner accordingly.

Verification of the unique features of the cascade system, i.e., their signal amplification potential and capacity
to generate sigmoidal responses to change in effector concentrations, requires a more rigorous kinetic analysis of
the properties of the individual cascade enzymes. A com-
prehensive analysis of the GS cascade is not yet possible
because of the instability of the UR/UT complex and the
inability to obtain it in a reasonably pure state. Nevertheless, some kinetic properties of the bicyclic cascade
have been determined by taking advantage of the fact that
the sole function of the uridylylation–deuridylylation
cycle is to control the steady state ratios of P_{IID} and
P_{IIA}. Indeed a steady state analysis of P_{II} interconversion cycle yields the equation:

$$(P_{IIA})_{mf} = \frac{\alpha_{1f}[UR]}{\alpha_{1r}[UT] + \alpha_{1f}[UR]} \qquad (3)$$

in which $(P_{IIA})_{mf}$ is the mole fraction, $P_{IIA}/(P_{IIA} + P_{IID})$,
and α_{1f} and α_{1r} are each products of the 4 different
reaction constants that govern the forward and regeneration
steps in the uridylylation cycle, respectively (See Eq. 1).

It can also be shown that Eq. 1 may be rearranged to
the expression:

$$\bar{n} = \left[\frac{1}{12} + \frac{\alpha_{2r}}{12\alpha_{2f}} \left(\frac{\alpha_{1r}[UT] + \alpha_{1f}[UR]}{\alpha_{1f}[UR]} - 1 \right) \right]^{-1} \qquad (4)$$

in which α_{2f} and α_{2r} are products of the 4 constants that govern the adenylylation and deadenylylation reactions respectively, in the second cycle.

It therefore is evident that:

$$\bar{n} = \left[\frac{1}{12} + \frac{\alpha_{2r}}{12\alpha_{2f}} \left(\frac{1}{(P_{IIA})_{mf}} - 1 \right) \right]^{-1} \qquad (5)$$

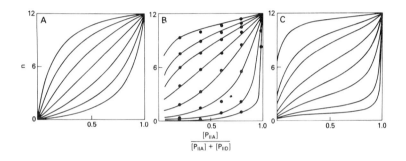

$$\frac{[P_{IIA}]}{[P_{IIA}] + [P_{IID}]}$$

Figure 8. The interdependence of the steady state level of adenylylation, the mole fraction of P_{IIA} and the ratio α_{2f}/α_{2r}. (A) Computer simulated curves with α_{2f}/α_{2r} set at 8, 4, 2, 1, 0.5, 0.25 and 0.125, from top to bottom, respectively. (B) Experimental data from studies in which $(P_{IIA})_{mf}$ was varied by varying the proportions of pure P_{IIA} and P_{IID}, and the ratio α_{2f}/α_{2r} was varied by varying the relative concentration of α-ketoglutarate and glutamine as described in reference (21). (C) Computer simulated curves based on the steady state equation describes the theoretical model in Fig. 10. For details see reference (21) from which this figure was taken.

Fig. 8A shows how, according to Eq. 5, the value of $\bar{n}$ will vary as a function of the $(P_{IIA})_{mf}$ and the α_{2f}/α_{2r} ratio. From this figure, it is evident that as the proportion of P_{IIA} is increased from 0 to 1.0, the value of $\bar{n}$ can increase in either a linear, a hyperbolic, or a parabolic manner, depending upon the α_{2f}/α_{2r} ratio. In view of the fact that the α_{2f}/α_{2r} can be modulated by changes in the concentrations of allosteric effectors, these theoretical results illustrate once again the remarkable flexibility of the GS cascade system to allosteric regulation.

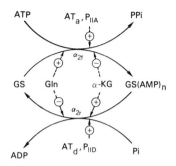

Figure 9. The modified monocyclic cascade illustrating the regulatory roles of P_{IIA}, P_{IID}, α-ketoglutarate and glutamine on the adenylylation of GS.

A comparison of Eqs. 4 and 5 shows that the response of $\bar{n}$ to perturbations in the activities of UR and UT can be mimicked by varying experimentally $(P_{IIA})_{mf}$ and the α_{2f}/α_{2r} ratio. The net effect of substituting mixtures of P_{IIA} and P_{IID} for UR and UT is to convert the bicyclic cascade system to the monocyclic cascade system depicted in Fig. 9. The figure illustrates also how the adenylylation and deadenylylation reactions are regulated in a reciprocal manner by the allosteric effectors L-glutamine and α-keto-glutarate.

With the isolation of GS, AT, P_{IIA} and P_{IID} in highly purified forms, it became feasible to test the validity of the cascade model by <u>in vitro</u> experiments. To this end, the steady state level of $\bar{n}$ was measured following the incuba-tion of GS with a purified preparation of AT in the presence of various mixtures of pure P_{IIA} and P_{IID} (to vary the $[P_{IIA}]_{mf}$) and various concentrations of α-ketoglutarate and glutamine (to vary the α_{2f}/α_{2r} ratio). As is shown in Fig. 8B, the experimental data yield a family of curves that is very similar to, but not identical with the theoretical family of curves in Fig. 8A.

An explanation for the imperfect agreement between the observed and theoretical curves was disclosed by kinetic experiments to determine the values of the 8 reaction con-stants that make up the complex constants α_{2r} and α_{2f}. these studies showed that the adenylylation and deadenylyla-

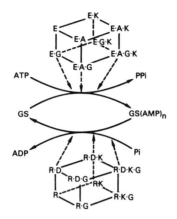

Figure 10. Mechanistic scheme for the adenylylation-deadenylylation cycle. See text for definition of notations. From reference (21).

tion reactions are catalyzed by three different complexes at both the AT_a and AT_d sites, and in addition, L-glutamine and α-ketoglutarate exhibit opposite and reciprocal effects on the adenylylation and deadenylylation reactions. Based on these studies it is evident that the adenylylation-deadenylylation cycle is more accurately described by the model depicted in Fig. 10, in which E and R denote AT_a and AT_d, respectively, A and D denote P_{IIA} and P_{IID}, respectively, and K and G represent α-ketoglutarate and L-glutamine. We will not describe here the steady state equation for this more realistic model. It has been published elsewhere, (21), suffice it to say that 10 different binding constants, 12 synergistic or antagonistic coefficients and 6 rate constants are required to describe the steady state for this model. Of these 28 constants, 22 have been estimated directly by a detailed kinetic analysis, 4 were determined by computer simulation analysis with rigorous constraints imposed by the other constants, and very reasonable values of only 2 were estimated solely by computer curve fitting (21).

By means of these 28 constants and the steady state equation of the complex model presented in Fig. 10, the steady state levels of $\bar{n}$ were computed as a function of $(P_{IIA})_{mf}$ at the various concentrations of α-ketoglutarate

and L–glutamine used in the experiments summarized in
Fig. 8B. The curves generated from the calculations (Fig.
8C) are in good agreement with the experimental curves in
Fig. 8B, showing that the steady state model depicted in
Fig. 10 can indeed explain the experimental findings. The
data show that for any given value of $(P_{IIA})_{mf}$ greater than
0 or less than 1.0, the state of adenylylation can be shifted from nearly 0 to almost 12 by varying the ratio of
α–ketoglutarate to glutamine from 0 to 0.4, and conversely,
for any given ratio of the allosteric effectors within the
range examined, the state of adenylylation can be shifted
from nearly 0 to 12 by changes in the proportions of P_{IIA}
and P_{IID}. In other words the specific activity of GS can
be varied continuously from almost zero ($\bar{n}$ = 12) to maximal
values ($\bar{n}$ = 0) merely by changes in the relative concentrations of positive and negative allosteric effectors. The
experimental data thus illustrate the extraordinary flexibility of the cyclic cascade system to metabolite
regulation.

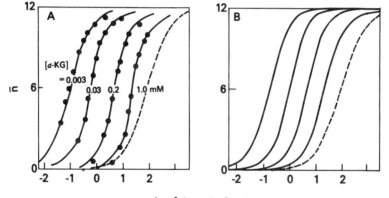

log [glutamine], mM

Figure 11.   (A) Steady state levels of $\bar{n}$ as a function
of glutamine concentration. $(P_{IIA})_{mf}$ was 0.6 and the concentration of α-ketoglutarate for curves from left to right
were 3 μM, 30 μM, 0.17 mM and 1.0 mM, respectively.   The
dashed curve is a calculated saturation curve based on the
experimentally determined K_d of 80 mM for the binding of
glutamine to AT in the absence of effectors and substrates.
(B) Computer simulated curves for the corresponding curves in
(A) using the steady state equation and the experimentally
determined reaction constants.   For details see reference
(21) from which this figure was taken.

Experimental verification of two other features of the cascade system is illustrated by the data in Fig. 11. Here the steady state value of $\bar{n}$ is plotted as a function of the concentrations of L-glutamine at several fixed levels of α-ketoglutarate, where the $(P_{IIA})_{mf}$ is held constant at a value of 0.6. The dashed line is the computed saturation curve for the binding of glutamine to AT, based on a K_d value of 80 mM, as was determined experimentally in the absence of other effectors. In agreement with a theoretical analysis of the cascade model the data show that the GS cascade is capable of high signal amplification. Whereas 80 mM glutamine is required to produce half saturation of AT in the absence of effectors, only 0.1 mM is required to obtain 50% adenylylation of GS ($\bar{n}$ = 6.0), when $(P_{IIA})_{mf}$ = 0.6 and α-ketoglutarate = 3 μM. This corresponds to a signal amplification of 800. However, as the concentration of α-ketoglutarate is increased from 3 μM to 1 mM, the concentration of glutamine required to achieve an $\bar{n}$ value of 6.0 is increased from 0.1 mM to 21 mM. The signal amplification is therefore decreased from 800 to 4.

Furthermore, it can be seen that the slopes of nearly linear portions of the $\bar{n}$ vs. glutamine concentration curves in Fig. 11A are greater than the dashed line which describes non-cooperative binding of glutamine to AT in the absence of effectors and substrate. These greater slopes show that the value of $\bar{n}$ is a sigmoidal function of the glutamine concentrations. As pointed out earlier, the cooperative-type of behavior, (yielding in this case a Hill number of 1.5) is predicted by the cascade model and reflects the fact that glutamine is a positive effector of the adenylylation reaction and also inhibits the deadenylylation reaction. Utilizing the more rigorous steady state equation, (not shown) and the 28 experimentally determined reaction constants needed to describe the cascade model depicted in Fig. 10, computer-simulated curves were obtained for the experimental condition used in obtaining the data summarized in Fig. 11A. The excellent agreement between the experimental curves in Fig. 11A and the corresponding computer-simulated curves in Fig. 11B support the validity of the cascade model for GS regulation.

V. ROLE OF ATP DECOMPOSITION

Finally, it should be noted that in the steady state analysis developed here, the roles of ATP and UTP as nucleotidyl group donors in the two interconversion cycles

has been ignored. This does not invalidate the theoretical
analysis, because for a given physiological state the concentrations of ATP and UTP are maintained at nearly constant
levels that are several orders of magnitude greater than the
concentrations of the cascade enzymes. Nevertheless, with
each interconversion cycle one equivalent of nucleoside
triphosphate is consumed. This triphosphate consumption is
absolutely necessary since it provides the energy that is
needed to sustain steady state levels of the covalently
modified enzyme that are far removed from the thermodyna-
mically stable states.

VI.   SUMMARY AND CONCLUSIONS

To summarize, it has been demonstrated that the
regulation of glutamine synthetase in E. coli is mediated
by a cascade system composed of two interconvertible protein
cycles. The uridylylation and deuridylylation of a small
regulatory protein in the first cycle modulates the
adenylylation and deadenylylation of glutamine synthetase
in the second cycle. For any given metabolic condition a
steady state is established in which the rate of adenylylation of glutamine synthetase and the rate of deadenylylation
are equal. The fraction of adenylylated subunits in this
steady state determines the specific activity of glutamine
synthetase and is specified by the magnitudes of 16 reaction
constants that govern the activities of the converter enzymes
that catalyze the covalent modification reactions. Through
multiple interactions with allosteric and substrate effectors
the four converter enzyme activities are programmed to sense
changes in the concentrations of a multitude of metabolites.
By causing changes in the catalytic constants of the cascade
enzymes, these multiple interactions lead continuously to
rapid shifts in the steady state distribution of adenylylated
and unadenylylated GS subunits, and consequently to changes
in the specific activity of glutamine synthetase commensurate
with the ever changing metabolic need for glutamine. In
addition to its role as a metabolic integration system, the
cascade is endowed with an enormous signal amplification
potential, making it possible to obtain large shifts in the
specific activity of GS in response to changes in primary
effector concentration that are orders of magnitude lower
than the dissociation constants of the effector-converter
enzyme complexes. Furthermore, the cascade exhibits unusual
flexibility with respect to allosteric control patterns.
Depending upon the metabolic situation it can not only
respond differently to changes in the concentration of a

given metabolite, but it can also modulate the amplitude of
the metabolite effect, and can generate a cooperative-type
of response to one particular allosteric effector and not
another. These unique characteristics provide the versatil-
ity and sensitivity needed in the regulation of one of the
most important enzymes in metabolism.

REFERENCES

1. Stadtman, E. R., in "The Enzymes of Glutamine
   Metabolism" (S. Prusiner and E. R. Stadtman, eds.)
   p. 1, Academic Press, New York, (1973)

2. Magasanik, B., Prival, M.J., Brenchley, J. E., Tyler,
   B. M., DeLeo, A. B., Streicher, S. L., Bender, R. A.,
   and Paris, C. G., in, "Current Topics in Cellular
   Regulation," Vol. 8, (B. L. Horecker and E. R. Stadtman,
   ets.), p. 119, Academic Press, New York (1974)

3. Stadtman, E. R. and Ginsburg, A., in "The Enzymes,"
   3rd Ed., (P. D. Boyer, ed.), p. 755, Academic Press,
   New York (1974)

4. Kingdon, H. S., Shapiro, B. M., and Stadtman, E. R.
   Proc. Nat. Acad. Sci., U.S., 58, 1703 (1967)

5. Wulff, K., Mecke, D., and Holzer, H. Biochem. Biophys.
   Res. Commun. 28, 740 (1967)

6. Shapiro, B. M., and Stadtman, E. R. J. Biol. Chem.
   243, 3769 (1968)

7. Shapiro, B. M., and Ginsburg, A. Biochemistry, 7,
   2153 (1968)

8. Valentine, R. C., Shapiro, B. M., and Stadtman, E. R.
   Biochemistry, 7, 2143 (1968)

9. Anderson, W. B., and Stadtman, E. R. Biochem. Biophys.
   Res. Commun. 41, 704 (1970)

10. Shapiro, B. M. Biochemistry, 8, 659 (1969)

11. Brown, M. S., Segal, A., and Stadtman, E. R. Proc. Nat.
    Acad. Sci. U.S., 68, 2949 (1971)

12. Mangum, J. H., Magni, G., and Stadtman, E. R. <u>Arch</u>.
     <u>Biochem</u>. <u>Biophys</u>. <u>158</u>, 514 (1973)

13. Adler, S. P., Purich, D., and Stadtman, E. R. <u>J</u>. <u>Biol</u>.
     <u>Chem</u>. <u>250</u>, 6264 (1975)

14. Bancroft, S., Rhee, S. G., Neumann, C., and Kustu, S.
     <u>J</u>. <u>Bacteriol</u>. <u>134</u>, 1046 (1978)

15. Stadtman, E. R., and Chock, P. B., <u>in</u> "Current Topics
     in Cellular Regulation," Vol. 13, (B. L. Horecker and
     E. R. Stadtman, eds.), p. 53, Academic Press, New York
     (1978)

16. Segal, A., Brown, M. S., and Stadtman, E. R. <u>Arch</u>.
     <u>Biochem</u>. <u>Biophys</u>. <u>161</u>, 319, (1974)

17. Pettit, F. H., Pelley, J. W., and Reed, L. J. <u>Biochem</u>.
     <u>Biophys</u>. <u>Res</u>. <u>Commun</u>. <u>65</u>, 575 (1975)

18. Stadtman, E. R., and Chock, P. B. <u>Proc</u>. <u>Nat</u>. <u>Acad</u>. <u>Sci</u>.
     <u>U.S</u>. <u>74</u>, 2761 (1977)

19. Chock, P. B. and Stadtman, E. R. <u>Proc</u>. <u>Nat</u>. <u>Acad</u>. <u>Sci</u>.
     <u>U.S</u>. <u>74</u>, 2766 (1977)

20. Stadtman, E. R., and Chock P. B. <u>in</u> "The Neurosciences
     Fourth Study Program," (F. O. Schmitt, ed.) p. 801,
     MIT Press (1979)

21. Rhee, S. G., Chock P. B., and Stadtman, E. R. <u>Proc</u>.
     <u>Nat</u>. <u>Acad</u>. <u>Sci</u>., <u>U.S</u>. <u>75</u>, 3138 (1978).

DISCUSSION

P. COHEN: Could you just confirm, is this still the only system
in which adenylylation and uridylylation have been identified
and if this mechanism is still restricted to just the few
classes of bacteria. Is this still the position?

E.R. STADTMAN: Yes, that is correct. To my knowledge there is
no other adenylylation or uridylylation mechanism for
regulation of other enzymes. None have been demonstrated at

least. Slight adenylation of aspartokinase in E. coli has been observed, but its role in regulation has not been demonstrated. As far as the organisms are concerned, this form of regulation, seems to be restricted to a glutamine synthetase and all gram negative bacteria. It has not been observed in any gram positive organisms nor in the higher forms of life.

P. COHEN: Could you just do so briefly, review for us the current state of the protein chemistry of the system, particularly the glutamine synthetase molecule itself, as to how far advanced this is.

E.R. STADTMAN: The glutamine synthetase has been studied very extensively by Dr. Ginsburg and her collaborators, and in earlier studies by Shapiro and Woolfolk. As I have already mentioned, it is a dodecamer, composed of two hexagonal rings, of six subunits each of around 50,000 molecular weight. Each of these subunits, has the capacity to be adenylylated. Earlier studies by Woolfolk had indicated that on each subunit of the enzyme there is a separate allosteric site for six to eight different allosteric effectors. As a consequence of the fact that these effectors can react at separate sites on the enzyme their effects are more or less independent of one another, and their combined effects are cumulative. More recent studies of Chock and Rhee support the conclusion that there are multiple allosteric sites on each subunit.

P. COHEN: In terms of what we know about the three dimensional structure of interconvertible enzymes at the present time, the most advanced knowledge we have is about glycogen phosphorylase. Has any crystallographic studies on the glutamine synthetase molecule been initiated?

E.R. STADTMAN: Yes, and in addition to X-ray analysis there have been a lot of EPR and NMR studies that have mapped the positions of several allosteric binding sites with respect to the two metal binding sites of the enzyme. The actual distances between the metal ion binding sites and some of the allosteric sites have been measured directly by these physical chemical methods. In addition by the use of fluorometric techniques, it has been possible to learn something about the mechanism of the catalytic process when glutamine synthetase converts ammonia and glutamate to glutamine. Using this approach Chock and Rhee have shown that the catalytic mechanism involves at least three consecutive steps and that several feedback inhibitors exert their effects by inhibiting different steps in the catalytic process. So I would say that a great deal of detailed information is available on the properties of the enzyme, the catalytic mechanism and the mechanism of feedback inhibition as well.

A.R. PETERSON: Is it reasonable to expect that the regulation of pools of purine and pyrimidine nucleotides used in DNA synthesis could be profoundly affected by this particular cascade?

E.R. STADTMAN: I think the pools of pyrimidine and purine nucleotides would certainly be influenced by the activities of RNA and DNA synthesis and that the activity of glutamine synthetase would respond to the changes in those pool sizes so that glutamine could continue to be formed at a rate which would, in fact, be commensurate with the requirement for those biological processes.

POLY(ADP-RIBOSE) AND ADP-RIBOSYLATION OF PROTEINS[1]

O. Hayaishi, K. Ueda, M. Kawaichi, N. Ogata, J. Oka,
K. Ikai, S. Ito, Y. Shizuta, H. Kim and H. Okayama

Department of Medical Chemistry
Kyoto University Faculty of Medicine
Kyoto, Japan

Introduction

Poly(ADP-ribose) and the ADP-ribosylation of proteins constitute a novel type of covalent modification of proteins and
have recently attracted the attention of a number of molecular
biologists, because they are ubiquitously distributed in nature and are implicated in the regulation of cell proliferation, protein synthesis, and DNA as well as RNA metabolism (1).
This type of post-translational modification is unique because
NAD, nicotinamide adenine dinucleotide, whose primary function
is an electron carrier in biological oxidation, invariably
provides the ADP-ribosyl moiety which is transferred on to a
protein molecule. The ADP-ribosyl unit thus covalently attached to a protein acceptor is present as either a monomer or
a polymer as in the case of the nuclear system. In this article, I shall attempt to summarize recent developments in this
field of research and to present some of our current thoughts
and experimental results. By way of orientation, I will first
review briefly, mono ADP-ribosylation of proteins, in which
only a single ADP-ribosyl moiety is transferred to a protein
acceptor. Then I would like to cover a more complex reaction,
poly(ADP-ribose) in nuclei, in which the ADP-ribosyl units are
polymerized. The structure, the biosynthesis and degradation
and the possible function of poly(ADP-ribose) will be discussed.

---

[1]Supported in part by grants-in-aid for Cancer Research
from the Ministry of Education, Science and Culture, Japan.

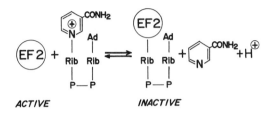

FIGURE 1. Mono ADP-ribosylation of elongation factor 2
by diphtheria toxin.

Mono ADP-ribosylation

In 1968, we reported that diphtheria toxin catalyzed the
transfer of the ADP-ribosyl moiety of NAD to elongation factor
2 (EF2) to form mono ADP-ribosyl EF2 with a concomitant release of nicotinamide and a proton (2) (Fig. 1). Diphtheria
toxin is an exotoxin produced by Corynebacterium diphtheriae,
and is the first example of an ADP-ribosyl transferase, which
catalyzes mono ADP-ribosylation of proteins. EF2 is an enzyme,
that is involved in protein synthesis in eukaryotic cells.
This factor catalyzes the translocation of peptidyl t-RNA on
the ribosome. The ADP-ribosylated EF2 thus formed was catalytically inactive. Diphtheria toxin, therefore, prevents protein synthesis and kills the host cells. In contrast to the
nuclear poly(ADP-ribose) synthetase system, which will be discussed later, only one ADP-ribosyl unit is transferred specifically onto each molecule of EF2 and the reaction is reversible.
Subsequent to our original finding of diphtheria toxin,
a number of similar mono ADP-ribosylation reactions has been
reported. The enzymes and the acceptors presently reported to
be involved in mono ADP-ribosylation reactions are summarized
in Table I.
Pseudomonas toxin was recently shown to catalyze the
reaction identical to that catalyzed by diphtheria toxin (3).
Pseudomonas exoenzyme S was reported to catalyze a similar reaction, but the acceptor protein has not been exactly identified (4). Furthermore, the virions of E. coli phage T_4 (5)
and N_4 (6) contain NAD : protein ADP-ribosyl transferases with
a broad acceptor specificity. Infection of E. coli with T_4
phage induces another enzyme that catalyzes a more specific
ADP-ribosylation of arginyl residues of α-subunits of RNA polymerase (7, 8). More recently, several enzymes such as cholera
toxin (9, 10), E. coli enterotoxin (11) and avian erythrocyte
protein (12) were reported to catalyze mono ADP-ribosylation

TABLE I.   Mono ADP-ribosylation reactions

Enzyme	Acceptor
Diphtheria toxin	Elongation factor 2
Pseudomonas toxin	Elongation factor 2
Pseudomonas exoenzyme S	Elongation factor I (?)
T4 phage { Viral	*E.coli* proteins
{ Induced	RNA polymerase
N4 phage	*E.coli* proteins
Cholera toxin	Adenylate cyclase
E. coli enterotoxin	Adenylate cyclase
Avian erythrocyte protein	Adenylate cyclase

of a GTP binding component of adenylate cyclase. It is worthy
of note that most enzymes that catalyze mono ADP-ribosylation
reactions are found in prokaryotes rather than in eukaryotes.

Poly ADP-ribosylation

Now I would like to move on to the main topic, the poly-
(ADP-ribose) system in nuclei and talk about first the structure of this unique macromolecule and its acceptor proteins.
The basic structure of poly ADP-ribosyl protein is schematically illustrated in Fig. 2. It is a unique homopolymer
composed of a linear sequence of repeating ADP-ribose units
linked together by ribose 1" to ribose 2' glycosidic bonds and
covalently attached to an acceptor protein through a terminal
ribose. The chain length of this polymer, synthesized either
in vitro or in vivo, ranges from 1 up to about 100 ADP-ribose

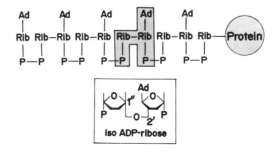

FIGURE 2.   Basic structure of poly ADP-ribosyl protein.

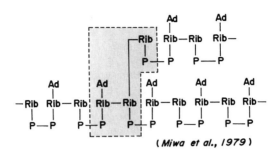

(Miwa et al., 1979)

FIGURE 3.   Proposed branched structure.

units depending on the conditions.   Recently, several investigators have suggested a branching structure of poly(ADP-ribose).
For example, Miwa and coworkers isolated a compound, with a structure shown in Fig. 3 from a snake venom phosphodiesterase digest of high molecular weight poly(ADP-ribose) (13).   Although the final structure is yet to be established, these authors suggested that the frequency of branching may be about one per 20-30 ADP-ribose residues of poly(ADP-ribose) of high molecular weight.

Now I should like to discuss, in some detail, the nature of the acceptor protein and the linkage through which the terminal ribose of poly(ADP-ribose) is attached to the acceptor molecule.

As early as 1968, we reported that ADP-ribose might be covalently bound to histones through an ester linkage (14).
Several other proteins and peptides in chromatin have also been reported to be ADP-ribosylated.   However, definitive characterization of these acceptor proteins has not been successful because both acceptor proteins and the polymer size are hetero-

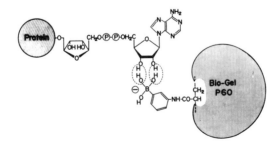

FIGURE 4.   Affinity chromatography of ADP-ribosyl protein on borate-polyacrylamide gel.

genous, they tend to aggregate and the ADP-ribosyl protein
linkages are rather unstable.

Recently, we made use of affinity chromatography on a borate polyacrylamide gel column and were able to make more definitive identification of the acceptor proteins (15). As
shown in Fig. 4, borate interacts specifically with cis-diol-
containing compounds at alkaline pH's. Therefore, rat liver
nuclei were incubated with radioactive NAD and extracted with
dilute HCl. When the extract was subjected to borate gel chromatography in the presence of 6 M guanidine·HCl, the modified
proteins were bound and separated from the bulk protein (Fig.
5). The broken line represents the amount of protein and the
solid line ^{14}C-ADP-ribosylated molecules. When pH was maintained at 8.2 where boric acid was negatively charged, more
than 95% of the protein passed through the column but most
ADP-ribosylated proteins were retained on the gel. At pH 6.0
these compounds were eluted because there was no interaction
when boric acid was uncharged. The ADP-ribosylated proteins
thus isolated were further purified by CM-cellulose column
chromatography in the presence of 7 M urea, and separated into
four fractions A, B, C, and D, and then analyzed by SDS-gel
slab electrophoresis (Fig. 6). Under these conditions ADP-
ribose was completely removed from the proteins; proteins were
therefore visualized by Coomassie Blue staining. Here NC stands
for crude rat liver nuclear proteins and STD stands for standard proteins of known molecular weights for reference. Fractions A and B contain several non-histone proteins. Fraction
C contains H2A, H2B, H3 or
H4 plus small amounts of non-
histone proteins. Fraction
D appears to contain H1 almost exclusively. Further
analysis by acid urea polyacrylamide gel electrophoresis and amino acid analysis
provided definitive evidence
that the major acceptor proteins in fractions C and D
were histones H2B and H1,
respectively.

The use of a borate gel
column thus provided a simple
and convenient method for the
isolation of ADP-ribosylated
proteins and enabled a more
definitive characterization
of the acceptor proteins.
The results are schematically

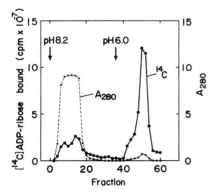

FIGURE 5. Isolation of
ADP-ribosylated histones with
borate gel chromatography.

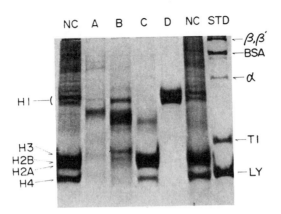

FIGURE 6.   SDS-gel electrophoresis of ADP-ribosylated proteins.

shown in Fig. 7.   We were able to demonstrate that the major acceptor proteins were histones H1 and H2B, but non-histone proteins such as A24 (16) and so-called "high mobility group proteins" (17) were also ADP-ribosylated under these conditions.

Then, we investigated the linkage and site of poly ADP-ribosylation on histone H1 (18).   ADP-ribosylated histone H1 was extracted with 5% perchloric acid and bisected with N-bromosuccinimide (Fig. 8).   The two fragments referred to as NBS_N and NBS_C were separated by gel filtration and then the stability of the linkages in the whole histone and these two fragments was investigated under different pH's and also in the presence of neutral hydroxylamine.   These fragments were then subjected to treatments using various proteases and peptidases.   The results of stability experiments and enzymatic degradation are presented in Figs. 9 and 10, respectively.

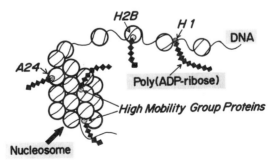

FIGURE 7.   Chromatin structure and poly ADP-ribosylation.

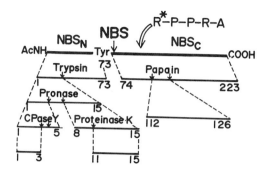

FIGURE 8. Fragmentation of poly ADP-ribosyl histone H1.

Fig. 9 shows the stability of the ADP-ribosyl histone
linkages of whole H1, NBS$_C$ and NBS$_N$ fragments. It can be seen
that the stability of poly ADP-ribosyl histone linkages under
different pH values or in the presence of neutral hydroxylamine is essentially identical in the whole histone H1 as well
as in the NBS$_C$ and NBS$_N$ fragments, suggesting that these linkages are homogeneous and involve an ester bond with a carboxyl
group of either a glutamate or an aspartate residue of histone because other known linkages of glycoproteins such as N-
and O-glycosides are much more stable under these conditions.
     The results of fragmentation experiments and amino acid
analysis revealed that glutamate residues No.2, 14 and 116 are
ADP-ribosylated in histone H1 (Fig. 10). It has been suggested that the N-terminal region and the carboxyl half of the
molecule play an important role in the DNA binding of H1 while

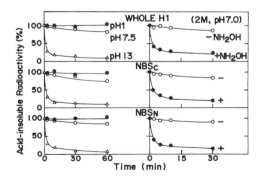

FIGURE 9. Stability of poly ADP-ribosyl histone.

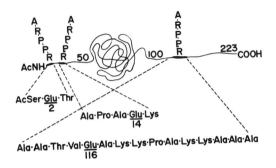

FIGURE 10. Poly ADP-ribosylation sites of histone H1.

the apolar central region is required for other functions.   The
well-defined apolar central region folds up with a precise
globular conformation, while the N-terminal and C-terminal
regions remain unstructured.   I would like to call your attention to the fact that all three ADP-ribosylated glutamate residues are present in the so-called N-terminal and the carboxyl
terminal regions, and appear to play a role in the binding to
DNA.

So much for the structure of poly(ADP-ribose) and acceptor
proteins; now I would like to turn to the enzymatic synthesis
of poly(ADP-ribose).   Enzymatic synthesis of poly ADP-ribosyl
protein is catalyzed by poly(ADP-ribose) synthetase and two
seemingly different reactions are involved.   (1) The initial
attachment of the ADP-ribose unit of NAD to an acceptor protein and (2) subsequent elongation of the chain to produce a
polymer (Fig. 11).   Poly(ADP-ribose) synthetase has been extensively purified and investigated from several sources.   Purification of calf thymus poly(ADP-ribose) synthetase is shown
in Table II.   The enzyme was purified from crude extracts of

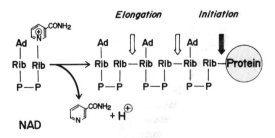

FIGURE 11.   Enzymatic synthesis of poly ADP-ribosyl
protein.

TABLE II.   Purification of calf thymus poly(ADP-ribose) synthetase

Fraction	Total protein	Specific activity	Yield
	mg	nmol/min/mg	%
Crude extract	29,800	1	100
DNA-agarose	137	146	69
Hydroxyapatite	20	522	37
Sephadex G-150	3	1,250	14

calf thymus about 1,250-fold, with an over-all yield of 14% by three steps (19). Gel electrophoretic patterns of these four preparations are illustrated in Fig. 12. The final preparation is apparently homogeneous. The properties of this final preparation are shown in Table III. The molecular weight was estimated to be about 110,000. $S_{20,w}$ is 5.8 and the frictional ratio about 1.4. It is a basic protein but DNA contents are negligible. Some catalytic parameters are also shown here. This is the best preparation of poly(ADP-ribose) synthetase ever reported in terms of both molecular and catalytic properties.

As is the case with other poly(ADP-ribose) synthetase preparations, the enzyme required the presence of DNA for activity. When DNA was omitted from the incubation mixture, little activity was detected (Table IV).

TABLE III. Properties of calf thymus poly(ADP-ribose) synthetase

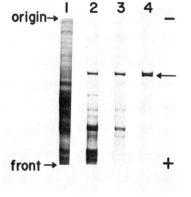

FIGURE 12. SDS-gel electrophoresis at each stage of purification.

M_r	11.0×10^4
$s^0_{20,w}$	5.80 S
f/f_0	1.39
pI	9.8
DNA content	N.D.
V_{max}	1,430 nmol/min/mg
K_m for NAD	55 μM
for DNA	2.5 μg/ml

TABLE IV.   Requirements of calf thymus poly(ADP-ribose) synthetase

Conditions	Enzyme activity	
	nmol/min/mg	%
Complete	990	100
−DNA	15	1
−Histone	430	43
−DNA, −Histone	15	1
−MgCl$_2$	470	47
−Dithiothreitol	600	61
Boiled enzyme	3	0

Histone stimulated the activity several fold in the presence of DNA.

The role of DNA in this reaction is not fully understood. Recently we were able to demonstrate that a mixture of decamers of poly(dA) and poly(dT) was as effective as native DNA with a molecular weight of several millions (Fig. 13)(20). However, a mixture of poly(dC) and poly(dG) was less than half as active and poly(dA) alone was almost ineffective, indicating that the chain length of DNA was not so important, while the presence of dA and dT was required. On the other hand, histones serve for two functions; one, as an activator and the other, as an acceptor protein.

When a highly purified synthetase was incubated with about 5 μg each of DNA and histone H1 per ml in the presence of radioactive NAD, and the product of the reaction was extracted with dilute acid and analyzed by SDS-polyacrylamide gel electrophoresis, there was no radioactive peak corresponding to the stained band of histone H1, as shown in the top figure (Fig. 14) (21). Under these conditions, histone did not serve as an acceptor protein but stimulated the synthetase activity several fold.

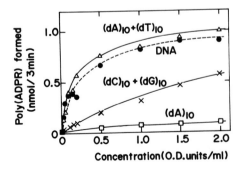

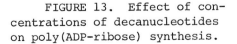

FIGURE 13.   Effect of concentrations of decanucleotides on poly(ADP-ribose) synthesis.

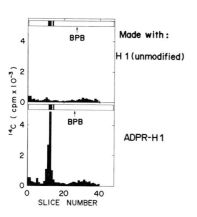

FIGURE 14. SDS-polyacryl-
amide gel electrophoresis of
poly ADP-ribosylated products.

However, when ADP-ribosyl
histone H1 was prepared by
chemical synthesis or by preincubating rat liver nuclei
with cold NAD, and ADP-
ribosyl histone H1 thus prepared was used in place of
unmodified histone H1, a
large amount of radioactive
ADP-ribosyl moiety was shown
to be associated with the
histone band suggesting that
the chain was elongated under
these conditions.

The above experiments
were carried out with about
5 μg each of DNA and histone
per ml of a reaction mixture.
When the concentrations of
native histone H1 and
DNA were increased to about
50 μg, then a large radioactive peak was observed that
corresponded exactly to his-
tone H1, indicating that initiation of the poly(ADP-ribose)
synthesis had occurred (Fig. 15). When the sample was treated
with proteinase K, the radio-
activity migrated along with
the indicator dye bromo-
phenol blue indicating
that the radioactive poly-
(ADP-ribose) was covalently
bound to the protein. In
order to ascertain that initiation reaction took place
under these conditions, the
following experiment was carried out.

In this experiment, ^{14}C-
ribose containing NAD was
used as substrate and poly
ADP-ribosylated protein was
isolated. Then, the protein
was removed by alkali treatment and the free poly(ADP-
ribose) chains were sub-
jected to the action of
snake venom phosphodiesterase.

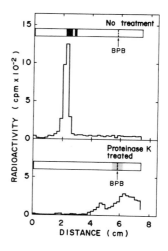

FIGURE 15. Association
of poly(ADP-ribose) with
histone H1.

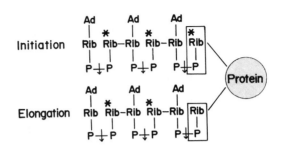

FIGURE 16.   Design of initiation experiment.   (*, ^{14}C)

The terminal ribose 5-phosphate was isolated and radioactivity
determined.   If initiation had occurred, the ribose 5-phos-
phate thus isolated should have the same specific radioactivity as the substrate, NAD, whereas if only elongation had occurred on the preexisting ADP-ribose moiety, the ribose 5-phos-
phate should not contain any radioactivity (Fig. 16).   Under
the conditions mentioned above, more than 80% of the radioactivity was retained in ribose 5-phosphate indicating that initiation was the predominant type of the reaction observed under
these conditions.

    So much for the biosynthesis of this polymer.   Fig. 17
summarizes the enzymatic reactions involved in the degradation
of poly ADP-ribosyl proteins.   Until recently, two different
types of enzymes have been reported.   Poly(ADP-ribose) glycohydrolase cleaves the ribose-ribose linkage yielding ADP-
ribose as the reaction product (22, 23).   This enzyme appears
to be responsible for the degradation of poly(ADP-ribose)

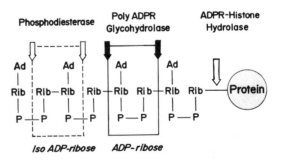

FIGURE 17.   Enzymatic degradation of poly ADP-ribosyl
protein.

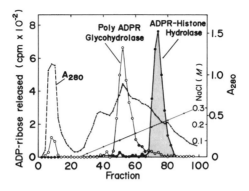

FIGURE 18. DEAE-cellulose column chromatography of ADP-ribosyl histone hydrolase.

in vivo and has been partially purified from calf thymus, rat liver and testis. The other type of enzyme is snake venom phosphodiesterase which cleaves the pyrophosphate linkage yielding isoADP-ribose. A similar enzyme was isolated from rat liver by Futai and Mizuno (24) but the physiological function is at present unknown. The enzyme which catalyzes the cleavage of the ADP-ribosyl histone linkage has been looked for by a number of investigators. We recently found an enzyme which liberates ADP-ribose from mono ADP-ribosyl histones and tentatively termed it "ADP-ribosyl histone hydrolase" (25). Purification of this enzyme is illustrated in Fig. 18. The high-speed supernatant fraction of rat liver homogenate was precipitated with 45% saturated ammonium sulfate and then chromatographed on a DEAE-cellulose column. The broken line represents the amount of protein. The ADP-ribosyl histone hydrolase activity was eluted in a single peak and was clearly separated from the peak of the well-known poly(ADP-ribose) glycohydrolase. In contrast

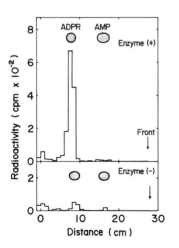

FIGURE 19. Paper chromatography of split products from ADP-ribosyl histone H1.

TABLE V. Poly ADP-ribosylation reactions

Enzyme	Acceptor
Nuclear	Chromosomal proteins
	Histones
	Non-histone proteins
	Mg^{++}, Ca^{++}-endonuclease
	RNA polymerase
Mitochondrial	Mitochondrial proteins
Cytoplasmic	Histones

to the latter enzyme, hydrolase catalyzes the splitting of a bond between ADP-ribose and the protein portion in mono ADP-ribosylated histones H1 and H2B but little, if any, of the glycosidic bonds within poly(ADP-ribose) chains. Analysis of the reaction product is shown in Fig. 19. As shown here, the split product was tentatively identified as ADP-ribose on paper chromatography. However, it was not exactly identical with ADP-ribose on Dowex 1 column chromatography. Preliminary evidence suggested that it was either an altered ADP-ribose molecule produced by structural rearrangement, or ADP-ribose bound to an unidentified small molecule and identification studies are now under way.

Finally, I should like to say a few words about the possible physiological function of poly(ADP-ribose). Table V summarizes the enzymes and acceptors reportedly involved in poly ADP-ribosylation reaction in eukaryotes. The nuclear enzymes catalyze the ADP-ribosylation of histones as well as several non-histone proteins. A Mg^{++}, Ca^{++}-dependent endonuclease (26) and RNA polymerase (27) were also reported to serve as acceptor proteins. In addition, Kun reported a mitochondrial enzyme which ADP-ribosylated a mitochondrial protein (28). Smulson and coworkers described a cytoplasmic, probably ribosomal, enzyme, which ADP-ribosylated histones (29).

A number of investigators suggested that poly(ADP-ribose) may play a role in the regulation of (1) DNA and RNA metabolism, (2) cell proliferation, differentiation and transformation, (3) structural integrity and DNA binding of nuclear proteins and (4) the NAD level in the cell. These inferences are certainly provocative but in most cases based on circumstantial evidence. In an attempt to answer this question, we recently prepared specific immune sera against poly(ADP-ribose) and studied intracellular localization, natural distribution and

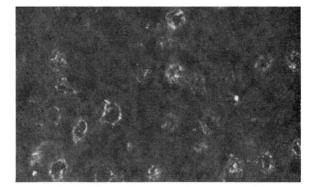

FIGURE 20.   Localization of poly(ADP-ribose) in liver
cells of adult rat, X 480.

possible biological function of poly(ADP-ribose).   In the
remainder of my talk I should like to say a few words concerning some preliminary results of these immunohistochemical
studies.
    Fig. 20 shows a rat liver section prepared on a Cryostat
and stained by indirect immunofluorescence technique. As this
section is approximately 4 μ thick, most cells were cut by
this technique. As you can see, poly(ADP-ribose) is localized
in the peripheral region that coincides with the localization
of heterochromatin. When a control serum was used, the presence of poly(ADP-ribose) could hardly be demonstrated. Similar
but not exactly identical pictures were obtained with slices of
all other tissues including brain, heart and so forth, indicating that poly(ADP-ribose) is ubiquitously distributed in
nuclei of most tissues in the rat and probably other mammalian
species as well.

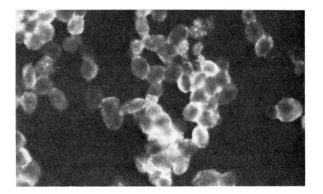

FIGURE 21.   Immunofluorescent staining of poly(ADP-ribose)
synthesized in rat liver nuclei, X 480.

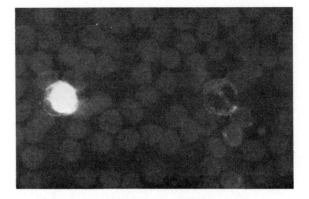

FIGURE 22. Immunostaining of poly(ADP-ribose) in normal peripheral blood preincubated with NAD (left, lymphocyte; right, polymorphonuclear leukocyte), X 600.

When nuclei were isolated from rat liver and stained by the same procedure, poly(ADP-ribose) is hardly detectable because poly(ADP-ribose) present in vivo is probably degraded

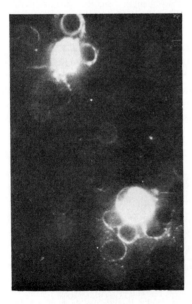

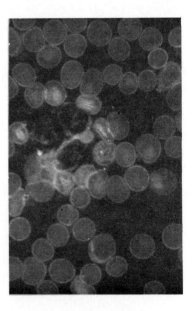

FIGURE 23. Demonstration of poly(ADP-ribose) synthesis in peripheral blood of a patient with acute myeloblastic leukemia, X 600.

FIGURE 24. Absence of immunofluorescence of poly-(ADP-ribose) in peripheral blood of a patient with chronic myelocytic leukemia, X 600.

during the isolation procedures. However, when isolated rat liver nuclei were preincubated with NAD, poly(ADP-ribose) is clearly seen in the nuclei as shown in Fig. 21.

Poly(ADP-ribose) is widely distributed in almost all types of cell nuclei, with only one exception. We were not able to detect its presence in polymorphonuclear leukocytes, while lymphocytes contain a significant quantity (Fig. 22). In agreement with this finding, the poly(ADP-ribose) synthetase activity was not detectable in the nuclei of polymorphonuclear leukocytes. While poly(ADP-ribose) and the synthetase activity were not detectable in normal polymorphonuclear leukocytes, a considerable amount of poly(ADP-ribose) was demonstrated in nuclei of myeloblasts, the immature leukocytes seen in cases of acute myeloblastic leukemia (Fig. 23).

In contrast, premature leukocytes in chronic patients do not contain poly(ADP-ribose) just like normal leukocytes (Fig. 24). Studies are now in progress to determine whether or not poly(ADP-ribose) plays a role in the differentiation of leukocytes and this finding is of clinical value for the differential diagnosis of acute and chronic leukemias.

In summary, the increasing number of poly and mono ADP-ribosylation reactions in both eukaryotes and prokaryotes reported during the last few years, indicates that these reactions constitute a new and rather general way by which proteins are modified covalently and such reactions may play a major role in the regulation of various activities of the cell, particularly the chromatin function and the metabolism of proteins and nucleic acids.

REFERENCES

1. Hayaishi, O., and Ueda, K., Ann. Rev. Biochem. 46, 95-116
    (1977).
2. Honjo, T., Nishizuka, Y., Hayaishi, O., and Kato, I., J.
    Biol. Chem. 243, 3553-3555 (1968).
3. Iglewski, B. H., and Kabat, D., Proc. Nat. Acad. Sci. USA
    72, 2284-2288 (1975).
4. Iglewski, B. H., Sadoff, J., Bjorn, M. H., and Maxwell,
    E., Proc. Nat. Acad. Sci. USA 75, 3211-3215 (1978).
5. Zillig, W., Mailhammer, R., Storko, R., and Rohrer, H.,
    Curr. Top. Cell. Regul. 12, 263-271 (1977).
6. Pesce, A., Casoli, C., and Schito, G. G., Nature 262,
    412-414 (1976).
7. Goff, C. G., J. Biol. Chem. 249, 6181-6190 (1974).
8. Rohrer, H., Zillig, W., and Mailhammer, R., Eur. J. Bio-
    chem. 60, 227-238 (1975).
9. Cassel, D., and Pfeuffer, W., Proc. Nat. Acad. Sci. USA
    75, 2669-2673 (1978).
10. Gill, D. M., and Meren, R., Proc. Nat. Acad. Sci. USA
    75, 3050-3054 (1978).
11. Moss, J., and Richardson, S. H., J. Clin. Invest. 62,
    281-285 (1978).
12. Moss, J., and Vaughan, M., Proc. Nat. Acad. Sci. USA 75,
    3621-3624 (1978).
13. Miwa, M., Saikawa, N., Ohashi, Z., Nishimura, S., and
    Sugimura, T., Proc. Nat. Acad. Sci. USA in press.
14. Nishizuka, Y., Ueda, K., Honjo, T., and Hayaishi, O., J.
    Biol. Chem. 243, 3765-3767 (1968).
15. Okayama, H., Ueda, K., and Hayaishi, O., Proc. Nat. Acad.
    Sci. USA 75, 1111-1115 (1978).
16. Okayama, H., and Hayaishi, O., Biochem. Biophys. Res.
    Commun. 84, 755-762 (1978).
17. Kawaichi, M., Ueda, K., and Hayaishi, O., Seikagaku 50,
    920 (1978).
18. Ogata, N., Ueda, K., and Hayaishi, O., Seikagaku 50, 920
    (1978).
19. Ito, S., Shizuta, Y., and Hayaishi, O., J. Biol. Chem.
    in press.
20. Shizuta, Y., Ito, S., Shizuta, H., and Hayaishi, O.,
    Seikagaku 50, 919 (1978).
21. Ueda, K., Kawaichi, M., Okayama, H., and Hayaishi, O., J
    Biol. Chem. in press.
22. Ueda, K., Oka, J., Narumiya, S., Miyakawa, N., and Haya-
    ishi, O., Biochem. Biophys. Res. Commun. 46, 516-523
    (1972).
23. Miwa, M., Tanaka, M., Matsushima, T., and Sugimura, T.,

J. Biol. Chem. 249, 3475-3482 (1974).

24. Futai, M., and Mizuno, D., J. Biol. Chem. 242, 5301-5307
     (1967).

25. Okayama, H., Honda, M., and Hayaishi, O., Proc. Nat. Acad.
     Sci. USA 75, 2254-2257 (1978).

26. Yoshihara, K., Tanigawa, Y., Burzio, L., and Koide, S. S.
     Proc. Nat. Acad. Sci. USA 72, 289-293 (1975).

27. Müller, W. E. G., and Zahn, R. K., Mol. Cell. Biochem.
     12, 147-159 (1976).

28. Kun, E., Zimber, P. A., Chang, A. C. Y., Puschendorf, B.,
     and Grunicke, H., Proc. Nat. Acad. Sci. USA 72, 1436-1440
     (1975).

29. Roberts, J. H., Stark, P., Giri, C. P., and Smulson, M.,
     Arch. Biochem. Biophys. 171, 305-315 (1975).

DISCUSSION

R. LEIF: The fluorescence distribution of the nuclei shown
in your slides is similar to heterochromatin, or coarse
chromatin. I presume that your antibody is active with
undenatured double-stranded DNA. If this is the case your
fluorescence technique may have considerable use in exfoliative cytology because the detection of the presence of a
specific molecular constituent is preferable to quantitative
measurements of DNA since the former is not susceptible to
artifacts due to cell clumping, which has been one of the
major problems in exfoliative cytology.

O. HAYAISHI: I quite agree with you. However, I must
emphasize that this is a very preliminary result and I think
it has to be confirmed with more caution.

E. KUN: I would like to make a brief addition to this
elegant presentation of Professor Hayaishi. If one is
interested in the quantitative distribution and macro-molecular association of poly-ADP-ribose in normal animal tissues
with the aid of quick freezing and with the application of a
quantitative radioimmuno- assay which is specific for larger
than tetrameric ADP-ribose then one finds that 99.4% of
macro-molecular poly-ADP-ribose is associated with non-
histone proteins, which means that the DNA and RNA metabolism
controlling proteins of chromatin seem to be the primary
target of poly-ADP-ribosylation in vivo. I think that the
apparent discrepancy between radiochemistry and quantitative
analysis combined with radiochemistry is based on the fact
that a primary weight determination of the polymer has to

be done in conjunction with radiochemistry. In addition,
if we interrupt in vivo the activity of poly(ADP-ribose)
polymerase, in other words we inhibit the biosynthesis of the
polymer with drugs, specifically inhibiting this enzyme, we
find a rapid induction of inducible proteins suggesting that
this poly-ADP-ribose in the non-chromatin nuclear fraction
may play a repressor role. The kinetics of ribose labelling
indicates that the half life of ribose in NAD and poly-ADP-
ribose is between 2-1/2 and 3 hours, which coincides in
kinetic sense with the time constants of enzyme induction.

O. HAYAISHI: This is an extremely interesting finding, but
both Dr. Kun's procedure and our procedure suffer from one
drawback, that is these immune sera do not pick up mono or
diADP ribose; so I just wondered whether your finding
indicates that histones are either mono or di-ADP-ribosylated
and non-histones have long chain poly ADP-ribose.

E. KUN: I would be tempted to agree with you that
perhaps the mono-ADP-ribosylation of histones as you have
shown, has a direct role in the polymerization process
itself. On the other hand the long chain poly-ADP-ribose
moieties have an additional control function.

E. STADTMAN: What really is known about the linkage of the
mono-ADP-ribosyl group to proteins other than the one you
mentioned in which it is attached to an arginine residue.
Also does mono-ribosylation of histones involve a glutamic
acid residue or some other residue.

O. HAYAISHI: Well there are two cases where the linkages of
mono-ADP-ribosylated proteins have been studied in great
detail. One is the elongation of factor II, in that case
the results were not very conclusive and Maxwell at NIH
suggested either histidine or some new amino acid resembling
histidine as a site of ADP ribosylation. The second case is
the alpha subunit of RNA polymerase of E. coli in that case I
think they are very fortunate to be able to chop off all the
amino acids to be able to isolate the ADP ribosylated
arginine; in that case I think it is quite definite.

About the histones, many people have confirmed our
observations including Dixson and Moore, Koide etc. So there
is no doubt the carboxyl group of perhaps glutamate is the
main site of ADP ribosylation, but there are also other
possibilities such as phosphoserine reported by Stockton in
England. So, there could be some other sites.

W. WHELAN: I have two questions. First of all where is the
position of equilibrium in the polymerization? Is the
reaction reversible?

O. HAYAISHI: In the poly-ADP-ribose system we have tried
very hard to reverse it for obvious reasons because we
thought first this might be a sort of a storage house of NAD
in nuclei. But this reaction is completely irreversible as
far as we can tell, whereas the elongation factor II system
is easily reversible. One can generate NAD from ADP
ribosylated elongation factor II under proper conditions.

W. WHELAN: If it is not reversible and if the first addition
step is rate limiting one could predict that a few rather
long chains would be formed as opposed to a population of
many short chains with a Gaussian distribution that you might
get if the reaction was reversible and the initiation step
was an easy one. Have you taken a look at the distribution
of chain length in an in vitro synthesis?

O. HAYAISHI: No we have not done this sort of experiment.
But I did not quite understand your question. Why should it
elongate easily if there is plenty of glycohydrase or
hydrolase present in the chromatin?

W. WHELAN: If it is difficult to initiate the first one site,
but once initiated then the reaction proceeds easily, then
you will preferentially elongate the first site initiated,
rather than initiate new sites and only build short chains.

O. HAYISHI: Well no one seems to know what determines the
chain lengths and even in in vitro systems we could sometimes
get a very long chain - as long as 40, 50 or 100; sometimes
it just does not elongate more than a few units. There must
be some other factors which control the length of the chain,
in addition to the equilibrium.

B. LESSER: With regard to the role of DNA in the ADP-
ribosylation, is it possible that the histone must be bound
to the DNA in order for the reaction to occur and if that is
a possibility, have you examined the affinity of your histone
substrate for the various DNAs you were using?

O. HAYAISHI: Well I must say that all the previous experiments are very inconclusive. Apparently the order of addition and mixing is quite important, but no one seems to know
what exactly is going on. At this moment I cannot answer
your question.

E. KUN: I would like to respond more specifically to Dr. Stadtman's question. I think we have to discriminate more specifically between mono-ADP-ribosylation and oligo-ADP-ribosylation. What has been shown by several people is that there is a histone dimer formation connected by 17 or 16 unit ADP ribose oligomers. The mono-ADP- ribosylation of histones are Schiff bases and they function in Dr. Hayaishi's system. This would suggest that the binding of ADP-R to the histone, that is the monomer Schiff base regulates the polymerase activity. On the other hand there is evidence that other types of bonds are also formed with ADP-R oligomers.

P. COHEN: The state of poly-ADP-ribosylation through the cell cycle in a synchronized cell system; has this not yet been looked at?

O. HAYAISHI: In the past, this has been done by a number of people. But it is very inconsistent. As yet we do not understand why the results are so different. It could be due to the types of cells, or the conditions and there is no consensus at this moment.

IDENTIFICATION OF TWO COLLAGEN RNAS LARGER THAN α1 AND α2
COLLAGEN MRNAS WHICH ARE DECREASED IN TRANSFORMED CELLS

Sherrill L. Adams, James C. Alwine,* V. Enrico Avvedimento,
Mark E. Sobel, Tadashi Yamamoto, Benoit de Crombrugghe and
Ira Pastan. Laboratories of Molecular Biology and *Molecular
Virology, National Cancer Institute, National Institutes
of Health, Bethesda, Maryland 20014.

We have constructed and cloned various recombinant
plasmids containing collagen DNA sequences. The hybrid
plasmid pCOL3 contains pro-α1 (Type I) collagen DNA, and
the plasmid pCOL1 contains pro-α2 collagen DNA sequences.
These two recombinant plasmids were bound covalently to
diazotized cellulose and used to purify pro-α1 and pro-α2
collagen mRNAs. They were also used as hybridization
probes to characterize these mRNAs in tissues which are
active in synthesis of Type I collagen (calvaria and long
bones of chick embryos), as well as in normal and Rous
sarcoma virus (RSV) transformed chick embryo fibroblasts.
RNAs from these sources were fractionated by size on
agarose gels containing methylmercuric hydroxide and
transferred to diazotized paper filters. The RNAs were
then identified by hybridization to [^{32}P]-labelled pro-α1
or pro-α2 collagen DNA.

RNA from calvaria and long bones and from primary
fibroblast cultures gave identical patterns of hybridization. The pro-α1 collagen DNA hybridized to a major
RNA species of 5000 bases and a minor one of 7100 bases.
The pro-α2 collagen DNA hybridized to a major RNA of 5200
bases and a minor one of 5700 bases. The two major RNAs
identified by this method are the mature, functional mRNAs
for pro-α1 and pro-α2 collagen. The minor species are
probably precursors of these mRNAs.

In chick embryo fibroblasts transformed with RSV,
both pro-α1 and pro-α2 collagen mRNA, as well as the two
larger RNAs, were reduced by more than 10-fold. Our
results indicate that collagen mRNA and collagen mRNA
precursors are present in decreased amounts in RSV
transformed cells. We suggest that RSV exerts its action
on collagen synthesis at the transcriptional level to
decrease collagen mRNA synthesis.

PROCESSING OF RNA IN BACTERIA

David Apirion, Basanta K. Ghora, Greg Plautz, Tapan K.Misra
and Peter Gegenheimer. Department of Microbiology and Immunology, Washington University Medical School, St. Louis,
Missouri 63110. U.S.A.

Primary endonucleolytic cleavage of ribosomal and transfer RNA transcripts in Escherichia coli is accomplished by
four major processing enzymes. Inactivation of one or more
of these enzymes results in the accumulation in vivo of RNA
processing intermediates which can be processed in vitro by
these enzymes. Most transfer RNAs are also produced by
the action of these enzymes, since in their absence few if
any monomeric tRNA species accumulate in vivo.

A variety of experiments lead to the following conclusions:

1. Transcription of RNA is independent of its processing.

2. Primary processing of rRNA and tRNA occurs during transcription.

3. Most RNA transcripts are processed by more than a single
endoribonuclease.

4. The four enzymes responsible for most endonucleolytic
processing of stable RNA transcripts in E. coli are
RNases III, E, F, and P. In processing of rRNA RNase III
removes the 16S (p16) and the 23S (p23) sequences from
the growing transcript, RNase E removes the 5S rRNA
(p5) from the growing transcript, RNase F cuts near the
3' ends of the spacer tRNAs while RNase P cuts at the
5' ends of the spacer tRNAs.

5. A given ribonuclease, by and large, fulfills either a
degradative or a processing function.

6. Processing endonucleases are highly specific and each
performs a unique function. Their recognition sites
may be composed of unique combinations of relatively
simple yet distinctive features of secondary and tertiary structure.

7. The efficiency but not the specificity of processing
cuts is affected by the size of the substrate.

8. A certain amount of flexibility exists in the order of
initial processing events, but the final steps must be
preserved.

THE PROTEIN MOIETIES OF THE NON-POLYSOMAL CYTOPLASMIC (FREE)
mRNP COMPLEXES OF EMBRYONIC CHICKEN MUSCLE:

J. Bag and B.H. Sells, Mol. Biol. Laboratory, Fac. Med.,
Memorial University of Newfoundland, Canada A1B 3V6.

Free mRNP complexes of embryonic chicken muscle were
isolated by oligo(dT)-cellulose chromatography. The bound
mRNP complex was eluted at 37° using a low ionic strength
buffer. The combination of increase in temperature and
change of salt concentration enabled us to elute 50-60% of
the bound mRNP. These mRNPs were free from proteins that
independently bind to oligo(dT)-cellulose. The protein
moieties of this mRNP complex were analyzed by two dimensional gel electrophoresis using separation according to
charge in the first dimension and separation according to
molecular weight in the second. Seventeen proteins of Mr
37,000 to 75,000 were present in the mRNP complex. The
majority of these proteins have migrated towards the positive electrode, only three proteins of Mr 75,000, 66,000
and 37,000 were found to migrate towards the negative
electrode. Comparison of the mRNP proteins by two dimensional gel electrophoresis with 0.5M KCl washed 40S and 60S
ribosomal subunit proteins showed that the mRNP proteins
are non-ribosomal in nature. However, four acidic proteins
of Mr 41,000, 42,000, 45,000 and 48,000 present in the high
salt washed proteins from polysomes and two basic proteins
of Mr 75,000 and 37,000 of the $^{pH}5.0$ enzyme fraction displayed electrophoretic migration properties similar to that
of mRNP proteins.

A protein kinase activity was found associated with
the mRNP. This activity was resistant to washing with 0.5M
KCl and was co-sedimented with the mRNP through a 50% sucrose
cushion. This enzyme can transfer phosphate group from ATP
to a large number of mRNP proteins. Two acidic proteins of
Mr 37,000 and 54,000 and one basic protein of Mr 75,000 were
the major phosphorylated mRNP proteins. These observations
therefore suggest the presence of the enzyme and the phosphate acceptor proteins in the mRNP complex.

(Supported by a grant from The Muscular Dystrophy Association
of Canada).

COMPLEX OLIGOSACCHARIDES ASSEMBLED ON LIPID INTERMEDIATES:
STUDIES WITH UDP-N-ACETYLGLUCOSAMINE AND GDP-MANNOSE

David S.Bailey, Mathias Dürr and Gordon Maclachlan
Biology Department, McGill University, Montreal, Canada.

Membrane preparations from growing regions of plants
(pea stems) catalyse the formation of lipid-linked
oligosaccharides from both uridine diphospho-N-acetyl-
glucosamine and guanosine diphospho-mannose. However,
the efficiency with which the membranes utilise the two
substrates differs: UDP-GlcNAc-labelling results in the
accumulation of a disaccharide which is converted to
higher oligosaccharides only when exogenous GDP-mannose
is added; GDP-mannose forms higher oligosaccharides very
efficiently without the addition of exogenous UDP-GlcNAc.

Structural analysis of the oligosaccharides using
glycosidases shows that there are internal GlcNAc sites
within the structure as well as at the reducing end.
This is true for oligosaccharides derived from both
GDP-mannose and UDP-GlcNAc. The lipid intermediates
formed under our incubation conditions are not homogeneous and represent structurally distinct products of
the biosynthetic mechanism. This may reflect a
heterogeneity amongst the glycoproteins produced in this
plant tissue.

The importance of the structure of the oligosaccharide
side-chains of glycoproteins in recognition phenomena is
well-documented (1). We propose that at least part of the
structural complexity evident in completed glycoproteins
(2) is determined at the level of the lipid intermediates.

(1) Hughes,R.C. & Sharon,N. Nature 274 637 (1978)
(2) Kornfeld,R. & Kornfeld,S. Ann. Rev. Biochem. 45
    217 (1976)

This work was supported by a NATO Fellowship to DSB,
a Swiss NSF Fellowship to MD and an NSERC(Canada) operating
grant to GM.

IS THERE A DIFFERENT MODE OF DNA SYNTHESIS IN NORMAL AND
TRANSFORMED HUMAN DIPLOID FIBROBLASTS?

David E. Berry, James W. Rawles and James M. Collins,
Department of Biochemistry and MCV/VCU Cancer Center,
Medical College of Virginia, Virginia Commonwealth Univer-
sity, Richmond, Virginia 23298

When normal human diploid fibroblasts (WI-38) are
stimulated to proliferate, the earliest detectable DNA
synthesis is in the form of small 4.5 S pieces. These
"Okazaki fragments" are converted, after about 10 hr, into
large chromosomal-sized DNA of 400 S*. We have detected
similar small pieces of DNA in transformed WI-38 cells
(called 2RA). When both cell lines are "pulsed" with
^{3}H-thymidine for times ranging from 8 sec to 90 sec. alkaline sucrose gradient sedimentation reveals both radioactive "Okazaki fragments" and radioactive DNA molecules
larger than Okazaki fragments. However, the distribution
of radioactivity in small and larger pieces of DNA is quite
different with the two cell lines. Extrapolation of the
data to zero time reveals that the normal cell line contains 50% of the ^{3}H in DNA pieces smaller than 6 S, and 50%
in DNA pieces larger than 6 S. However, the transformed
cell line contains 67% of the ^{3}H in pieces smaller than
6 S, and 33% of the ^{3}H in pieces larger than 6 S. These
data seem to argue that DNA synthesis in the normal cells
is semi-discontinuous, whereas in the transformed cells it
is totally discontinuous. An alternative explanation is
that both cell lines actually have a totally discontinuous
mode of synthesis, but in the normal cell line the rate of
ligation of one of the two elongating chains of the replication fork is much faster than the other. It should be
stressed that the data obtained with the transformed cell
line cannot be accounted for by postulating that both cell
lines actually have a semi-discontinuous mode of DNA synthesis.

Supported by grants CA-24158 and CA-16059 from the National
Cancer Institute.

*Rawles, J.W. and Collins, J.M., J. Biol. Chem. 252 4762-
4766 (1977).

A study of the DNA region coding for initiation of precursor rRNA transcription in Lytechinus variegatus.

D. Bieber, N. Blin, and D.W. Stafford

Dept. of Zoology, University of North Carolina at Chapel Hill, N.C., 27514

The ribosomal DNA from the sea urchin Lytechinus variegatus was mapped with various restriction enzymes and with the R-loop technique (1). Eco RI cuts a tandem rDNA repeat into 4 fragments. The largest Eco RI fragment (5120 bp) encompasses the majority of the non-transcribed spacer (NTS) and overlaps with the precursor rRNA, but does not hybridize with mature 18S rRNA. This NTS region was cloned using Bam HI and 2 fragments were obtained (2); the smaller one (1800 bp) hybridized with precursor rRNA.

From the size of the precursor molecule the origin of transcription can be placed about 900 bases from the 5' end of the 18S sequence (3). The 1800 bp Bam HI fragment overlaps the 18S rRNA sequence by ca. 300 bp and therefore extends in the other direction ca. 600 bp beyond the 5' end of the precursor. Hence, this 1800 bp Bam HI fragment, which includes the site of transcription initiation, should contain putative control sequences for rRNA transcription.

This Bam HI fragment was used as substrate for further restriction enzyme analysis and was cut by the endonucleases Sma I, Eco RI, Xho I, and Hpa II. Sequencing data have been obtained from the 4 Sma I fragments of this Bam HI fragment.

Supported in part by grants from NSF and Deutsche Forschungs-gemeinschaft.

References:
(1) F.E. Wilson, N. Blin, D.W. Stafford (1978) Chromosoma 65: 373-381.
(2) N. Blin et al. (1978) J. Biol. Chem. In press.
(3) N. Blin (1978) submitted for publication.

A LECTIN RESISTANT VARIANT OF MOUSE FIBROBLASTS WITH AL-
TERED CELL SURFACE GALACTOSE

D. A. Blake, W. S. Stanley, B. P. Peters, E. H. Y. Chu and
I. J. Goldstein. Departments of Biological Chemistry and
Human Genetics, The University of Michigan, Ann Arbor,
Michigan.

The isolation of the first lectin-resistant variant of
the Swiss 3T3 mouse cell line has recently been reported
(1). This variant is able to resist the cytotoxic effects
of the α-D-galactopyranosyl-binding isolectin, Bandieraea
simplicifolia I-B₄ and the α- and β-D-galactose-binding
lectin, abrin. Compared to the parental cell line, the
variant cell exhibits a reduced number of binding sites on
its cell surface for the Bandieraea simplicifolia I-B₄
isolectin. A second lectin isolated from Bandieraea
simplicifolia seeds but specific for terminal N-acetyl-D-
glucosaminyl groups is cytotoxic to the variant but not the
parental cells. These results suggest a lesion in the biosynthesis of cell surface structures which results in the
exposure of subterminal N-acetyl-D-glucosaminyl residues.

Crude membrane preparations from the variant cell
line were ten-fold more active than parental 3T3 membrane
preparations in the transfer of D-galactose from radiolabeled UDP-galactose to endogenous glycoprotein acceptors.
Parental and variant cells were shown to possess approximately equivalent levels of uridine diphospho-galactose-4-
epimerase activity, when this enzyme was assayed in vitro
with added substrate and cofactors. Thus, the metabolic
lesion in the variant cells would appear to occur at a
step preceding galactosyl transfer to glycoprotein.

Supported by grants GB 34302 from NSF and AM 10171, CA
20424 and GM 20608 from USPH. D.A.B. was supported by
NRSA Postdoctoral Fellowship CA 06172.

(1) W. S. Stanley, B. P. Peters, D. A. Blake, D. Yep,
E. H. Y. Chu, and I. J. Goldstein. Proc. Nat. Acad. Sci.
Dec., 1978, in press.

ALTERATIONS IN THE RATE OF PROTEIN SYNTHESIS OF POLYSOMES
EXTRACTED FROM FIBROBLASTS OF PATIENTS WITH DUCHENNE MUSCU-
LAR DYSTROPHY AND CARRIERS OF THE DISEASE.

Léa Brakier-Gingras, Marie Boulé et Michel Vanasse.
Département de Biochimie, Université de Montréal et Section
de Neurologie, Hôpital Sainte-Justine, Montréal, Québec,
Canada.

Fibroblasts from patients suffering from Duchenne muscular dystrophy (DMD), from carriers of the disease and
from normal controls were grown in tissue culture. Polysomes were extracted and pelleted by centrifugation through
a 50% w/v sucrose cushion. Their protein synthesizing activity was assayed in a cell-free heterologous system, using
wheat germ as a source of soluble enzymes. The rate of
protein synthesis was measured by the amount of $[^{14}C]$-leu-
cine incorporated into trichloroacetic acid insoluble material.

A preliminary screening of 7 patients and 7 carriers
revealed that the rate of protein synthesis was decreased
about 2-fold for the carriers and about 3-fold for the patients when compared to controls. The choice of this system of assay focuses on the influence of alterations in polysomes on the rate of protein synthesis; this is in contrast
to the homologous systems used previously[1]. The potential use of such assays for carrier detection, and their significance for the understanding of the molecular causes of
Duchenne muscular dystrophy will be discussed.

Supported by the Muscular Dystrophy Association of Canada.

(1) Ionasescu, V., Zellweger, H., Ionasescu, R. and Lara-
Braud, C., Protein synthesis in human muscular dystrophies,
pp 362-375, in Pathogenesis of Human Muscular Dystrophies,
ed. L. Rowland, 1977.

BIOSYNTHESIS OF THE SMALL NUCLEAR RNAs IN A CELL-FREE
SYSTEM FROM MOUSE MYELOMA CELLS

Anne Brown, Chi-Jiunn Pan and William F. Marzluff
Dept. of Chemistry, Florida State University, Tallahassee,
Florida 32306

Small nuclear RNAs are the most abundant RNAs in animal cells, after ribosomal RNAs, on a molar basis. There
are 4 major species (120-250 nucleotides) which contain
"capped" 5' termini, *Gppp A^mp U^m_p Cp with the modified
guanosine a 2,2,7 trimethyl guanosine. These RNAs are
synthesized at 10-20% the rate of 5S rRNA, are stable and
are synthesized throughout the cell cycle.

Isolated nuclei from mouse myeloma cells did not synthesize significant amounts of the small nuclear RNAs, as
analyzed by gel electrophoresis. However, on addition of a
concentrated soluble protein extract (derived from both
nuclei and cytoplasm) these RNAs accumulated at a rate similar to that found in whole cells. These RNAs are transcribed by RNA polymerase III, as their synthesis was sensitive to 100 µgm/ml α-amanitin but not to 1 µgm/ml.

When isolated nuclei were incubated with ^{3}H-S-adenosyl
methionine in the presence of soluble protein extract small
amounts of labeled RNA with identical electrophoretic mobility with the small nucelar RNAs was found. The bulk of
the label was found on the 4.5S tRNA precursor and ribosomal rRNA precursors. After digestion with RNase T2, 70%
of the label in small nuclear RNAs was found on oligonucleotides with a charge of -5—-6.5. These oligonucleotides
were cap I and cap II structures with label in ^{2,2,7}tri-
methyl guanosine and 2'-OCH₃ adenosine. In contrast the
4.5S tRNA precursor gave predominately labeled mononucleotides and the rRNA precursor labeled dinucleotides.

Methylation of the small nuclear RNAs and the 4.5S
tRNA precursor was dependent on the presence of all 4 ribo-
triphosphates, sensitive to actinomycin D and 100 µgm/ml
α-amanitin, while rRNA methylation was largely (>80%) independent of transcription.

We conclude that the small nuclear RNAs are probably
synthesized from a larger precursor, not processed in isolated nuclei but processed and capped in the presence of
the protein extract. The capping is intimately related to

Supported by NIH grant AG 00413 and NSF PCM 76-11681.

575

A SHARED CONTROL MECHANISM FOR REGULATION OF
ADENOVIRUS AND SV40 GENE EXPRESSION

Timothy Carter[1], Rabecca Blanton and David Cosvin[2]
Department of Biological Sciences, St. John's Uni-
versity, Jamaica, NY[1] and The Milton S. Hershey
Medical Center, Penn State University, Hershey, PA[2]

Temperature-sensitive mutations in the Ad5 gene
encoding a 72K dalton DNA-binding protein allow prolonged accumulation of all early virus RNA at $40.5^{\circ}C$
when DNA synthesis is blocked by ara-C (1).
We have measured the capacity of isolated nuclei
to synthesize virus RNA during a short pulse of
^{3}H-UTP at various times after infection by wt Ad5
and ts125. The rate of virus RNA synthesis was
severely depressed in nuclei infected by wt for
more than 6 hours at $40.5^{\circ}C$, whereas the rate of
virus transcription remained high in ts125-infected
nuclei. A temperature-shift experiment showed that
the decrease in capacity to synthesize virus RNA was
dependent on a functional 72K protein.
A similar regulatory function has been shown for
SV40 T-antigen: tsA mutants accumulate early virus
RNA at the restrictive temperature $(41^{\circ}C)$. In cells
infected at $32^{\circ}C$ and then shifted to $41^{\circ}C$ late in
infection, early virus transcription increased,
whereas late transcription decreased (2).
We have measured the rates of Ad5 and SV40
transcription in monkey cells coinfected by both
viruses, in a similar temperature shift protocol
employing combinations of wt and ts mutants. We
find that under these conditions, the Ad5 72K
protein can replace SV40 T-antigen function in regulating SV40 transcription.

(1) Carter, T.H. and R.A. Blanton, J.Virol.25:664-674
    (1978)
(2) Reed, S.I. et al., Proc.Nat.Acad.Sci. U.S.A. 73:
    3083-3087 (1976)

IN VITRO SYNTHESIS OF VIRUS-SPECIFIC RNA USING TRANSCRIP-
TIONAL COMPLEX ISOLATED FROM ADENOVIRUS-INFECTED CELLS.

N. K. Chatterjee and M. H. Sarma
N.Y. State Department of Health, Empire State Plaza, Albany
New York, 12201

The control mechanisms of regulation of transcription
and post-transcriptional modifications of RNA in eukaryotes
are largely undefined at present. As an approach to understanding these mechanisms, we have developed a cell-free
system containing an active viral-transcription complex
from adenovirus-infected HeLa cells. The complex contains
viral DNA, cellular RNA polymerase II which has initiated
transcription, methylase(s) and probably a poly(A) polymerase and some other enzymes of processing. It appears
that by proper manipulations, such a system could be used
to transcribe and process RNA to generate translatable
mRNA. Some of the characteristics of this system are
described.

In the presence of four nucleoside triphosphates, the
complex synthesizes adenovirus-specific RNA in high yield.
Depending on the experimental conditions, RNA synthesis
could be carried out by elongation of already initiated
RNA chains present in the complex, or by de novo initiation
and elongation of RNA chains in the presence of exogenous
viral DNA. The complex associated RNA chains (at the time
of isolation) are "capped" and the majority of them are of
small molecular weight. Under both conditions of synthesis, incorporation of label from UTP, ATP or SAM proceeds
at a linear rate for at least 2 hrs. The newly synthesized
RNA molecules appear to be methylated and polyadenylated.
Although most of the RNA molecules fall within the size
range of virus-specific cytoplasmic mRNA, a consistent
amount of it is larger than 28S. In SDS polyacrylamide
gels, the complex displays a number of proteins of MW
>120,000 to 12,000.

Supported by NIH Grant #CA19707

GENE EXPRESSION OF MAMMALIAN LENS TISSUE: COEXISTENCE AND
PROPERTIES OF POLY(A)-CONTAINING AND POLY(A)-LACKING 10S
and 14S mRNA SPECIES.

John H. Chen
College of Physicians and Surgeons, Columbia University,
New York, New York 10032

Calf lens 10S and 14S mRNAs have been shown to contain both poly(A)-containing (poly(A)$^+$) and poly(A)-lacking
(poly(A)$^-$) populations. Experimental results indicate
that 47-58% of the 10S and 14S mRNA lack polyadenylated
sequences. In vitro translation studies indicate that
both species of the poly(A)$^-$ RNA can direct the synthesis
of a heterogeneous population of lens polypeptides, including putative alpha-crystallin chains. In contrast,
the poly(A)$^+$ mRNAs direct only the synthesis of alpha-
crystallin polypeptides. On the transcriptional level,
poly(A)$^+$ mRNA can be used as a template for the synthesis
of complementary DNA while poly(A)$^-$ mRNA fraction lacks
the template activity with Avian myeloblastosis versus reverse transcriptase. Based upon molecular hybridization
data, the poly(A)$^-$ mRNAs possess some complementary polynucleotide sequences to the cDNA synthesized from the
poly(A)$^+$ alpha-crystallin mRNA. The over-all data suggest that the native mRNAs for alpha-crystallin polypeptides may contain two molecular forms, with and without
polyadenylation.

GENE EXPRESSION OF MAMMALIAN LENS TISSUE: IS LENS mRNA BI-
CISTRONIC?

John H. Chen, College of Physicians and Surgeons,
Columbia University, New York, N.Y.

Lens 10S and 14S poly(A)$^+$ mRNAs have been shown to
code for the synthesis of the 175-amino acid B_2 chain
and the 173-amino acid A_2 chain of alpha-crystallin, respectively. The 10S mRNA contains 735 nucleotide, a
number which is theoretically compatible with the size
of the B_2 chain. The 14S mRNA of 1,500 nucleotides contains more than twice the required number of nucleotides
necessary for the synthesis of A_2 chain. The following
experiments concerning the functional role of the extra
nucleotide of the 14S mRNA suggest that it may be bicistronic: (1) Molecular hybridization with cDNA from 10S
mRNA indicates that 14S mRNA possesses two binding sites;
(2) Examination of the nucleotide sequences of 10S and
14S mRNA by two dimensional maps of T_1 and pancreatic
RNase digests shows a strikingly similar complexity;
(3) Sequence analysis of the initiation sites for protein
synthesis suggests that 14S mRNA contains a greater
abundance of initiation codans than that of the 10S mRNA;
(4) Alkaline-urea polyacrylamide gel analysis of the cell-
free protein products indicates that in addition to the
A_2 chain 14S mRNA codes for a putative B_2 polypeptide;
(5) Preliminary immunological studies reveal that the
polypeptide end-products directed by the 14S mRNA contain antigenic sites which bind to antibodies of alpha-
crystallin. The over-all data suggest that lens 14S
mRNA contains a second cistronic unit and this additional
cistron may code for a B_2 chain of alpha-crystallin or a
closely-related lens polypeptide.

All rights of reproduction in any form reserved.
ISBN 0-12-604450-3

THE EXPRESSION OF α-FETOPROTEIN GENE IN NORMAL AND NEOPLAS-
TIC RAT LIVER.

J.F. Chiu[1], L. Belanger[2], P. Commer[1], and C. Schwartz[1] [1]Dept.
of Biochemistry, University of Vermont College of Medicine
Burlington, Vt. 05405 and[2]Laboratoire de Biologie Molecu-
laire Centre de Rocherche de l'Hotel-Dieu de Quebec,
Quebec, Canada.

α-Fetoprotein (AFP) is a glycoprotein which is synthesized mainly by the fetal liver and yolk sac. AFP is present only in trace amounts in the serum of adult rat and
human, but increases dramatically in animals with chemically induced hepatocarcinoma and in human patients with primary liver tumors. Therefore AFP may provide a unique
model for the understanding of the relationship between embryonic development and malignant transformation.

Rat AFP mRNA was isolated from Morris hepatoma 7777 by
means of indirect immunoprecipitation of AFP-synthesizing
polysomes followed by oligo (dT)-cellulose affinity column
chromatography for poly A-containing mRNA. Purified AFP
mRNA migrated as a single band at 21S on SDS polyacrylamide
gels. In a cell free translation system, it yielded a peptide product immunoprecipitable by anti-rat AFP antiserum
and co-migrating with serum AFP on SDS-urea polyacrylamide
gels.

AFP mRNA was used for the synthesis of a radioactive com-
plimentary DNA. In hybridization assays, the cDNA reasso-
ciated with its purified template at a $C_rot_{1/2}$ of 1.1 x
10^{-2}. Molecular hybridization assays also showed that the
amount of AFP mRNA sequences decreased dramatically in rat
liver polysomes with an increase in age.

The effect of hormones on AFP synthesis in developing
rat liver was also studied. The results indicate that exogenously administered pharmacological doses of glucocorticoids, ACTH and thyroxin cause serum AFP levels to fall
prematurely, with glucocorticoid being the most effective.
From the hybridization assay of AFP mRNA content in liver
cytoplasm of rats treated with dexamethasone, it appears
that the effect of the hormone is at the level of AFP gene
expression but not at the level of secretion of AFP.

This investigation was supported by Grant No. CA-25098
awarded by the National Cancer Institute, DHEW.

NUCLEAR PHOSPHOPROTEIN PHOSPHATASE FROM CALF LIVER

Ming W. Chou, Eng L. Tan and Chung S. Yang
Department of Biochemistry, CMDNJ- New Jersey Medical
School, Newark, New Jersey   07103

Calf liver nuclear phosphoprotein phosphatase has been purified approximately 850 fold. The enzyme has a molecular weight of 34,000 daltons as determined by SDS polyacrylamide gel electrophoresis. The purified enzyme has a pH optimum between 7.0 and 7.5 with phosphophosphorylase, phosphohistones f_1 and f_{2b}, and phosphoprotamine as substrates. The enzyme activity towards these substrates follows the order, phosphophosphorylase > phosphohistone f_1 > phosphohistone f_{2b} > phosphoprotamine. The K_m values toward phosphophosphorylase and phosphohistone f_1 are 17 and 28 μM phosphate, respectively. Dephosphorylated histone f_1 and orthophosphate are competitive inhibitors of the enzyme with respective K_i values of 11 μM and 4.1 mM. Divatent metal ions and NaCl inhibit the enzyme, but $CaCl_2$ is slightly stimulatory. The results of Dixon plots suggested that metal ion inhibition occurs at two sites, one on the enzyme and the other on the substrate. The enzyme is also inhibited by NaF and EDTA. Nucleotides bearing the pyrophosphate structure are potent inhibitors of the enzyme while mononucleotides are slightly inhibitory. Low concentrations of cyclic AMP and cyclic GMP have little effect on the phosphatase activity, but high concentrations of these cyclic nucleotides inhibit the activity. DNA and other polyions also inhibit the enzyme. The enzyme appears to require free sulfhydryl groups for activity since it is inhibited by N-ethylmaleimide and p-hydroxymercuribenzoate; the latter inhibition can be reversed by mercaptoethanol and dithiothreitol. The purified nuclear phosphoprotein phosphatase is similar to the cytosol phosphoprotein phosphatase of calf liver, although the two enzymes have different properties in the crude state (1).

Supported by USPHS Grant GM-23164.

(1) Chou, M.W., Tan, E.L., and Yang, C.S. Int. J. Biochem. 9 27-32 (1978).

CHEMICALLY INDUCED GENE EXPRESSION: THE ROLE OF HISTONE
HYPER-ACETYLATION IN TRANSFORMATION BY SV40 MINICHROMATIN

Bennett N. Cohen and Thomas E. Wagner, Dept. of Chemistry,
Ohio University, Athens, Ohio 45701

We have developed an in vivo-like, chemical acetylating
procedure, utilizing acetyl adenylate, that mimics enzymatic
acetylation of histones; i.e., there is an increase in DNAse
I digestion and no increase in micrococcal nuclease digestion
of the chemically acetylated chromatin (1). This procedure
has been utilized to acetylate SV40 minichromatin in vitro,
in order to study the effect of acetylated vs. non-acetylated
minichromatin in cellular transformation.

All transformation studies were carried out with cultured Balb/3T3 cells. SV40 minichromatin was reconstituted
from SV40 form I DNA and calf thymus histone octamer. Cellular uptake of acetylated and non-acetylated minichromatin
was accomplished at high Ca^{++} concentration. Cells were sub-
cultured 24 hrs after infection with acetylated or non-
acetylated minichromatin.

After 5-6 days transformed cell foci appeared only in the
cultures infected with acetylated minichromatin. To select
for transformed cells, foci were transferred to agar-lined
culture flasks and grown in a suspension of 1% methylcellu-
lose. The selected cells were plated on slides and analyzed
for the presence of T-antigen by indirect immunofluorescence.
The acetylated minichromatin infected cells exhibited positive
T-antigen specific fluorescence, while non-acetylated minichromatin treated cells and normal cells showed no specific
fluorescence.

Schaffhausen and Benjamin have isolated non-transforming
mutants of SV40 and found them to be deficient in multi-
acetylated histones H3 and H4 (2). Our results tend to
support the proposed function of hyper-acetylation in cellular
modulation of gene activation.

Supported by American Cancer Society Grant No. NP 123.

(1) Shewmaker, C.K., Cohen, B.N. and Wagner, T.E. Biochem.
    Biophys. Res. Commun. (1978) in press.
(2) Schaffhausen, B.S. and Benjamin, T.L. Proc. Nat. Acad.
    Sci., 73, 1092-1096 (1976).

TRANSCRIPTION OF THE HISTONE mRNA PRECURSOR DURING THE CELL CYCLE.

Dave Cooper, Chi-Jiunn Pan, William F. Marzluff
Dept. of Chemistry, Florida State University, Tallahassee, Florida

Radiolabeled histone mRNA from mouse myeloma cells was quantitatively isolated by hybridization to plasmid DNA, carrying the histone genes from sea urchin[1], covalently bound to phosphocellulose. Non-polyadenylated 9S RNA from the cytoplasm of S-phase mouse myeloma cells bound to the resin, but not cytoplasmic RNA from cells in G1 or G2. Different RNA species bound to the plasmids containing the gene for histone 3 (pSp17[1]) and the gene for histone 4 (pS p2[1]). When cells were synchronized by isoleucine starvation these RNAs accumulated in the cytoplasm only during early S phase.

The labeled histone RNA found after short pulses (5-30 min) with ^{3}H-uridine had very different properties from the cytoplasmic histone mRNA. The same RNA molecules bound to plasmids containing the histone genes H3 and H4, indicating the RNA was potentially polycistronic. The RNA was larger than 28S rRNA on formamide-sucrose gradient centrifugation. With increasing length of labeling time, the proportion of 9S RNA increased and the proportion of histone RNA molecules containing sequences for both histone genes decreased indicating a precursor-product relationship between the large RNA and 9S histone mRNA.

Nuclei isolated from S phase cells synthesized histone RNA molecules which hybridized to both plasmids. The RNAs were similar in size to those formed in vivo but accumulated in the cell-free system and were not processed. Thus similar molecules were formed in vivo and on completion of nascent transcripts in isolated nuclei.

Synthesis during the cell cycle was monitored by 5 min pulses with ^{3}H-uridine and by the ability of isolated nuclei to synthesize histone RNA. Transcription of the histone genes was 5-10 fold higher in early S phase, compared to late S, G1 or G2. We conclude histone gene transcription yields a large, potentially polycistronic, primary transcript. Transcription of histone genes is activated during early S phase.

Supported by NIH Grant Ag 00413.

1. Kedes, L. Cell **8** 321 (1976)

CHARACTERIZATION OF A CYTOPASMIC POLY(A)·PROTEIN COMPLEX:
END-PRODUCT OR BY-PRODUCT OF MESSENGER RNA METABOLISM?

Peter L. Davies. The Group in Eukaryotic Molecular Biology
and Evolution, Department of Biochemistry, Queen's University, Kingston, Ontario, Canada.

A 10-13S ribonucleoprotein (RNP) particle containing
poly(A) was present in cytoplasmic extracts from wheat germ,
chick liver and chick embryonic muscle. A similar particle
has been reported in wheat germ (1) and Artemia salina (2).
Dissociation of this RNP particle with SDS released a poly
(A) tract which sedimented at 3-4S. Over 90% of the poly(A)
tracts in the postmitochondrial supernatant fraction from
wheat germ were present in this particle. Artefactual generation of the particle during extraction of the wheat germ
is unlikely because the postmitochondrial supernatant fraction supported the translation of added mRNA.

The average length of the poly(A) tracts in the RNP
particles from wheat germ and chick liver was 60 AMP residues, similar to the length determined for the poly(A)
tracts in globin messenger RNP particles from rabbit reticulocyte polysomes. The similarity between the 10-13S RNP
particle and the poly(A)-containing portion of messenger
RNP particles was extended by the observation that a 73-
78,000d protein copurified with the free poly(A) tracts
from chick liver through three successive CsCl isopycnic
centrifugations.

The cytoplasmic poly(A)·protein complex might simply
be an end-product of mRNA metabolism. However, if the junction between the poly(A) and the rest of the mRNA retains
the sensitivity that it has in the hnRNP particle, then an
initial, endonucleolytic cleavage at this point could generate a portion of the cellular poly(A)⁻mRNA population in
addition to the free, cytoplasmic poly(A)·protein complex
reported here.

Supported by the Medical Research Council of Canada.

1) Filimonov, N.G. et al. (1977) FEBS lett 79 348-352.

2) Slegers, H. and Kondo, M. (1977) N.A.R. 4 625-639.

FACTORS WHICH INFLUENCE THE N^{α}-ACETYLATION OF ACTH AND
FRAGMENTS OF ACTH

Jack E. Dixon and Terry A. Woodford
Department of Biochemistry
Purdue University
West Lafayette, Indiana

The rat pituitary contains an enzyme which acetylates
the α-amino group of the NH_2-terminal serine residue of
corticotropin (ACTH) and the structurally related peptide,
des-N^{α}-acetyl α-melanotropin, using acetyl CoA. The enzyme
works equally well with $ACTH_{(1-39)}$, $ACTH_{(1-24)}$, and
$ACTH_{(1-10)}$, but will not utilize Ser (the N-terminal amino
acid of ACTH) or Ser·Tyr, suggesting that the information
required for N^{α}-acetylation by the enzyme resides in the
first 10 amino acids. The product of the acetylation of
des-acetyl α-melanotropin was shown to be identical to
α-melanotropin (α-MSH). Removal of the NH_2-terminal serine
residue of des-N^{α}-acetyl α-MSH by Edman degradation results
in the formation of a peptide which is no longer a sub-
strate for the enzyme. The reaction of the enzyme with
$ACTH_{(1-10)}$ is also specific for the α-amino group of the
NH_2-terminal amino acid as determined by identification
of N-[^{3}H]acetyl serine. The N^{α}-acetyltransferase is localized predominantly in the subcellular fractions sedimenting
at forces greater than 10,000 x g. In comparison to other
organs, the rat pituitary exhibits the highest N^{α}-acetyl-
transferase-specific activity using $ACTH_{(1-10)}$ as a substrate. Incorporation of acetate into $ACTH_{(1-10)}$ was not
appreciably altered by ATP, GTP, AMP, or $MgCl_2$, but was
strongly inhibited by 1.0 mM CoA.

Supported by USPHS Grant #AM 18024.

INSTABILITY OF HIGHER PLANT CHROMOSOMAL DNA MOLECULARLY
CLONED IN E. COLI VIA RECOMBINANT PLASMIDS

Virginia K. Eckenrode[1], Harald Friedrich[2], Ron T. Nagao[2],
Sidney R. Kushner[1], Richard B. Meagher[2,3]. Departments of
Biochemistry[1], Botany[2], and Microbiology[3], University of
Georgia, Athens, Georgia 30602

High molecular weight DNA from Glycine max (soybean)
and Arabidopsis thaliana was highly purified to remove
polysaccharide contamination by banding in CsCl-PdI
gradients. Restriction fragments were enriched 60-fold
for the soybean rDNA genes by preparative agarose gel
electrophoresis. Restriction fragments of total DNA from
both plants and from fractions enriched for rDNA were
inserted into the HindIII and BamHI sites (tetracycline
resistance) of plasmid pBR322 and the EcoRI site (chlor-
amphenicol resistance) of plasmid pBR325. Hundreds of
independently isolated transformants containing recombinant plasmids were screened. The vast majority of the
plant DNA inserts were less than 10^6 daltons. The size of
the input DNA varied from 10^6 to 10^7 daltons with a MW_{av}
of 4×10^6 daltons. We have shown that the cloned rDNA
fragments result from a 60% loss of DNA from within the
cloned rDNA fragment. When unique restriction fragment
size classes of non-rDNA (e.g., 3.8×10^6 daltons) were used,
the majority of cloned fragments obtained were also less
than 2×10^6. These results were true in both HB101 and
C600 r^-m^-, two commonly used E. coli recipient strains.
However, we have subsequently cloned and maintained full
length plant DNA fragments in a different host strain,
SK1592. The ability of this strain to maintain plant DNA
is currently under investigation.

Supported by NSF Grant #PCM77-15011, USDA/SEA Grants #
5901-0410-8-0068-0 and 5901-0410-8-0045, NIH Grants #
GM21454 and GM00048.

TRANSLATION OF ENDOGENOUS NON-IgG and IgG MESSENGER RNA IN
HUMAN LEUKEMIC PLASMA CELLS AFTER TREATMENT WITH SEVERAL
ALKYLATING AGENTS

B. Emmerich, R. Maurer, G. Schmidt and J. Rastetter
Department of Haematology and Oncology, Technical University
Munic, Munic, GFR

During the clinical course of a plasma cell leukemia developed from a IgG (κ) plasmocytoma the cellular protein synthesis and immunoglobulin synthesis were measured in cytoplasmic
extracts from peripheral plasma cells.

The observation that protein synthesis in these extracts
could be followed up to 60 minutes and is blocked by aurintri-
carboxylic acid after a lag period of 5 minutes suggest that
endogenous messenger RNAs are relative stable and were reinitiated
under cell-free conditions.

The cell-free products were analysed on SDS slab gels and
the relative synthesis of individual proteins was determined
by their labelling pattern. The ratio of radioactivity in the
immunoglobulin polypeptide bands was 2:1, indicating that an
equivalent amount of heavy- and light chain messenger RNA was
translated. Additionally IgG synthesis was examined by binding
to solid phase antibodies and protein A. The data from gel
analysis and from binding to specific antibodies and protein A
showed that IgG synthesis shares 10 to 15 % of total protein
synthesis.

After treatment with cyclophosphamid and a combination of
this drug with melphalan and vincristine no chance in the
activity of total protein synthesis was found. In contrast,
a decrease of total protein synthesis could by observed after
treatment with 1,3- bis (2-chloroethyl)-1-nitrosoures (BCNU)
combined with cyclophosphamid and cytosins-arabinoside. The
relative amount of 10-15 % Ig synthesis kept unchanged under
the different chemotherapeutic regims.

The results suggest a primary effect of BCNU on cellular
protein synthesis in human myeloma cells similar to that reported in HeLa cell cultures (1).

Supported by the Deutsche Forschungsgemeinschaft, Grant Em 20/3

(1) Penman, M., Huffman R., Kumar, A., Biochemistry 15 2661-
2668 (1976)

THE IN SITU PHOSPHORYLATION OF THE α-SUBUNIT OF eIF-2 IN
RETICULOCYTE LYSATES INHIBITED BY HEME DEFICIENCY, dsRNA,
GSSG, OR THE HEME-REGULATED INHIBITOR.

Vivian Ernst, Daniel Levin and Irving M. London.

Dept. of Biology, M.I.T., and Harvard-M.I.T. Division of
Health Sciences and Technology, M.I.T., Cambridge, MA 02139.

The initiation of protein synthesis in reticulocyte lysates is inhibited in the absence of hemin, or in the presence of GSSG (50–500 μM) or dsRNA (1–50 ng/ml). All three
conditions activate cAMP-independent protein kinases that
phosphorylate the α-subunit (38,000 daltons) of eIF-2 (eIF-
2α). We have examined this phosphorylation directly in lysates by labelling for brief periods with a pulse of high
specific activity $[\gamma-^{32}P]$ATP. The $[^{32}P]$phosphoprotein profiles were analyzed by one-dimensional SDS-PAGE and autoradiography under conditions in which the eIF-2α polypeptide
can be clearly distinguished. All modes of inhibition produced a rapid and significant increase in the phosphorylation of eIF-2α, compared to uninhibited control incubations
which displayed little or no phosphorylation of eIF-2α. In
heme-deficient lysates, phosphorylation of eIF-2α was detectable within 30 sec after the start of incubation, and
at all times subsequent both before and after the shut-off
of protein synthesis. The restoration of synthesis by the
delayed addition of hemin is accompanied by a decrease in
the phosphorylation of eIF-2α. Similar kinetics of phosphorylation of eIF-2α are observed in hemin-supplemented
lysates inhibited by GSSG (0.5 mM), dsRNA (20 ng/ml), or
purified HRI, the heme-regulated protein kinase. These
data provide further evidence that phosphorylation of eIF-
2α is a critical event in the inhibition of protein synthesis and indicate that a rapid turnover of phosphate
occurs at the site(s) of phosphorylation. Lysates inhibited
by heme-deficiency, GSSG, or purified HRI also show a significant phosphorylation of an 80,000-dalton polypeptide
which co-migrates with purified HRI. Lysates inhibited by
dsRNA display phosphorylation of a 67,000-dalton polypeptide,
but no labelling of the 80,000-dalton polypeptide. (Supported by USPHS Grants AM-16272-07 and GM-24825-01.)

PURIFICATION OF ARGINYL- AND LYSYL-tRNA SYNTHETASES FROM
RAT LIVER: EVIDENCE OF GLYCOPROTEINS

Raymant L. Glinski, Jr.[†], Philip C. Gainey[†], Thomas P.
Mawhinney[*] and Richard H. Hilderman[†]. [†]Department of Biochemistry, Clemson University, Clemson, SC. [*]Department
of Biochemistry, University of Missouri, Columbia, MO.

Multiple forms, in terms of molecular weight, of
arginyl- and lysyl-tRNA synthetases can be isolated from
rat liver. The small molecular weight forms of arginyl-
and lysyl-tRNA synthetases have been purified to apparent
homogeneity. Throughout the entire purification scheme
arginyl- and lysyl-tRNA synthetases co-chromatograph on
DEAE-cellulose, hydroxylapatite, phosphocellulose and Blue
Sepharose affinity chromatography. In the pure state they
also copurify on gel filtration chromatography. Thus
these synthetases may be tightly associated as a complex.
Other workers have also found a close association of
arginyl- and lysyl-tRNA synthetases in rat liver (1,2).

Gas chromatographic analysis of the purified complex
demonstrates that it contains about 12% carbohydrate by
weight. The sugars found associated with the synthetases
were mannose, fucose, glucose, galactose, and N-acetyl-
glycosamine. This is the first report that mammalian
aminoacyl-tRNA synthetases are glycoproteins.

Supported by   USPHS Grant GM 27718.

1. Goto, T. and Schweiger, A. 1973. Z. Physiol. Chem.
    354: 1027-1033.

2. Dang, C. V. and Yang, C. H. 1978. Biochem. Biophys.
    Res. Commun. 80: 709-714.

MAMMALIAN CELL NUCLEAR RNA LIGASE ACTIVITY

Norma Ornstein Goldstein
Department of Animal Biology, University of Pennsylvania
School of Veterinary Medicine, Philadelphia, PA 19104

A longtime interest of mine in a RNA ligase, going as far
back as 1968 before any such enzyme was known, was restimu-
lated by reports of split β-globin genes and their split
pre-mRNA transcripts in mouse cells (1,2). Earlier work
about such enzyme activity in animal cells grown in tissue
culture (3,4) had been open to question (5), long before an
implicit requirement for RNA ligase had been proposed for a
joining step(s) in the processing of β-globin pre-mRNA (1).
I now report finding RNA ligase activity in crude nuclear
extracts of 16 day fetal mouse liver cells which synthesize
β-globin pre-mRNA and some adult globin mRNA (6).

Extracts were prepared as by others (3) except that nuclei
were freeze-thawed twice and Dounce homogenized in place of
sonication. A simple oligoribonucleotide synthetic assay
was designed and analyzed with paper chromatography, similar
to methods used for T4 RNA ligase (7). Commercially available unlabeled and labeled 3'-OH acceptors such as ApApA,
IpIpI etc. and 5'-P donors such as pAp, pCp etc. served as
model substrates at relatively high concentrations (0.1-1 mM)
with T4 RNA ligase as control enzyme. Assay volumes (50-
100 µl) were incubated in sealed microcaps at 27° and 37° C
for 10-120 min. Long incubation at 37° favored substrate
degradation by cell nucleases. A single ligated product,
e.g., XpXpXbYp with 5'-OH and 3'-P termini resulted from any
one assay. Chromatogrammed molecules were monitored with
UV light or radioactivity counting. Other cell extracts are
being tested as possible sources of enzyme(s) for purification and joining studies.

Supported in part by USPHS Grant #CA05295 to Dr. R. J.
Rutman and by NIH Biomedical Research funds to N.O.G. from
the University. Fetal mouse livers were a gift from Drs.
R. L. Brinster and E. A. Merz in our Department.

1 Tilghman SM Tiemeir DC & Leder P PNAS USA 75:725 (1978)
2 Tilghman SM Curtis PJ Tiemeir DC Leder P & Weissmann C
  PNAS USA 75:1309 (1978)
3 Linne T Oberg B & Philipson L Eur J Biochem 42:157 (1974)
4 Cranston JW Silber R Malathi VG & Hurwitz J JBC 249:7447 (1974)
5 Bedows E Wachsman JT & Gumport RI BBRC 67:1100 (1975)
6 Kinniburgh AJ Mertz JE & Ross J Cell 14:681 (1978)
7 Uhlenbeck OC & Cameron V Nucleic Acids Res. 4:85 (1977)

MULTIPLE FORMS OF α2u GLOBULIN IN THE RAT STUDIED IN VIVO
AND IN PRIMARY CULTURES OF ADULT HEPATOCYTES

Laura Haars and Henry Pitot
McArdle Laboratory, Department of Oncology and Pathology,
University of Wisconsin, Madison, WI

The synthesis and secretion of α2u globulin, a major
urinary protein of adult male rats (1), was examined in primary cultures of hepatocytes. Antibody specific for α2u
globulin was used to immunoprecipitate α2u globulin from
cell homogenates and media. A single protein species of
approximately 24,000 daltons was found in homogenates prepared from cultured adult hepatocytes. The corresponding
culture medium also contained this protein species as well
as two or more higher molecular weight (HMW) forms. Only
the HMW forms were labeled isotopically with glucosamine or
mannose. The appearance of the HMW forms but not the 24,000
dalton form could be completely inhibited by the incubation
of cell cultures with tunicamycin, an inhibitor of the first
step in the lipid-linked pathway of glycoprotein synthesis.

Vitamin A deficiency has been shown to reduce glycosylation of proteins (2), presumably due to reduced pools
of mannosylretinylphosphate. Cultured hepatocytes derived
from vitamin A deficient rats synthesized α2u globulin at
10% of the level of control cultures, and the corresponding
culture medium contained only the HMW forms. The HMW forms
could be labeled with 14C-glucosamine but not 14C-mannose.

α2u globulin immunoprecipitated from homogenates of
liver and kidney was found only in the form of a 24,000
dalton species. In contrast, serum α2u globulin was
present exclusively in HMW forms.

α2u globulin in vivo is rapidly cleared from the circulation by the kidney and appears as a single species in the
urine (3). The data presented here suggest one possible
function for the carbohydrate moiety of α2u globulin. Glycosylation of the globulin may produce a molecule that is
too large and negatively charged to pass readily through
the glomerular capillary membranes, thereby maintaining a
critical level of α2u globulin in the serum.

Supported by NCI Grants No. CA-07175, P01CA-22484, and
T32-09135.

(1) Roy, A.K. et al. (1966) Biochim. Biophys. Acta 127,
    72-81.
(2) DeLuca, L.M. et al. (1975) Biochim. Biophys. Acta 409,
    342-359.
(3) Roy, A.K. and Neuhaus, O.W. (1966) Biochim. Biophys.
    Acta 127, 82-87.

BIOSYNTHESIS OF LYSOSOMAL ENZYMES IN DIPLOID HUMAN FIBRO-
BLASTS. Andrej Hasilik and Elizabeth F. Neufeld. National
Institute of Arthritis, Metabolism & Digestive Diseases,
National Institutes of Health, Bethesda, Maryland 20014

Cultured fibroblasts were labeled with ^{3}H-leucine or ^{33}P$_i$.
Extracts from labeled cells and concentrates of medium were
subjected to immunoprecipitation with goat antisera raised
against human placental β-N-acetylhexosaminidase. The
immunoprecipitates were solubilized in SDS and dithiothreitol, and the radioactive components detected by fluorography
after polyacrylamide gel electrophoresis. After prolonged
periods (> 24 hr) the ^{3}H-labeled material precipitated from
cells showed the same polypeptide pattern as a mixture of
purified hexosaminidase A and B isozymes (54,000, 29,000,
as well as some minor bands). However, in a 3 hr pulse, the
label was found predominantly in two bands of higher molecular weight (67,000, 63,000), which were identified as precursors of the α and β subunits, respectively, by means of a
second immunoprecipitation as follows: the primary immunoprecipitates were solubilized, diluted and reprecipitated
with antisera prepared against denatured α and β subunits of
hexosaminidase. In the media, the label was found only in
the precursor polypeptides even after a 3-day chase. The
secretion of precursors was greatly enhanced by the presence
of ammonia or chloroquine, which also inhibited the intracellular processing of the precursors. Increased secretions
of the precursor polypeptides and decreased intracellular
processing were also seen in fibroblasts from patients with
I-cell disease. Similar observations were made for α-
glucosidase, α-L-fucosidase and cathepsin D. In the
presence of ^{33}P$_i$ the hexosaminidase polypeptides were
labeled in normal fibroblasts and their secretions but not
in I-cell fibroblasts. These data are in agreement with
the defect in posttranslational processing which has been
proposed for the I-cell mutation.

A.H. is a recipient of a fellowship from Deutsche
Forschungsgemeinschaft.

ISOLATION AND PARTIAL CHARACTERIZATION OF β-LIKE GLOBIN
GENES FROM TOTAL GENOMIC GOAT DNA

Joel R. Haynes, Paul Rosteck, Jr., Jeffrey Robbins, Greg
Freyer, Douglas J. Burks, Michael Cleary, and Jerry B.
Lingrel. Department of Biological Chemistry, University
of Cincinnati College of Medicine, Cincinnati, Ohio   45267.

Total genomic DNA from the spleen of a juvenile goat
was partially digested with the restriction endonuclease
EcoRI and the 12-22 kbp size class was isolated by sucrose
density gradient centrifugation. This DNA was ligated to
Charon 4A DNA following removal of the internal phage DNA
fragments produced by EcoRI digestion, and the resulting
recombinant DNA was packaged into infectious phage particles and grown on *E. coli* DP50 supF.

Recombinant clones containing globin gene sequences
were identified by hybridization to ^{32}P-labeled globin complementary DNA. Four different clones were isolated.

Characterization of the clones involved localization of
the globin gene sequences by restriction enzyme mapping,
Southern blotting techniques, and electron microscopic observation of R-loops formed by hybridization to goat globin
mRNA. Cell-free translation of globin mRNA isolated by
hybridization to the recombinant DNAs demonstrated that
only β-like sequences were present in the clones.

These studies and additional restriction mapping will
be discussed.

We are grateful to Dr. T. Maniatis for communicating
the detailed methodology of construction of libraries from
eukaryotic DNA before publication and for his helpful discussions. This work was supported by grant PCM 76-80222
from the National Science Foundation, and grants AM 20119
and GM 10999 from the U.S. Public Health Service.

REGULATION OF METALLOTHIONEIN SYNTHESIS IN CULTURED MAMMA-
LIAN CELLS: INDUCTION, DEINDUCTION, AND SUPERINDUCTION

C. E. Hildebrand and M. D. Enger. Cellular and Molecular
Biology Group, Los Alamos 'Scientific Laboratory, University
of California, Los Alamos, New Mexico 87545

Metallothioneins (MTs), low-molecular weight, cysteine-
rich, metal-binding proteins, are induced in animal tissues
and in cultured cells upon exposure to Cd^{2+} or Zn^{2+}. The
ubiquity of this response in higher eukaryotes and the generality of the proposed roles for MTs in regulation of
essential trace metals (Zn^{2+}), as well as in protection
against the toxicity of nonessential trace metals (Cd^{2+}),
make the MT induction system unique for studying the
regulation of gene expression. A cloned, stable, variant
subline (CdR) of the Chinese hamster cell (CHO), selected
for its resistance to Cd^{2+}-mediated cytotoxicity, was found
to have increased capacity to induce MT upon exposure to
Cd^{2+}. The CdR cell was used to study the regulation of MT
induction and deinduction. Maximal induction of MT in CdR
was achieved by exposure of cells in suspension culture to
2 μM $CdCl_2$. Following induction, the rate of MT synthesis
increased above the low uninduced rate within 1.5-2.0 h,
reached a maximal rate (30-fold > basal rate) by 8 h, and
decreased to a constant rate (50% of maximal) by 24 h. MT
synthesis attains the maximal rate in the absence of any
comparable stimulation of general protein or RNA synthesis.
Actinomycin D (AM, 2-5 μg/ml) added 0-0.5 h postinduction
inhibits MT synthesis assayed 8 h postinduction, suggesting
that primary induction is regulated at the level of transcription. AM (2 μg/ml) added to maximally induced cultures
(8 h) results in an increased rate of MT synthesis (10% >
the maximal 8-h rate) during the first 2 h after AM addition
and then maintains an approximately constant rate during the
following 4 h, while the MT synthesis rate in control, non-
AM-treated cells decreases. Transfer of maximally induced
(8-h) cells to Cd^{2+}-free medium results in a rapid decline
in the rate of MT synthesis (17% of maximal by 8 h following removal from Cd^{2+}). The rapid deinduction is markedly
reduced by addition of AM (2 μg/ml) at the time of removal
from Cd^{2+} so that, by 8 h after Cd^{2+} removal, the rate of
MT synthesis in AM-treated cells is 85% of normal. Qualitatively, AM effects on induction and deinduction mimic the
behavior of several hormonally controlled inducible systems,
suggesting that the regulatory mechanisms involved in
controlling gene expression are similar in these quite different inducible systems.

This work performed under the auspices of the U.S.D.O.E.

A CLASS OF HIGHLY REPEATED DNA SEQUENCES UBIQUITOUS TO THE
HUMAN GENOME

Catherine M. Houck, Frank P. Rinehart and Carl W. Schmid.
Department of Chemistry, University of California, Davis.

A highly repetitive family of 300 nucleotide long
sequences has been isolated from human DNA. These sequences appear to be interspersed with single copy DNA sequences and are characterized by a site for the restriction enzyme Alu I located 170 nucleotides from one end.
The same sequence family has been found in 300 nucleotide
long inverted repeated human DNA.

This Alu family of sequences contains over half of
the 300 nucleotide long repeated sequences and constitutes
at least three percent of the human genome. This corresponds to a 300 nucleotide long sequence repeated more
than 200,000 times. The Alu family appears to be interspersed throughout 30 to 60 percent of the human genome.

The Alu family does not appear to be related to the
tandemly repeated sequences which have been isolated from
human DNA by restriction enzyme cleavage (1). We therefore, conclude that the well known interspersion pattern
of repetitious and single copy sequences in human DNA is
dominated by one family of repeated sequences.

These sequences may be related to repeated sequences
which have been found to be interspersed throughout the
hnRNA (2). These hnRNA sequences have a very low complexity as indicated by fingerprint analysis (2). In
addition, the inverted repeated DNA sequences have been
shown to contain the same sequences as the repeated hnRNA
sequences (3). If the Alu family of DNA sequences is
related to the repeated RNA sequences, it may be that they
function at the level of hnRNA processing.

Supported by grant GM 21346 from the National Institutes
of Health.

(1) Manuelidis, L., Chromasoma, 66, 1-21 (1978).

(2) Robertson, H. D., Dickson, E., and Jelinek, W.,
J. Mol. Biol., 115, 571-589 (1977).

(3) Jelinek, W. R., J. Mol. Biol., 115, 591-601 (1977).

REGULATION OF HMG-CoA REDUCTASE AND REDUCTASE KINASE
BY REVERSIBLE PHOSPHORYLATION

T. S. Ingebritsen, R. A. Parker and D. M. Gibson,
Department of Biochemistry, Indiana University School of
Medicine, Indianapolis, Indiana 46223.

Rat liver microsomal HMG-CoA reductase (R) is readily
interconverted between an active (dephosphorylated) and
an inactive (phosphorylated) state. Inactivation of R is
catalyzed by R kinase (RK). This enzyme is present both
in the cytosol and in microsomes. Reactivation of R is
catalyzed by a second cytosolic enzyme P which appears to
be identical to the 35,000 dalton liver phosphorylase
phosphatase (1,2).

In the present studies, RK was extracted from microsomes by repeatedly washing with neutral buffer. RK activity was assayed using R in RK-deficient microsomes as
substrate. The apparent K_m for ATP was .2 mM. RK activity was not enhanced by cAMP, cGMP, cCMP, or cIMP. Substantial purification of soluble RK was achieved by chromatography on DEAE-Cellulose. RK was found to have a molecular weight of 360,000 daltons by chromatography on
Agarose A 0.5 m.

Incubation of either soluble or microsomal RK at 37°
resulted in a time dependent inactivation of the enzyme.
The rate of inactivation was greatly enhanced by purified
P and inhibited by fluoride. RK was reactivated in a time
dependent manner by incubation with 2 mM MgATP in a low
ionic strength medium. Reactivation was blocked in the
high ionic strength medium routinely used to assay R and
RK. The RK activating enzyme was found both in the soluble RK preparation and tightly bound to the particulate
glycogen-protein complex. These data suggest that R is
regulated by a bicyclic system in which both R and RK are
interconvertible enzymes.

Research supported by grants from NIH (AM21278, AM19299),
the American Heart Assoc., Indiana Affiliate and the
Showalter Foundation.

(1) Ingebritsen, T.S., Lee, H.S., Parker, R.A. and Gibson,
    D.M., Biochem. Biophys. Res. Comm. 81 1268-1277 (1978).

(2) Gibson, D.M. and Ingebritsen, T.S., Review, Life
    Sciences (1978), in press.

STRUCTURE AND EXPRESSION OF THE ALPHA FETOPROTEIN GENE.

M.A. Innis and D.L. Miller. Department of Biochemistry,
Roche Institute of Molecular Biology, Nutley, New Jersey,
07110.

Alpha fetoprotein (AFP), a 72,000 dalton serum protein, is secreted during embryonic development by the
fetal liver and yolk sac. After birth serum levels of AFP
decline by a factor of 10^5; however, AFP reappears in the
sera of animals bearing hepatomas and teratocarcinomas.
We have shown that the lack of AFP synthesis in normal
liver is due to the absence of mRNA-AFP in this tissue
(1); therefore, the expression of the gene is controlled
by processes that regulate the level of mRNA-AFP.

We have studied the regulation of mRNA-AFP levels in
a cloned cell culture derived from Morris Hepatoma 7777,
which shows a density dependent variation in the AFP synthesis rate. By hybridization of pulse-labeled mRNA-AFP
to cDNA-AFP produced by recombinant DNA techniques, we
have measured the half-life of mRNA-AFP in the hepatoma to
be 40 hours, about five times as long as the half-life of
total mRNA. The rate of secretion of AFP by the hepatoma
is found to be determined by the level of mRNA-AFP.
Changes in the level of mRNA-AFP are controlled by changes
in the rate of release of mRNA-AFP from the nucleus.

The transcriptional unit of the AFP gene was found by
U.V. mapping to be very large, probably 20-24 kilobase
pairs, or nearly ten times as large as the cytoplasmic
mRNA. This was confirmed by the discovery of a very large
precursor nuclear RNA whose size is also 20-24 kilobases.
Several discrete intermediates appear during density gradient sedimentation of pulse-labeled nuclear RNA, which
suggests that the removal of intervening sequences determines the rate of processing of the nuclear precursor.
The existence of several such sequences is demonstrated
by restriction mapping of the AFP gene. Restriction
analysis has thus far revealed no differences between the
structure of the AFP genes in liver and hepatoma 7777,
which suggests that other regulatory genes are involved
in the expression of AFP.

(1)   Innis, M.A. and Miller, D.L., J. Biol. Chem. 252,
      8469-8475 (1977).

CONTROL OF eIF-2 PHOSPHORYLATION IN RABBIT RETICULOCYTE LYSATE

Rosemary Jagus and Brian Safer. Laboratory of Molecular Hematology, NHLBI, NIH, Bethesda, Md.

An increased phosphorylation state of the 38,000 dalton subunit (α) of eIF-2 is associated with translational inhibition resulting from hemin deficiency, double-stranded RNA and oxidised glutathione. Since it has been suggested that eIF-2α phosphorylation directly or indirectly inhibits initiator met t-RNA$_f$ binding by eIF-2, studies have previously been made on eIF-2α kinase and conditions which modulate its activity (1-2).

We have demonstrated, however, that the steady state level of eIF-2 phosphorylation is determined by a combination of kinase and phosphatase activities. Using (^{32}P)eIF-2, we have found that dephosphorylation of eIF-2α in reticulocyte lysate is extremely rapid (t$\frac{1}{2}$=20 secs) and occurs at the same rate in both hemin-supplemented and hemin-depleted lysate. In contrast, the phosphorylation state of eIF-2β remains constant under all conditions studied.

eIF-2α phosphatase activity can be modulated by the energy charge of the guanine nucleotide pool. At concentrations required to produce a transient inhibition of protein synthesis, GDP reduced the extent of phosphorylation by 50%. This may reflect a change in the accessibility of a site on eIF-2α to its phosphatase.

Highly active eIF-2α phosphatase and kinase activities allow a rapid adjustment of the steady-state level of phosphorylation over a wider range than either activity alone would allow. In addition, control of eIF-2α phosphatase activity by the energy charge expands the response of eIF-2 phosphorylation to a wider variety of conditions known to affect translational activity.

(1)   Revel, M. and Groner, Y., Ann. Rev. Biochem. 47 1079-1126 (1978).

(2)   Safer, B. and Anderson, W. F., Crit. Rev. Biochem. In Press (1978).

ORGANIZATION OF IMMUNOGLOBULIN KAPPA LIGHT CHAIN GENES IN
GERM LINE AND SOMATIC CELLS

Rolf Joho[1], Irving Weissman[1], Philip Early[2], and
Leroy Hood[2].

[1]Department of Pathology, Stanford University School of
Medicine, Stanford, CA. [2]Division of Biology, California
Institute of Technology, Pasadena, CA.

    The origin of antibody diversity is still unclear.
In order to investigate this problem we studied the organization of Ig genes in several tissues and most importantly in the germ line. Mouse sperm and embryo DNA
were digested with several restriction enzymes and the
resulting fragments were analyzed for their content of
kappa light chain sequences by hybridization techniques
as described by Southern. A kappa light chain probe specific for the constant region yielded the same restriction
map for sperm and embryo DNA; a probe specific for a
kappa variable region sequence (MOPC 167) revealed several
V_k genes located on the same restriction fragments in
sperm and embryo DNA. Hybridization with v and c region
probes also revealed differences in the intensity of the
corresponding bands on Southern gels. These results will
be discussed in the light of different hypotheses that
have been put forward in order to explain the origin of
antibody diversity.

CLONING OF FRAGMENTS OF THE SEA URCHIN HISTONE GENE CLUSTER WITH SV40 DNA AS A VECTOR

John S. Kaptein and George C. Fareed. Molecular Biology Institute, University of California, Los Angeles, California 90024.

DNA from the sea urchin (Strongylocentrotus purpuratus) histone gene cluster was digested with restriction enzymes. Selected fragments were linked at one end to portions of the SV40 genome via ligation of cohesive termini. Hybrid molecules formed consisted of the large Taq I-Pst I (0.28) fragment of SV40 linked to a Pst I-Eco R1 fragment containing the start of the H1 histone gene, and the large Bam H1-Hae II fragment of SV40 linked to a Hae II-Eco R1 fragment containing the entire H2B histone gene. Permissive African Green Monkey kidney cells (CV1-P) were coinfected with this DNA and with DNA from temperature sensitive mutants of SV40, tsA58 for insertions of sea urchin DNA into the late region of SV40, and tsBC11 for insertions into the early regions of SV40. Individual plaques appearing at the non-permissive temperature were expanded and screened for the presence of virus containing DNA from the sea urchin histone gene cluster, using a filter hybridization assay developed for this purpose. Circularization of the hybrid linear DNA molecules occurred <u>in vivo</u> during the transfection procedure as evidenced by various restriction endonuclease analyses. During the circularization there was no detectable change in the size of the DNA; however, loss of restriction endonuclease sites at the open ends did occur. Marker rescue of SV40 genomes, and genetic rearrangements (primarily of the SV40 helper DNA) were also observed. Propagation of the hybrid genomes was achieved through lytic infections. Expression of genetic information in cytoplasmic RNA transcripts was examined for selected clones.

Supported in part by a Medical Research Council of Canada Fellowship and USPHS grant #CA 06091.

Regulation by Metabolites of Inactivating and Additional
Phosphorylation Reactions in the Pig Heart Pyruvate Dehydrogenase Complex.

Alan L. Kerbey, Peter H. Sugden and Philip J. Randle

Nuffield Department of Clinical Biochemistry, Radcliffe
Infirmary, Oxford OX2 6HE, U.K.

The pyruvate dehydrogenase (PDH) complex of mammalian
tissues is phosphorylated and inactivated by an $ATPMg^{2-}$-
dependent kinase (PDH kinase) intrinsic to the complex.
The kinase reaction is inhibited by ADP (competitive with
ATP), pyruvate (uncompetitive with ATP), CoA and NAD^+,
and is activated by acetyl-CoA and NADH. Inactivation of
PDH is accomplished by incorporation of a single phosphate
into the tetrameric decarboxylase component of the complex.
This is followed in sequence by incorporation of two
further phosphates without any change in enzyme activity.
The amino acid sequences around the three phosphorylation
sites have been determined. These additional phosphorylations interfere with re-formation of active PDH complex
by PDH phosphate phosphatase. Using PDH complex purified
from pig heart the relative rates of the inactivating and
additional phosphorylation reactions have been investigated. The effects of inhibitors and activators of the
kinase reaction on the rate of inactivating and additional
phosphorylation reactions have also been investigated. The
rate of the additional phosphorylation reactions was
markedly slower than that of the inactivating phosphorylation. The Km for ATP in the inactivating kinase reaction
was $25.45 \pm 4.67\mu M$. The corresponding Km value for the
additional kinase reactions was significantly lower
($9.67 \pm 1.46\mu M$). ADP was found to be a competitive inhibitor with respect to ATP of both inactivating and additional phosphorylation reactions. The Ki values were
$69.84 \pm 11.63\mu M$ and $31.47 \pm 6.70\mu M$ respectively. Pyruvate
was an uncompetitive inhibitor with respect to ATP of both
inactivating and additional phosphorylation reactions. The
Ki values were $2.81 \pm 0.67mM$ and $1.14 \pm 0.08mM$ respectively. Increasing concentration ratios of $NADH/NAD^+$ and
acetyl-CoA/CoA increased the rate of both the inactivating
and additional phosphorylation reactions. Stimulation of
these reactions by increasing concentration ratios of
acetyl-CoA/CoA was more marked when a partially stimulating concentration ratio of $NADH/NAD^+$ was used.

CYCLIC AMP STIMULATION OF RIBOSOMAL PHOSPHOPROTEIN
PHOSPHATASE ACTIVITY

Robert Kisilevsky, and Margaret Treloar
Department of Pathology and Biochemistry, Queen's Univer-
sity and Kingston General Hospital, Kingston, Ontario,
Canada K7L 3N6

Within minutes after female rats are treated with
ethionine (E) one can observe the phosphorylation of
ribosomal protein S6. This phosphorylation persists
until 2 hrs following E administration at which point it
is gradually lost. By 4 hrs after E no phosphate remains
on S6. The phosphate loss is correlated with a marked
increase in liver cAMP. In contrast control animals treated with saline remain phosphorylated for long periods of
time, without any rise in cAMP. These results suggest that
a ribosomal phosphoprotein phosphatase exists in the liver
which can be stimulated by cAMP. This possibility was
examined in vitro.

^{32}P labelled ribosomes with virtually completely phosphorylated S6, and no other phosphorylated protein, were prepared by first treating animals with E for 4 hrs followed by
the adenine induced reversal of E intoxication in the
presence of ^{32}P ortho phosphate. These ribosomes were
used in an in vitro assay for phosphatase activity in the
presence and absence of cAMP. The phosphatase was prepared
from rat liver 100,000 x g supernatant using a 35% ethanol
cut. The precipitate was resuspended in 20mM Tris, 5mM
$MnCl_2$, 2mM βME, 0.5mM EDTA at pH 7.4. The solubilized
material was dialysed extensively against the same buffer
and used as the phosphatase preparation. Cyclic AMP led
to a 1.5-2 fold stimulation of ribosomal phosphoprotein
phosphastase activity.

Supported by Grant MT-4133 from the M.R.C. of Canada.

602

UNUSUAL GENOME ORGANIZATION OF NEUROSPORA CRASSA

Robb Krumlauf and George Marzluf, Dept. Biochem. Ohio State
University, Columbus, Ohio 43210

The genome of the fungus Neurospora crassa was analyzed
and found to have a sequence complexity relative to E. coli
of $2x10^7$BP. The genome is composed of 90% unique, 8% repetitive and 2% foldback sequences. The complexity of the repetitive component is 15,400 BP with a repeat frequency of
140. Genetic evidence indicates that many non-linked structural genes are coordinately and positively controled. Therefore, we examined the organization of the repetitive components
in the genome since they may play a role in gene regulation.

A variety of studies all indicate that the repetitive
sequences are organized in streches 10,000 BP or longer. At
most only 2-3% of the unique sequences are adjacent to the
repetitive sequences at fragment lengths up to 10,000 BP.
Ribisomal DNA represents 5-6% of the total DNA and about 80%
of the repetitive component. Since the rDNA units are believed
to be tandemly repeated on a single chromosome, as is the case
in yeast, their clustering can largely account for the large
repetitive streches and lack of interpersion of the repetitive
sequences.

These results indicate that Neurospora crassa does not conform to either the short period or long period interspersion
pattern for repetitive sequences seen in other eucaryotes.
This outcome implies that repetitive DNA sequences do not serve
a regulatory role in Neurospora crassa. However, since the
techniques used in these experiments would not detect double
stranded regions smaller than 45-50 Bp, it is conceivable that
additional repetitive sequences shorter than 45 BP might exist
and occur in an interspersed pattern. These sequences might
provide a regulatory role.

Supported by NIH grant GM-23367

CLONING AND CHARACTERIZATION OF ALLELES OF THE NATURAL OVAL-
BUMIN GENE CREATED BY MUTATIONS IN THE INTERVENING SEQUENCES

Eugene C. Lai, Savio L.C. Woo, Achilles Dugaiczyk, Donald
A. Colbert and Bert W. O'Malley. Department of Cell Biology
Baylor College of Medicine, Houston, Texas 77030.

Two allelic forms of the natural chicken ovalbumin gene
have been independently cloned. These alleles differ from
each other by an Eco RI restriction cleavage site in one of
the seven intervening sequences within the natural ovalbumin
gene. Detailed restriction mapping and sequence analyses of
these cloned genotypic alleles have shown identical sequence
organization and molecular structures of the interspersed
structural and intervening sequences except for the particular Eco RI cleavage site. Sequencing data of the cloned DNA
suggest that this Eco RI site may be created or eliminated
by a single base mutation in the intervening sequence of the
ovalbumin gene. The occurrence of apparent homozygous and
heterozygous allelic forms of the ovalbumin gene in individual hens and roosters within the same breed has been
observed. Out of 40 chickens examined, 40% of them are
homozygous for the ovalbumin gene without the extra Eco RI
site, while 10% of them are homozygous possessing this site.
The other half of the population are heterozygous for this
Eco RI site. Further analysis of individual chicken DNA
cleaved by restriction endonuclease Hae III has revealed
that there may be a series of such mutational variations
within the ovalbumin gene. We have identified two Hae III
cleavage sites that do not occur in all of the chickens.
The extent of divergence in intervening DNA sequences can
be fully appreciated only if the same gene could be cloned
and the DNA sequenced from a series of individual animals.
This type of mutation within the intervening sequences of
an eucaryotic gene has no known phenotypic manifestation
and represents a new type of a extrastructural silent
mutation. It should be taken into consideration in
estimating relative rates of evolutionary sequence
divergence of eucaryotic genes.

DIFFERENTIAL INFLUENCE OF mRNA CAP METABOLISM ON PROTEIN
SYNTHESIS IN DIFFERENTIATING EMBRYONIC CHICK LENS CELLS.

Gene C. Lavers, Gregory Gang and E.M. Ciccone
Department of Biochemistry, New York University Dental
Center, New York, N.Y. 10010

A homologous cell-free translation system, derived
from differentiating cells obtained from 15-day embryonic
chick lenses, efficiently utilizes endogenous lens mRNA
for the synthesis of lens polypeptides. The incorporation
of [^{35}S]-methionine into lens polypeptides indicates that
differential synthesis of lens proteins is altered in
response to the accumulation of mRNA caps. m^{7}GMP,
specifically released from short m^{7}GMP-containing capped
oligonucleotides, but not from m^{7}GMP-containing capped
mRNA by a m^{7}GpppN-pyrophosphatase activity, is associated
with a differential increase in lens polypeptide chain
synthesis. Product inhibition of the m^{7}GpppN-pyro-
phosphatase by m^{7}GMP is relieved by conversion of m^{7}GMP
to m^{7}Guanine and ribose-5-phosphate. Decreased levels of
m^{7}GMP is associated with increased levels of endogenous
mRNA translation. The binding of [^{3}H-methyl]-labeled
delta-crystallin mRNA or [^{35}S]-met-tRNA$_f^{met}$ to lens ribosomes
is suppressed by m^{7}GpppG. The results suggest that
translation of functional lens mRNA can be effected by the
metabolism of m^{7}GMP-containing capped mRNA fragments in
differentiating epithelial cells during lens development.

Supported by USPHS Grant EY02352.

MOLECULAR STUDIES OF RETRODIFFERENTIATION PHENOMENA IN PROLIFERATION-COMPETENT ADULT HEPATOCYTE CULTURES.

H. Leffert[*], J. Bonner[**], J. Brown[*+], T. Moran[*], J. Sala-Trepat[**], S. Sell[+], H. Skelly[*+], and K. Thomas[+]. [*]Cell Biology Lab, Salk Institute, LaJolla, CA; [+]Pathology Dept., School of Medicine, University of California at San Diego, LaJolla, CA; and[**]Biology Division, California Institute of Technology, Pasadena, CA.

Developmental studies of growth-state dependent phenotypes in normal epithelial cells are possible using primary monolayer cultures of adult rat hepatocytes (H. Leffert et al., PNAS 75:1834-1838, 1978). These studies have focussed upon the control of albumin and alpha$_1$-fetoprotein (AFP) production. To quantitate the numbers of total cellular mRNA-containing sequences coding for these proteins, total cellular RNA was isolated during the 12-day growth cycle and hybridized under conditions of RNA excess to purified albumin and AFP-specific [^{3}H]- or [^{32}P]-labelled cDNA probes (J. Sala-Trepat et al., PNAS, in press). Specific protein synthesis rates were measured by immunoprecipitation of [^{3}H]-leucine pulse-labelled antigens. Overall protein synthesis rates were determined by [^{3}H]-leucine uptake into hot acid-insoluble material.

Total (cellular and secreted) albumin synthesis and total albumin mRNA-containing sequences per cell followed U-shaped curves during the growth cycle. By contrast, total AFP synthesis per cell was undetectable during lag-phase (day 2) but then increased by at least 10-fold during log-phase (day 8), whereas total numbers of AFP mRNA-containing sequences remained constant throughout (days 2-12) at ca. 5-10 sequences per cell. Overall protein synthesis rates followed a bell-shaped curve with maximal increases (between days 8-10) of ca. 2-3 fold.

The combined observations suggest that transcriptional and post-transcriptional processes contribute to mechanisms which control retrodifferentiation phenomena during "normal" hepatoproliferative transitions.

Supported by USPHS Grants CA21230, GM13762, P50 AA03504 and the French Research Council.

ROLE OF PROTEIN SYNTHESIS IN THE REPLICATION OF THE "KILLER VIRUS" OF YEAST

Michael J. Leibowitz. Department of Microbiology, CMDNJ-Rutgers Medical School, Piscataway, NJ 08854.

The "killer virus" is a cytoplasmically-inherited double'stranded RNA virus of Saccharomyces cerevisiae which confers upon host cells the ability to secrete a toxin which kills strains not harboring this virus (or plasmid). Growth of killer yeast strains in the presence of sublethal concentrations of cycloheximide (1) or at elevated temperature (2) cures these strains of the killer virus.

The protein-synthesis inhibitor cryptopleurine resembles cycloheximide in its ability to cure the killer plasmid. A yeast strain bearing the mutation cryl contains ribosomes which are resistant to cryptopleurine inhibition in vitro (3). This mutant is also resistant to curing of the killer virus by cryptopleurine. This suggests that the mechanism of cryptopleurine-induced curing is secondary to inhibition of protein synthesis.

Supported by grants from the Foundation of CMDNJ and General Research Support-Rutgers Medical School.

(1) Fink, G.R., and Styles, C.A. Proc. Nat. Acad. Sci. 69 2846-2849 (1972).

(2) Wickner, R.B., J. Bacteriol. 117 1356-1357 (1974).

(3) Skogerson, L., McLaughlin, C., and Wakatama, E. J. Bacteriol. 116 818-822 (1973).

EFFECTS OF THE CATALYTIC SUBUNIT OF PROTEIN KINASE II FROM
RETICULOCYTES AND BOVINE HEART MUSCLE ON PROTEIN PHOSPHORY-
LATION AND PROTEIN SYNTHESIS IN RETICULOCYTE LYSATES.

Daniel Levin, Vivian Ernst and Irving M. London.
Dept. of Biology, M.I.T., and Harvard-M.I.T. Division of
Health Sciences and Technology, M.I.T., Cambridge, MA 02139.

The inhibition of protein synthesis initiation in heme-
deficient lysates is due to the activation of a heme-regulated cAMP-independent protein kinase (HRI) which phosphorylates the α-subunit of eIF-2 (eIF-2α). Recent findings
suggest that the phosphorylation of HRI is related to its
activation and is correlated with the phosphorylation of
eIF-2α both in vitro and in situ. In a previous study, we
observed that high levels of BHM protein kinase II (PK)
(50 μg/ml) plus cAMP (10^{-4}M) partially inhibited protein
synthesis in lysates. Since cAMP alone (10^{-7}M to 10^{-2}M)
has no effect on protein synthesis in lysates, we examined
the effects of highly purified catalytic subunits of PK II
from reticulocytes (rC) and BHM (bC). Both rC and bC migrate as 43,000 dalton polypeptides in SDS-PAGE and are
autophosphorylated when preincubated with $[\gamma-^{32}P]$ATP. The
addition of rC (or bC) to lysates at physiological concentrations does not affect protein synthesis, but at 5 to 10-
fold higher levels, some inhibition (20-25%) is observed;
by comparison, heme-deficiency produced an 83% inhibition.
Incubation of the proinhibitor form of HRI with $[\gamma-^{32}P]$ATP
and rC (or bC) results in the phosphorylation of proinhibitor, but this does not result in phosphorylation of eIF-2α
or inhibition of protein synthesis. Lysates treated with
rC, bC or 100 μM cAMP display a phosphoprotein profile
different from that of heme-deficiency: 5-10 polypeptides
are phosphorylated but not eIF-2α or HRI. In heme-deficiency, HRI and eIF-2α are rapidly phosphorylated. (see
Ernst et al, these Abstracts). (Supported by USPHS Grants
AM-16272-07 and GM-24825-01).

FUNCTIONAL HETEROGENEITY IN THE RNAS OF TYMV VIRIONS

D. Lightfoot and R. Clark. Department of Biochemistry
and Nutrition, Virginia Polytechnic Institute and State
University, Blacksburg, Virginia

Turnip Yellow Mosaic Virus (TYMV) RNA occurs as a
complex mixture of different size RNAs in native virions
and only under fully denaturing conditions can these RNAs
be resolved. On pH 3.5, agarose urea gel electrophoresis
six major RNAs are discernable among a 1.9×10^6 D to
0.22×10^6 range of RNA fragments. A variety of isolation
procedures do not change this pattern. Virions of very
young (5 day) infections, compared to the usual 24 day infections, contain a simpler array of RNA including $1.9 \times
10^6$ D, 0.39×10^6 D, and 0.22×10^6 D species and no de-
graded RNA. In virions of 8 day and progressively older
infections 1×10^6 D, 0.46×10^6 D, and 0.41×10^6 D RNAs
also are prominent but accompanied by increasing levels of
heterogeneous RNA. The mechanism producing this complex
array of RNA species is not known.

The smallest of the six RNAs is the coat protein
mRNA which, like the full genome RNA, is capped but, unlike the large RNA, contains its coat protein mRNA
sequence in an active form. The purified coat protein
mRNA stimulates TCA precipitable amino acid incorporation
four fold in an E. coli cell-free protein synthesizing
system. This is 1/3 the stimulation observed with added
MS 2 RNA. Neither are these high levels of incorporation
observed using the whole genome RNA nor is the coat protein detectible among the products.

TYMV RNA 3'terminus is an active valine tRNA. The
valylation activity requires, for 79% of the total RNA,
prior adenosylation of the terminal ACC_{OH} sequence to
$ACCA_{OH}$ while 8% of the native RNA already has the terminal
A. This low level of native 3' adenosine is a constant,
even for viral RNA from 5 day infections or for any of the
size classes of RNA. On the other hand, the proportion of
viral RNA which can be activated by adenosylation prior to
aminoacylation falls from 79% to only 35% in 5 day infections. Aminoacylation kinetics show that these two
independent forms of the viral tRNA function have the same
K_m and, presumably, the same structure. The nonadenosy-
lated form is a competitive inhibitor (K_i = 32 nM) of the
aminoacylation reaction. However, the very low K_m of the
aminoacylatable form is 7 nM and these data indicate that
at least some TYMV RNA may be esterified to valine in vivo.
The RNAs of TYMV may serve several functions at different
stages of infection essential to the infection process, in
addition to the usual role of mRNA.

SUBCELLULAR DISTRIBUTION OF SV40 T ANTIGEN SPECIES IN
VARIOUS CELL LINES

Samuel W. Luborsky and K. Chandrasekaran.
National Cancer Institute, National Institutes of Health,
Bethesda, Maryland 20014

We investigated the presence in various SV40 virus
transformed cell lines of the SV40 T antigen (TA) on the
plasma membrane, as well as the subcellular size distribution of TA species. The cells were labelled with ^{35}S-
methionine, suspended in hypotonic solution, dounced, and
the nuclei, membranes and cytoplasm fractionated in a two-
phase system. TA was immunoprecipitated with anti TA serum
in the presence of Sepharose linked to protein A of Staph.
aureus, and analyzed by electrophoresis on SDS-polyacrylamide gels, and fluorography. We examined SV40 transformed
cells from normal rat kidney, Balb/c and AL/N mice, cloned
and mass cultures.

Both 94K and 56K molecular weight (MW) species were
found in the nuclei of the four cell lines examined, while
the membrane fractions contained the 94K dalton species and
only low levels of the 56K species. The cytoplasm fractions contained little radioactive label specifically immunoprecipitated by the anti TA serum. Detection of the small
TA (∿17K) species was variable, and rendered more difficult
by generally containing least label. Work is in progress
to improve detection of this species. Control mixing experiments were carried out to clarify the nature and purity of
the cell fractions. Cell surface labelling experiments are
in progress to measure more directly the plasma membrane TA.

These results showing the presence of TA on the plasma
membrane, and previous reports that TA can function as a
tumor specific transplantation antigen (TSTA) fit a model in
which plasma membrane bound TA can function in vivo as .
TSTA, while nuclear TA binds to DNA and controls some unknown
functions. Thus the genetic information contained in the
early region of the SV40 virus genome is enabled to code
for protein(s) capable of multiple diverse functions.

HORMONAL REGULATION OF PROTEIN BIOSYNTHESIS IN RAT EPIDIDY-
MAL FAT CELLS: STIMULATION BY INSULIN, INHIBITION BY FAST-
ING, AND PERMISSIVE EFFECTS OF ADRENAL GLUCOCORTICOIDS.

Robert T. Lyons and Donald A. Young. E. Henry Keutman Lab.
Endocrine-Metabolism Unit, Department of Medicine, Univ. of
Rochester Medical Center; Rochester, New York 14642.

A cell-free protein synthesis assay was set up which
used the incorporation of ^{3}H-leucine into TCA-insoluble
protein as a measure of the run-off of nascent polypeptide
chains from polysomes of fat cell lysates. The reaction
was performed in the presence of a large excess (relative
to internal pools) of 20 unlabeled amino acids and a de-
salted post-microsomal supernatant fraction from rat liver
in order to ensure that the number of initiated fat cell
ribosomes was the rate-limiting factor in each assay.
Simultaneous sucrose density gradient analysis of fat cell
lysates was performed in order to quantitate changes in
pre-polysomal (ribosomal subunits, monomers, dimers and
disomes) and in polysomal components which may occur in
response to a variety of hormonal and nutritional manipulations of intact and adrenalectomized rats.

Our results indicate that: 1) polysomes from rat epididymal fat pad tissue may be recovered in a native, essentially undegraded state using a procedure involving freezing and grinding at liquid N_2 temperature; 2) the extent of
ribosomal aggregation into polysomes in vivo and the activity of these polysomes in vitro may be reliably and reproducibly compared between groups of rats undergoing a
variety of hormonal and nutritional manipulations; 3) either
fasting of intact rats or adrenalectomy results in a marked
loss of fat pad polysomes in vivo, and a parallel decline
in protein synthetic activity in vitro; 4) brief (30 min)
insulin treatment of starved intact rats results in extensive polysome formation in vivo and in a doubling of protein synthetic activity of fat pad polysomes in vitro;
5) starved, adrenalectomized rats require dexamethasone
pre-treatment in order to exhibit the insulin effect mentioned above.
    We have observed a similar insulin effect on protein
synthesis in isolated fat cells and are now seeking verification in vitro of the possible "permissive" action of
glucocorticoids that was observed in vivo.

Supported by NIH Grants # 5 T32 AM 07092 and AM 16177, and
by an award from the United Cancer Council of Monroe County.

611

REVERSAL OF POST TRANSLATIONAL TYROSYLATION OF TUBULIN

Todd M. Martensen and Martin Flavin.   Laboratory of Cell
Biology, NHLBI, NIH, Bethesda, Md. 20014

A proteolytic enzyme with carboxypeptidase activity
has been purified from sheep brain extracts.   Enzymatic
properties were similar to tubulin carboxypeptidase
activity in rat brain extracts described by Argarana et
al. (1978) Mol. Cell. Biochem. 19, 17-21.   The enzyme
reverses the reaction catalyzed by tyrosine tubulin
ligase which attaches tyrosine to the c-terminal glutamyl
residue of the α subunit of tubulin by the following
reaction:

$$\text{tyrosine} + \text{tubulin} \underset{\text{Cpase}}{\overset{\text{Ligase}}{\rightleftharpoons}} \text{tyrosyl-tubulin}$$

$$\text{MgATP} \rightharpoonup \text{MgADP} + \text{Pi}$$

After homogenization of sheep brain the majority of
carboxypeptidase activity resided in the soluble fraction.
Some activity copolymerized with tubulin with an increase
in specific activity during three polymerization cycles.
Purification by NH_4SO_4 fractionation and DEAE cellulose
chromatography removed ligase and tubulin.   Further
chromatography on CM cellulose gave a 500-fold purification with 60% recovery of soluble activity.

The purified exoprotease activity was inhibited by
EDTA, o-phenanthroline, and β-phenylpropionate.   Dithiothreitol and a carboxypeptidase inhibitor isolated from
potato tubers was not inhibitory.   Calcium ion was several
fold more inhibitory than Mg^{++} at mM levels.   Complete
inhibition was seen with 100 mM KCl or NaCl.

Kinetic studies gave a Km for ^{14}C-tyrosyltubulin of
approximately 50 μg/ml ($5 \times 10^{-7}M$).   Tubulin which was
reduced and alkylated could also serve as a substrate.
Carbobenzoxy-glu-tyr was inhibitory at mM levels.
Colchicine which inhibits tubulin polymerization and GTP
which promotes polymerization were inhibitory.   Utilization of a regenerating system for GTP with pyruvate kinase
and phosphoenolpyruvate could completely inhibit the
reaction.

SALT-DEPENDENT TRANSCRIPTIONAL SPECIFICITY OF T7 RNA POLY-MERASE

William T. McAllister and Anthony D. Carter. Department of Microbiology, CMDNJ-Rutgers Medical School, Piscataway, NJ 08854.

The late genes of bacteriophage T7 are transcribed by a phage-specified RNA polymerase. We have found that the appearance of T7 late messenger RNAs in vivo is temporally regulated: class II mRNAs are synthesized from 4 until 16 min after infection, class III mRNAs are synthesized from 6-8 min after infection until lysis (at 30 min). The regions of T7 DNA that are transcribed by the phage RNA polymerase in vitro depend upon the ionic conditions: whereas synthesis of class II RNAs is inhibited at KCl concentrations greater than 150mM, synthesis of class III RNAs proceeds efficiently at this salt concentration. The dependence of transcriptional specificity upon ionic conditions in vitro suggests that this phenomenon may be important in regulating transcription during T7 infection. T7-infected cells undergo marked permeability changes at about 12 min after infection as a result of the expression of T7 gene 3.5 (lysozyme). In cells infected with phage defective in gene 3.5, class II RNA synthesis is not shut off with normal kinetics. These observations indicate that phage-induced alterations in intracellular ionic conditions are important in the regulation of T7 late transcription. We have mapped six salt-resistant class II promoters between 13.5 and 35.4 units on the T7 map, and six salt-resistant class III promoters between 46.5 and 97.2 T7 units.

This research was supported by NIH Grant GM-21783 to William T. McAllister. Anthony D. Carter is a predoctoral trainee supported by NCI Training Grant 27-1241.

DIFFERENTIAL SENSITIVITY OF INHIBITION OF CYCLIC AMP PHOS-
PHODIESTERASE IN THE THYMUS DURING MURINE LEUKEMOGENESIS.

Lawrence A. Menahan
Department of Pharmacology, The Medical College of
Wisconsin, Milwaukee 53226

A large proportion of AKR mice develop lymphoid leukemia at the age of 7 to 9 months with the initial oncogenic
transformation occurring in the thymus. Significant changes
in thymic cyclic AMP (cAMP) metabolism during the transition from the non-leukemic to the leukemic state have been
reported (1). It was shown in these studies that a sharp
decrease in the concentration of cAMP was found if an overt
thymic lymphoma was present. In spite of an elevated thymic adenylate cyclase activity in the leukemic AKR mice,
the higher activity of cAMP phosphodiesterase in the thymic
lymphoma could account for the drop in cAMP content observed. More recently, it was shown that much of high affinity
cAMP phosphodiesterase in the thymus of normal and leukemic
mice was associated with the plasma membrane (2).

In the present study, the inhibitory effects of methyl-
xanthines [theophylline, isobutylmethylxanthine (MIX)] and
SQ 20,009 on thymic cAMP phosphodiesterase were compared.
SQ 20,009 (10^{-5} M) in the presence of 1 µM cAMP inhibited
60% of the phosphodiesterase activity in a membrane preparation from normal mice while the inhibition found under
comparable conditions with thymic lymphoma tissue from AKR
mice was only 30%. Yet, theophylline (2×10^{-4} M) inhibited
thymic cAMP phosphodiesterase to the same extent (40%) in
both normal and leukemic mice. Parallel studies with out-
bred Swiss albino mice have indicated that the sensitivity
of cAMP phosphodiesterase in thymus to inhibition by both
MIX (5×10^{-5} M) and SQ 20,009 (5×10^{-5} M) decreases progressively with increasing age. However, the pattern of
inhibition by the two agents at all ages up to 24 months
was similar.

In summary, there is not only an increase in membrane-
bound cAMP phosphodiesterase in leukemic AKR mice but there
is also a differential loss of inhibition by certain drugs
suggesting an enrichment of certain phosphodiesterase
isozyme(s) during murine leukemogenesis.

Supported by Grant CA21631 awarded by National Cancer
Institute, DHEW.

(1) Kemp, R. G., Hsu, P. -Y. and Duguesnoy, R. J., Cancer
Res. 32 2440 (1975).
(2) Menahan, L. A. and Kemp, R. J., J. Cyclic Nucleotide
Res. 2 417 (1976).

IN VIVO SUBNUCLEAR DISTRIBUTION OF LARGER THAN TETRAMERIC
POLYADENOSINE DIPHOSPHORIBOSE IN RAT LIVER

Takeyoshi Minaga and Ernest Kun
Departments of Pharmacology, Biochemistry and Biophysics, and
the Cardiovascular Research Institute, University of California,
San Francisco, CA 94143

The occurence of polyadenosine diphosphoribose in rat
liver in vivo has been established by pulse labeling with
^{14}C-ribose (ref. 1). Although this experiment demonstrated
the existence of the polymer in liver, the estimation of
absolute amounts of polyadenosine diphosphoribose and the
establishment of in vivo macromolecular polydispersity
were not possible because quantitation was necessarily
dependent on tracer analysis only.

Recently a specific radioimmunoassay has been developed
(ref. 2) which is suitable for the quantitative determination
of larger than tetramers. Further modification of this
method enables the specific extraction and assay of poly
$(ADP-R)_{n>}$ 4 from whole tissues and the determination of the
polymer as associated with histone and non-histone proteins of
the chromatin. It was found that more than 99% of the larger
than tetrameric polymer is associated with non-histone nuclear
proteins, whereas histone fractions (less than 1%) contain an
approximately even distribution between all fractions. The
macromolecular profile of the polymer as present in subnuclear
protein fractions was determined by the combination of molecular
filtration and radioimmunoassay. From specific activities of
in vivo ^{14}C-ribose labeling $t1/2$ of poly (ADP-R) and of NAD^+
were of the same magnitude (2.7 to 3.1 h). A time dependent
variation in ^{14}C labeling between long and short chain
polymers was also observed.

1. Ueda, K., Omachi, A., Kawaichi, M. and Hayaishi, O. Proc.
Natl. Acad. Sci. USA 72. 205-209, 1975.

2. Ferro, A. M., Minaga, T., Piper, W.N. and Kun. E. Biochim.
Biophys. Acta 519, 291-305, 1978.

Supported by USPHS Program Project Grant HL-6285 and a grant
of the U.S. Air Force Office of Scientific Research (AFOSR
78-3698).

PROCESSING AND TRANSPORT OF RNA FOLLOWING MITOGENIC
STIMULATION OF HUMAN LYMPHOCYTES

M.F. Mitchell, E. Bard and J.G. Kaplan. Department of
Biology, University of Ottawa, Ottawa, Canada.

The kinetics of appearance of labeled RNA in the cytoplasm of human peripheral blood lymphocytes stimulated with
Concanavalin A (Con A) was studied using an improved method
that afforded clean separation of nuclei from cytoplasm (1).
We observed no significant increase over resting cell levels
in labeled total cell RNA before 10 h exposure to Con A; we
detected increases in ^{3}H-uridine-labeled non-polyadenylated
RNA in the cytoplasm following 6 h exposure to Con A (1). In
stimulated cells prelabeled with ^{3}H-uridine, labeled cytoplasmic non-polyadenylated RNA levels increased after 3 h
exposure to Con A (2). ^{3}H-uridine-labeled polyadenylated RNA
required at least 13 h before it appeared in cytoplasmic
extracts of stimulated cells and it did not appear in those
of resting cells at least for the 8-9 h duration of our
experiments (1). Increased levels of ^{3}H-adenine-labeled polyadenylated RNA were detectable in cytoplasm after 5 h exposure
to mitogen (2). The lag between addition of isotope and
appearance of labeled RNA in cytoplasm was inversely related
to the duration of incubation of cells with mitogen. However,
once the label appeared in the cytoplasm its rate of
accumulation was similar (1). Preliminary double-labeling
experiments suggest that transcripts from resting cells are
polyadenylated and transported to the cytoplasm very soon
after stimulation. These data suggest that an important early
effect of mitogen is to increase processing and transport of
RNA, especially polyadenylation of pre-mRNA. Similar conclusions have been reported by Schäfer and colleagues (3).

Thus, we do not regard extensive derepression of the
genome to be an early manifestation of lymphocyte activation.

(1) Mitchell, M.F., Bard, E., L'Anglais, R.L. and Kaplan, J.G.
Can. J. Biochem. 56 659-666 (1978).

(2) Kaplan, J.G., Mitchell, M. and Bard, E. Biochemistry and
Molecular Biology of Lymphocyte Transformation (M.R.
Quastel, ed.) Academic Press, New York. (1978).

(3) Land, H. and Schäfer, K.P. Biochem. Biophys. Res. Comm.
79 947 (1977).

IDENTIFICATION OF DROSOPHILA 2S rRNA AS THE 3'-PART OF 5.8S rRNA

George Pavlakis, Regina Wurst and John Vournakis
Department of Biology, Syracuse University, Syracuse, N.Y. 13210

Drosophila melanogaster 5.8S rRNA was end-labeled with [^{32}P] at either the 5'- or 3'-end and was sequenced using the rapid RNA sequencing method. It is 122 nucleotides long and homologous to the 5'-part of the sequenced 5.8S molecules from other species. We refer to this molecule as "mature 5.8S rRNA" (m5.8S rRNA). The 3'-end of m5.8S rRNA is able to form extensive base-pairing with the 5'-end of Drosophila 2S rRNA at positions equivalent to a GC-rich hairpin that has been found in all sequenced 5.8S rRNA molecules. Structure-mapping (1) of the m5.8S rRNA with S1 nuclease verifies its similarity to the other 5.8S rRNAs. End-labeled 2S rRNA is shown to form a complex with m5.8S rRNA and 26S rRNA separately, and with both together, as does m5.8S rRNA with 26S rRNA.

These experiments lead us to conclude that:
a) The 5.8S rRNA molecule in Drosophila is very similar both in sequence and structure to the other 5.8S rRNAs but is split in two, the 2S rRNA being the 3'-part of the molecule.
b) The hairpin loop of the "GC rich arm" is not essential for ribosome function and must be exposed in the ribosome.
c) The 2S rRNA anchors the m5.8S rRNA on the 26S rRNA but there are additional sites of interaction between the m5.8S rRNA and the 26S rRNA.

These results are in agreement with the known relative positions of m5.8S rRNA and 2S rRNA within the spacer separating the 18S DNA and 26S DNA (2) in the ribosomal DNA repeat of Drosophila.

Supported by NIH Grant GM 22280, and grants from the Alton-Jones Foundation and the Syracuse University Senate Research Fund.

(1) Wurst, R., Vournakis, J.N. and Maxam, A.M. (1978) Biochem., in press.
(2) Jordan, B.R., and Glover, D.M. (1977) FEBS Letters 78, 271-274.

FACILITATION BY PYRIMIDINE NUCLEOSIDES AND HYPOXANTHINE OF MNNG MUTAGENESIS IN CHINESE HAMSTER CELLS.

A.R. Peterson, Hazel Peterson, J.R. Landolph and Charles Heidelberger.
University of Southern California Cancer Center, 1303 N. Mission Road, Los Angeles, CA 90033.

Hypoxanthine (Hx), thymidine (TdR) and deoxycytidine (CdR), at concentrations of 10^{-5}M increased the yield of 8-azaguanine resistant (AzG^r) mutants induced by $\underline{N}$-methyl-$\underline{N}'$-nitro-N-nitrosoguanidine (MNNG) in cultured Chinese hamster V79 cells. The cytotoxicity of MNNG was increased 2 fold in the presence of Hx, and 10 fold in the presence of TdR. This effect of TdR was abolished by equal concentrations of CdR, which by itself did not increase the cytotoxicity of MNNG. However, the yield of MNNG–induced AzG^r colonies was increased 2-10 fold in the presence of both CdR and TdR. The AzG^r colonies displayed the AzG^r,HAT^s (10^{-5}M Hx, 10^{-6}M aminopterin, 5×10^{-5}M TdR) phenotype characteristic of hypoxanthine:guanine phosphoribosyl transferase deficient ($HGPRT^-$) mutants, or the AzG^r, HAT^r or AzG^s, HAT^s phenotypes, indicative of mutant HGPRT with altered substrate affinities. The nucleosides did not affect the growth or expression time of the $HGPRT^-$ mutants; the same extent of alkali-labile DNA damage occurred in cells treated with alkylating agents in the presence and absence of TdR and CdR. The increase in mutation frequency in the presence of these nucleosides occurred not only with MNNG, but also with ethylating agents, and not only at the HGPRT locus but also at that of the ouabain sensitive sodium–potassium ATPase. Nucleosides treated with MNNG were not mutagenic, and treatment of the cells with TdR and CdR only prior to treatment with MNNG did not increase the induced mutation frequency. Therefore, we consider that Hx, TdR, and CdR produce nucleotide pool imbalances that increase the probability of errant replication of damaged DNA, and thus facilitate mutagenesis.

Supported by grant CA-21,036 from the National Cancer Institute and by American Cancer Society Postdoctoral Fellowship (to J.R.L.) PF-1331.

NONCODING INSERTS IN HEMOGOLOBIN AND SIMIAN VIRUS 40 GENES ARE BOUNDED BY SELF-COMPLEMENTARY REGIONS

Manfred Philipp, Robert Bernholc, Avi Rubinsztejn, and Darrell Ballinger, Department of Chemistry, Lehman College of the City University of New York, Bronx, New York 10468

Recent studies have shown that many eukaryotic genes contain noncoding internal spacer regions called introns. These are, at least in some cases, transcribed into RNA. We have prepared several programs which analyze DNA and RNA sequences. These programs help determine whether these primary sequences contain self-complementarities that might aid in ligation of coding, exon regions that remain in the final mRNA's. The presence of complementary regions in exons that are eventually ligated to form mRNA's may facilitate such ligation, by bringing areas to be ligated into close physical proximity, and by providing recognition sufficient to ensure that appropriate exons are ligated.

Extensive complementary regions were found in both rabbit β-globin mRNA and SV-40 DNA. The potentially most stable self-complementarity found in the rabbit β-globin mRNA extends from nucleotide number 231 to nucleotide number 488. The known rabbit intron is excised at the central, non-base-paired loop which may be formed by this self-complementarity.

In the early genes of SV-40, there are at least two overlapping introns of different extent. Using the sequence indicated for a precursor RNA containing both introns, we found a self-complementarity that originates in exon regions and extends into intron regions. The self-complementarity extends from nucleotide positon 4293 to 4956. Deletion of either of the two introns reveals other self-complementarities which could act to bring exon region termini together, in a manner similiar to that in the rabbit β-globin mRNA.

A VARIANT OF POLYOMA VIRUS ALTERED IN ITS PROCESSING OF
LATE VIRAL RNA

B.J. Pomerantz, C.E. Blackburn, E.K. Thomas, V.A. Merchant,
J.D. Hare; Department of Microbiology, School of Medicine
& Dentistry, University of Rochester, Rochester, New York.

We have evidence that the 3049 variant of Py virus is
altered in a step associated with the processing of its
primary late transcript. This variant was originally isolated from a spontaneous hamster tumor in the lab of Dr.
Eddy at the N.I.H. It is recognized by its unusual phenotype in an indirect immunofluorescent test using antibody
directed against capsid antigens.[1] Cells infected with w.t.
virus fluoresce only in the nucleus, whereas 3049 infected
cells fluoresce in both nucleus and cytoplasm, a phenotype
related to overproduction of viral proteins.[5]

The evidence that 3049 is a processing variant is
summarized below:

(1) In all early lytic events and in the amount and timing
of viral DNA replication, 3049 and w.t. are the same.

(2) In the nucleus, at late times, the amount of high
molecular weight and total nuclear viral RNA is the same.
The amount of 3049 nuclear poly (A) RNA is increased,
though.[3,4]

(3) 3049 RNA is more abundant in cytoplasmic poly (A)
selected or total RNA pools at late times.

(4) In 3049 infected cells, there is a 3-5 fold increase
in the amount of viral capsid proteins[5] and a difference
in their relative distribution.

(5) Burst size and assembly are the same.

(6) There are structural differences in the viral genome
as seen in restriction enzyme analysis.

The difference in relative distribution of 3049 capsid
proteins suggests the possibility of a difference in the
relative distribution of 3049 messages. Experiments are in
progress to quantify the messages, map their leaders, and
investigate the relationship between the amounts of viral
message and their respective protein products.

REFERENCES:

(1) Hare, J.D., Virology, 40: 684, 1970.
(2) Betts, R.F., Tachovsky, T.E. and Hare, J.D., J. Gen.
     Virology, 16: 29, 1972.
(3) Rutherford, R.B. and Hare, J.D., B.B.R.C. 58: 839,
     1974.
(4) Rutherford, R.B. and Hare, J.D., B.B.R.C. 62: 789,
     1975.
(5) Tachovsky, T.G. and Hare, J.D., J. Virol. 16: 116,
     1975.

PROTEASE-MEDIATED MORPHOLOGICAL CHANGES INDUCED BY TREATMENT OF TRANSFORMED CHICK FIBROBLASTS WITH A TUMOR PROMOTER: EVIDENCE FOR DIRECT CATALYTIC INVOLVEMENT OF PLASMINOGEN ACTIVATOR

James P. Quigley. Department of Microbiology and Immunology, SUNY, Downstate Medical Center, Brooklyn, New York

The tumor promoter phorbol myristate acetate (PMA) induces the production of the serine protease plasminogen activator (PA) in cultures of normal chick embryo fibroblasts (CEF) and synergistically enhances PA production in Rous sarcoma virus transformed chick embryo fibroblasts (RSVCEF). Concomitant with the PA induction, distinct morphological changes occur in serum-free RSVCEF cultures upon PMA treatment. The morphological changes include enhanced cell clustering and the formation of colonial aggregates of cells. These alterations in the morphology of the PMA-treated transformed cells are inhibited by a number of protease inhibitors including DFP, leupeptin, NPGB and benzamidine. A number of protease inhibitors were ineffective in preventing the PMA-induced morphological changes and these included inhibitors of trypsin, chymotrypsin, thrombin and most importantly plasmin. The use of a fluorescent substrate to assay PA directly and the use of ^{3}H DFP to label and characterize serine enzymes in the culture fluid from PMA-treated cells indicate that PA is the serine enzyme responsible for the morphological changes. Thus, PA itself can catalytically alter cellular behavior in transformed cultures independent of plasminogen, up to now its only known substrate.

RECOVERY OF A TRANSFORMING VIRUS FROM AKR MOUSE TUMOR
TISSUE BY DNA TRANSFECTION

John M. Rice, Anna D. Barker and Anthony J. Dennis
Battelle's Columbus Laboratories
Columbus, Ohio 43201

Infectious proviral DNA for MULV was recovered directly
from the tissues of both aged overtly leukemic and young
non-leukemic AKR mice. Infectious DNA of a mink cell transforming virus was recovered only from the tumor tissue of
leukemic AKR mice and not from the control (liver) tissue
from the same leukemic mice nor from the pooled lymphoid
tissue of young non-leukemic mice. These mink cell transforming viruses (MCT viruses) were isolated from both mink
and mouse cells transfected with tumor DNA and exhibited
properties similar to previously described transforming
isolates. Our MCT virus isolates produced foci or morpho-
logic transformation but not cytopathology when tested on
mink cells and exhibited an amphotropic host range. The DNA
transfection technique proved to be an efficient approach
for the isolation of MCT viruses.

SYNTHESIS AND POST-TRANSLATIONAL MODIFICATION OF RETROVIRUS
PRECURSOR POLYPROTEINS ENCODED BY gag AND env GENES

A.M. Schultz and S. Oroszlan. Molecular Oncology Program,
Frederick Cancer Research Center, Frederick, Maryland 21701

The structural proteins of type-C RNA tumor viruses
(family Retroviridae) are synthesized as large precursor
polyproteins (Pr), products of the gag (group specific antigen)- and env (envelope)-genes. These polyproteins are
accurately cleaved by proteolytic enzymes. Other posttranslational modifications important for virus maturation
and assembly are glycosylation and phosphorylation. We have
studied these processess in cells infected with Rauscher
murine leukemia virus using radiolabeled precursors in pulse-
chase experiments, immune precipitation with antibodies
directed against specific antigenic determinants of virion
component polypeptides, and sodium dodecyl sulfate-polyacryl-
amide gel electrophoresis.

We have found that gag-gene encoded proteins are processed from a common precursor (Pr75) by two divergent pathways: one leading to internal virion components (p15, ppl2,
p30, p10) via cleavage and phosphorylation, the other to non-
virion, cell-associated proteins via glycosylation. In addition, env-gene encoded envelope protein p15E and glycoprotein
gp70 arise from Pr70 by glycosylation and subsequent cleavage
as summarized in the following scheme:

I.   mRNA^{gag} $\xrightarrow{}$ Pr75 $\Big\langle$ $\xrightarrow{g}$ gPr80 $\xrightarrow{gt}$ gP94

   $\xrightarrow{nc}$ Pr65 $\xrightarrow{p}$ pPr65 $\xrightarrow{mc}$ p15+ppl2+p30+p10

II. mRNA^{env} $\xrightarrow{}$ Pr70 $\xrightarrow{g}$ gPr85 $\xrightarrow{mc}$ gp70 + p15E

nc=nascent cleavage, p=phosphorylation,
mc=morphological cleavage, g=glycosylation
(tunicamycin sensitive), and gt=glycosyla-
tion and transport; numbers represent
molecular weights in thousands

Chemical cleavage of Pr65, Pr75 and gPr80 followed by
peptide mapping demonstrated that all three polyproteins
possess similar carboxyl-terminal regions whereas Pr75 and
gPr80 are distinguished from Pr65 by a long amino-terminal
"leader" sequence. The effect of each modification step on
subsequent processing leading to virus assembly will be
discussed. Supported by NCI Contract #N01-CO-75380.

AVIAN MYELOBLASTOSIS VIRUS (AMV) LINEAR DUPLEX DNA: IN VITRO
ENZYMATIC SYNTHESIS AND STRUCTURAL ORGANIZATION ANALYSIS

Robert A. Schulz, Jack G. Chirikjian+ and Takis S. Papas
Laboratory of Tumor Virus Genetics, National Cancer Institute
NIH, Bethesda, Md. 20014, and +Department of Biochemistry,
Georgetown University, Washington, DC 20007.

Avian retroviruses contain a unique genome composed of
a diploid single-stranded RNA. AMV RNA consists of two 35S
subunits which code for at least three viral gene products.
These include "gag", "pol" and "env" which code for viral
structural proteins, a polymerase, and glycoproteins of the
viral envelope, respectively. The identification and location of the "leuk" gene responsible for leukemogenic transformation remains unknown.

We have utilized a product of the viral genome, reverse
transcriptase, to synthesize both the negative and positive
DNA strands from AMV RNA. Full length cDNA transcripts of
intact RNA were isolated after alkaline sucrose gradient
centrifugation (1). The complete copy is 11% double stranded
after nuclease S_1 digestion. A hairpin structure at the 3'
end of the cDNA selfprimes synthesis of the second strand.
The product of this two step reaction, catalyzed by reverse
transcriptase, is a linear duplex of 5.2 x 10^6 daltons. The
duplex is covalently linked and 90-100% S_1 resistant. The
mechanism of hairpin formation by reverse transcriptase and
its possible involvement in proviral synthesis in vivo will
be discussed.

Analysis of the viral RNA template, genomic length
cDNA, and linear duplex on agarose gels demonstrates two
distinct species of each. RNAs of 7600 and 7000 ribonucle-
otides, cDNAs of 2.6 and 2.3 x 10^6 daltons, and duplexes
of 5.2 and 4.0 x 10^6 daltons have been identified. Since
AMV is defective and requires Myeloblastosis Associated
Virus for replication, the additional RNA and DNA species
may be helper virus, rather than AMV, related. Ongoing
restriction mapping of the two duplex DNAs should help us
understand the nature of the two species and may provide
further insight into the nature and organization of the
"leuk" gene.

(1) Myers, J.C., Spiegelman, S., and Kacian, D.L., Proc.
Nat. Acad. Sci., USA 74, (1977) 2840.

STUDIES ON THE DNA-DIRECTED IN VITRO SYNTHESIS OF BACTERIAL
ELONGATION FACTOR TU.

Tanya Schulz, Carlos Spears and Herbert Weissbach.
Department of Biochemistry, Roche Institute of Molecular
Biology, Nutley, New Jersey, 07110.

We have studied the synthesis of biologically active
EF-Tu in a DNA-directed in vitro system using the transduc-
ing phage λrif^{d}18 as template and a fractionated lysate of
E. coli. Immunoprecipitation and gel electrophoresis of the
radioactively labelled products synthesized in vitro showed
a major peak (>80%) which co-migrated with authentic EF-Tu.
Additionally, minor peaks at 12,000 M.W. and 25-30,000 M.W.
were observed. The cell-free system was optimized for this
gene product and the characteristic in vivo effect of a
putative regulatory factor, ppGpp, has been observed in
vitro. The inhibition of EF-Tu synthesis by ppGpp suggests
that the regulation of this protein seen in this in vitro
system is similar to at least one physiological regulatory
mechanism in the whole cell.

In conjunction with similar studies on the synthesis of
β-galactosidase being done in the laboratory, attempts are
being made to achieve synthesis in a more defined system.
In this system, the crude E. coli fractions are replaced by
purified preparations of all the known factors that are
thought to be essential for transcription and translation of
genetic information. Analysis of the incubations using the
more defined system shows that although amino acid incorporation into total protein decreases only two-fold, the yield
of EF-Tu in this system is significantly lower (<15%) compared to that synthesized in the crude in vitro system. The
minor peaks at 12,000 M.W. and 25-30,000 M.W. are still present in amounts similar to those observed in the crude in
vitro system. These results suggest that the crude extracts
contain additional factors necessary for EF-Tu synthesis
that are missing in the partially purified system. It has
been possible to partially restore the synthesis of EF-Tu by
the addition of a fraction from DEAE-cellulose chromatography. The active factors in this preparation are being
studied.

(1) Chu, F., Miller, D.L., Schulz, T., Weissbach, H. and
    Brot, N. Biochem. Biophys. Res. Commun. 73, 917-927
    (1976).
(2) Treadwell, B.V., Kung, H-F., Redfield, B., Spears, C.,
    Eskin, B. and Weissbach, H. Fed. Proc. 36, 1988 (1977).

IN VITRO TRANSCRIPTION OF Ad2 DNA BY EUKARYOTIC RNA POLY-
MERASES II. I. KINETICS OF FORMATION OF STABLE BINARY
COMPLEXES AT SPECIFIC SITES

S. Seidman, F. Witney and S. Surzycki.
Department of Biology, Indiana University, Bloomington, IN

The equilibrium constant (K_{ass}) for the formation of
stable binary complexes between Adenovirus DNA (Ad2) and
wheat germ RNA polymerase II and human RNA polymerase II
(isolated from placenta) were determined using the filter
binding technique of Hinkle and Chamberlin (1). The rate
constant for the formation of the stable binary complex
with either enzyme was 10^7-$10^8 M^{-1}sec^{-1}$. The rate constant
for dissociation of either enzyme was 10^{-3}-$10^{-4}sec^{-1}$.
Thus the calculated K_{ass} is 10^{10}-$10^{12}M^{-1}$, which indicated
that these complexes are as stable as the binary complexes
formed between E. coli RNA polymerase and promoters.

The position of these stable binary complexes were
mapped by EM using the protein-free BAC spreading technique of Vollenweider et al (2). Polymerases were bound
to both total Ad2 DNA and Eco RI fragment A in order to
orient the ends of the molecule. We mapped 9-11 strong
binding sites, at least 5 of which correspond to in vivo
promoters for Ad2 transcription. In addition the 9 sites
mapped for wheat germ RNA polymerase are identical to 9
of the sites observed for human RNA polymerase.

(1) Hinkle, D.C., and Chamberlin, M.J. (1972). J. Mol.
     Biol. 70 187-195.

(2) Vollenweider, H.J., Sogo, J.M., and Koller, T. (1975).
     Proc. Nat. Acad. Sci. USA 72 83-87.

CHARACTERIZATION OF HUMAN FETAL γ GLOBIN DNA

Jerry L. Slightom and Philip W. Tucker
Laboratory of Genetics, University of Wisconsin
Madison, Wisconsin 53711

Using a cDNA transcript of adult human globin mRNA as hybridization probe, we have cloned from the human genome an Eco RI fragment (2.7 kbp) of fetal γ globin gene into Charon phage vector 3AΔlac (1). The clone, Hs 51.1, has been characterized with respect to size, type of gene sequence contained, and position of these sequences (2).

Detailed DNA sequencing and restriction enzyme mapping will be presented documenting the fact that Hs 51.1 contains the major portion of the γ globin amino acid coding region (residues 1-121) and 1.25 kbp of adjacent 5' non-translated DNA. The coding region is divided into three blocks by two intervening sequences designated IVS 1 (about 100 bp between residues 30 and 31) and IVS 2 (about 900 bp between residues 104 and 105). The location of these intervening sequences are precisely the same as those found in human, rabbit, and mouse β globin, and the junction between them and the corresponding coding regions show nucleotide sequence homology.

Supported by NIH Grants GM21812, AM20120, GM20069, GM06526, and GM07131.

(1) Blattner, F. R., Blechl, A. E., Denniston-Thompson, K., Faber, H. E., Richards, J. E., Slightom, J. L., Tucker, P.W., and Smithies, O. Science, In Press (1978).

(2) Smithies, O., Blechl, A. E., Denniston-Thompson, K., Newell, N., Richards, J. E., Slightom, J. L., Tucker, P. T., and Blattner, F. R. Science, In Press (1978).

PHOSPHORYLATION AND INACTIVATION OF GLYCOGEN SYNTHASE BY A
Ca^{++} CDR-DEPENDENT KINASE. Thomas R. Soderling and Ashok
Srivastava, Dept. Physiology, Vanderbilt Medical School, Nash-
ville, TN, 37232.

Phosphorylation and a to b conversion of skeletal muscle
glycogen synthase by a protein kinase is stimulated up to 10-
fold by addition of the calcium-dependent regulator (CDR) protein. CDR is the calcium binding protein which is responsible
for the Ca^{++}-stimulation of phosphodiestease, adenylate cyclase,
myosin light chain kinase, and phosphorylase kinase. Stimulation of synthase phosphorylation is half-maximal with 0.06 μM
CDR and can be blocked by EGTA (1mM) or trifluoperazine (0.1 mM).
Synthase phosphorylation by this CDR-dependent kinase has a pH
8.6/6.8 activity ratio of about 10. Approximately 60-70% of
the ^{32}P is incorporated into the trypsin-insensitive region.
Phosphorylation of synthase by cAMP-dependent kinase is not
affected by CDR, EGTA, or trifluoperazine and has a pH 8.6/6.8
activity ratio of 1. Synthase phosphorylation by cAMP-independent synthase kinase (J. Biol. Chem. 252, 7517) is not
affected by EGTA or trifluoperazine, is stimulated 30-50% by
CDR, and has a pH 8.6/6.8 activity ratio of about 1.5. The
CDR-dependent synthase kinase is not myosin light chain kinase
as this enzyme does not phosphorylate glycogen synthase. A
possible physiological role for the CDR-dependent synthase
kinase in -adrenergic mediated synthase a to b conversion
will be discussed. Supported by NIH grant AM 17808 and the
Howard Hughes Medical Institute.

A POLYPEPTIDE IN INITIATION FACTOR PREPARATIONS THAT
RECOGNIZES THE 5'-TERMINAL CAP IN EUKARYOTIC mRNA

Nahum Sonenberg and Aaron J. Shatkin
Roche Institute of Molecular Biology, Nutley, NJ   07110.

Most eukaryotic mRNAs contain a 5'-terminal cap,
$m^7G5'pppN$ which promotes translation in vitro at the level
of initiation. To determine if the cap structure is recognized in the process of translation, we assayed rabbit reticulocyte initiation factors for the ability to interact specifically with the 5'-end of oxidized, ^3H-methyl-labeled
mRNA by chemical cross-linking (1). All of the factors
tested (except eIF-4A) could be cross-linked to the 5'-ends
of reovirus, vesicular stomatitis virus (VSV) and vaccinia
virus mRNAs. However, cross-linking of a single polypeptide
of apparent mol. wt. 24,000 by SDS-polyacrylamide gel electrophoresis was prevented by cap analogs such as m^7GDP,
indicating that this polypeptide interacted specifically
with the cap. The 24,000 polypeptide was detected mainly in
eIF-3 (90% pure) and to a lesser extent in eIF-4B (85% pure).
Analysis of cross-linked eIF-3 by gel electrophoresis under
non-denaturing conditions revealed that the cross-linked
protein was not part of the multi-subunit eIF-3 complex.
Fractions from sucrose gradients that resolved eIF-3 and -4B
activities and which contained the 24,000 polypeptide were
found to stimulate preferentially the translation in reticulocyte lysates of uncapped VSV mRNA relative to capped mRNA,
suggesting that it may be a new protein synthesis factor
which has a higher affinity for capped mRNA and is present
in limiting amounts in the cell-free system (J.E. Bergmann,
H. Trachsel, N. Sonenberg, A.J. Shatkin and H.F. Lodish,
submitted). Affinity chromatography procedures (in collaboration with S. Hecht) were used to purify putative cap
binding proteins from high salt wash fractions of rabbit
reticulocyte and mouse ascites cell ribosomes. From these
mixtures of proteins, a polypeptide of apparent mol. wt.
24,000 was bound preferentially to m^7GDP-Sepharose as compared to GDP-Sepharose. It was eluted from the resin upon
addition of excess (5 mM) m^7GDP, consistent with the suggestion that a 24,000 polypeptide recognizes and binds to the
cap structure in mRNA. Studies with the purified protein
should help to elucidate its possible roles in protein synthesis and regulation of eukaryotic genetic expression, for
example in the differential translation of viral mRNAs in
picornavirus-infected cells.

(1) Sonenberg, N., Morgan, M.A., Merrick, W.C. and Shatkin,
    A.J., Proc. Nat. Acad. Sci. 75 4843-4847 (1978).

LOCATION AND SEQUENCE OF A PUTATIVE PROMOTER FOR THE B GENE OF BACTERIOPHAGE S13.

J.H. Spencer, E. Rassart and K. Harbers. Department of Biochemistry, McGill University, Montreal and Department of Biochemistry, Queen's University, Kingston, Ontario.

The binding of *E. coli* RNA polymerase to *Hin*d II + III and *Hae* III restriction fragments of bacteriophage S13 has been used to localize binding sites on the phage genome. Three major binding sites were located which correlate with promoter sites at the beginning of genes A and B and a site overlapping gene D and the beginning of gene E. Two less definite binding sites were also localized, one in gene F and one at the gene G-H junction. Complexes between *E. coli* RNA polymerase and S13 RF DNA have also been visualized in the electron microscope. Two major locations were mapped corresponding to the beginning of genes B and D. Minor sites were also localized corresponding to positions iden- tified in the restriction fragment binding experiments. From the overall binding data putative promoters have been placed at the start of genes A, B and D. The site at the start of gene B has been sequenced, using T4 endonuclease IV hydrolysis, a methylmethane sulfonate alkylation and alkaline hydrolysis procedure and the Maxim-Gilbert rapid sequencing method. The area sequenced is bounded by a *Hae* III site and a *Hin*d III site and contains a *Hin*d II site and three *Alu* I sites. The sequence has an identifiable Pribnow box at the *Hin*d II site and at this position there is a one base sequence difference from the corresponding sequence in φX174 DNA. This putative promoter position is approximately 50 nucleotides from the predicted translation initiation site on the B gene protein of S13, which is identical in site to the translation site predicted for the B gene protein of φX174 and is approximately 100 nucleotides in the 3' direction from the equivalent promoter site sug- gested for the B gene of φX174. A comparison of the region sequenced with that of φX174 shows two differences, the one close to the *Hin*d II site and the other at the *Hin*d III site A comparison of the A gene protein readout shows no amino acid changes compared to φX174 since the two changes φX174 → S13 are C → T and G → T and both are in the third position of the reading frame. The sequence also contains sites for translation of a small protein of approximately 35 residues in a second reading frame with the B protein in the third frame.

Supported by the MRC of Canada, Grant #MT 1453 to J.H.S.

REGULATION OF PROTEIN SYNTHESIS: ROLE OF NAD^+ AND SUGAR PHOSPHATES IN LYSED RABBIT RETICULOCYTES

R. J. Suhadolnik, J. M. Wu, C. P. Cheung, A. Bruzel, and J. Mosca. Department of Biochemistry, Temple University School of Medicine, Philadelphia, PA 19140

NAD^+, NAD^+ analogs, glucose-6-phosphate, fructose-6-phosphate, fructose-1,6-diphosphate, and cAMP have been shown to stimulate and/or inhibit protein synthesis in lysed rabbit reticulocytes (Lennon, Wu, and Suhadolnik, Biochem. Biophys. Res. Commun. 72, 530 (1976); Lennon, Wu, and Suhadolnik, Arch. Biochem. Biophys. 184, 42 (1977); Wu, Cheung, and Suhadolnik, Biochem. Biophys. Res. Commun. 82, 921 (1978); Wu, Cheung, and Suhadolnik, J. Biol. Chem. 253, 7295 (1978)). NAD^+ and NAD^+ analogs can replace the energy requirement for CP/CPK and PEP/PK in cell-free synthesizing systems of lysed rabbit reticulocytes. NAD^+ at 160 μM stimulates the initiation step of the protein synthetic process as determined by (i) the transfer of formyl-methionine from $f[^{35}S]Met\text{-}tRNA_f^{met}$ into polypeptide, (ii) the accumulation of methionyl-valine in the presence of pactamycin, and (iii) the formation of the 40S initiation complex.

The effect of NAD^+ on protein synthesis is closely correlated with the activation of glycolysis. Activation of glycolysis by NAD^+ generates ATP, thereby maintaining a high energy charge. The conversion of glycolytic intermediates to lactate may trigger a signal which stimulates the initiation of protein synthesis. Moreover, the addition of hexose phosphates and NAD^+ stimulates protein synthesis to a greater extent than that observed with NAD^+ alone. Using the gel-filtered lysates, we have demonstrated a synergistic stimulation of polypeptide synthesis by hexose phosphates and cAMP. Furthermore, fructose-6-phosphate prevents the formation of a protein synthesis inhibitor induced by double stranded RNA. The opposite is observed with cAMP, i.e., the expression of the double stranded RNA induced protein synthesis inhibitor is suppressed by cAMP.

Supported by USPHS grant #AI12066.

THE EFFECT OF CORTISOL ON GLUCOCORTICOID "RESISTANT" THYMOCYTES

E. Aubrey Thompson, Jr., Department of Biology,
University of South Carolina, Columbia, SC 29208.

In the sexually immature mouse, the thymus is composed primarily of undifferentiated cortical thymocytes. These cells proliferate rapidly; and, in the presence of glucocorticoids, they undergo a cytolytic response which leads to thymic involution. Cortical thymocytes are therefore designated as glucocorticoid sensitive cells.

Under normal circumstances, cortical thymocytes migrate into the medullary region of the tissue where they differentiate to form the immediate precursors of the peripheral T lymphocytes. Medullary lymphocytes are mitotically quiescent and do not undergo cytolysis in the presence of cortisol or related steroids. For this reason, medullary thymocytes have been designated as glucocorticoid resistant.

Studies have been undertaken to identify isozyme markers of thymocyte differentiation. α-Glycerolphosphate dehydrogenase (GPDH) is apparently such a marker. Cortisol-induced destruction of undifferentiated thymocytes does not cause a decrease in the amount of this enzyme/thymus. Instead total GPDH/thymus increases 3 fold under the influence of cortisol. The "induced" enzyme and the enzyme of the untreated thymus are similar in chemical properties. Moreover, thymic GPDH resembles liver GPDH in kinetic and chromatographic properties, although liver GPDH activity is not increased in the cortisol-treated mouse.

On this basis we conclude that the so called resistant medullary cells are, in fact, target cells for glucocorticoids. They differ from the undifferentiated cells in that medullary cells respond, not by undergoing cytolysis, but by increasing the activity of at least one enzyme. It remains to be determined if this results from an alteration in the receptor(s) or an alteration in the genome.

THE EFFECTS OF STREPTOCOCCUS PNEUMONIAE INFECTION ON HEPATIC
MESSENGER RNA PRODUCTION

William L. Thompson and Robert W. Wannemacher, Jr.
U.S. Army Medical Research Institute of Infectious Diseases,
Fort Detrick, Frederick, Md.

A rise in the concentration of certain serum proteins
has been observed during the early stages of various infections (1). Studies in our laboratory on the regulation of
production of these "acute-phase" proteins in rat hepatic
cells have shown an increase in synthesis and accumulation
of RNA in response to infection, primarily directed to the
bound ribosomes (2), presumably resulting in the subsequent
increase in these serum proteins. Further studies have
been conducted on messenger RNA (mRNA) from control and S.
pneumoniae-infected rats at the peak time of RNA response.
The mRNA from both free and bound ribosomes was extracted
and isolated on oligo dT cellulose. The specific activity
of the isolated mRNA was determined after in vivo labeling
with [^{14}C]orotic acid. A significant increase was seen in
both total mRNA concentration and specific activity in the
bound ribosome fraction from infected rat livers when compared to controls, while no significant changes in mRNA from
the free ribosomes were noted. In order to determine if
increased concentrations of mRNA in cytoplasm were due to
production and/or movement from nuclei, in vitro transport
studies were conducted on the nuclei isolated from control
and infected rats. A 30-min, in vivo, pulse label of
[^{14}C]orotic acid served as a marker for rapidly transcribed
mRNA. Although higher levels of labeled RNA were present
in the infected nuclei, the percent mRNA transported across
the nuclear membrane was the same as in nuclei from controls. Also, a soluble fraction taken from the high-speed
supernatant (2×10^7 g·min) of infected and control liver
homogenates had no effect on transport rates when tested in
the incubation medium against both infected and control
nuclei. These observations suggest that the regulation of
acute-phase protein production is at least in part due to
the transcriptional regulation of specific mRNA.

(1) Bostian, K.A., Blackburn, B.S., Wannemacher, Jr., R.W.,
McGann, V.G., Beisel, W.R. and DuPont, H.L., J. Lab. Clin.
Med. 87 577-585 (1976).

(2) Thompson, W.R. and Wannemacher, Jr., R.W., Biochem. J.
134 79-87 (1973).

IDENTIFICATION OF PUTATIVE PRECURSORS OF OVALBUMIN AND OVO-
MUCOID MESSENGER RNAs.

Ming-Jer Tsai, J.L. Nordstrom, D.R. Roop and B.W. O'Malley.
Department of Cell Biology, Baylor College of Medicine,
Houston, Texas 77030.

Specific DNA probes prepared from structural and intervening sequences within the natural ovalbumin gene were
used to identify the precursors of ovalbumin mRNA. Nuclear
RNA from hormone-stimulated chick oviducts was electrophoresed under denaturing conditions (in the presence of
methylmercury hydroxide) followed by transfer to diazobenzyloxymethyl paper and hybridization to ^{32}P-DNA probes
for structural and intervening sequences. Hybridization of
structural sequence probes to cytoplasmic RNA gave a single
low molecular weight band at the expected size of mature
ovalbumin mRNA (2050 nucleotides). However, hybridization
of the same probe to nuclear RNA demonstrated multiple
species of RNA that are higher in molecular weight than
mature ovalbumin mRNA. The largest molecule has a molecular
weight of more than 8000 nucleotides in length which could
accomodate the entire natural ovalbumin gene sequence.
These high molecular weight species contain RNA complementary to both structural and intervening sequences of the
ovalbumin gene and thus appear to represent ovalbumin gene
transcripts at various processing stages.
   A DNA probe prepared from ovomucoid cDNA was also used to
search for a potential precursor(s) of ovomucoid mRNA.
After electrophoresis of nuclear RNA followed by transfer
to diazobenzyloxymethyl paper and hybridization to the ovomucoid mRNA probes, once again, we observed multiple species
of RNA that are higher in molecular weight than mature ovomucoid mRNA. The largest one has a molecular weight > 7000
nucleotides in length. None of these bands have the same
molecular weight as that of the high molecular weight ovalbumin RNA.
   These results are consistent with the entire ovalbumin and
ovomucoid genes being transcribed into large precursor
molecules, followed by excision and turnover of the intervening sequence RNA and consecutive ligation of the
structural sequences to form mature $mRNA_{ov}$ and $mRNA_{om}$.

PURIFICATION OF mRNA GUANYLYL TRANSFERASE FROM HELA CELLS.
S. Venkatesan and B. Moss. NIH, Bethesda, Maryland

An enzyme that transfers a GMP residue from GTP to
the 5'-end of RNA to form a cap structure characteristic
of eukaryotic mRNA has been detected in HeLa cell nuclei.
This enzyme, which now has been purified approximately 500
fold, catalyzes the following reaction, in which ppN-
represents a polyribonucleotide containing a 5'- diphosphate:

$$\text{GTP} + \text{ppN-} \rightarrow \text{G(5')pppN-} + \text{PPi}$$

In view of the low levels of activity of this enzyme the
enzyme could be monitored only by isolating and characterizing the "capped" structure. For this reason we have
utilized as substrates oligoadenylate sequences derived
by partial hydrolytic cleavage of commercial poly(A) with
nuclease P1 and subsequent chemical addition of a second
phosphate moiety at the 5' end of the oligomers. The
resulting oligomers with 5' diphosphate ends were excel-
lent acceptors of $[\alpha\text{-}^{32}\text{P}]$GTP in the enzymatic reaction.
The specificity of the 5' moiety was established by compar-
ing the "capping" efficiency of poly (A) synthesized using
purified RNA polymerase with calf thymus DNA as template.
The poly (A) was labeled with $[\beta\text{-}^{32}\text{P}]$ATP and the 5' triphosphate end was converted to diphosphate by hydrolysis
with vaccinia virus polynucleotide triphosphatase. Although crude cellular guanylyltransferase preparations can
use RNA acceptors with a tri- or diphosphate terminus, the
latter is strongly preferred by the purified enzyme. No
transfer occurs to RNA containing a 5'-monophosphate or a
5'-hydroxyl. With saturating concentrations of acceptor,
the Km for GTP is approximately 1 μM. Unlike the guanylyltransferase previously purified from vaccinia virus, the
cellular enzyme has no associated RNA (guaninine-7-)-
methyltransferase or polynucleotide triphosphatase activity. The purified enzyme has a pH optimum of 7.5 and
requires divalent cations for activity. Optimal concentrations of Mg^{2+} and Mn^{2+} are 5 mM and 0.5 mM, respectively. The molecular weight of the enzyme is approximately
37,500 as determined by velocity sedimentation through
linear sucrose gradients.

EMERGENCE OF RESISTANCE TO GLUCOCORTICOID HORMONES IN CANCER CELLS: DETECTION OF DIFFERENCES IN PROTEINS SYNTHESIZED BY GLUCOCORTICOID-SENSITIVE AND -RESISTANT MOUSE P1798 LYMPHOSARCOMA CELLS.

Bruce P. Voris, Mary L. Nicholson, and Donald A. Young.
E. Henry Keutman Lab. Endocrine-Metabolism Unit, Departments of Medicine and Radiation Biology and Biophysics, Univ. of Rochester Medical Ctr., Rochester, New York 14642.

Previous studies from this laboratory (1) have shown that the emergence of strains of P1798 lymphosarcoma cells that are resistant to glucocorticoid killing in vivo (in the mouse) may occur via the selection of cells with hardier membranes; membranes (cytoplasmic and/or nuclear) that are better able to withstand the hormone's intracellular destructive actions.

In this study, we have looked for changes in the kinds of proteins made that accompany the emergence of resistance to glucocorticoids by examining individual proteins separated by two-dimensional polyacrylamide gel electrophoresis. By increasing the size of the first and second dimension gel systems by 2.5 fold, the resolution is approximately doubled, allowing the detection of >2000 separate protein peaks. In comparing glucocorticoid-sensitive versus -resistant cells, there are at least 20 proteins whose synthetic rates differ significantly (as measured by densitometry). These differences are retained through at least three tumor passages, suggesting that the changes are relatively stable. Some proteins are detectable only in the sensitive line; whereas, others are detectable only in the resistant line. Thus, the possibility remains open that resistance may be conferred through either an increase or a decrease in the amounts of one or more proteins.

Continuing investigations are aimed at localizing the protein differences to specific subcellular compartments with particular interest in determining which proteins are membrane-associated.

Supported by NIH Grants: 5 T32 GM 07136, 5 F32 CA05381 and AM 16177; and by an award from the United Cancer Council of Monroe County.

(1) Nicholson, M.L. and Young, D.A. Cancer Res. 38: 3673-3680 November, 1978.

COMPLEXITY AND DIVERSITY OF POLYSOMAL AND INFORMOSOMAL
mRNA FROM CHINESE HAMSTER CELLS

Ronald A. Walters, Paul M. Yandell, and M. Duane Enger.
Cellular and Molecular Biology Group, Los Alamos Scientific
Laboratory, University of California, Los Alamos, New Mexico
87545

In eucaryotic cells, a fraction of the cytoplasmic
polyadenylated mRNA population is not associated with ribosomes. These ribosome-free ribonucleoproteins (informosomes) are thought to play a role in control at the translational level in that (1) early development of the fer-
tilized egg involves programmed utilization of informosomes,
(2) distribution of mRNAs between polysomal and informosomal
fractions can change more rapidly than the rate of mRNA
synthesis, and (3) some mRNAs can shift from one fraction to
the other under the appropriate stimulus. There is no
information, however, on the sequence complexity of informosomal mRNA and little data on the degree of sequence
diversity between polysomal and informosomal mRNAs. We
present here the results of our work obtained from annealing
cDNA copies of polyadenylated mRNA to a vast excess of template mRNA from Chinese hamster cells (line CHO). Homologous annealing reactions showed that polysomal and informosomal mRNA populations had similar complexities (9000 to
12,000 mRNA species). However, the most abundant informosomal mRNA contained only ~ 19% the number of different
species comprising the most abundant polysomal mRNA. The
diversity of mRNA species between polysomal and informosomal
mRNA was determined by heterologous annealing reactions with
either cDNA copied from total cytoplasmic mRNA (informosomal
plus polysomal mRNA) or cDNAs prepared from purified polysomal and informosomal mRNA. The results indicated that all
of the different sequences detected were present in both the
polysomal and informosomal mRNA populations. However, comparisons of the rate of annealing of fractionated cDNA
complementary to the most abundant polysomal and informosomal mRNAs showed that the most abundant informosomal
mRNAs were found in the polysomal mRNA at a considerably
lower relative frequency. Our results are not compatible
with models postulating translational control by the complete (or nearly so) sequestering of many mRNA sequences in
an untranslatable form in the cytoplasm. The data are
consistent, however, with models encompassing differential
rates of initiation on the polysome or preferential affinity of some mRNAs for initiation factors.

Supported by the Division of Biomedical and Environmental
Research of the U. S. Department of Energy.

CHARACTERIZATION OF A VIRION-ASSOCIATED RNA POLYMERASE FROM KILLER YEAST

J. Douglas Welsh and Michael J. Leibowitz. Department of Microbiology, CMDNJ-Rutgers Medical School, Piscataway, NJ 08854.

Killer strains of Saccharomyces cerevisiae contain a 1.7 x 10^6 dalton double-stranded RNA plasmid (denoted M); such strains secrete a protein toxin which kills strains lacking the plasmid (sensitives) but to which the killers are immune. The killer plasmid and two larger double-stranded RNA species [3.0 x 10^6 (L) and 3.8 x 10^6 (XL) daltons] are present in intracellular virus-like particles. Yeast strains in the stationary phase of growth with ethanol as the carbon source contain a DNA-independent RNA polymerase activity which co-purifies with these virions. This activity requires all four ribonucleotide triphosphates and catalyzes the incorporation of all four into single-stranded RNA (based on nuclease sensitivity). The reaction requires Mg^{++}, is sensitive to sulfhydryl reagents and high levels of monovalent cation, and is insensitive to DNase (0.5 mg/ml), α-amanitin (100 µg/ml), actinomycin D (100 µg/ml), and thiolutin (50 µg/ml). Pyrophosphate (5mM) completely blocks incorporation, which is also inhibited by the intercalating agent ethidium bromide. Exogenous nucleic acids fail to stimulate incorporation. Deoxythimidine triphosphate is not a substrate in this reaction in the presence or absence of the other deoxynucleotide triphosphates. Incorporation into RNA is linear with respect to time for at least one hour under the assay conditions. The majority of the single-stranded RNA product is released from the virions. The size distributions of the polymerase products indicate that the polymerase may be a virion-bound "transcriptase" acting on all of the virion-encapsidated double-stranded RNA species. The polymerase reaction and its products are being investigated in particles from mutants which are defective in genes required for plasmid maintenance (mak), or expression of the plasmid functions of toxin production (kex) or toxin-resistance (rex).

Supported by grants from the Foundation of CMDNJ, General Research Support-Rutgers Medical School, and National Cancer Institutional Research Service Award CA-09069.

IN VITRO TRANSCRIPTION OF Ad2 DNA BY EUKARYOTIC RNA POLY-
MERASES II. II. TRANSCRIPTION FROM STABLE BINARY COM-
PLEXES

F. Witney, S. Seidman, and S. Surzycki.
Department of Biology, Indiana University, Bloomington,
IN

The mechanism of initiation of transcription from
stable binary complexes between eukaryotic RNA polymerase
II (wheat germ and human) has been studied. The mechanism for initiation resembles the three-step process proposed by Chamberlin (1). 1. RNA polymerase binds stably
to DNA, forming "I" (closed binary complexes); 2. "I"
complexes are converted into "RS" (open binary complexes);
3. In the presence of ribonucleotide triphosphates, each
"RS" complex can catalyze the rapid formation of the
first phosphodiester bond in the nascent RNA chains. All
of our 9 mapped stable binary complexes (Seidman et al,
this symposium) are able to initiate RNA chains.

The transition from "I" to "RS" complex exhibits a
sigmoidal dependence on temperature for which a melting
temperature (T_m) can be determined. The T_m for these
polymerases is 22-25 C. A Van't Hoff plot of the temperature dependence suggests the opening of 4-8 nucleotide
base pairs on Ad2 DNA by these polymerases. This is the
same as the number of base pairs opened by E. coli RNA
polymerase.

The direction of synthesis (strand selection) from
the stable binding sites has been determined by two independent methods. 1. Electron microscopic mapping of
the position of DNA:RNA polymerase:RNA tertiary complexes
as a function of time of RNA chain synthesis. 2. Hybridization of synthesized RNA to DNA blots of separated
strands of various Ad2 DNA fragments.

(1) Chamberlin, M.J., Mangel, W., Rhodes, G., and
     Stahl, S. (1976). in Control of Ribosome Bio-
     synthesis, O. Kjeldgaard, O. Maaloe, eds., pp.
     22-39, Academic Press, New York.

MOUSE FETAL HEMOGLOBIN SYNTHESIS IN MURINE
ERYTHROLEUKEMIC CELLS

N.C. Wu and R.M. Zucker
Papanicolaou Cancer Research Institute, Miami, Fl. 33123

Murine erythroleukemic cells (MELC, clone T3Cl2) derived
from the spleen of adult DDD mouse were induced to differentiate by butyric acid (BA) or dimethylsulfoxide (DMSO).
MELC synthesized both DDD mouse adult and fetal hemoglobin by
the induction of BA and DMSO; however, BA was more potent to
induce the synthesis of fetal hemoglobin. At 0.5 mM BA for 5
days, 20% of the total hemoglobin which the cells synthesized
was the fetal hemoglobin; at 1 mM BA, 40% of the total was
the fetal hemoglobin; and at 2 mM BA, 60% of the total was
the fetal hemoglobin. When the cells were in 2 mM BA for 2
days, then in fresh medium without inducer for 3 days, only
30% of total hemoglobin was the fetal hemoglobin. It
suggests that the synthesis of the fetal and adult hemoglobin
may be dependent on either the post-transcriptional regulation or the existence of two cell subpopulations, i.e., F
cells and A cells.

SPECTROSCOPIC PROBES FOR STUDY OF GENE TRANSCRIPTION

L.R. Yarbrough, Michael Baughman, and Joseph G. Schlageck
Department of Biochemistry, Kansas University Medical Center, Kansas City, KS 66103.

Gene transcription is a complex process involving at least the following events: 1. template binding and promoter melting, 2. nucleoside triphosphate (NTP) binding 3. formation of the first phosphodiester bond (initiation), 4. chain elongation, and 5. termination. Although the first two steps have been studied in some detail, much less in known about subsequent steps. We have recently described the synthesis and properties of a new fluorescent nucleotide analog, adenosine-5'-triphosphoro-γ-1-(5-sulfonic acid) napthylamidate or (γ-AmNS)ATP which can be used as substrate by DNA-dependent RNA polymerase of E. coli (Yarbrough, Biochem. Biophys. Res. Comm. 81: 35-41, 1978). The comparable UTP analog has now also been synthesized. Both (γ-AmNS)ATP and (γ-AmNS)UTP are good substrates for E. coli RNA polymerase. V_{max} values are 50-70% that found for unmodified NTPs; K_m values are increased by only 50% or less. (γ-AmNS)ATP can be used to initiate RNA chains. The conformational properties of the analogs have been studied by absorption, fluorescence and circular dichroic techques. There studies suggest that the napthyl ring and the nitrogen base are in close proximity, perhaps involved in stacking interactions. Both the absorption and fluorescence properties of the (γ-AmNS)NTPs are sensitive to binding of divalent cations such as Mg or Mn^{2+}. This property has been used to obtain binding constants for the metal·(γ-AmNS)NTP complexes. The presence of the substituted napthyl group decreases the binding affinity of (γ-AmNS)ATP for Mg by about twenty fold. Measurements of excited state lifetimes and quantum yields show that binding of Mg to (γ-AmNS)UTP produces conformational alterations which decrease the interaction between the uridine and napthyl ring. These analogs therefore provide a potential means of exploring the conformation of nucleotides during binding by the polymerase and subsequent events leading to phosphodiester bond formation.

Supported by a grant from the National Foundation-March of Dimes.

POSITIVE TRANSCRIPTIONAL CONTROL OF INDUCIBLE GALACTOSE
PATHWAY ENZYMES IN YEAST

James G. Yarger and James E. Hopper
Rosenstiel Basic Medical Sciences Research Center, Brandeis
University, Waltham, Massachusetts 02154

The genetically well-defined galactose-melibiose regulatory system is being studied in Saccharomyces cerevisiae
toward an understanding of how the interplay between
positive and negative regulatory gene activity controls the
induction of structural gene expression in a eukaryote.
Genetic control of the inducible galactose-melibiose pathway enzymes involves the five structural genes: Meli, GAL1,
GAL7, GAL10, and GAL2, and four regulatory genes: GAL4, c
or GAL81, i or GAL80, and GAL3.

The regulatory protein, GAL4, has been previously shown
to have a key role in the galactose-specific induction of
functional GAL1 and GAL7 mRNAs through the use of nonsense
mutations in GAL4 gene along with in vitro translation and
immunoprecipitation assays for the GAL1 and GAL7 mRNAs (1,
2). However, the involvement of GAL4 gene product could be
with the induction of mRNA transcription, transcription itself, or in an immediate post-transcriptional event.

Therefore, in order to ascertain more precisely where
GAL4 gene product exerts its control, we have developed
physical probes for galactose-specific mRNAs in the form of
galactose-specific, yeast-pMB9 recombinant DNA plasmids.
These plasmids have been identified by Southern hybridization to galV RNA and by HART analysis (3). Using these
probes, we have determined that GAL4 gene product is required for the induction of de novo galactose gene transcripts.

Since the structural genes Meli and GAL2 are on separate
chromosomes from the three linked genes GAL1, GAL7 and
GAL10 and require coordinate regulatory control from GAL4
gene product, it is reasonable to expect GAL4 gene product
recognition sites to be adjacent to the structural genes.
Current studies are aimed at determining the existence,
distribution, and nature of regulatory recognition sites
adjacent to the cloned galactose structural genes.

1) Hopper, J.E., Broach, J.R., and Rowe, L.B. (1978),
   PNAS (USA) 75:2878.
2) Hopper, J.E. and Rowe, L.B. (1978), J. Biol. Chem.,
   in press.
3) Patterson, B.M., Roberts, B.E., and Huff, E.L. (1977),
   PNAS (USA) 74:4370

LOCALIZATION AND CHARACTERIZATION OF THE 5' END OF ADENOVIRUS-2 FIBER mRNA.

B.S. Zain, J. Sambrook, A. Dunn, and R.J. Roberts.
Cold Spring Harbor Laboratory, Cold Spring Harbor, New York 11724.

Adenovirus-2 mRNAs are composed of RNA segments transcribed from noncontiguous regions of the Ad2 genome. In the case of late mRNA sequences encoded at map positions 16.6, 19.6 and 26.6 become joined to form a common leader sequence at the 5' end of the major mRNAs. One of these, from which the major structural protein, fiber, is translated has been studied in detail. The 5' end of the mRNA (main body) has been precisely localized on the Ad2 genome, and the nucleotide sequences of segments of the genome and mRNA containing the 5' end has been derived.

Fiber mRNAs are heterogeneous at their 5' ends (1); i.e., in addition to the common tripartite leader sequence, species contain additional leader components (X, Y, Z). cDNA was prepared by reverse transcription of the fiber mRNA and cloned into pBR322 plasmid. Analysis of DNA from one of this clone helped us derive the nucleotide sequences of the entire leader region (tripartite and "Y" leader components). Comparison with the genomic sequences lead to the localization of various splice points.

Supported by Cancer Center Grant (CA 13106) from NCI and by grants to B.S.Z. (ACS-VC-267, NSF #PCN 77-16480) and NIH (1 RO1 CA22698-01).

(1) Chow, L.T. and Broker, T.R. (1978) submitted to Cell.